Animal Models for Study of Pathophysiological Mechanisms of Hypertension and Its Complications

Animal Models for Study of Pathophysiological Mechanisms of Hypertension and Its Complications

Editors

Josef Zicha

Ivana Vaněčková

MDPI • Basel • Beijing • Wuhan • Barcelona • Belgrade • Manchester • Tokyo • Cluj • Tianjin

Editors
Josef Zicha
Institute of Physiology of the Academy of Sciences of the Czech Republic
Czech Republic

Ivana Vaněčková
Institute of Physiology of the Academy of Sciences of the Czech Republic
Czech Republic

Editorial Office
MDPI
St. Alban-Anlage 66
4052 Basel, Switzerland

This is a reprint of articles from the Special Issue published online in the open access journal *Biomedicines* (ISSN 2227-9059) (available at: https://www.mdpi.com/journal/biomedicines/special_issues/AM_Hypertension).

For citation purposes, cite each article independently as indicated on the article page online and as indicated below:

LastName, A.A.; LastName, B.B.; LastName, C.C. Article Title. *Journal Name* **Year**, *Volume Number*, Page Range.

ISBN 978-3-0365-6318-3 (Hbk)
ISBN 978-3-0365-6319-0 (PDF)

Cover image courtesy of Josef Zicha.

Contents

About the Editors

Josef Zicha

Josef Zicha (born in 1950) studied at the Faculty of Pediatric Medicine, Charles University, Prague. Since the end of his study he was employed at the Institute of Physiology, Czech Academy of Sciences (IP CAS) in Prague. His early research was focused on experimental salt hypertension for which he received Young Investigator Award of the International Society of Hypertension (ISH 1984). He continued his studies within the frame of Alexander von Humboldtstiftung (Munich 1987-1988) and Franz Volhard Fellowship (Paris 1990-1992). In 1994-2002 he was the Secretary of the Organizing Committee and the Chair of Program Committee of the Joint Meeting of International and European Societies of Hypertension held in Prague 2002. Thereafter he became a member of the Scientific Council of ISH (2002-2010). Since 2008 he is a Member of the International Committee of SHR Meetings. In 2007-2015 he was the head of the Laboratory of Experimental Hypertension (IP CAS) where he is working up to now.

Ivana Vaněčková

Ivana Vaněčková (born in 1964) is currently a research professor and the Head of the Department of Experimental Hypertension at the Institute of Physiology, Prague, Czech Republic. Dr. Vaněčková graduated from Charles University in 1987. She pursued her postgraduate studies at the Institute of Physiology, Czech Academy of Sciences, where she obtained her Ph.D. Degree. Dr. Vaněčková became a postdoctoral fellow at the renal physiology laboratory at the Institute of Physiology, Prague. In 2001 she became assistant research professor at the Department of Experimental Medicine, Institute for Clinical and Experimental Medicine, Prague, where she was Head of the Department for Experimental Hypertension. As a Senior Investigator Dr. Vaněčková continued her work at the Institute of Physiology AS CR, in the Laboratory of Experimental Hypertension where she is interested in the role of renin-angiotensin and endothelin systems in hypertension and chronic kidney disease.

Preface to "Animal Models for Study of Pathophysiological Mechanisms of Hypertension and Its Complications"

Essential hypertension is still an important health care problem. It is necessary to investigate its mechanisms especially in animal models. The potential clinical importance of such experimental research might be expected. This Special Issue is concerned with several important topics. First, several studies were focused on the pathophysiological mechanisms responsible for (a) blood pressure elevation during hypertension development [1–3], (b) organ damage in chronic hypertension [4,5], and (c) drugs targeting hypertension and/or its complications [6–8]. Other studies were interested in the participation of (a) central and peripheral blood pressure control [9], (b) changes in vascular structure and function [3,10], and (c) neural, humoral, and endocrine factors [11–13]. Furthermore, the contribution of (a) altered redox signaling [9,14], (b) chronic inflammation (6), (c) microbiome changes (14), and (d) interactions of genetic and environmental factors [3,15] were evaluated in multiple papers. Finally, special attention was paid to the progress in (a) pharmacological tools for the control of hypertension and associated organ damage [4–6], (b) genetic modifications to alter blood pressure levels [16], and (c) non-pharmacological interventions attenuating hypertension or its complications [17]. The original articles or reviews covered the interesting aspects of the pathophysiology of hypertension and associated end-organ damage, the use of various experimental hypertensive models, and the importance of specific environmental factors acting in distinct phases of the ontogeny. We especially appreciate the presentation of new ideas and the critical discussion of traditional theories.

1. Latic N, Zupcic A, Frauenstein D, Erben RG: Activation of RAAS signaling contributes to hypertension in aged hyp mice. *Biomedicines* **2022**, *10*, 1691.
2. Maeng L, Lee K: Role of Translationally Controlled Tumor Protein (TCTP) in the development of hypertension and related diseases in mouse models. *Biomedicines* **2022**, *10*, 2722.
3. Vieira-Rocha MS, Sousa JB, Rodriguez-Rodríguez P, Arribas SM, Diniz C: Elevated vascular sympathetic neurotransmission and remodeling is a common feature in a rat model of fetal programming of hypertension and SHR. *Biomedicines* **2022**, *10*, 1902.
4. Šimko F, Baka T, Stanko P, Repova K, Krajcirovicova K, Aziriova S, Domenig O, Zorad S, Adamcova M, Paulis L: Sacubitril/valsartan and ivabradine attenuate left ventricular remodelling and dysfunction in spontaneously hypertensive rats: different interactions with the renin-angiotensin-aldosterone system. *Biomedicines* **2022**, *10*, 1844.
5. Hojná J, Kotsaridou Z Vaňourková Z, Rauchová H, Behuliak M, Kujal P, Kadlecová M, Zicha J, Vaněčková I: Empagliflozin is not renoprotective in non-diabetic rat models of chronic kidney disease. *Biomedicines* **2022**, *10*, 2509.
6. Malínská H, Hüttl M, Marková I, Miklánková D, Hojná S, Papoušek F, Šilhavý J, Mlejnek P, Zicha J, Hrdlička J, Pravenec P, Vaněčková I: Beneficial effects of empagliflozin are mediated by reduced renal inflammation and oxidative stress in spontaneously hypertensive rats expressing human C-reactive protein. *Biomedicines* **2022**, *10*, 2066.
7. Hawlitschek C, Brendel J, Gabriel P, Schierle K, Salameh A, Zimmer H-G, Rassler B: How effective is a late-onset antihypertensive treatment? Studies with captopril as monotherapy and in combination with nifedipine in old spontaneously hypertensive rats. *Biomedicines* **2022**, *10*, 1904.

8. Rassler B, Hawlitschek C, Brendel J, Zimmer H-G: How do young and old spontaneously hypertensive rats respond to antihypertensive therapy? Comparative studies on the effects of combined captopril and nifedipine treatment. *Biomedicines* **2022**, *10*, 3059.
9. Chao YM, Rauchová H, Chan JYH: Disparate roles of oxidative stress in rostral ventrolateral medulla in age-dependent susceptibility to hypertension induced by systemic L-NAME treatment in rats. *Biomedicines* **2022**, *10*, 2232.
10. Goto K, Kitazono T: Chloride ions, vascular function and hypertension. *Biomedicines 10*, 2316.
11. Latic N, Peitzsch M, Pietzsch J, Erben RG: Long-term excessive dietary phosphate intake causes hypertension via activation of RAAS and stimulation of the sympathetic tone in mice. *Biomedicines* **2022**, *10*, 2510.
12. Andelova K, Szeiffova Bacova B, Sykora M, Pavelka S, Rauchova H, Tribulova N: Cardiac Cx43 signaling is enhanced and TGF-β1/SMAD2/3 suppressed in response to cold acclimation and modulated by thyroid status in hairless SHRM. *Biomedicines* **2022**, *10*, 1707.
13. Castoldi G, Carletti R, Ippolito S, Villa I, Palmisano B, Bolamperti S, Rubinacci A, Zerbini G, Meani M, Zatti G, Di Gioia CRT: Angiotensin II modulates calcium/phosphate excretion in experimental model of hypertension: Focus on bone. *Biomedicines* **2022**, *10*, 2928.
14. Tain Y-L, Hsu C-N: Hypertension of developmental origins: consideration of gut microbiome in animal models. *Biomedicines* **2022**, *10*, 875.
15. Vieira-Rocha MS, Sousa JB, Rodriguez-Rodríguez P, Arribas SM, Diniz C : Elevated vascular sympathetic neurotransmission and remodeling is a common feature in a rat model of fetal programming of hypertension and SHR. *Biomedicines* **2022**, *10*, 1902.
16. Ohara H, Nabika T: Genetic modifications to alter blood pressure level. *Biomedicines 10*, 1855.
17. Yamori Y, Sagara M, Mori H: Stroke-prone SHR as experimental models for cardiovascular disease risk reduction in humans. *Biomedicines* **2022**, *10*, 2974.

Josef Zicha and Ivana Vaněčková

Editors

biomedicines

Review

Hypertension of Developmental Origins: Consideration of Gut Microbiome in Animal Models

You-Lin Tain [1,2] and Chien-Ning Hsu [3,4,*]

1 Department of Pediatrics, Kaohsiung Chang Gung Memorial Hospital and Chang Gung University College of Medicine, Kaohsiung 833, Taiwan; tainyl@cgmh.org.tw
2 Institute for Translational Research in Biomedicine, Kaohsiung Chang Gung Memorial Hospital and Chang Gung University College of Medicine, Kaohsiung 833, Taiwan
3 Department of Pharmacy, Kaohsiung Chang Gung Memorial Hospital, Kaohsiung 833, Taiwan
4 School of Pharmacy, Kaohsiung Medical University, Kaohsiung 807, Taiwan
* Correspondence: cnhsu@cgmh.org.tw; Tel.: +886-975-368-975; Fax: +886-7733-8009

Abstract: Hypertension is the leading cause of global disease burden. Hypertension can arise from early life. Animal models are valuable for giving cogent evidence of a causal relationship between various environmental insults in early life and the hypertension of developmental origins in later life. These insults consist of maternal malnutrition, maternal medical conditions, medication use, and exposure to environmental chemicals/toxins. There is a burgeoning body of evidence on maternal insults can shift gut microbiota, resulting in adverse offspring outcomes later in life. Emerging evidence suggests that gut microbiota dysbiosis is involved in hypertension of developmental origins, while gut microbiota-targeted therapy, if applied early, is able to help prevent hypertension in later life. This review discusses the innovative use of animal models in addressing the mechanisms behind hypertension of developmental origins. We will also highlight the application of animal models to elucidate how the gut microbiota connects with other core mechanisms, and the potential of gut microbiota-targeted therapy as a novel preventive strategy to prevent hypertension of developmental origins. These animal models have certainly enhanced our understanding of hypertension of developmental origins, closing the knowledge gap between animal models and future clinical translation.

Keywords: developmental origins of health and disease (DOHaD); gut microbiota; hypertension; short chain fatty acid; oxidative stress; probiotics; prebiotics; renin–angiotensin system

Citation: Tain, Y.-L.; Hsu, C.-N. Hypertension of Developmental Origins: Consideration of Gut Microbiome in Animal Models. *Biomedicines* **2022**, *10*, 875. https://doi.org/10.3390/biomedicines10040875

Academic Editors: Josef Zicha and Ivana Vaněčková

Received: 10 March 2022
Accepted: 8 April 2022
Published: 9 April 2022

1. Introduction

Hypertension is the most common chronic disease and yields considerable morbidity and mortality globally [1]. Because of the multifactorial nature of hypertension, the use of various animal models, which evoke hypertension by different mechanisms, is advantageous for unraveling disease pathogenesis and developing novel antihypertensive drugs [2–4]. Though we are seeing tremendous progress on experimental hypertension, the prevalence of hypertension is still high and continues to rise worldwide [5].

Epidemiological and animal studies support that hypertension may be programmed in utero [6–9]. The association between fetal development and increased risk of adult disease has emerged as the concept of developmental origins of health and disease (DOHaD) [10]. A wide spectrum of early-life insults can evoke developmental programming resulting hypertension later in life. These insult stimuli include, but are not limited to, maternal malnutrition (both under- and overnutrition), maternal medical conditions, environmental exposure to toxins/chemicals, lifestyle changes, and medicines taken during pregnancy [7–9,11–14].

Over the past decade, the pathogenesis behind hypertension of developmental origins has not been fully elucidated, but data from animal models have proposed several key mechanisms [14]. Until now, the proposed mechanisms consist of aberrant renin–angiotensin system (RAS), oxidative stress, reduced nephron numbers, gut microbiota dysbiosis, dysregulated nutrient-sensing signals, sex differences, epigenetic regulation, etc. [7–9,11–14]. Among them, the interaction between the gut microbiota and the host implicated in hypertension has received significant interest [15–18]. Despite gut microbiota dysbiosis being observed in multiple animal models of hypertension [15,16], too little attention has been paid to its role in hypertension of developmental origins.

Although blood pressure (BP) is considered with a multifactorial pattern of inheritance, genome-wide association studies cumulatively could only explain ~3.5% of BP trait variability [19]. Accordingly, it is likely that the influence of environmental and epigenetic factors on the developmental programming of hypertension should receive wider recognition. Notably, maternal insults can impair gut microbiota composition and function, leading to adverse offspring outcomes later in life [20]. Conversely, review elsewhere indicted that early-life gut microbiota-targeted therapies have benefits on the prevention of the developmental programming of adult disease, including hypertension [21]. All this raises the notion that we need to pay more attention to prevent and not just treat hypertension, with a focus on the influence of dysbiotic gut microbiota on hypertension of developmental origins. Accordingly, animal models would likely be very useful in unraveling these actions.

In this review, we describe the role of gut microbiota implicated in animal models used for studying the developmental programming of hypertension. Therefore, we summarize the contributions of animal models linking the gut microbiota to developmental programming of hypertension, which helps in developing valuable strategies to prevent hypertension from happening. We specifically focus on addressing gut microbiota-targeted therapies such as probiotics, prebiotics, and postbiotics as a reprogramming strategy for prevention of hypertension of developmental origins.

In view of the above, a search was performed in the electronic bibliographic database PubMed/MEDLINE. Search terms were as follows: "developmental programming", "DOHaD", "animal model", "pregnancy", "gestation", "offspring", "progeny", "prenatal", "perinatal", "mother", "maternal", "reprogramming", "gut microbiota", "probiotics", "prebiotics", "postbiotics", "synbiotics", "blood pressure", and "hypertension." Relevant abstracts were identified and reviewed to identify appropriate studies. Suitable published articles in English were included, without restriction of the time of publication.

2. Hypertension of Developmental Origins: Choice of Animal Models

Compared to animal models of essential hypertension established in prior research [2,3], the etiologies of hypertension of developmental origins and underlying pathogenic mechanisms are more complex [14]. Animal models of hypertension of developmental origins can be categorized in different ways (Figure 1).

Firstly, these models can be classified according to early-life adverse conditions. Nutritional programming is the most common type of animal model being studied in the field of DOHaD research [22]. Dietary caloric or protein restriction in animals mimics the starvation linked to famine in human cohorts [23,24]. Imbalance of maternal nutrition can have long-term changes in BP, resulting hypertension in adult offspring [25]. Inadequate or excessive intake of a specific nutrient has been established to induce hypertension of developmental origins in animal models, as reviewed elsewhere [11,26]. These models of undernutrition related to hypertension of developmental origins include, but are not limited to, caloric restriction [27], protein restriction [28], and deficiencies in sodium [29], calcium [30], zinc [31], iron [32], methyl donor nutrients (choline; vitamins B2, B6, and B12; folic acid; and methionine) [33], and vitamin D [34]. On the other hand, overnutrition characterized by the consumption of a high-fat [35,36], high-fructose [37,38], or high-protein diet [39] by rodent mothers also leads to early programming of hypertension in the offspring. Additionally, animal models resembling maternal medical conditions have also been evaluated in developmental programming of hypertension. These models include hypertensive disorders of pregnancy [40], preeclampsia [41], diabetes [42], chronic kidney disease (CKD) [43], maternal hypoxia [44], etc. Furthermore, chemical and medication exposures during pregnancy increase the risk of developing hypertension in offspring [13,14]. Prenatal exposure to 2,3,7,8-tetrachlorodibenzo-p-dioxin (TCDD) [45], bisphenol A [46], nicotine [47], caffeine [48], cyclosporine [49], gentamicin [50], tenofovir [51], minocycline [52], or glucocorticoids [53] has been reported to induce hypertension of developmental origins in various animal models.

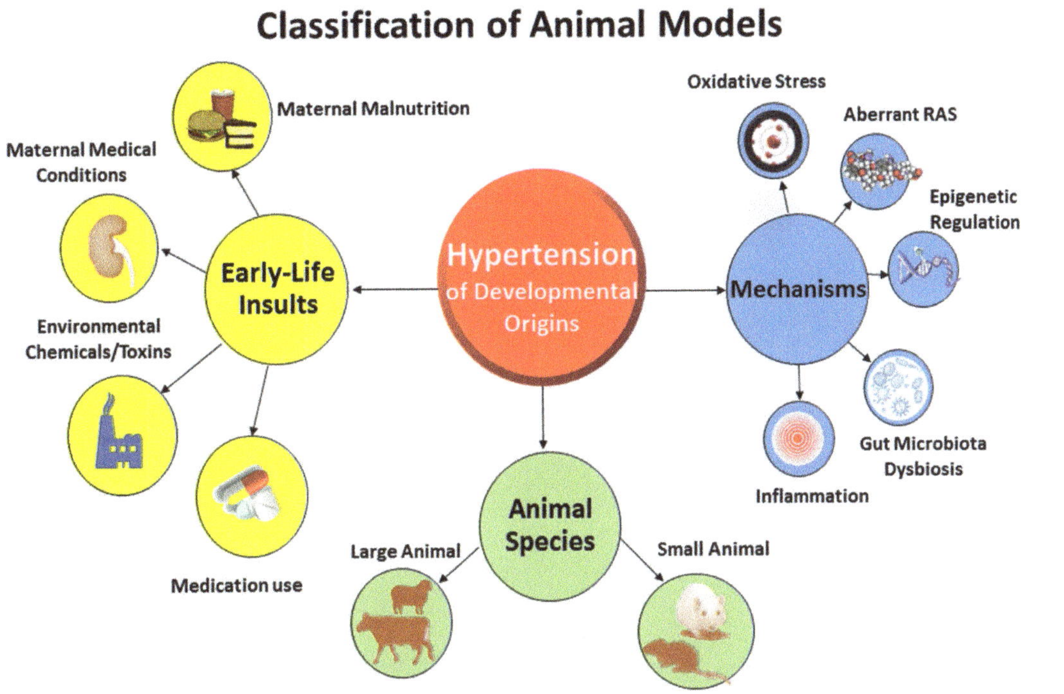

Figure 1. The classification of major animal models for studying hypertension of developmental origins.

Secondly, animal models can be classified based on molecular mechanisms. In view of different early-life adverse environmental factors producing the same outcome, that is to say hypertension in adult offspring, there might be core mechanisms underlying hypertension of developmental origins. These mechanisms include gut microbiota dysbiosis [21], oxidative stress [12], aberrant RAS [54], reduced nephron numbers [7], dysregulated nutrient-sensing signals [55], sex differences [56,57], epigenetic regulation [58], inflammation [9,14], nitric oxide (NO) deficiency [59], etc. Up to date, various animal models have been developed to test such proposed mechanisms. Because of the multifactorial nature of developmental hypertension, the use of various animal models, each of which induces hypertension by a different mechanism yet with the same end result, is advantageous. This approach would allow for a novel and effective reprogramming intervention targeting a specific molecular pathway to be adopted for preventions and therapies.

Lastly, animal models in DOHaD research can be classified according to species [60]. Diverse large- and small-animal models have been used, each with its own natural advantages and disadvantages [8]. Former reviews demonstrated that cow [61], sheep [62], rat [27], and mice [63] have be used to study hypertension of developmental origins [14]. Considering that rat models are cost-effective and easy to maintain and breed, they became the most common species used in the research field of DOHaD-related hypertension [14]. Although nonhuman primates [64], swine [65], rabbits [66], and guinea pig [67] have been studied for cardiovascular outcomes induced by maternal insult stimuli, none of them have been reported for examining hypertension of developmental origins.

Rats are by far the most often used species in the field of primary hypertension research. Of these, the spontaneously hypertensive rat (SHR) without any doubt is the most popular strain [2]. However, the majority of the rat strains used for studying developmental hypertension are Sprague-Dawley (SD) or Wistar [14]. In view of the genetic background of SHR, offspring develops hypertension spontaneously without programming induced by early-life insults, weakening its application in studying hypertension of developmental origins. Hence, the choices of animal models between essential hypertension and hypertension of developmental origins are quite different. Many more aspects of animal models need to be taken into further consideration, such as the timing of organogenesis [68], life cycle [69], gestation period [70], litter size [71], offspring outcomes, than in human studies [72], and valuable therapeutic interventions need to be evaluated and validated [14].

Together, it is noted that remarkable advances in hypertension of developmental origins have been originated from animal models. However, what is missing in the literature is animal models used for studying hypertension-related complications. Although elevated BP is the core feature of human hypertension, its morbidity and mortality occur with complications in the heart, brain, kidneys, and vessels. The contributions of early-life insults to these hypertension-related complications later in life in an organ-dependent manner have not yet been well-studied in the above-mentioned animal models.

3. Gut Microbiota: Choice of Animal Models

Trillions of bacteria living in the gut—the gut microbiota—coexist with the host in a mutually beneficial relationship [73]. Microbiota refers to all the microorganisms found in the environment, while the term microbiome refers to the collection of genomes from all microorganisms in a given environment. A variety of environmental factors can cause the disturbance of gut microbiota (i.e., dysbiosis), which in turn can influence human health and disease. Although the influence of gut microbiota in hypertension has been extensively reviewed elsewhere [15–18], less attention was paid to exploring its role in hypertension of developmental origins.

Directly after birth, microbes colonize the neonatal gut immediately [74]. These alterations continue until three years of age and mediate the transition toward an adult-like gut microbiota [75]. During pregnancy and lactation, the mothers share gut microbes and microbial metabolites with their offspring, which highlights the importance of maternal influences in the development of early-life gut microbiota [76]. A diversity of early-life factors governs the establishment of the gut microbiota, such as maternal medical conditions, gestational age, types of delivery, antibiotic exposure, formula feeding, and ecological factors [74–77].

So far, animal models have been broadly established to investigate human diseases in gut microbiota research [78]. Figure 2 illustrates various approaches to alter the gut microbiota in animal models of disease. Several gut microbiota-targeted therapies have been used to alter gut microbiota compositions and its derived metabolites. These interventions consist of probiotics, prebiotics, synbiotics, postbiotics, etc. [14]. The embryo transfer (ET) method is considered the gold standard for gut microbiota transfer. Additionally, researchers often use other methods to transfer the gut microbiota, such as fecal microbiota transfer (FMT), co-housing (CH), or cross-fostering (CF) [78].

Several gut microbiota-targeted therapies have shown to alter the gut microbiome. Probiotics (i.e., live beneficial microbes) and prebiotics (i.e., substances in foods that promote the growth of healthy microbes) are the most commonly used gut microbiota-targeted modalities in clinical practice [79]. Synbiotics refer to a mixture comprising probiotics and prebiotics that also confers a health benefit [79]. In addition, the use of substances leased or produced through gut microbial metabolism, namely postbiotics, have shown an influence on gut microbiota compositions and metabolites [80].

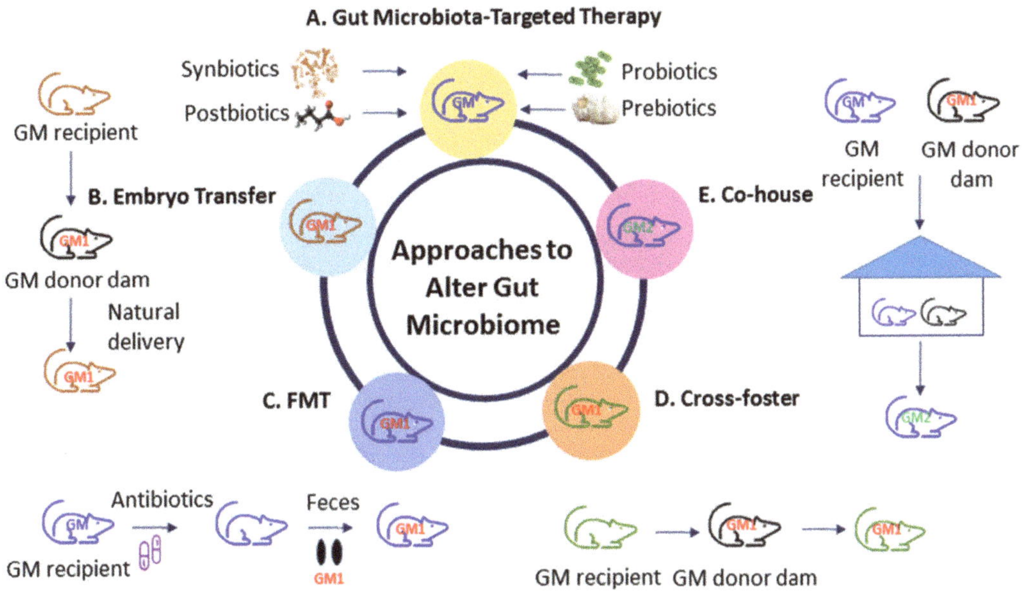

Figure 2. Different approaches to altering the gut microbiota. (**A**) Gut microbiota-target therapy; (**B**) embryo transfer; (**C**) fecal microbiota transfer; (**D**) cross-foster; (**E**) co-house. FMT = fecal microbiota transfer; GM = gut microbiota; GM1 = transferred gut microbiota; GM2 = GM + GM1.

In the approach to transfer embryos, they are collected from the gut microbiome recipient and surgically transferred to a pseudopregnant donor dam [78]. Accordingly, the recipient pups obtain the vaginal microbiota from the donor dam through vaginal delivery. Nevertheless, this method needs considerable costs and expertise, making it inaccessible for many labs. Using the FMT approach, feces or fecal contents from donors are transferred to recipient animals via gastric gavage. Germ-free mice or antibiotics-treated depleted microbiota animals are commonly used as recipients [81].

Another commonly used method is CH, wherein recipients are co-housed with a donor after weaning [82], leading to the transfer of the donor gut microbiome through coprophagy and grooming [82]. Although the co-housing approach is easy and low-cost, the transfer of the gut microbiota after the critical developmental period results an incomplete transfer as well as a hybridized gut microbiome. When the recipient pups are housed in cages with the donor dam within 24 h after birth, the CF method allows the recipients to obtain most of their gut microbiota from the donor dam [78]. Compared to CH, the CF approach transfers the gut microbiota from an early age during the maternal care process.

All these methods each carry certain advantages and limitations. Researchers should thus be mindful of these method-related differences in the context of the transfer methods used for studying the role of the gut microbiota on hypertension of developmental origins.

4. Gut Microbiota in Hypertension of Developmental Origins

There is mounting evidence to support the pathogenic interconnection between the gut microbiome and hypertension [15–18]. However, there is paucity of information regarding the influence of the gut microbiota on the developmental programming of hypertension later in life. Therefore, most data obtained from patients with established hypertension and knowledge received from animal models of essential hypertension might be extrapolated to hypertension of developmental origins.

4.1. Gut Microbiota and BP Regulation

A great deal of work on the influence of the gut microbiota and its derived metabolites on BP regulation has been conducted. First, data from several genetic hypertensive rat models (e.g., SHR) indicated that the gut microbiota of hypertensive rats is dysbiotic and significantly different from the microbiota of normotensive control rats [15]. Gut microbiota dysbiosis was also noted for other hypertension models such as animals treated with high salt [83], angiotensin II [84], and deoxycorticosterone acetate-salt [85]. Another line of evidence comes from germ-free animals. The absence of microbiota in germ-free rats resulted with relative hypotension compared with their conventionalized counterparts, suggesting an essential role of gut microbiota in BP regulation [86]. Additionally, germ-free mice that received FMT from a hypertensive human donor developed a gut microbiota similar to that of their donor, as well as elevated BP [87]. There are observations that microbial metabolites are involved in BP homeostasis. Short chain fatty acids (SCFAs) are the main metabolites produced during bacterial fermentation of carbohydrates. SCFAs are generally known to regulate BP via activating their SCFA receptor, including olfactory receptor 78 (Olfr78), G protein-coupled receptors (GPR) GPR41, GPR43, and GRP109A [88]. Another example is trimethylamine-N-oxide (TMAO). TMAO is a small colorless amine oxide produced by gut microbiota metabolism [89]. A high TMAO level correlates with CVD mortality [90]. Fourth, the uses of probiotics [91] or prebiotics [92] have shown benefits on hypertensive patients.

4.2. Animal Models Linking Gut Microbiota Dysbiosis to Hypertension of Developmental Origins

Much work investigating the actions of the gut microbiome has directly studied the hypertension models, yet relatively little data exists on its programming effect related to hypertension of developmental origins. A summary of animal studies indicating the association between dysbiotic gut microbiota and developmental hypertension in adult offspring is provided in Table 1 [40,52,93–106].

The current review is only restricted to early-life insults starting in the pregnancy and/or lactation period. Table 1 shows that rats are the most common species being used. A variety of early-life insults have been reported to induce developmental hypertension, accompanying alterations of the gut microbiota, including a maternal high-fructose diet [93–95], maternal high-fructose diet plus TCDD exposure [96], maternal high-fat/high-cholesterol diet [97], maternal high-fat and/or post-weaning high-fat diet [98,99], gestational hypertension [40,100], maternal CKD [43], maternal dyslipidemia [101], maternal N^G-nitro-L-arginine-methyl ester (L-NAME) administration plus postnatal high-fat diet [102], maternal administration of minocycline [52], maternal TMAO and asymmetric dimethylarginine (ADMA) exposure [103], maternal TCDD exposure [104,105], and prenatal androgen exposure [106].

Table 1 lists the timing of hypertension determined from rat models, with age ranging from 12 weeks to four months. As every month of an adult rat corresponds to three human years [55], the observed periods correspond with humans from childhood to early adulthood.

4.3. Gut Microbiota Dysbiosis in Hypertension of Developmental Origins

The study of the gut microbiome in animal models of developmental hypertension mainly focuses on four types of dysbiosis: loss of diversity, decreases in beneficial microbes, shifts in key taxa, and alterations of microbial metabolites. A schematic summarizing the gut microbiota and a possible molecular pathway linked to hypertension of developmental origins is presented in Figure 3.

Table 1. Animal models reporting hypertension of developmental origins associated with dysbiotic gut microbiota.

Animal Models	Species/Gender	Age at Measure	Alterations of Gut Microbiota	Ref.
Maternal high-fructose diet	SD rat/M	12 weeks	Decreased renal GPR41 and GPR43 expression	[93]
Maternal high-fructose diet	SD rat/M	12 weeks	Decreased plasma TMA level; reduced abundance of genus *Akkermansia* and phylum *Verrucomicrobia*	[94]
Maternal plus post-weaning high-fructose diet	SD rat/M	12 weeks	Decreased abundance of genera *Bacteroides*, *Dysgonomonas*, and *Turicibacter*	[95]
Maternal high-fructose diet and TCDD exposure	SD rat/M	12 weeks	Increased abundance of genus *Gordonibacter*	[96]
Maternal high-fat and high-cholesterol diet	Wistar rat/M	90 days	Decreased α-diversity	[97]
Maternal plus post-weaning high-fat diet	SD rat/M	16 weeks	An increased F/B ratio; a reduction of genera *Lactobacillus* and *Akkermansia*	[98,99]
Maternal hypertension	SHR/M	12 weeks	An increased abundance of the genera *Bifidobacterium*, *Lactobacillus*, *Turicibacter*, and *Akkermansia*	[40]
Maternal hypertension	SHR/M	12 weeks	An increased F/B ratio	[100]
Maternal CKD	SD rat/M	12 weeks	An increased F/B ratio; a reduction of genera *Bifidobacterium*, *Ruminococcus*, *Alistipes*; decreased acetate and butyrate in the plasma; and increased plasma TMAO level.	[43]
Maternal dyslipidemia	Wistar rat/M and F	24 weeks	A decrease of genera *Lactobacillus* abundance	[101]
Maternal L-NAME administration plus post-weaning high-fat diet	SD rat/M	16 weeks	An increased F/B ratio	[102]
Maternal minocycline administration	SD rat/M	12 weeks	An increase F/B ratio, and decreased genera *Lactobacillus*, *Ruminococcus*, and *Odoribacter* abundance	[52]
Maternal TMAO and ADMA exposure	SD rat/M	12 weeks	Decreased abundance of *Erysipelotrichaceae* family	[103]
Maternal TCDD exposure	SD rat/M	12 weeks	Decreased α-diversity, and increased F/B ratio, and a decreased abundance of genera *Ruminococcus*, *Roseburia*, and *Odoribacter*	[104,105]
Prenatal androgen exposure	Wistar rat/F	4 months	An increased abundance of bacteria associated with production of SCFAs.	[106]

Studies tabulated according to animal models, and age at measure; SD = Sprague-Dawley; SHR = spontaneously hypertensive rat; M = male; F = female; CKD = chronic kidney disease; TCDD = 2,3,7,8-tetrachlorodibenzo-p-dioxin; ADMA = asymmetric dimethylarginine; GPR41 = G protein-coupled receptor 41; GPR43 = G protein-coupled receptor 43; TNA = trimethylamine; TMAO = trimethylamine N-oxide; L-NAME = N^G-nitro-L-arginine-methyl ester; F/B ratio = *Firmicutes* to *Bacteroidetes* (F/B) ratio; SCFA = short chain fatty acid.

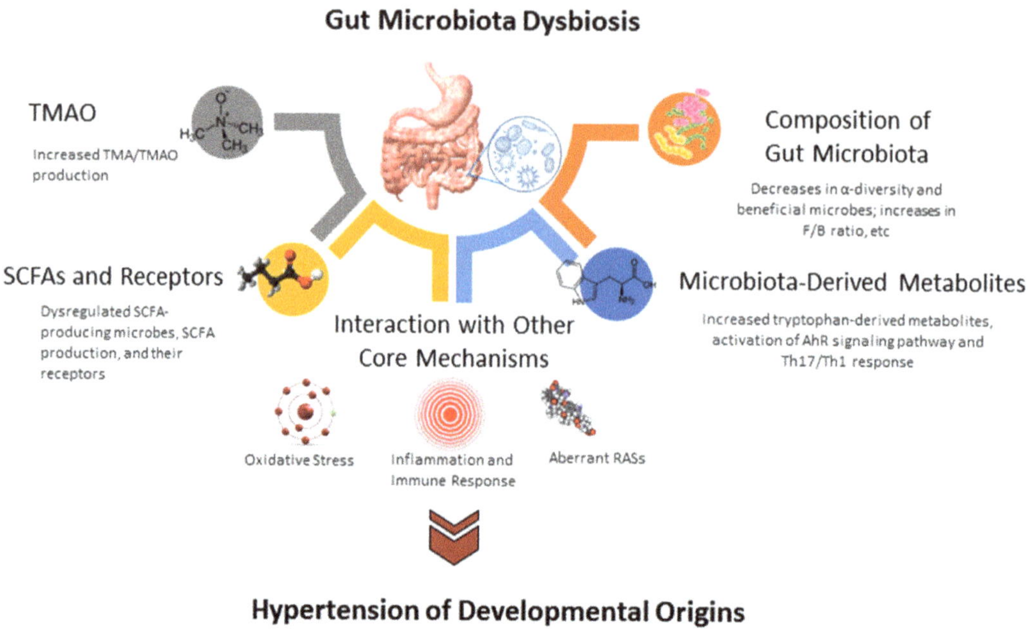

Figure 3. Overview of the gut microbiota and potential molecular mechanisms related to hypertension of developmental origins. SCFA short chain fatty acid. TMAO = trimethylamine N-oxide; TMA = trimethylamine; SCFA = short chain fatty acid; RAS = renin-angiotensin system; TH17 = T helper 17 cells; TH1 = T helper 1 cells; F/B ratio = *Firmicutes* to *Bacteroidetes* ratio; AhR = aryl hydrocarbon receptor.

4.3.1. Alterations in Gut Microbiota Compositions

First, α-diversity is decreased in models of maternal high-fat and high-cholesterol diet [97] and maternal TCDD exposure [104,105]. A similar pattern of gut dysbiosis was reported in several hypertensive animal models [15]. Second, a maternal plus post-weaning high-fat diet programming offspring's hypertension coincides with an increased *Firmicutes* to *Bacteroidetes* (F/B) ratio and a reduction of genera *Lactobacillus* and *Akkermansia* [98,99]. This was found to be consistent with hypertension models showing the F/B ratio was increased and served as a microbial marker of hypertension [15]. Likewise, the increase of the F/B ratio is noted in other models of developmental hypertension programmed by a variety of maternal insults, including CKD [43], minocycline administration [52], hypertension [100], L-NAME administration plus high-fat diet [102], and TCDD exposure [104,105]. Both *Akkermansia* and *Lactobacillus* are known as one of the beneficial probiotic bacterial strains [107,108]. Decreases of certain beneficial microbes were also found in developmental models of hypertension, like maternal minocycline administration [52], maternal high-fructose diet [94], maternal hypertension [40], maternal dyslipidemia [101], and maternal TCDD exposure [104,105].

4.3.2. SCFAs and Their Receptors

Notably, an association between microbiota-derived metabolites and hypertension has been found in several models of developmental hypertension [43,93,94,106].

SCFAs, the main metabolites produced by the gut microbiota, have one to six carbon atoms (C1–C6), mainly consisting of acetic acid (C2), propionic acid (C3), and butyric acid (C4) [88]. In SHR, hypertension is associated with decreased abundance of acetate- and butyrate-producing bacteria [15]. Similarly, SCFAs and their receptors are involved in hypertension of developmental origins, as reported in several animal models [52,93,106].

In a model of maternal administration of minocycline, minocycline-induced hypertension is associated with a reduction of plasma acetate and butyrate [52]. Another report demonstrated that dam rats receiving a 60% fructose diet caused offspring's hypertension, coinciding with an increased plasma acetate level and a reduction of renal GPR41 and GPR43 expression [93]. As acetate is a ligand for GPR41 to induce vasodilatation, and Olfr78 exhibits vasoconstrictive action [109], these findings support the notion that SCFAs and their receptors contribute to maternal high-fructose-diet-induced hypertension in adult offspring. Additionally, maternal garlic oil therapy protects against offspring hypertension programmed by a high-fat diet, which is related to increased acetate, butyrate, and propionate, as well as their producing microorganisms [110]. Moreover, maternal SCFA supplementation have been reported to be protective on hypertension of developmental origins [85,94]. These findings support the notion that SCFAs and their receptors might be a crucial mechanism behind developmental programming of hypertension.

4.3.3. TMAO

TMAO is an end-product of microbial carnitine and choline metabolism [89]. TMAO is converted from trimethylamine (TMA) by flavin-containing monooxygenase (FMO). TMAO is able to activate nuclear factor-κB (NF-κB) signaling, enhance leukocyte-endothelial cell adhesion, and induce inflammatory gene expression, all of which are related to the development of hypertension [111].

Maternal exposure to TMAO results in hypertension in adult male offspring [103]. Conversely, microbe-dependent TMA and TMAO formation can be inhibited by 3,3-dimethyl-1-butanol (DMB), a structural analogue of choline [112].

In a maternal high-fructose diet model, maternal DMB therapy showed protection against hypertension in adult rat offspring, which was relevant to the reduction of TMA and TMAO levels [94]. Another study demonstrated that perinatal resveratrol therapy protected adult rat offspring against maternal CKD-induced hypertension, which was associated with a decrease of the TMAO-to-TMA ratio [113]. These observations suggest a pathogenic link between the TMAO metabolic pathway and hypertension of developmental origins.

4.4. *Core Mechanisms Linking to Gut Microbiota*

Considering that various early-life insults during fetal development produce the same outcome—hypertension in adulthood—there might be some core mechanisms involved in the pathogenesis of hypertension of developmental origins. A number of mechanisms so far have been proposed, such as aberrant RAS, oxidative stress, reduced nephron numbers, dysregulated nutrient-sensing signals, inflammation, sex differences, epigenetic regulation [7–9,11–14]. Among them, some are interconnected to gut microbiota dysbiosis and will be discussed in turn.

4.4.1. Oxidative Stress

During pregnancy, the presence of excessive reactive oxygen species (ROS) under suboptimal in utero conditions may prevail over the defensive antioxidant system and compromise fetal development, leading to oxidative stress damage [114]. A review elsewhere indicated that there are various types of in utero insult stimuli linked to oxidative stress in mediating hypertension of developmental origins [115]. The main mechanisms underlying the actions of oxidative stress-related hypertension of developmental origins include increased ROS-producing enzyme expression [116], increased ROS formation [117], decreased antioxidant capacity [118], impaired NO signaling pathway [27], increased lipid peroxidation [119], increased oxidative DNA damage [43], and increased peroxynitrite production [120].

Data from several animal models listed in Table 1 shows that the connections between gut microbiota dysbiosis and oxidative stress may be involved in the pathogenesis of programmed hypertension, including maternal CKD [43], high-fructose diet [95], and high-fat diet models [110]. Gut microbial communities are able to elicit redox signaling and maintain host–microbiota homeostasis [121]. An imbalance in the redox state can lead to inflammatory responses and gut damage, resulting in gut microbiota dysbiosis. In a maternal CKD model, offspring developed hypertension related to increased oxidative stress and impaired NO signaling [43]. In a subsequent study, perinatal resveratrol therapy protected adult offspring against hypertension programmed by maternal CKD, accompanied with reshaping the gut microbiota and reducing oxidative stress concurrently [113].

Together, oxidative stress may work together with the gut microbiota under hypertension of developmental origins. More attention needs to be paid to evaluate how the gut microbiota interconnects with oxidative stress to elicit organ-specific programming processes behind hypertension, and whether antioxidant therapy in pregnancy may also benefit the gut microbiota to protect adult offspring against hypertension of developmental origins.

4.4.2. Aberrant RAS

The RAS is a major regulatory network that maintains BP, and blockade of the RAS has emerged as a therapeutic option for hypertension [122]. An increasing number of animal models linked to aberrant RAS are now being developed to evaluate hypertension of developmental programming, as reviewed elsewhere [54].

Within the RAS, regulation is achieved through a cascade of proteases generating some bioactive peptides [122]. The classical RAS consists of angiotensin-converting enzyme (ACE), angiotensin (ANG) II, and angiotensin II type 1 receptor (AT1R). Activation of the classical RAS elicits vasoconstriction and inflammation under pathophysiological conditions, consequently resulting in hypertension and its related complications [123].

On the other hand, the nonclassical RAS is composed of ACE2/angiotensin-(1-7) (Ang-(1-7))/Mas receptor/ANG II type 2 receptor (AT2R), by which it can counterbalance the adverse effects of ANG II [124]. A growing body of evidence supports that aberrant RAS plays a key role in developmental hypertension, and RAS-based interventions can be used as a reprogramming strategy to prevent hypertension [54].

Mounting evidence suggests a bidirectional interaction between the gut microbiota and RAS; alterations in RAS shift microbiota composition and metabolic activity, while gut microbiota-derived metabolites can modulate the gut RAS [125]. Through regulation of intestinal amino acid transport, prior research reported that ACE2 plays a key non-catalytic role in gut biology and modulation of the gut microbiota [126].

Adult rat progeny of CKD mothers developed hypertension, coinciding with decreased expression of Mas receptor and AT2R [43]. In another high-fructose diet plus TCDD exposure model, 3,3-dimethyl-1-butanol (DMB) therapy protected against hypertension, coinciding with a reduction of AT1R but an increase of AT2R protein abundance, as well as reshaping the gut microbiota [96]. Other developmental hypertension models such as a perinatal high-fat diet [99], maternal administration of minocycline [52], and maternal TMAO plus ADMA exposure [103] also interfere with aberrant RAS and gut microbiota dysbiosis.

As gut microbiota dysbiosis has been linked to hypertension by modulating the systemic and local RAS [127], these observations endorse the idea that the interaction between the RAS and the gut microbiota implicates the pathogenesis of developmental programming of hypertension, although this remains speculative.

4.4.3. Inflammation and Immune Response

Pregnancy is considered a physiologic systemic inflammatory response; compromised pregnancies and associated complications may be attributed to inflammation [128]. The accumulation of T cells, monocyte/macrophages, and T cell-derived cytokines is involved in the pathogenesis of hypertension [129]. An imbalance of T regulatory cells (Treg) and T helper 17 (TH17) cells has been linked to hypertension [129], which can be restored by postbiotic therapy [130]. In CKD, the interplay between Treg/TH17 balance and inflammation has also been related to hypertension [131]. Treg and TH17 cells can both be regulated by aryl hydrocarbon receptor (AhR) [132].

It is noted that several microbial tryptophan metabolites are uremic toxins as well as AhR ligands. AhR signaling can initiate inflammation through increasing monocyte adhesion, upregulating proinflammatory gene expression, reducing NO bioavailability, and inducing the expression of endothelial adhesion molecules [133]. Several gut microbiota-derived uremic toxins, like indoleacetic acid and indoxyl sulfate, have pro-oxidant, proinflammatory, procoagulant, and proapoptotic effects, all of which are involved in the pathogenesis of hypertension [134].

Using a rat model of maternal CKD-induced hypertension, we observed that maternal tryptophan therapy preventing offspring's hypertension coincides with restoration of the AhR signaling pathway and several tryptophan-metabolizing microbes [134]. Another study showed that TCDD-induced programming hypertension is related to TH17-induced renal inflammation, the activation of AhR signaling, and alterations of gut microbiota compositions [105]. In contrast, TCDD-induced activation of AhR signaling and TH17 responses can be restored by perinatal supplementation with resveratrol, an AhR antagonist. Additionally, resveratrol is reported to have benefits on offspring hypertension in several developmental hypertension models [45,46,95,102].

Although results from animal models support the role of inflammation and immunity on hypertension of developmental origins, more research is required to gain comprehensive insight into their interconnections with the gut microbiota and develop therapeutic potential of inflammation- or immune-targeted therapies in hypertension of developmental origins and associated organ damage.

5. Reprogramming Strategy: Gut Microbiota-Targeted Therapy

The idea from DOHaD research creates opportunities to reverse the programming process, namely reprogramming, by early intervention aiming to prevent hypertension of developmental origins later in life [135]. Current literature on animal studies for hypertension of developmental origins supports that gut microbiota-targeted therapy can work as a reprogramming strategy to prevent hypertension induced by various early-life insult stimuli.

Animal Models Used for Reprogramming

Here, we show Table 2 that summarizes studies documenting microbiota-targeted reprogramming therapies in animal models of developmental hypertension, restricting those starting before or upon disease onset [93,94,96,99,103,105,110,113,134,136,137]. The therapeutic duration is from pregnancy through lactation, which cover the periods of organogenesis. The literature review states that gut microbiota-targeted strategies used to prevent hypertension include probiotics, prebiotics, postbiotics, and dietary nutrients.

Table 2 illustrates that the most commonly used species are rats. Several models of developmental hypertension have been used to examine gut microbiota-targeted interventions, such as maternal high-fructose diet [93,94], perinatal high-fat diet [99,110,136], perinatal TCDD exposure [105], maternal adenine-induced CKD [113,134,137], maternal TMAO and ADMA exposure [103], and maternal high-fructose intake plus TCDD exposure [96].

Despite probiotics and prebiotics showing benefits in hypertension [91,92], there was very limited evidence in regard to their role on hypertension of developmental origins. Probiotic treatment with *Lactobacillus casei* in pregnancy and lactation prevents the development of hypertension in adult male rat offspring raised on a maternal high-fructose diet [93] or perinatal high-fat diet model [99].

As a prebiotic, inulin has a protective effect in hypertension of developmental origins [93,99]. A previous study using a maternal high-fructose model demonstrated that perinatal inulin treatment protects against offspring's hypertension, accompanied by increased abundance of *Lactobacillus*, the most-known probiotic strain [93]. Another study demonstrated that perinatal supplementing with inulin protects against maternal high-fructose-diet-induced hypertension, accompanied by increases of plasma propionate concentrations [99].

Table 2. Summary of animal models documenting gut microbiota-targeted therapies for hypertension of developmental origins.

Gut Microbiota-Targeted Therapies	Animal Models	Species/Gender	Age at Evaluation	Ref.
Probiotics				
Daily oral gavage of *Lactobacillus casei* (2×10^8 CFU/day)	Maternal high-fructose diet	SD rat/M	12 weeks	[93]
Daily oral gavage of *Lactobacillus casei* (2×10^8 CFU/day)	Perinatal high-fat diet	SD rat/M	16 weeks	[99]
Prebiotics				
5% *w/w* long chain inulin	Maternal high-fructose diet	SD rat/M	12 weeks	[93]
5% *w/w* long chain inulin	Perinatal high-fat diet	SD rat/M	16 weeks	[99]
Resveratrol (50 mg/L) in drinking water	Maternal TMAO and ADMA exposure	SD rat/M	12 weeks	[103]
Resveratrol (50 mg/L) in drinking water	Perinatal TCDD exposure model	SD rat/M	12 weeks	[105]
Resveratrol (50 mg/L) in drinking water	Maternal adenine-induced CKD	SD rat/M	12 weeks	[113]
Daily oral gavage of garlic oil (100 mg/kg/day)	Perinatal high-fat diet	SD rat/M	16 weeks	[110]
Postbiotics				
Magnesium acetate (200 mmol/L) in drinking water	Maternal high-fructose diet	SD rat/M	12 weeks	[94]
1% DMB in drinking water	Maternal high-fructose diet	SD rat/M	12 weeks	[94]
1% DMB in drinking water	Maternal high-fructose diet and TCDD exposure	SD rat/M	12 weeks	[96]
1% conjugated linoleic acid	Maternal high-fat diet	SD rat/M	18 weeks	[136]
Dietary Nutrients				
Daily oral gavage of tryptophan (200 mg/kg/day)	Maternal adenine-induced CKD	SD rat/M	12 weeks	[134]
Daily oral gavage of L- or D-cysteine (8 mmol/kg/day)	Maternal adenine-induced CKD	SD rat/M	12 weeks	[137]

Studies tabulated based on types of intervention and animal models. TCDD = 2,3,7,8-tetrachlorodibenzo-p-dioxin; CKD = chronic kidney disease; TMAO = trimethylamine-N-oxide; ADMA = asymmetric dimethylarginine; SD = Sprague-Dawley rat; DMB = 3,3-maternal dimethyl-1-butanol.

Additionally, resveratrol could be used to protect against adult disease of developmental origins due to its probiotic properties [138]. Studies using a maternal TMAO plus ADMA exposure rat model indicate that adult rat progeny born to dams exposed to uremic toxins develop hypertension [103]. Nonetheless, maternal resveratrol therapy rescues from hypertension programmed by TMAO plus ADMA exposure, accompanied by increased butyrate-producing bacteria and fecal butyrate level.

Another study demonstrated that adult rat progeny born to dams exposed to TCDD have hypertension [105], which is related to activation of AhR signaling, induction of TH17-dependent renal inflammation, and shifts of gut microbiota compositions [105]. Conversely, the induction of TH17- and AhR-mediated inflammation can be counterbalanced by perinatal resveratrol supplementation. The beneficial effects of resveratrol are also relevant to reshaping the gut microbiota by augmenting microbes that can inhibit TH17 responses and reducing the F/B ratio. In a maternal CKD model, adult rat progeny developed renal hypertrophy and hypertension [82]. Perinatal resveratrol therapy protects from hypertension, accompanied by restoration of microbial richness and diversity and an increase in beneficial microbes, *Lactobacillus* and *Bifidobacterium* [113]. Nevertheless, the low bioavailability of resveratrol limits its clinical translation [139]. In this regard, we synthesized resveratrol butyrate ester

(RBE) from resveratrol and butyrate by esterification to improve the efficacy [140]. We recently found that low-dose RBE (25 mg/L) is as effective as resveratrol (50 mg/L) in protecting against CKD-induced hypertension [141], despite the beneficial effect of RBE in models of developmental hypertension awaiting further evaluation.

Although there are many other prebiotic foods, only garlic oil has shown benefits on protection of perinatal high-fat-diet-induced hypertension in adult progeny [110]. The beneficial effects of garlic oil include increased α-diversity; increased plasma acetate, butyrate, and propionate; and increased beneficial bacteria *Lactobacillus* and *Bifidobacterium*.

In addition to probiotics and prebiotics, postbiotics are another gut microbiota-targeted modality. Postbiotics include various components like microbial cell fractions, extracellular vesicles, cell lysates, extracellular polysaccharides, functional proteins, cell wall-derived muropeptides, etc. [80]. Nevertheless, very limited information exists regarding the use of postbiotics in hypertension. As a postbiotic, acetate for perinatal supplementing showed benefits on maternal high-fructose-diet-induced hypertension [94].

Additionally, two studies demonstrated that maternal DMB therapy, an inhibitor of TMA formation, protects from hypertension in adult rat progeny exposed to maternal high-fructose diet with or without TCDD exposure [94,96]. Its beneficial effect was accompanied by affection of the metabolic pathway of TMA-TMAO and reshaping the gut microbiota. Another example of postbiotics use for hypertension of developmental origins is conjugated linoleic acid. One study showed the benefit of conjugated linoleic acid, a gut microbial metabolite derived from dietary polyunsaturated fatty acids, on high-fat-diet-induced hypertension [136].

Moreover, dietary nutrients have also been applied as gut microbiota-targeted therapies for hypertension of developmental origins. Former reviews have sufficiently illustrated the impact of diet on the gut microbiome [142,143]. Prior research demonstrating that specific nutrient intake can be beneficial to protecting from hypertension of developmental origins in various animal models [26,144]. These nutrients include folic acid [145], vitamin E [146], polyunsaturated fatty acids [147], and certain amino acids [27,148,149]. However, very few of them have been studied about their impact in gut microbiota related to developmental hypertension. One recent study showed that L-cysteine therapy protects adult offspring against maternal CKD-induced hypertension, associated with enhancement of beneficial genera *Oscillibacter* and *Butyricicoccus*, as well as depletion of indole-producing genera *Alistipes* and *Akkermansia* [137]. Another study reported that tryptophan supplementation during gestation prevents maternal CKD-induced offspring hypertension. The protective effect of tryptophan supplementation is related to alterations to several tryptophan-metabolizing microbes and the AHR signaling pathway [134].

Notably, increasing evidence supports the notion that altered gut microbial composition and function is evident in hypertension that can be evoked with diets high in salt [83,150–152]. Although maternal high salt consumption has been associated with offspring hypertension [29], much remains to be elucidated about the interplay between gut microbiota with high-sodium diets and the role of low-salt diet as a reprogramming strategy for hypertension of developmental origins.

Together, current evidence from animal models supports that modulation of gut microbiota compositions and its derived metabolites through gut microbiota-targeted therapies, in the long term, may enable the capacity to prevent the development of hypertension in a desired favorable direction.

6. Conclusions and Perspectives

Animal models have made significant contributions to research in hypertension of developmental origins, giving rise to substantial evidence of an interconnection between early-life insult stimuli, gut microbiota dysbiosis, and hypertension in adulthood. Our review provides insights into the importance of animal models not only in investigating underlying mechanisms behind hypertension of developmental origins, but also in developing early-life gut microbiota-targeted therapy to help prevent hypertension in later life. Considering that hypertension is a major hallmark of metabolic syndrome, aforementioned animal models may be potential models to evaluate developmental programming of metabolic syndrome-related disorders, not to mention that the link between gut microbiota and metabolic syndrome have been extensively discussed in many research papers [153,154].

Animal models are generally considered an intermediate step between bench and human trials. While animal studies are a regulatory requirement for validating preliminary experimental data, animals will remain indispensable in research for some time [155]. However, alternative approaches to animal models need to be investigated and adopted. The integration of computer models with modeling in vitro tissues and organs should be considered as alternative protocols to reduce the use of animals in scientific research [156].

To move the field forward, some unsolved aspects toward clinical translation need to be considered. Despite abovementioned early-life insults having been identified in animal models of developmental hypertension, there may be more risk factors in nature that can adversely influence the BP of adult progeny awaiting to be discovered.

Another important aspect is that the implication of the gut microbiota transfer early in life to the development of hypertension is still unknown, although FMT has been extensively studied in microbiome-associated pathologies, including hypertension [78,81,82]. However, currently, little information exists about their potential application in hypertension of developmental origins.

Animal studies suggest that early use of certain prebiotics, probiotics, or postbiotics may prevent the developmental programming of hypertension, while the exact mechanisms have not been entirely elucidated. What is absent in the literature is whether other prebiotic-rich foods or prebiotic-like components, either individually or in combination, in pregnancy and lactation can also change the gut microbiota to protect adult progeny against hypertension in various animal models.

In conclusion, gut microbiota dysbiosis is a meaningfully pathogenetic link for hypertension of developmental origins. Each of the aforementioned animal models was applied to examine a specific hypothesis, and neither can be considered superior in respect of all aspects of research on hypertension of developmental origins. After all this greater understanding of animal models used for DOHaD research and remarkable growth in gut microbiota-targeted therapies, we believe that translating this growing body of evidence into clinical practice is a valuable strategy to reduce the global hypertension-related burden.

Author Contributions: Conceptualization, C.-N.H. and Y.-L.T.; data curation, C.-N.H. and Y.-L.T.; funding acquisition, Y.-L.T.; project administration, C.-N.H. and Y.-L.T.; writing-original draft, C.-N.H. and Y.-L.T.; writing-review and editing, C.-N.H. and Y.-L.T. All authors have read and agreed to the published version of the manuscript.

Funding: This work was supported by the Chang Gung Memorial Hospital, Kaohsiung, Taiwan, grants CORPG8M0201, CORPG8M0151, and CORPG8M0081, and the Ministry of Science and Technology, Taiwan, grants MOST 110-2314-B-182-020-MY3 (Y.-L.T.) and MOST 110-2314-B-182A-029 (C.-N.H.).

Institutional Review Board Statement: Not applicable.

Informed Consent Statement: Not applicable.

Data Availability Statement: Data are contained within the article.

Conflicts of Interest: The authors declare no conflict of interest.

References

1. World Health Organization. Hypertension. 2022. Available online: https://www.who.int/health-topics/hypertension#tab=tab_1 (accessed on 25 February 2022).
2. Pinto, Y.M.; Paul, M.; Ganten, D. Lessons from rat models of hypertension: From Goldblatt to genetic engineering. *Cardiovasc. Res.* **1998**, *39*, 77–88. [CrossRef]
3. Lerman, L.O.; Kurtz, T.W.; Touyz, R.M.; Ellison, D.H.; Chade, A.R.; Crowley, S.D.; Mattson, D.L.; Mullins, J.J.; Osborn, J.; Eirin, A.; et al. Animal Models of Hypertension: A Scientific Statement From the American Heart Association. *Hypertension* **2019**, *73*, e87–e120. [CrossRef] [PubMed]

4. Padmanabhan, S.; Joe, B. Towards precision medicine for hypertension: A review of genomic, epigenomic, and microbiomic effects on blood pressure in experimental rat models and humans. *Physiol. Rev.* **2017**, *97*, 1469–1528. [CrossRef] [PubMed]
5. Mills, K.T.; Bundy, J.D.; Kelly, T.N.; Reed, J.E.; Kearney, P.M.; Reynolds, K.; Chen, J.; He, J. Global Disparities of Hypertension Prevalence and Control: A Systematic Analysis of Population-Based Studies From 90 Countries. *Circulation* **2016**, *134*, 441–450. [CrossRef]
6. Luyckx, V.A.; Bertram, J.F.; Brenner, B.M.; Fall, C.; E Hoy, W.; E Ozanne, S.; E Vikse, B. Effect of fetal and child health on kidney development and long-term risk of hypertension and kidney disease. *Lancet* **2013**, *382*, 273–283. [CrossRef]
7. Bagby, S.P. Maternal nutrition, low nephron number, and hypertension in later life: Pathways of nutritional programming. *J. Nutr.* **2007**, *137*, 1066–1072. [CrossRef]
8. Ojeda, N.B.; Grigore, D.; Alexander, B.T. Developmental programming of hypertension: Insight from animal models of nutritional manipulation. *Hypertension* **2008**, *52*, 44–50. [CrossRef]
9. Paauw, N.D.; Van Rijn, B.B.; Lely, A.T.; Joles, J.A. Pregnancy as a critical window for blood pressure regulation in mother and child: Programming and reprogramming. *Acta Physiol.* **2016**, *219*, 241–259. [CrossRef]
10. Hanson, M. The birth and future health of DOHaD. *J. Dev. Orig. Health Dis.* **2015**, *6*, 434–437. [CrossRef]
11. Hsu, C.N.; Tain, Y.L. The Double-Edged Sword Effects of Maternal Nutrition in the Developmental Programming of Hypertension. *Nutrients* **2018**, *10*, 1917. [CrossRef]
12. Hsu, C.N.; Tain, Y.L. Early Origins of Hypertension: Should Prevention Start Before Birth Using Natural Antioxidants? *Antioxidants* **2020**, *9*, 1034. [CrossRef] [PubMed]
13. Hsu, C.N.; Tain, Y.L. Adverse Impact of Environmental Chemicals on Developmental Origins of Kidney Disease and Hypertension. *Front. Endocrinol.* **2021**, *12*, 745716. [CrossRef] [PubMed]
14. Hsu, C.N.; Tain, Y.L. Animal Models for DOHaD Research: Focus on Hypertension of Developmental Origins. *Biomedicines* **2021**, *9*, 623. [CrossRef] [PubMed]
15. Yang, T.; Richards, E.M.; Pepine, C.J.; Raizada, M.K. The gut microbiota and the brain-gut-kidney axis in hypertension and chronic kidney disease. *Nat. Rev. Nephrol.* **2018**, *14*, 442–456. [CrossRef] [PubMed]
16. Marques, F.Z.; Mackay, C.R.; Kaye, D.M. Beyond gut feelings: How the gut microbiota regulates blood pressure. *Nat. Rev. Cardiol.* **2018**, *15*, 20–32. [CrossRef] [PubMed]
17. Palmu, J.; Lahti, L.; Niiranen, T. Targeting Gut Microbiota to Treat Hypertension: A Systematic Review. *Int. J. Environ. Res. Public Health* **2021**, *18*, 1248. [CrossRef]
18. Avery, E.G.; Bartolomaeus, H.; Maifeld, A.; Marko, L.; Wiig, H.; Wilck, N.; Rosshart, S.P.; Forslund, S.K.; Müller, D.N. The Gut Microbiome in Hypertension: Recent Advances and Future Perspectives. *Circ. Res.* **2021**, *128*, 934–950. [CrossRef]
19. Seidel, E.; Scholl, U.I. Genetic mechanisms of human hypertension and their implications for blood pressure physiology. *Physiol. Genom.* **2017**, *49*, 630–652. [CrossRef]
20. Chu, D.M.; Meyer, K.M.; Prince, A.L.; Aagaard, K.M. Impact of maternal nutrition in pregnancy and lactation on offspring gut microbial composition and function. *Gut Microbes* **2016**, *7*, 459–470. [CrossRef]
21. Hsu, C.N.; Hou, C.Y.; Hsu, W.H.; Tain, Y.L. Cardiovascular Diseases of Developmental Origins: Preventive Aspects of Gut Microbiota-Targeted Therapy. *Nutrients* **2021**, *13*, 2290. [CrossRef]
22. McMullen, S.; Mostyn, A. Animal models for the study of the developmental origins of health and disease. *Proc. Nutr. Soc.* **2009**, *68*, 306–320. [CrossRef] [PubMed]
23. Painter, R.C.; Roseboom, T.J.; Bleker, O.P. Prenatal exposure to the Dutch famine and disease in later life: An overview. *Reprod. Toxicol.* **2005**, *20*, 345–352. [CrossRef] [PubMed]
24. Stein, A.D.; Zybert, P.A.; van der Pal-de Bruin, K.; Lumey, L.H. Exposure to famine during gestation, size at birth, and blood pressure at age 59 y: Evidence from the Dutch Famine. *Eur. J. Epidemiol.* **2006**, *21*, 759–765. [CrossRef]
25. Langley-Evans, S.C.; Langley-Evans, A.J.; Marchand, M.C. Nutritional programming of blood pressure and renal morphology. *Arch. Physiol. Biochem.* **2003**, *111*, 8–16. [CrossRef]
26. Hsu, C.N.; Tain, Y.L. The Good, the Bad, and the Ugly of Pregnancy Nutrients and Developmental Programming of Adult Disease. *Nutrients* **2019**, *11*, 894. [CrossRef] [PubMed]
27. Tain, Y.L.; Hsieh, C.S.; Lin, I.C.; Chen, C.C.; Sheen, J.M.; Huang, L.T. Effects of maternal L-citrulline supplementation on renal function and blood pressure in offspring exposed to maternal caloric restriction: The impact of nitric oxide pathway. *Nitric Oxide* **2010**, *23*, 34–41. [CrossRef] [PubMed]
28. Zohdi, V.; Lim, K.; Pearson, J.T.; Black, M.J. Developmental programming of cardiovascular disease following intra uterine growth restriction: Findings utilising a rat model of maternal protein restriction. *Nutrients* **2014**, *7*, 119–152. [CrossRef]
29. Koleganova, N.; Piecha, G.; Ritz, E.; Becker, L.E.; Müller, A.; Weckbach, M.; Nyengaard, J.R.; Schirmacher, P.; Gross-Weissmann, M.L. Both high and low maternal salt intake in pregnancy alter kidney development in the offspring. *Am. J. Physiol. Ren. Physiol.* **2011**, *301*, F344–F354. [CrossRef]
30. Bergel, E.; Belizán, J.M. A deficient maternal calcium intake during pregnancy increases blood pressure of the offspring in adult rats. *BJOG* **2002**, *109*, 540–545. [CrossRef]
31. Tomat, A.; Elesgaray, R.; Zago, V.; Fasoli, H.; Fellet, A.; Balaszczuk, A.M.; Schreier, L.; Costa, M.A.; Arranz, C. Exposure to zinc deficiency in fetal and postnatal life determines nitric oxide system activity and arterial blood pressure levels in adult rats. *Br. J. Nutr.* **2010**, *104*, 382–389. [CrossRef]

32. Gambling, L.; Dunford, S.; Wallace, D.I.; Zuur, G.; Solanky, N.; Srai, K.S.; McArdle, H.J. Iron deficiency during pregnancy affects post-natal blood pressure in the rat. *J. Physiol.* **2003**, *552*, 603–610. [CrossRef] [PubMed]
33. Tain, Y.L.; Chan, J.Y.H.; Lee, C.T.; Hsu, C.N. Maternal melatonin therapy attenuates methyl-donor diet-induced programmed hypertension in male adult rat offspring. *Nutrients* **2018**, *10*, 1407. [CrossRef] [PubMed]
34. Tare, M.; Emmett, S.J.; Coleman, H.A.; Skordilis, C.; Eyles, D.W.; Morley, R.; Parkington, H.C. Vitamin D insufficiency is associated with impaired vascular endothelial and smooth muscle function and hypertension in young rats. *J. Physiol.* **2011**, *589*, 4777–4786. [CrossRef] [PubMed]
35. Williams, L.; Seki, Y.; Vuguin, P.M.; Charron, M.J. Animal models of in utero exposure to a high fat diet: A review. *Biochim. Biophys. Acta* **2014**, *1842*, 507–519. [CrossRef] [PubMed]
36. Tain, Y.L.; Lin, Y.J.; Sheen, J.M.; Yu, H.R.; Tiao, M.M.; Chen, C.C.; Tsai, C.C.; Huang, L.T.; Hsu, C.N. High fat diets sex-specifically affect the renal transcriptome and program obesity, kidney injury, and hypertension in the offspring. *Nutrients* **2017**, *9*, 357. [CrossRef]
37. Seong, H.Y.; Cho, H.M.; Kim, M.; Kim, I. Maternal High-Fructose Intake Induces Multigenerational Activation of the Renin Angiotensin-Aldosterone System. *Hypertension* **2019**, *74*, 518–525. [CrossRef] [PubMed]
38. Tain, Y.L.; Chan, J.Y.; Hsu, C.N. Maternal Fructose Intake Affects Transcriptome Changes and Programmed Hypertension in Offspring in Later Life. *Nutrients* **2016**, *8*, 757. [CrossRef]
39. Thone-Reineke, C.; Kalk, P.; Dorn, M.; Klaus, S.; Simon, K.; Pfab, T.; Godes, M.; Persson, P.; Unger, T.; Hocher, B. High-protein nutrition during pregnancy and lactation programs blood pressure, food efficiency, and body weight of the offspring in a sex-dependent manner. *Am. J. Physiol. Regul. Integr. Comp. Physiol.* **2006**, *291*, R1025–R1030. [CrossRef]
40. Hsu, C.N.; Hou, C.Y.; Chang-Chien, G.P.; Lin, S.; Tain, Y.L. Maternal N-Acetylcysteine Therapy Prevents Hypertension in Spontaneously Hypertensive Rat Offspring: Implications of Hydrogen Sulfide-Generating Pathway and Gut Microbiota. *Antioxidants* **2020**, *9*, 856. [CrossRef]
41. Tain, Y.L.; Lee, C.T.; Chan, J.Y.; Hsu, C.N. Maternal melatonin or N-acetylcysteine therapy regulates hydrogen sulfide-generating pathway and renal transcriptome to prevent prenatal N(G)-Nitro-L-arginine methyl ester (L-NAME)-induced fetal programming of hypertension in adult male offspring. *Am. J. Obstet. Gynecol.* **2016**, *215*, 636. [CrossRef]
42. Tain, Y.L.; Lee, W.C.; Hsu, C.N.; Lee, W.C.; Huang, L.T.; Lee, C.T.; Lin, C.Y. Asymmetric dimethylarginine is associated with developmental programming of adult kidney disease and hypertension in offspring of streptozotocin-treated mothers. *PLoS ONE* **2013**, *8*, e55420. [CrossRef]
43. Hsu, C.N.; Yang, H.W.; Hou, C.Y.; Chang-Chien, G.P.; Lin, S.; Tain, Y.L. Maternal Adenine-Induced Chronic Kidney Disease Programs Hypertension in Adult Male Rat Offspring: Implications of Nitric Oxide and Gut Microbiome Derived Metabolites. *Int. J. Mol. Sci.* **2020**, *21*, 7237. [CrossRef] [PubMed]
44. Giussani, D.A.; Camm, E.J.; Niu, Y.; Richter, H.G.; Blanco, C.E.; Gottschalk, R.; Blake, E.Z.; Horder, K.A.; Thakor, A.S.; Hansell, J.A.; et al. Developmental programming of cardiovascular dysfunction by prenatal hypoxia and oxidative stress. *PLoS ONE* **2012**, *7*, e31017. [CrossRef] [PubMed]
45. Hsu, C.N.; Lin, Y.J.; Lu, P.C.; Tain, Y.L. Maternal resveratrol therapy protects male rat offspring against programmed hypertension induced by TCDD and dexamethasone exposures: Is it relevant to aryl hydrocarbon receptor? *Int. J. Mol. Sci.* **2018**, *19*, 2459. [CrossRef] [PubMed]
46. Hsu, C.N.; Lin, Y.J.; Tain, Y.L. Maternal exposure to bisphenol A combined with high-fat diet-induced programmed hypertension in adult male rat offspring: Effects of resveratrol. *Int. J. Mol. Sci.* **2019**, *20*, 4382. [CrossRef] [PubMed]
47. Xiao, D.; Huang, X.; Li, Y.; Dasgupta, C.; Wang, L.; Zhang, L. Antenatal Antioxidant Prevents Nicotine-Mediated Hypertensive Response in Rat Adult Offspring. *Biol. Reprod.* **2015**, *93*, 66. [CrossRef]
48. Serapiao-Moraes, D.F.; Souza-Mello, V.; Aguila, M.B.; Mandarim-de-Lacerda, C.A.; Faria, T.S. Maternal caffeine administration leads to adverse effects on adult mice offspring. *Eur. J. Nutr.* **2013**, *52*, 1891–1900. [CrossRef]
49. Slabiak-Blaz, N.; Adamczak, M.; Gut, N.; Grajoszek, A.; Nyengaard, J.R.; Ritz, E.; Wiecek, A. Administration of cyclosporine a in pregnant rats—The effect on blood pressure and on the glomerular number in their offspring. *Kidney Blood Press. Res.* **2015**, *40*, 413–423. [CrossRef]
50. Chahoud, I.; Stahlmann, R.; Merker, H.J.; Neubert, D. Hypertension and nephrotoxic lesions in rats 1 year after prenatal exposure to gentamicin. *Arch. Toxicol.* **1988**, *62*, 274–284. [CrossRef]
51. Gois, P.H.; Canale, D.; Luchi, W.M.; Volpini, R.A.; Veras, M.M.; Costa Nde, S.; Shimizu, M.H.; Seguro, A.C. Tenofovir during pregnancy in rats: A novel pathway for programmed hypertension in the offspring. *J. Antimicrob. Chemother.* **2015**, *70*, 1094–1105. [CrossRef]
52. Hsu, C.N.; Chan, J.Y.H.; Wu, K.L.H.; Yu, H.R.; Lee, W.C.; Hou, C.Y.; Tain, Y.L. Altered Gut Microbiota and Its Metabolites in Hypertension of Developmental Origins: Exploring Differences between Fructose and Antibiotics Exposure. *Int. J. Mol. Sci.* **2021**, *22*, 2674. [CrossRef] [PubMed]
53. Tain, Y.L.; Sheen, J.M.; Chen, C.C.; Yu, H.R.; Tiao, M.M.; Kuo, H.C.; Huang, L.T. Maternal citrulline supplementation prevents prenatal dexamethasone-induced programmed hypertension. *Free Radic. Res.* **2014**, *48*, 580–586. [CrossRef] [PubMed]
54. Hsu, C.N.; Tain, Y.L. Targeting the Renin-Angiotensin-Aldosterone System to Prevent Hypertension and Kidney Disease of Developmental Origins. *Int. J. Mol. Sci.* **2021**, *22*, 2298. [CrossRef] [PubMed]

55. Tain, Y.L.; Hsu, C.N. Interplay between Oxidative Stress and Nutrient Sensing Signaling in the Developmental Origins of Cardiovascular Disease. *Int. J. Mol. Sci.* **2017**, *18*, 841. [CrossRef]
56. Ojeda, N.B.; Intapad, S.; Alexander, B.T. Sex differences in the developmental programming of hypertension. *Acta Physiol.* **2014**, *210*, 307–316. [CrossRef] [PubMed]
57. Tomat, A.L.; Salazar, F.J. Mechanisms involved in developmental programming of hypertension and renal diseases. Gender differences. *Horm. Mol. Biol. Clin. Investig.* **2014**, *18*, 63–77. [CrossRef]
58. Scherrer, U.; Rimoldi, S.F.; Sartori, C.; Messerli, F.H.; Rexhaj, E. Fetal programming and epigenetic mechanisms in arterial hypertension. *Curr. Opin. Cardiol.* **2015**, *30*, 393–397. [CrossRef] [PubMed]
59. Hsu, C.N.; Tain, Y.L. Regulation of Nitric Oxide Production in the Developmental Programming of Hypertension and Kidney Disease. *Int. J. Mol. Sci.* **2019**, *20*, 681. [CrossRef]
60. Dickinson, H.; Moss, T.J.; Gatford, K.L.; Moritz, K.M.; Akison, L.; Fullston, T.; Hryciw, D.H.; Maloney, C.A.; Morris, M.J.; Wooldridge, A.L.; et al. A review of fundamental principles for animal models of DOHaD research: An Australian perspective. *J. Dev. Orig. Health Dis.* **2016**, *7*, 449–472. [CrossRef]
61. Mossa, F.; Carter, F.; Walsh, S.W.; Kenny, D.A.; Smith, G.W.; Ireland, J.L.; Hildebrandt, T.B.; Lonergan, P.; Ireland, J.J.; Evans, A.C. Maternal undernutrition in cows impairs ovarian and cardiovascular systems in their offspring. *Biol. Reprod.* **2013**, *88*, 92. [CrossRef] [PubMed]
62. Gopalakrishnan, G.S.; Gardner, D.S.; Rhind, S.M.; Rae, M.T.; Kyle, C.E.; Brooks, A.N.; Walker, R.M.; Ramsay, M.M.; Keisler, D.H.; Stephenson, T.; et al. Programming of adult cardiovascular function after early maternal undernutrition in sheep. *Am. J. Physiol. Regul. Integr. Comp. Physiol.* **2004**, *287*, R12–R20. [CrossRef] [PubMed]
63. Brain, K.L.; Allison, B.J.; Niu, Y.; Cross, C.M.; Itani, N.; Kane, A.D.; Herrera, E.A.; Skeffington, K.L.; Botting, K.J.; Giussani, D.A. Intervention against hypertension in the next generation programmed by developmental hypoxia. *PLoS Biol.* **2019**, *17*, e2006552. [CrossRef] [PubMed]
64. Kuo, A.H.; Li, C.; Li, J.; Huber, H.F.; Nathanielsz, P.W.; Clarke, G.D. Cardiac remodeling in a baboon model of intrauterinegrowth restriction mimics accelerated ageing. *J. Physiol.* **2017**, *595*, 1093–1110. [CrossRef] [PubMed]
65. Gonzalez-Bulnes, A.; Astiz, S.; Ovilo, C.; Lopez-Bote, C.J.; Torres-Rovira, L.; Barbero, A.; Ayuso, M.; Garcia-Contreras, C.; Vazquez-Gomez, M. Developmental Origins of Health and Disease in swine: Implications for animal production and biomedical research. *Theriogenology* **2016**, *86*, 110–119. [CrossRef]
66. Chavatte-Palmer, P.; Tarrade, A.; Rousseau-Ralliard, D. Diet before and during Pregnancy and Offspring Health: The Importance of Animal Models and What Can Be Learned from Them. *Int. J. Environ. Res. Public Health* **2016**, *13*, 586. [CrossRef]
67. Morrison, J.L.; Botting, K.J.; Darby, J.R.T.; David, A.L.; Dyson, R.M.; Gatford, K.L.; Gray, C.; Herrera, E.A.; Hirst, J.J.; Kim, B.; et al. Guinea pig models for translation of the developmental origins of health and disease hypothesis into the clinic. *J. Physiol.* **2018**, *596*, 5535–5569. [CrossRef]
68. Hartman, H.A.; Lai, H.L.; Patterson, L.T. Cessation of renal morphogenesis in mice. *Dev. Biol.* **2007**, *310*, 379–387. [CrossRef]
69. Sengupta, P. The Laboratory Rat: Relating Its Age with Human's. *Int. J. Prev. Med.* **2013**, *4*, 624–630.
70. Barry, J.S.; Anthony, R.V. The pregnant sheep as a model for human pregnancy. *Theriogenology* **2008**, *69*, 55–67. [CrossRef]
71. Chahoud, I.; Paumgartten, F.J.R. Influence of litter size on the postnatal growth of rat pups: Is there a rationale for litter-size standardization in toxicity studies? *Environ. Res.* **2009**, *109*, 1021–1027. [CrossRef]
72. Krege, J.H.; Hodgin, J.B.; Hagaman, J.R.; Smithies, O. A noninvasive computerized tail-cuff system for measuring blood pressure in mice. *Hypertension* **1995**, *25*, 1111–1115. [CrossRef] [PubMed]
73. Lynch, S.V.; Pedersen, O. The Human Intestinal Microbiome in Health and Disease. *N. Engl. J. Med.* **2016**, *375*, 2369–2379. [CrossRef] [PubMed]
74. Milani, C.; Duranti, S.; Bottacini, F.; Casey, E.; Turroni, F.; Mahony, J.; Belzer, C.; Delgado Palacio, S.; Arboleya Montes, S.; Mancabelli, L.; et al. The First Microbial Colonizers of the Human Gut: Composition, Activities, and Health Implications of the Infant Gut Microbiota. *Microbiol. Mol. Biol. Rev.* **2017**, *81*, e00036-17. [CrossRef]
75. Matamoros, S.; Gras-Leguen, C.; Le Vacon, F.; Potel, G.; De La Cochetiere, M.-F. Development of intestinal microbiota in infants and its impact on health. *Trends Microbiol.* **2013**, *21*, 167–173. [CrossRef] [PubMed]
76. Arrieta, M.C.; Stiemsma, L.T.; Amenyogbe, N.; Brown, E.M.; Finlay, B. The intestinal microbiome in early life: Health and disease. *Front. Immunol.* **2014**, *5*, 427. [CrossRef]
77. Vandenplas, Y.; Carnielli, V.P.; Ksiazyk, J.; Luna, M.S.; Migacheva, N.; Mosselmans, J.M.; Picaud, J.C.; Possner, M.; Singhal, A.; Wabitsch, M. Factors affecting early-life intestinal microbiota development. *Nutrition* **2020**, *78*, 110812. [CrossRef]
78. Zhang, C.; Franklin, C.L.; Ericsson, A.C. Consideration of Gut Microbiome in Murine Models of Diseases. *Microorganisms* **2021**, *9*, 1062. [CrossRef]
79. Pandey, K.R.; Naik, S.R.; Vakil, B.V. Probiotics, prebiotics and synbiotics—A review. *J. Food Sci. Technol.* **2015**, *52*, 7577–7587. [CrossRef]
80. Żółkiewicz, J.; Marzec, A.; Ruszczyn'ski, M.; Feleszko, W. Postbiotics—A step beyond pre- and probiotics. *Nutrients* **2020**, *12*, 2189. [CrossRef]
81. Ericsson, A.C.; Personett, A.R.; Turner, G.; Dorfmeyer, R.A.; Franklin, C.L. Variable Colonization after reciprocal fecal microbiota transfer between mice with low and high richness microbiota. *Front. Microbiol.* **2017**, *8*, 196. [CrossRef]

82. Robertson, S.J.; Lemire, P.; Maughan, H.; Goethel, A.; Turpin, W.; Bedrani, L.; Guttman, D.S.; Croitoru, K.; Girardin, S.E.; Philpott, D.J. Comparison of co-housing and littermate methods for microbiota standardization in mouse models. *Cell Rep.* **2019**, *27*, 1910–1919.e2. [CrossRef] [PubMed]
83. Wilck, N.; Matus, M.G.; Kearney, S.M.; Olesen, S.W.; Forslund, K.; Bartolomaeus, H.; Haase, S.; Mähler, A.; Balogh, A.; Markó, L.; et al. Salt-responsive gut commensal modulates TH17 axis and disease. *Nature* **2017**, *551*, 585–589. [CrossRef] [PubMed]
84. Robles-Vera, I.; de la Visitación, N.; Toral, M.; Sánchez, M.; Romero, M.; Gómez-Guzmán, M.; Yang, T.; Izquierdo-García, J.L.; Guerra-Hernández, E.; Ruiz-Cabello, J.; et al. Probiotic Bifidobacterium breve prevents DOCA-salt hypertension. *FASEB J.* **2020**, *34*, 13626–13640. [CrossRef]
85. Marques, F.Z.; Nelson, E.; Chu, P.Y.; Horlock, D.; Fiedler, A.; Ziemann, M.; Tan, J.K.; Kuruppu, S.; Rajapakse, N.W.; El-Osta, A.; et al. High-fiber diet and acetate supplementation change the gut microbiota and prevent the development of hypertension and heart failure in hypertensive mice. *Circulation* **2017**, *135*, 964–977. [CrossRef]
86. Joe, B.; McCarthy, C.G.; Edwards, J.M.; Cheng, X.; Chakraborty, S.; Yang, T.; Golonka, R.M.; Mell, B.; Yeo, J.Y.; Bearss, N.R.; et al. Microbiota Introduced to Germ-Free Rats Restores Vascular Contractility and Blood Pressure. *Hypertension* **2020**, *76*, 1847–1855. [CrossRef] [PubMed]
87. Li, J.; Zhao, F.; Wang, Y.; Chen, J.; Tao, J.; Tian, G.; Wu, S.; Liu, W.; Cui, Q.; Geng, B.; et al. Gut microbiota dysbiosis contributes to the development of hypertension. *Microbiome* **2017**, *5*, 14. [CrossRef] [PubMed]
88. Pluznick, J.L. Microbial short-chain fatty acids and blood pressure regulation. *Curr. Hypertens. Rep.* **2017**, *19*, 25. [CrossRef]
89. Velasquez, M.T.; Ramezani, A.; Manal, A.; Raj, D.S. Trimethylamine N-Oxide: The good, the bad and the unknown. *Toxins* **2016**, *8*, 326. [CrossRef]
90. Schiattarella, G.G.; Sannino, A.; Toscano, E.; Giugliano, G.; Gargiulo, G.; Franzone, A.; Trimarco, B.; Esposito, G.; Perrino, C. Gut microbe-generated metabolite trimethylamine-N-oxide as cardiovascular risk biomarker: A systematic review and dose-response meta-analysis. *Eur. Heart J.* **2017**, *38*, 2948–2956. [CrossRef]
91. Chi, C.; Li, C.; Wu, D.; Buys, N.; Wang, W.; Fan, H.; Sun, J. Effects of Probiotics on Patients with Hypertension: A Systematic Review and Meta-Analysis. *Curr. Hypertens. Rep.* **2020**, *22*, 34. [CrossRef]
92. Ried, K. Garlic lowers blood pressure in hypertensive subjects, improves arterial stiffness and gut microbiota: A review and meta-analysis. *Exp. Ther. Med.* **2020**, *19*, 1472–1478. [CrossRef] [PubMed]
93. Hsu, C.N.; Lin, Y.J.; Hou, C.Y.; Tain, Y.L. Maternal Administration of Probiotic or Prebiotic Prevents Male Adult Rat Offspring against Developmental Programming of Hypertension Induced by High Fructose Consumption in Pregnancy and Lactation. *Nutrients* **2018**, *10*, 1229. [CrossRef] [PubMed]
94. Hsu, C.N.; Chang-Chien, G.P.; Lin, S.; Hou, C.Y.; Tain, Y.L. Targeting on Gut Microbial Metabolite Trimethylamine-N-Oxide and Short-Chain Fatty Acid to Prevent Maternal High-Fructose-Diet-Induced Developmental Programming of Hypertension in Adult Male Offspring. *Mol. Nutr. Food Res.* **2019**, *63*, e1900073. [CrossRef] [PubMed]
95. Tain, Y.L.; Lee, W.C.; Wu, K.L.H.; Leu, S.; Chan, J.Y.H. Resveratrol Prevents the Development of Hypertension Programmed by Maternal Plus Post-Weaning High-Fructose Consumption Through Modulation of Oxidative Stress, Nutrient-Sensing Signals, and Gut Microbiota. *Mol. Nutr. Food Res.* **2018**, *62*, e1800066. [CrossRef]
96. Hsu, C.N.; Chan, J.Y.H.; Yu, H.R.; Lee, W.C.; Wu, K.L.H.; Chang-Chien, G.P.; Lin, S.; Hou, C.Y.; Tain, Y.L. Targeting on Gut Microbiota-Derived Metabolite Trimethylamine to Protect Adult Male Rat Offspring against Hypertension Programmed by Combined Maternal High-Fructose Intake and Dioxin Exposure. *Int. J. Mol. Sci.* **2020**, *21*, 5488. [CrossRef]
97. Guimarães, K.S.L.; Braga, V.A.; Noronha, S.I.S.R.; Costa, W.K.A.D.; Makki, K.; Cruz, J.C.; Brandão, L.R.; Chianca Junior, D.A.; Meugnier, E.; Leulier, F.; et al. Lactiplantibacillus plantarum WJL administration during pregnancy and lactation improves lipid profile, insulin sensitivity and gut microbiota diversity in dyslipidemic dams and protects male offspring against cardiovascular dysfunction in later life. *Food Funct.* **2020**, *11*, 8939–8950. [CrossRef]
98. Hsu, C.N.; Hou, C.Y.; Lee, C.T.; Chan, J.Y.H.; Tain, Y.L. The Interplay between Maternal and Post-Weaning High-Fat Diet and Gut Microbiota in the Developmental Programming of Hypertension. *Nutrients* **2019**, *11*, 1982. [CrossRef]
99. Hsu, C.N.; Hou, C.; Chan, J.Y.H.; Lee, C.T.; Tain, Y.L. Hypertension Programmed by Perinatal High-Fat Diet: Effect of Maternal Gut Microbiota-Targeted Therapy. *Nutrients* **2019**, *11*, 2908. [CrossRef]
100. Li, H.B.; Yang, T.; Richards, E.M.; Pepine, C.J.; Raizada, M.K. Maternal Treatment with Captopril Persistently Alters Gut-Brain Communication and Attenuates Hypertension of Male Offspring. *Hypertension* **2020**, *75*, 1315–1324. [CrossRef]
101. De Oliveira, Y.; Cavalcante, R.G.S.; Cavalcanti Neto, M.P.; Magnani, M.; Braga, V.A.; de Souza, E.L.; de Brito Alves, J.L. Oral administration of Lactobacillus fermentum post-weaning improves the lipid profile and autonomic dysfunction in rat offspring exposed to maternal dyslipidemia. *Food Funct.* **2020**, *11*, 5581–5594. [CrossRef]
102. Chen, H.E.; Lin, Y.J.; Lin, I.C.; Yu, H.R.; Sheen, J.M.; Tsai, C.C.; Huang, L.T.; Tain, Y.L. Resveratrol prevents combined prenatal N^G-nitro-L-arginine-methyl ester (L-NAME) treatment plus postnatal high-fat diet induced programmed hypertension in adult rat offspring: Interplay between nutrient-sensing signals, oxidative stress and gut microbiota. *J. Nutr. Biochem.* **2019**, *70*, 28–37. [CrossRef] [PubMed]
103. Hsu, C.N.; Hou, C.Y.; Chang-Chien, G.P.; Lin, S.; Chan, J.Y.H.; Lee, C.T.; Tain, Y.L. Maternal resveratrol therapy protected adult rat offspring against hypertension programmed by combined exposures to asymmetric dimethylarginine and trimethylamine-N-oxide. *J. Nutr. Biochem.* **2021**, *93*, 108630. [CrossRef] [PubMed]

104. Hsu, C.N.; Hou, C.Y.; Lee, C.T.; Chang-Chien, G.P.; Lin, S.; Tain, Y.L. Maternal 3,3-Dimethyl-1-Butanol Therapy Protects Adult Male Rat Offspring against Hypertension Programmed by Perinatal TCDD Exposure. *Nutrients* **2021**, *13*, 3041. [CrossRef] [PubMed]
105. Hsu, C.N.; Hung, C.H.; Hou, C.Y.; Chang, C.I.; Tain, Y.L. Perinatal Resveratrol Therapy to Dioxin-Exposed Dams Prevents the Programming of Hypertension in Adult Rat Offspring. *Antioxidants* **2021**, *10*, 1393. [CrossRef] [PubMed]
106. Sherman, S.B.; Sarsour, N.; Salehi, M.; Schroering, A.; Mell, B.; Joe, B.; Hill, J.W. Prenatal androgen exposure causes hypertension and gut microbiota dysbiosis. *Gut Microbes* **2018**, *9*, 400–421. [CrossRef] [PubMed]
107. Cani, P.D.; de Vos, W.M. Next-Generation Beneficial Microbes: The Case of Akkermansia muciniphila. *Front. Microbiol.* **2017**, *8*, 1765. [CrossRef]
108. DiRienzo, D.B. Effect of probiotics on biomarkers of cardiovascular disease: Implications for heart-healthy diets. *Nutr. Rev.* **2014**, *72*, 18–29. [CrossRef]
109. Pluznick, J.L. Renal and cardiovascular sensory receptors and blood pressure regulation. *Am. J. Physiol. Ren. Physiol.* **2013**, *305*, F439–F444. [CrossRef]
110. Hsu, C.N.; Hou, C.Y.; Chang-Chien, G.P.; Lin, S.; Tain, Y.L. Maternal Garlic Oil Supplementation Prevents High-Fat Diet-Induced Hypertension in Adult Rat Offspring: Implications of H2S-Generating Pathway in the Gut and Kidneys. *Mol. Nutr. Food Res.* **2021**, *65*, e2001116. [CrossRef]
111. Seldin, M.M.; Meng, Y.; Qi, H.; Zhu, W.; Wang, Z.; Hazen, S.L.; Lusis, A.J.; Shih, D.M. Trimethylamine N-Oxide promotes vascular inflammation through signaling of mitogen-activated protein kinase and nuclear factor-κB. *J. Am. Heart Assoc.* **2016**, *5*, e002767. [CrossRef]
112. Wang, Z.; Roberts, A.B.; Buffa, J.A.; Levison, B.S.; Zhu, W.; Org, E.; Gu, X.; Huang, Y.; Zamanian-Daryoush, M.; Culley, M.K.; et al. Non-lethal Inhibition of Gut Microbial Trimethylamine Production for the Treatment of Atherosclerosis. *Cell* **2015**, *163*, 1585–1595. [CrossRef] [PubMed]
113. Hsu, C.N.; Hou, C.Y.; Chang-Chien, G.P.; Lin, S.; Yang, H.W.; Tain, Y.L. Perinatal Resveratrol Therapy Prevents Hypertension Programmed by Maternal Chronic Kidney Disease in Adult Male Offspring: Implications of the Gut Microbiome and Their Metabolites. *Biomedicines* **2020**, *8*, 567. [CrossRef] [PubMed]
114. Dennery, P.A. Oxidative stress in development: Nature or nurture? *Free Radic. Biol. Med.* **2010**, *49*, 1147–1151. [CrossRef] [PubMed]
115. Tain, Y.L.; Hsu, C.N. Oxidative Stress-Induced Hypertension of Developmental Origins: Preventive Aspects of Antioxidant Therapy. *Antioxidants* **2022**, *11*, 511. [CrossRef] [PubMed]
116. Liu, Y.; Qi, L.; Wu, J.; Xu, T.; Yang, C.; Chen, X.; Lv, J.; Xu, Z. Prenatal high-salt diet impaired vasodilatation with reprogrammed renin-angiotensin system in offspring rats. *J. Hypertens.* **2018**, *36*, 2369–2379. [CrossRef]
117. Svitok, P.; Okuliarova, M.; Varga, I.; Zeman, M. Renal impairment induced by prenatal exposure to angiotensin II in male rat offspring. *Exp. Biol. Med.* **2019**, *244*, 923–931. [CrossRef]
118. Do Nascimento, L.C.P.; Neto, J.P.R.C.; de Andrade Braga, V.; Lagranha, C.J.; de Brito Alves, J.L. Maternal exposure to high-fat and high-cholesterol diet induces arterial hypertension and oxidative stress along the gut-kidney axis in rat offspring. *Life Sci.* **2020**, *261*, 118367. [CrossRef]
119. Chen, L.; Zadi, Z.H.; Zhang, J.; Scharf, S.M.; Pae, E.K. Intermittent hypoxia in utero damages postnatal growth and cardiovascular function in rats. *J. Appl. Physiol.* **2018**, *124*, 821–830. [CrossRef]
120. Piecha, G.; Koleganova, N.; Ritz, E.; Müller, A.; Fedorova, O.V.; Bagrov, A.Y.; Lutz, D.; Schirmacher, P.; Gross-Weissmann, M.L. High salt intake causes adverse fetal programming–vascular effects beyond blood pressure. *Nephrol. Dial. Transplant.* **2012**, *27*, 3464–3476. [CrossRef]
121. Campbell, E.L.; Colgan, S.P. Control and dysregulation of redox signalling in the gastrointestinal tract. *Nat. Rev. Gastroenterol. Hepatol.* **2019**, *16*, 106–120. [CrossRef]
122. Te Riet, L.; van Esch, J.H.; Roks, A.J.; van den Meiracker, A.H.; Danser, A.H. Hypertension: Renin-angiotensin-aldosterone system alterations. *Circ. Res.* **2015**, *116*, 960–975. [CrossRef]
123. Forrester, S.J.; Booz, G.W.; Sigmund, C.D.; Coffman, T.M.; Kawai, T.; Rizzo, V.; Scalia, R.; Eguchi, S. Angiotensin II Signal Transduction: An Update on Mechanisms of Physiology and Pathophysiology. *Physiol. Rev.* **2018**, *98*, 1627–1738. [CrossRef] [PubMed]
124. Paz Ocaranza, M.; Riquelme, J.A.; García, L.; Jalil, J.E.; Chiong, M.; Santos, R.A.S.; Lavandero, S. Counter-regulatory renin angiotensin system in cardiovascular disease. *Nat. Rev. Cardiol.* **2020**, *17*, 116–129. [CrossRef] [PubMed]
125. Jaworska, K.; Koper, M.; Ufnal, M. Gut microbiota and renin-angiotensin system: A complex interplay at local and systemic levels. *Am. J. Physiol. Liver Physiol.* **2021**, *321*, G355–G366. [CrossRef] [PubMed]
126. Oliveira Andrade, J.M.; de Farias Lelis, D.; Mafra, V.; Cota, J. The angiotensin converting enzyme 2 (ACE2), gut microbiota, and cardiovascular health. *Protein Pept. Lett.* **2017**, *24*, 827–832. [CrossRef]
127. Richards, E.M.; Pepine, C.J.; Raizada, M.K.; Kim, S. The gut, its microbiome, and hypertension. *Curr. Hypertens. Rep.* **2017**, *19*, 36. [CrossRef]
128. Challis, J.R.; Lockwood, C.J.; Myatt, L.; Norman, J.E.; Strauss, J.F.; Petraglia, F. Inflammation and pregnancy. *Reprod. Sci.* **2009**, *16*, 206–215. [CrossRef]

129. McMaster, W.G.; Kirabo, A.; Madhur, M.S.; Harrison, D.G. Inflammation, immunity, and hypertensive end-organ damage. *Circ. Res.* **2015**, *116*, 1022–1033. [CrossRef]
130. Ren, J.; Crowley, S.D. Role of T-cell activation in salt-sensitive hypertension. *Am. J. Physiol. Heart Circ. Physiol.* **2019**, *316*, H1345–H1353. [CrossRef]
131. Zhang, J.; Hua, G.; Zhang, X.; Tong, R.; Du, X.; Li, Z. Regulatory T cells/T-helper cell 17 functional imbalance in uraemic patients on maintenance haemodialysis: A pivotal link between microinflammation and adverse cardiovascular events. *Nephrology* **2010**, *15*, 33–41. [CrossRef]
132. Stevens, E.A.; Mezrich, J.D.; Bradfield, C.A. The aryl hydrocarbon receptor: A perspective on potential roles in the immune system. *Immunology* **2009**, *127*, 299–311. [CrossRef]
133. Sallée, M.; Dou, L.; Cerini, C.; Poitevin, S.; Brunet, P.; Burtey, S. The aryl hydrocarbon receptor-activating effect of uremic toxins from tryptophan metabolism: A new concept to understand cardiovascular complications of chronic kidney disease. *Toxins* **2014**, *6*, 934–949. [CrossRef]
134. Hsu, C.N.; Lin, I.C.; Yu, H.R.; Huang, L.T.; Tiao, M.M.; Tain, Y.L. Maternal Tryptophan Supplementation Protects Adult Rat Offspring against Hypertension Programmed by Maternal Chronic Kidney Disease: Implication of Tryptophan-Metabolizing Microbiome and Aryl Hydrocarbon Receptor. *Int. J. Mol. Sci.* **2020**, *21*, 4552. [CrossRef] [PubMed]
135. Tain, Y.L.; Joles, J.A. Reprogramming: A preventive strategy in hypertension focusing on the kidney. *Int. J. Mol. Sci.* **2016**, *17*, 23. [CrossRef]
136. Gray, C.; Vickers, M.H.; Segovia, S.A.; Zhang, X.D.; Reynolds, C.M. A maternal high fat diet programmes endothelial function and cardiovascular status in adult male offspring independent of body weight, which is reversed by maternal conjugated linoleic acid (CLA) supplementation. *PLoS ONE* **2015**, *10*, e0115994.
137. Hsu, C.N.; Hou, C.Y.; Chang-Chien, G.P.; Lin, S.; Tain, Y.L. Dietary Supplementation with Cysteine during Pregnancy Rescues Maternal Chronic Kidney Disease-Induced Hypertension in Male Rat Offspring: The Impact of Hydrogen Sulfide and Microbiota-Derived Tryptophan Metabolites. *Antioxidants* **2022**, *28*, 483. [CrossRef] [PubMed]
138. Hsu, C.N.; Hou, C.Y.; Tain, Y.L. Preventive Aspects of Early Resveratrol Supplementation in Cardiovascular and Kidney Disease of Developmental Origins. *Int. J. Mol. Sci.* **2021**, *22*, 4210. [CrossRef]
139. Walle, T.; Hsieh, F.; DeLegge, M.H.; Oatis, J.E., Jr.; Walle, U.K. High absorption but very low bioavailability of oral resveratrol in humans. *Drug Metab. Dispos.* **2004**, *32*, 1377–1382. [CrossRef]
140. Tain, Y.L.; Chang, S.K.C.; Liao, J.X.; Chen, Y.W.; Huang, H.T.; Li, Y.L.; Hou, C.Y. Synthesis of Short-Chain-Fatty-Acid Resveratrol Esters and Their Antioxidant Properties. *Antioxidants* **2021**, *10*, 420. [CrossRef]
141. Hsu, C.N.; Hou, C.Y.; Chang, C.I.; Tain, Y.L. Resveratrol Butyrate Ester Protects Adenine-Treated Rats against Hypertension and Kidney Disease by Regulating the Gut-Kidney Axis. *Antioxidants* **2021**, *11*, 83. [CrossRef]
142. Quigley, E.M.M. Nutraceuticals as modulators of gut microbiota: Role in therapy. *Br. J. Pharmacol.* **2020**, *177*, 1351–1362. [CrossRef] [PubMed]
143. Zmora, N.; Suez, J.; Elinav, E. You are what you eat: Diet, health and the gut microbiota. *Nat. Rev. Gastroenterol. Hepatol.* **2019**, *16*, 35–56. [CrossRef] [PubMed]
144. Hsu, C.N.; Tain, Y.L. Amino Acids and Developmental Origins of Hypertension. *Nutrients* **2020**, *12*, 1763. [CrossRef] [PubMed]
145. Torrens, C.; Brawley, L.; Anthony, F.W.; Dance, C.S.; Dunn, R.; Jackson, A.A.; Poston, L.; Hanson, M.A. Folate supplementation during pregnancy improves offspring cardiovascular dysfunction induced by protein restriction. *Hypertension* **2006**, *47*, 982–987. [CrossRef] [PubMed]
146. Palinski, W.; D'Armiento, F.P.; Witztum, J.L.; de Nigris, F.; Casanada, F.; Condorelli, M.; Silvestre, M.; Napoli, C. Maternal hypercholesterolemia and treatment during pregnancy influence the long-term progression of atherosclerosis in offspring of rabbits. *Circ. Res.* **2001**, *89*, 991–996. [CrossRef] [PubMed]
147. Gregório, B.M.; Souza-Mello, V.; Mandarim-de-Lacerda, C.A.; Aguila, M.B. Maternal fish oil supplementation benefits programmed offspring from rat dams fed low-protein diet. *Am. J. Obstet. Gynecol.* **2008**, *199*, e1–e7. [CrossRef]
148. Fujii, T.; Yura, S.; Tatsumi, K.; Kondoh, E.; Mogami, H.; Fujita, K.; Kakui, K.; Aoe, S.; Itoh, H.; Sagawa, N.; et al. Branched-chain amino acid supplemented diet during maternal food restriction prevents developmental hypertension in adult rat offspring. *J. Dev. Orig. Health Dis.* **2011**, *2*, 176–183. [CrossRef]
149. Thaeomor, A.; Teangphuck, P.; Chaisakul, J.; Seanthaweesuk, S.; Somparn, N.; Roysommuti, S. Perinatal Taurine Supplementation Prevents Metabolic and Cardiovascular Effects of Maternal Diabetes in Adult Rat Offspring. *Adv. Exp. Med. Biol.* **2017**, *975*, 295–305. [PubMed]
150. Smiljanec, K.; Lennon, S.L. Sodium, hypertension, and the gut: Does the gut microbiota go salty? *Am. J. Physiol. Heart Circ. Physiol.* **2019**, *317*, H1173–H1182. [CrossRef]
151. Miranda, P.M.; De Palma, G.; Serkis, V.; Lu, J.; Louis-Auguste, M.P.; McCarville, J.L.; Verdu, E.F.; Collins, S.M.; Bercik, P. High salt diet exacerbates colitis in mice by decreasing Lactobacillus levels and butyrate production. *Microbiome* **2018**, *6*, 57. [CrossRef]
152. Bier, A.; Braun, T.; Khasbab, R.; Di Segni, A.; Grossman, E.; Haberman, Y.; Leibowitz, A. A high salt diet modulates the gut microbiota and short chain fatty acids production in a salt-sensitive hypertension rat model. *Nutrients* **2018**, *10*, 1154. [CrossRef] [PubMed]
153. Sikalidis, A.K.; Maykish, A. The Gut Microbiome and Type 2 Diabetes Mellitus: Discussing a Complex Relationship. *Biomedicines* **2020**, *8*, 8. [CrossRef] [PubMed]

154. Moszak, M.; Szulińska, M.; Bogdański, P. You Are What You Eat-The Relationship between Diet, Microbiota, and Metabolic Disorders-A Review. *Nutrients* **2020**, *12*, 1096. [CrossRef] [PubMed]
155. Fontana, F.; Figueiredo, P.; Martins, J.P.; Santos, H.A. Requirements for Animal Experiments: Problems and Challenges. *Small* **2021**, *17*, e2004182. [CrossRef] [PubMed]
156. Doke, S.K.; Dhawale, S.C. Alternatives to animal testing: A review. *Saudi Pharm. J.* **2015**, *23*, 223–229. [CrossRef]

biomedicines

MDPI

Review

Genetic Modifications to Alter Blood Pressure Level

Hiroki Ohara * and Toru Nabika

Department of Functional Pathology, Faculty of Medicine, Shimane University, Izumo 693-8501, Japan; nabika@med.shimane-u.ac.jp
* Correspondence: oharah@med.shimane-u.ac.jp

Abstract: Genetic manipulation is one of the indispensable techniques to examine gene functions both in vitro and in vivo. In particular, cardiovascular phenotypes such as blood pressure cannot be evaluated in vitro system, necessitating the creation of transgenic or gene-targeted knock-out and knock-in experimental animals to understand the pathophysiological roles of specific genes on the disease conditions. Although genome-wide association studies (GWAS) in various human populations have identified multiple genetic variations associated with increased risk for hypertension and/or its complications, the causal links remain unresolved. Genome-editing technologies can be applied to many different types of cells and organisms for creation of knock-out/knock-in models. In the post-GWAS era, it may be more worthwhile to validate pathophysiological implications of the risk variants and/or candidate genes by creating genome-edited organisms.

Keywords: knock-out; genome-editing; SHR; SHRSP; Dahl SS

Citation: Ohara, H.; Nabika, T. Genetic Modifications to Alter Blood Pressure Level. *Biomedicines* **2022**, *10*, 1855. https://doi.org/10.3390/biomedicines10081855

Academic Editors: Josef Zicha and Ivana Vaněčková

Received: 30 June 2022
Accepted: 29 July 2022
Published: 1 August 2022

1. Introduction

Hypertension is the leading preventable risk factor for cerebro-cardiovascular complications, including heart failure and stroke. Effective anti-hypertensive drugs with different pharmacological actions have been developed; nevertheless, it is deemed that there are 1.28 billion hypertensive patients globally and 0.7 billion or more patients are untreated [1]. Given the resulting mortality and disability as well as the high prevalence, hypertension is still a major public health burden in the world.

It is needless to say that gene-targeted knock-out (KO) and knock-in (KI) or transgenic rodent models have greatly contributed to understanding the pathophysiological basis of hypertension and its vascular complications. In particular, mice have been widely used as the best experimental animal since the gene engineering technique to create KO models was established for over 30 years ago. By contrast, it had been technically difficult to create KO rats for a long time because of the difficulty of rat ES cell culture. Recent advances in genome-editing technologies, however, have made it possible to easily create KO rats similar to mice [2,3]. Given that spontaneous cerebro-cardiovascular disease models, such as Dahl salt-sensitive (SS) and stroke-prone spontaneously hypertensive rats (SHRSP), have been commercially available, a genome-editing strategy using the rat disease models has much potential to clarify the novel pathogenesis of hypertension. In this review, we outlined recent advances in basic research for hypertension using KO and KI or transgenic rodent models to clarify the underlying mechanisms.

2. Mouse Models

Essential hypertension is a highly complex pathological condition that is formed by synergistic influences of multiple lifestyles, social, environmental, and genetic factors. Since blood pressure (BP) is collaboratively controlled by various organs and tissues, there are many studies that have investigated tissue (or cell)-specific roles of genes on BP regulation using conventional and conditional KO or transgenic mice. In contrast to rats, no spontaneous hypertensive mouse models have been established; accordingly, angiotensin

II (Ang II)-infused models have been widely used to investigate the pathogenesis of Ang II-related hypertension. Deoxycorticosterone acetate (DOCA)-salt or high-salt diet (usually containing 4% or 8% NaCl) models have been also used to investigate the pathogenesis of salt-sensitive hypertension. In this section, we overview proposed mechanisms for controlling BP found in the phenotyping of KO or transgenic mouse models, especially focusing on the findings in the recent decade.

2.1. Kidney

The kidney plays pivotal roles in arterial BP regulation by controlling blood volume and plasma electrolyte balance. Activities of the renin–angiotensin–aldosterone system (RAAS) and mineral transporters (Na^+/H^+ exchanger; NHE, Na^+–K^+–Cl^- co-transporter; NKCC, Na^+–Cl^- co-transporter; NCC, epithelial sodium channel; ENaC, etc.) distributed along with nephron are important for physiological BP regulation; thus, genes that may regulate those activities have been widely investigated (Table 1).

Ang II regulates BP via Ang II type 1 receptor (*Agtr1a*, AT1R). As BP lowering effects were observed in proximal tubules (PT) or collecting duct (CD)-specific KO mice [4,5], blockade of AT1R signaling in renal epithelial cells would be a pharmacological target for hypertension therapy. Of note, AT1R-associated protein (*Agtrap*), which is widely distributed along renal tubules, has been found to suppress AT1R signaling by facilitating internalization of AT1R resulting in decreased cell surface expression of AT1R [6,7], suggesting that activation of endogenous AGTRAP has potential to reduce BP. In fact, it has been reported that the renal-specific overexpression and conventional KO mice show lower and higher BP phenotype compared with the wild-type (WT) control, respectively [8–11]. In contrast to the results in mice, however, the deletion in Dahl SS rats exacerbated renal damage under a 4% NaCl diet condition with no change in BP [12]. Although AGTRAP may play double-edged roles in reno-cardiovascular functions in a context-specific manner, it is a potential candidate gene located in a GWAS loci for BP in humans [12].

Although Ang II is the most well-known bioactive peptide hormone in the RAAS, (pro)renin and Ang-(1-7) produced by angiotensin-converting enzyme 2 (ACE2) are also known to regulate BP via its specific receptors. The (Pro)renin receptor (PRR) that specifically recognizes both prorenin and renin was cloned by Nguyen et al. in 2002 [13]. In the kidney, PRR is mainly expressed in renal vasculature, PT and distal tubules (DT), and CD and enhances the catalytic activity of (pro)renin that converts Ang I to Ang II, resulting in an increase in Ang II production [14]. Consistent with the physiological function of PRR, decreases in BP elevation induced by Ang II infusion have been observed in both tubular- and CD-specific KO mice through inhibition of ENaC activation [15–17]. Ang-(1-7) generated by mainly ACE2 is a vasoactive peptide that induces a vasodilation response by binding to Mas receptor [18]. Therefore, ACE2-Ang-(1-7)-Mas axis exerts a counteracting effect on Ang II that causes BP elevation. Ni et al. reported that conventional double KO of both ACE2 and Mas receptor in mice caused greater Ang II-induced BP elevation when compared with the WT littermates [19]. In addition, they also showed that the dual deletion of ACE2 and Mas receptor worsened hypertensive nephropathy, suggesting that ACE2-Ang-(1-7)-Mas receptor axis has protective roles in both the development of hypertension and the resulting hypertensive kidney injury.

Tubuloglomerular feedback (TGF) is an important physiological system to regulate long-term BP by sensing blood volume and electrolyte balance at the level of juxtaglomerular apparatus in each nephron [20]. Accumulating evidence has shown that local activities of renal oxide synthases (NOS), which produce a major chemical vasodilator NO, play an important role in the regulation of the TGF system. NOS families are composed of three isoforms, i.e., neuronal NOS (nNOS, encoded by *Nos1*), inducible NOS (iNOS, *Nos2*), and endothelial NOS (eNOS, *Nos3*). Although all the three isoforms are expressed in the kidney, *Nos1* and *Nos3* are thought to be major isoforms that physiologically participate in the TGF because of low baseline expression of *Nos2*. Interestingly, Lu et al. showed that macula densa-specific deletion of *Nos1* exacerbated a high-salt diet-induced BP elevation under a condition of Ang II infusion accompanied by reduced glomerular filtration rate (GFR) and Na^+ excretion [21]. It was also reported that local NOS1 activity at the macula densa contributed to a sex difference in BP response to Ang II [22]. Moreover, Hyndman et al. and Gao et al. have investigated renal-specific roles of NOS1 and NOS3 on BP regulation using CD-specific and nephron-specific KO mice, respectively [23,24]. They suggested that deletion of the two isoforms caused greater high-salt-induced BP elevation by enhancing ENaC [25] and NCC activities in the tubular cells, respectively.

Pathophysiological roles of NEDD4-2 (encoded by *Nedd4l*) and with-no-lysine kinases 1 and 4 (*Wnk1* and *Wnk4*) in (salt-sensitive) hypertension have been well-investigated in humans as well as in rodent models. NEDD4-2 is an E3 ubiquitin ligase that ubiquitylates ENaC to down-regulate its cell surface expression and activity [26]. Although NEDD4-2 was initially found as a ENaC-specific regulator in the kidney [25], Ronzaud et al. reported that NEED4-2 also regulated NCC activity and its renal tubule-specific deletion caused salt-dependent hypertension [27]. Consequently, NEED4-2 is involved in the pathogenesis of salt-sensitive hypertension through the two-independent pathways that controls renal Na^+ homeostasis. WNK1 and WNK4 are known to be responsible genes of pseudohypoaldosteronism type 2 (PHA2) that is caused by large deletions in intron 1 of WNK1 or gain-of function mutations in WNK4 [28]. Mechanistically, WNKs phosphorylate SPAK/OSR1, thereby activating NCC in the DT and resulting in increased Na^+ reabsorption and salt-sensitive hypertension [28,29]; however, the molecular network may be a little complicated as a paradoxical role of kidney-specific WNK1 lacking a kinase domain on the development of salt-sensitive hypertension was reported [30]. Moreover, Mu et al. suggested a unique pathway involving salt-sensitive hypertension caused by epigenetic down-regulation of WNK4 [31]. In this context, kelch-like protein 3 (KLHL3) and cullin 3 (CUL3), which are the E3 ubiquitin ligase complex to degrade WNK, have also received much attention as target molecules to prevent salt-sensitive hypertension [28,29].

Unlike the local mechanisms in the kidney described above, Pan et al. uniquely identified the liver–kidney and liver–adipocytes axis to control BP via a hepatocytes-producing hormone, fibroblast growth factor 21 (FGF21), which has pleiotropic effects on glucose and lipid metabolism [32]. They found that FGF21 augmented peroxisome proliferator-activated receptor γ (PPARγ)-mediated activation of ACE2 in both the kidney and adipocytes; thereby, an increase in Ang-(1-7) production reduced both BP and vascular injury. Because FGF21 production was stimulated by Ang II, the FGF21–ACE2 axis may counteract Ang II-induced hypertension and the vascular injury. This might be a key mechanism in obesity-related hypertension.

Besides the above, multiple mechanisms have been proposed such as by circadian clock- [33,34], osmotic stress- [35], and genome-wide association study (GWAS)-related genes [36,37] as well.

Table 1. Target molecules in kidney.

Targets	Type of Genetic Modification	Models	Phenotypes	References
AGTRAP (angiotensin II receptor-associated protein, *Agtrap*)	Renal tubule-specific overexpression	Ang II	↓BP, ↓NCC and αENaC activities	Wakui et al. [8]
	Conventional KO	Ang II	↑BP, ↑ENaC activity	Ohsawa et al. [9]
	Conventional KO	5/6 nephrectomy	↑BP, ↑plasma volume, ↑αENaC and TNF-α expression	Kobayashi et al. [10]
	Proximal tubule-specific KO	Ang II	No differences in basal BP, pressor response to Ang II, and cardiac hypertrophy	Kinguchi et al. [11]
PRR ((Pro)renin receptor, *Atp6ap2*)	Tubular-specific KO	Ang II	↓BP, ↓Na^+ retention, ↓αENaC expression	Ramkumar et al. [15]
	Collecting duct-specific KO	Ang II	↓BP, ↓urinary renin and αENaC activities	Peng et al. [16]
	Collecting duct-specific KO	Ang II	↓BP (basal and Ang II), ↓α/γENaC activation, ↓urinary Ang II and renin levels	Prieto et al. [17]
ACE2 (angiotensin-converting enzyme-2, *Ace2*), Mas receptor (*Mas1*)	Conventional double KO	Ang II	↑BP, ↑ renal injury, ↑serum Cr, ↓Cr clearance	Ni et al. [19]
NOS1 (NO synthase 1, *Nos1*)	Macula densa-specific KO	Ang II + high-salt diet	↑BP, ↑ tubuloglomerular feedback response, ↓GFR, urine flow, and N^+ excretion	Lu et al. [21]
	Macula densa-specific KO	Ang II	Diminished sex difference in Ang II-induced BP, tubuloglomerular feedback response, and natriuretic response	Zhang et al. [22]
	Collecting duct-specific KO	High-salt diet	↑BP, ↓urine output, ↓Na^+, Cl^-, and NOx excretion	Hyndman et al. [23]
NOS3 (NO synthase 3, *Nos3*)	Doxycycline-inducible nephron-specific KO	High-salt diet	↑BP, ↑Na^+ retention, ↑NCC activation	Gao et al. [24]
NEDD4-2 (*Nedd4l*)	Tetracycline-inducible tubule-specific KO	High-salt diet	↑BP, ↑β/γENaC and ROMK expression, ↑NCC activation, hypercalciuria	Ronzaud et al. [27]
WNK1 (with-no-lysine kinase 1, *Wnk1*)	Kidney-specific overexpression of the kidney-specific isoform	No treatment	↓BP, ↑plasma Ang II and aldosterone, ↓NCC and NKCC2 activation	Liu et al. [30]
	Kidney-specific KO (targeted deletion of the first exon of the kidney-specific isoform)	High-salt diet	↑BP, ↑Na^+ retention, ↑NCC and NKCC2 activation	Liu et al. [30]

Table 1. *Cont.*

Targets	Type of Genetic Modification	Models	Phenotypes	References
FGF21 (fibroblast growth factor 21, *Fgf21*)	Conventional KO	Ang II	↑BP, ↑vascular hypertrophy and fibrosis, ↓vascular relaxation, ↓plasma/adipose ACE2 and Ang-(1-7), ↑plasma/adipose Ang II	Pan et al. [32]
BMAL1 (brain and muscle ARNT-like 1, *Arntl*)	Kidney-specific KO	No treatment (or K^+-restricted diet)	↑BP, ↓Na^+ retention under K^+-restricted diet	Crislip et al. [33]
Per1 (period 1, *Per1*)	Distal nephron-specific KO	DOCP-salt	↑BP, ↑Na^+ retention, ↑plasma aldosterone, ↑medullary endothelin-1	Douma et al. [34]
NFAT5 (nuclear factor of activated T-cells 5, *Nfat5*)	Doxycycline-inducible tubular cell-specific KO	High-salt diet	↑BP, hypernatremia, polyuria, ↓Na^+ excretion, ↑ENaC expression	Hiramatsu et al. [35]
HSD11β2 (11β-hydroxysteroid dehydrogenase, *Hsd11b2*)	Kidney-specific KO	No treatment	↑BP, ↑αENaC and NCC activation	Ueda et al. [36]
NPR-C (natriuretic peptide receptor-C, *Npr3*)	Conventional KO	Ang II	↓BP, ↑diuretic and natriuretic response, ↓NCC activation via WNK4/SPAK	Shao et al. [37]
	Tubule-specific KO	Ang II	↓BP, ↓NCC activation via WNK4/SPAK	Shao et al. [37]

2.2. Vasculatures

Table 2 summarizes target genes in vasculatures and the representative phenotypes described below. Peripheral vascular tone is one of the primary factors to control BP. Two primary cell types, i.e., endothelial cells (ECs) and vascular smooth muscle cells (VSMCs), play major roles in the regulation of the vascular tone mainly through production of vasodilators (NO, etc.) or vasoconstrictors (endothelin-1, etc.) and sympathetic vasoconstriction, respectively. Endothelial NOS (eNOS, NOS3) predominantly generates NO from L-arginine in ECs; thereby, the released NO activates NO-sensitive guanylyl cyclase (NO-GC) in VSMCs to increase cytosolic cGMP, then the activated cGMP-dependent protein kinase (PKG) induces smooth muscle relaxation. The NO-GC/cGMP/PKG signaling is indispensable for NO-dependent BP regulation as spontaneous BP elevation was found in VSMC-specific NO-GC deficient mice [38].

In addition, NO-independent pathways that stimulate cGMP/PKG also exist. Natriuretic peptides, which are composed of A- (atrial; ANP), B- (brain; BNP), and C-type (CNP), are well-studied vasoactive peptides that exert vasodilation via direct activation of the transmembrane receptor NPR1 (natriuretic peptide receptor 1, also known as guanylyl cyclase-A; GC-A) or NPR2 (also known as GC-B). Among the three members, CNP is secreted from ECs and specifically bind to NPR2, whereas ANP and BNP are cardiac peptides targeting NPR1 [39]. In addition, CNP is thought to be an autocrine/paracrine factor in the circulation system because of the relatively low plasma concentration compared with ANP and BNP [40]. Several recent reports have uncovered the detailed mechanisms of NPR1/2-mediated vasodilation.

Nakao et al. showed that EC-specific CNP KO mice had higher BP compared with WT control independently of NO production, whereas VSMC-specific NPR2 KO had unaltered BP [41]. On the other hand, Špiranec et al. thereafter reported that the deletion in 'precapillary arteriole SMCs and capillary pericytes' caused BP elevation in mice accompanied by an impaired CNP-induced vasodilatory response [42]. Collectively, these results indicate that EC-derived CNP acts on precapillary arteriole SMCs and capillary pericytes as well as ECs to lower peripheral vascular resistance and BP through an NO-independent manner. It is of note that a CNP-induced vasodilatory response in mesenteric arteries was also impaired in VSMC-specific KO mice by Nakao et al.; nevertheless, BP of the KO mice was compatible with that of WT [41]. Conflicting results for the BP phenotype between the two VSMC-specific KO models may be partly due to the difference in promoters driving Cre expression (*sm22* [41] or *Pdgf-rb* [42] promoter) to create the conditional NRP2 KO mice. Furthermore, it was very recently reported that EC-specific, but not VSMC-specific, deletion of NPR1 diminished BP reduction by intravenous ANP administration [43].

Intracellular Ca^{2+} mobilization is a key modulator to induce both NOS-mediated vascular relaxation by EC and VSMC contraction; thus, cell-type specific genetic modification is necessary to clarify functional implications of the target molecules on vascular responses. Stromal interaction molecule 1 (*Stim1*) is an endoplasmic reticulum (ER) resident transmembrane protein that senses Ca^{2+} store in ER lumen via its N-terminal EF hand motif. When the Ca^{2+} store is depleted, STIM1 moves toward the cytoplasmic membrane and opens the target Ca^{2+} channels, ORAI1, and transient receptor potential families (TRPs) to elicit Ca^{2+} entry into the cytosol (store-operated Ca^{2+} entry; SOCE) [44]. STIM1 is expressed in broad cell types including ECs and VSMCs and plays key roles in the maintenance of intracellular Ca^{2+} homeostasis [43]. Kassan et al. revealed that VSMC-specific deletion of *Stim1* ameliorated Ang II-induced hypertension with decreased vascular ER stress [45]. By contrast, a significant increase in nighttime BP was observed in EC-specific *Stim1* KO mice that showed decreased NO production and an EC-dependent vasodilation phenotype in vitro [46]. These studies suggest that STIM1 exerts an opposite role in the regulation of vascular tone in the two different cells. Interestingly, we found that the stroke-prone spontaneously hypertensive rat (SHRSP) had a premature stop codon in this gene that caused the expression of truncated STIM1 with decreased SOCE activity [47,48]. Although the recovery of STIM1 function in SHRSP by CRISPR-Cas9-mediated gene KI did not alter the BP [49], systemic impairment of SOCE activity would have important implications for the pathogenesis of hypertensive end-organ damage in SHRSP independently of the BP phenotype. Phenotyping of the KI rat model is currently in progress; the findings will be described elsewhere.

Fluid shear stress is an important mechanical stimulus that physiologically enhances NO production by ECs to maintain vascular integrity. Increasing evidence has shown that a mechanosensitive cation channel PIEZO1 on ECs mediates laminal flow-dependent activation of purinergic P2Y2 receptor, thereby activating PI3K/Akt signaling to phosphorylate NOS3 [50]. Recently, an alternative pathway mediated by the PIEZO1adrenomedullin (ADM) axis was reported [51]. ADM is a circulatory vasodilator and diuretic and natriuretic peptide that is mainly produced by ECs [52]. Iring et al. showed that PIEZO1 enhanced endothelial ADM secretion, then the secreted ADM bound to calcitonin receptor-like receptor (CALCRL) on ECs by an autocrine/paracrine fashion. ADM-CALCRL complex activates its adjacent adenylyl cyclase; thereby, cAMP-dependent protein kinase (PKA) phosphorylates and activates NOS3, resulting in NO-dependent vasorelaxation [50]. Actually, it is of interest that all EC-specific single KO of ADM, CALCRL, and Gαs, which is the downstream G protein of CALCRL, in mice caused apparent BP elevation. This finding indicates an essential role of PIEZO1-ADM signaling on controlling vascular tone and BP at a resting condition.

Prostaglandins (PGs) are endogenous lipid mediators that have multiple bioactivities such as uterine contraction, platelet aggregation, and bronchodilation and is generated from arachidonic acid by catalytic activities of cyclooxygenases (COXs). PGs are also involved in BP regulation as nonsteroidal anti-inflammatory drugs (NSAIDs), which block COX activity, have hypertensive side effects [53]. Among the known PGs, PGE_2 is a major prostanoid that affects BP both positively and negatively via its specific receptor EP1-4 [54,55]. Thus far, it was shown that EP1 and EP3 mediate vasoconstrictive response, while EP2 and EP4 lead to vasodilation [53]. The diverse effects of PGE_2 on vascular functions may be due to the characteristic of tissue distribution of the receptors. Recently, Xu et al. reported that EC-specific KO and overexpression of EP4 resulted in higher and lower BP compared with control mice, respectively [56], under both basal and high-salt diet conditions. Physiological roles of PGs on BP regulation may be still controversial; however, the recent report clearly indicates a hypotensive potential of PGE_2-EP4 signaling via enhancing NO production in ECs.

2.3. Immunity

A growing body of evidence has emerged in the last decade suggesting the pathogenic aspects of innate and adaptive immune responses on the development and progression of hypertension and hypertensive end-organ damages. The possible mechanisms have been well reviewed [57–59]; herein, we shortly highlight recent findings on this topic (Table 3). Among various subpopulations of immune cells, previous reports indicate that T cells especially have diverse contributions to the etiology [56]. In particular, $CD4^+$- and regulatory T cell (Treg)-mediated pathological cascades have been raised in several recent studies.

Table 2. Target molecules in vasculatures.

Targets (*Official Symbols*)	Type of Genetic Modification	Models	Phenotypes	References
NO-GC (NO-sensitive guanylyl cyclase, *Gucy1b1*)	Tamoxifen-inducible VSMC-specific KO	No treatment	↑BP, ↓NO-induced vasorelaxation	Groneberg et al. [38]
CNP (C-type natriuretic peptide, *Nppc*)	EC-specific KO	No treatment	↑BP, ↓acetylcholine- and endothelium-dependent relaxation, ↑Endothelin-1 and Ace expression in ECs	Nakao et al. [41]
NPR2 (natriuretic peptide receptor 2, *Npr2*)/Guanylyl cyclase-B (GC-B)	VSMC-specific KO	No treatment	No difference in BP, ↓CNP-induced relaxation in mesenteric arteries	Nakao et al. [41]
	Tamoxifen-inducible EC-specific KO	No treatment	↑BP, ↓cGMP production	Špiranec et al. [41]
NPR1 (natriuretic peptide receptor 1, *Npr1*)	EC-specific KO	No treatment	Loss of EC-dependent BP reduction by ANP, unaltered NO production, K^+ channel-mediated hyperpolarization in EC	Tokudome et al. [43]
STIM1 (stromal interaction molecule 1, *Stim1*)	VSMC-specific KO	Ang II	↓BP, ↓cardiac hypertrophy, ↓perivascular fibrosis, ↓endothelial dysfunction	Kassan et al. [45]
	EC-specific KO	No treatment	↑BP (nighttime), ↓NO production, ↓endothelium-dependent relaxation	Nishimoto et al. [46]
ADM (adrenomedullin, *Adm*), CALCRL (calcitonin receptor-like, *Calcrl*), $G_{\alpha s}$ (GNAS (guanine nucleotide binding protein, alpha stimulating) complex locus, *Gnas*)	Tamoxifen-inducible EC-specific KO	No treatment	↑BP, ↓eNOS activation, ↓flow-induced vasorelaxation	Iring et al. [51]
EP4 (prostaglandin E2 receptor, *Ptger4*)	EC-specific KO	No treatment, high-salt diet, Ang II	↑BP, ↓NO production, ↓vasorelaxation response	Xu et al. [56]
	EC-specific overexpression	No treatment, high-salt diet	↓BP, ↑eNOS activation, ↑NO production	Xu et al. [56]

Hydrogen sulfide (H_2S) is a cardioprotective endogenous gaseous mediator that is generated by three major enzymes: cystathionine beta synthase (CBS), cystathionine gamma lyase (CSE), or 3-mercaptopyruvate sulfurtransferase [60]. Although it was reported that a conventional CSE KO mice showed an age-dependent increase in BP [61], Cui et al. revealed that $CD4^+$ T cell-specific deletion of CSE was sufficient to induce greater BP in mice under both physiological and Ang II-treated conditions [62]. Mechanistically, they suggest that CSE-derived H_2S activates liver kinase B1 (LKB1)-PKA signaling and the resulting activation of Treg attenuates vascular and renal inflammation, thereby preventing BP elevation. In addition, Sun et al. reported that mineralocorticoid receptor (MR) deficiency in $CD4^+$ T cells ameliorated Ang II-induced BP elevation and vascular and renal damage in mice [63]. In contrast to the KO model, the MR overexpression exacerbated the increase in BP after Ang II infusion; however, IFN-γ-neutralizing antibodies could abolish the deleterious effect, suggesting that IFN-γ produced by infiltrated T cells was a key cytokine link between MR signaling in $CD4^+$ T cells and the resulting hypertension.

For Treg, a detailed pathological mechanism caused by a microRNA function has been proposed. MicroRNAs (miRs) are involved in numerous (patho)physiological conditions by controlling gene expression mainly at a translational level, and among the identified miRs, miR-31 has multifaceted roles in regulation of immune responses [64]. Interestingly, Li et al. reported that Ang II-induced BP elevation and vascular and renal damage were reduced in mice lacking miR-31 in Treg compared with control mice, which were accompanied by increased Treg differentiation [65]. Furthermore, the opposite phenotypes were observed in mice with Treg-specific deletion of protein phosphatase 6c (*Ppp6c*), a direct target of miR-31, suggesting that *Ppp6c* had potential to improve Ang II-induced hypertension [64]. This may be a novel posttranslational mechanism that worsens hypertensive phenotypes through an overexpression of a specific microRNA that regulates Treg functions.

AT1R, an Ang II receptor, is widely expressed in immune cells [66]. It has been shown that deletion of AT1R on T lymphocytes or macrophages do not affect BP even under an Ang II-infused condition [67,68]. In contrast to the previous findings, Lu et al. revealed that the deletion on $CD11c^+$ myeloid cells (dendritic cells; DCs) in mice with chronic Ang II infusion resulted in increased BP, renal infiltration of inflammatory cells (memory T, $CD40^+$ DCs), and Na^+ retention with greater β/γENaC expression [69]. It is of interest that AT1R on DCs exerts a cardioprotective role in spite of harmful effects of Ang II on renal and cardiovascular functions. Moreover, Sag et al. showed that mice with myeloid cell-, but not endothelial cell-, specific deletion of NADPH oxidase 2 (NOX2), which is a superoxide-generating enzyme, had lower basal BP compared with control mice [70].

Beside the above, the pathophysiological actions of C-C motif chemokine receptor 7 (CCR7) [71], toll-like receptor 3/4 (TLR3/4) [72], placental growth factor (PlGF) [73], complement C3a/C5a receptors (C3aR/C5aR) [74], T cell receptor delta chain (TCRδ) [75], and interleukin-1 receptor type 1 (IL-1R1) [76] have been also proposed using conventional KO mouse models. Overall, accumulated evidence has commonly suggested pathophysiological connections between immune responses and renal dysfunction on the development of hypertensive conditions. Clinical perspectives of anti-inflammatory therapies targeting specific cytokines were also discussed [77].

Table 3. Target molecules in immune system.

Targets (*Official Symbols*)	Type of Genetic Modification	Models	Phenotypes	References
CSE (Cystathionine γ lyase, *Cth*)	$CD4^+$ T cell-specific KO	Ang II	↑BP, ↑blood and renal Treg, ↑renal and peripheral adipose tissue $CD4^+/CD8^+$ T	Cui et al. [62]
MR (mineralocorticoid receptor)/nuclear receptor subfamily 3, group C, member 2 (*Nr3c2*)	$CD4^+$ T cell-specific KO	Ang II	↓BP, ↓renal/vascular damage, ↓IFNγ-producing T cell	Sun et al. [63]
	$CD4^+$ T cell-specific overexpression	Ang II	↑BP	Sun et al. [63]
MicroRNA-31 (miR-31, *Mir31*)	Conventional and Treg-specific KO	Ang II	↓BP, ↑Treg differentiation,↓renal and vascular injury	Li et al. [65]
Ppp6c (protein phosphatase 6c, *Ppp6c*)	Treg-specific KO	Ang II	↑BP, ↓Treg differentiation,↑renal injury	Li et al. [65]
AT1R (angiotensin II receptor type 1, *Agtr1a*)	$CD11c^+$ cell-specific KO	Ang II	↑BP, ↑renal memory T and $CD40^+$ DC, ↑ENaC	Lu et al. [69]
NOX2 (NADPH oxidase 2, *Cybb*)	Myeloid cells-specific KO	No treatment	↓BP, ↑NO bioavailability	Sag et al. [70]
		Ang II	No effect on BP	Sag et al. [70]
CCR7 (C-C motif chemokine receptor 7, *Ccr7*)	Conventional KO	Ang II	↓BP, ↑renal $CD8^+$ T, ↓renal draining lymph node $CD4^+$ T and $CD8^+$ T	Wen et al. [71]
TLR3/4 (toll-like receptor 3/4, *Tlr3/4*)	Conventional KO	Ang II	↓BP and cardiac hypertrophy in TLR3 KO, ↓cardiac hypertrophy in TLR4 KO	Singh et al. [72]
PIGF (placental growth factor, *Pgf*)	Conventional KO	DOCA-salt	↓BP, ↓renal damage and T cell infiltration	Perrotta et al. [73]
C3aR/C5aR (complement 3a and 5a receptors, C3ar1/C5ar1)	Conventional double KO	Ang II	↓BP, ↑renal Treg, ↓renal/vascular remodeling	Chen et al. [74]
TCRδ (T cell receptor delta chain, *Tcrd*)	Conventional KO	Ang II	↓BP, ↓endothelial dysfunction	Caillon et al. [75]
IL-1R1 (IL-1 receptor type 1, *Il1r1*)	Conventional KO	Ang II	↓BP, ↓NKCC2 activity	Zhang et al. [76]

2.4. Other Organs and Tissues (Brain, Adipocyte, and Adrenal Gland)

Table 4 summarizes target genes in other major organs and tissues and the representative phenotypes. Pathological implications of brain RAAS on hypertension have been well investigated [78]. Based on the distribution of RAAS components in the brain, it has been verified that brain RAAS activity can induce BP elevation independently of renal RAAS function by using cell type-specific transgenic mice targeting AGT and/or renin [79–81]. Consistent results were observed in DOCA-salt mice with neuron-specific deletion of PRR that exhibited decreases in BP and brain Ang II production [82].

Pathological relationships between salt intake and hypertension have been long suggested in humas as well as in rodents; however, the precise mechanism remains elusive. In this context, it is noteworthy that Nomura et al. reported that the Na_x channel expressed in specific glial cells in the organum vasculosum lamina terminalis (OVLT) functioned as the brain sensor detecting $[Na^+]$ increase in the body and that deletion of Na_x diminished salt-induced hypertensive phenotype [83]. Concerning this, neuronal 11β-hydroxysteroid dehydrogenase type2 (*Hsd11b2*), which encodes a corticosterone-producing enzyme, and PRR have been proposed to be involved in both the development of salt-sensitive hypertension and sodium appetite [84,85]. In addition, PRR deficiency in adipocytes was pathologically implicated in a high-fat diet-induced BP increase in male mice but not in female mice [86].

The adrenal gland is a major endocrine organ that plays a pivotal role in BP regulation and fluid and electrolyte homeostasis via production of steroid hormones and catecholamines. The two-pore domain K^+ channels (TASKs) expressed in zona glomerulosa (zG) cells down-regulate the production of aldosterone in the cells [87]. Guagliardo et al. showed that zG cell-specific deletion of TASK-1 and -3 caused autonomous hyperaldosteronism and chronic BP elevation in mice [88]. In addition, Mathar et al. reported that mice lacking transient receptor potential melastatin 4 (TRPM4) had chronically increased BP with exaggerated sympathetic tone [89]. TRP families are non-selective cation channels that are involved in many physiological processes and are regarded as potential targets for drug design for various diseases [50]. According to the report by Mathar et al., TRPM4 deficiency increases catecholamine release from chromaffin cells and thereby augments sympathetic tone resulting in a continuous BP elevation.

It is generally known that plasma concentrations of adrenal gland-derived steroid hormones are controlled by the physiological circadian rhythm of adrenocorticotropic hormone (ACTH) secretion. Circadian clock genes, cryptochrome-1 and -2 (Cry-1 and -2), play key roles in this mechanism by direct regulation of *Hsd3b6* expression encoding an aldosterone-producing enzyme, 3β-hydroxysteroid dehydrogenase-isomerase (3β-HSD). Therefore, Cry-1 and Cry-2 KO mice exhibited salt-sensitive hypertension under a high-salt diet condition due to constitutive activation of 3β-HSD such asthe DOCA-salt model [90].

Multiple genetic and physiological mechanisms as thus far described are complicatedly involved in the pathogenesis of hypertension. Furthermore, the pathogenic roles of epigenetic modifications [91–93], microbiota/metabolome [94–96], and sympathetic overactivity [97–99] in cardiovascular disease have been also discussed.

Table 4. Target molecules in brain, adipocyte, and adrenal gland.

Organs, Tissues	Targets (*Official Symbols*)	Type of Genetic Modification	Models	Phenotypes	References
Brain	Human AGT (angiotensinogen, *AGT*)	Glial-specific overexpression	No treatment	↑BP,↑salt preference	Morimoto et al. [79]
	Human REN (renin, *REN*)	Glial- and neuron-specific overexpression	No treatment	↑BP, ↑salt preference	Morimoto et al. [80]
	Human AGT (angiotensinogen, *AGT*)	Glial-specific KO	No treatment	↓BP	Sherrod et al. [81]
	PRR ((Pro)renin receptor, *Atp6ap2*)	Neuron-specific KO	DOCA-salt	↓BP, ↓brain Ang II production, ↓cardiac and vasomotor sympathetic tone	Li et al. [82]
	Na_x (sodium channel, voltage-gated, type VII, alpha, *Scn7a*)	Conventional KO	High-salt diet	↑BP	Nomura et al. [83]
	HSD11β2 (11β-hydroxysteroid dehydrogenase, *Hsd11b2*)	Neuron-specific KO	High-salt water	↑BP, ↑salt preference	Evans et al. [84]
Adipocytes	PRR ((Pro)renin receptor, *Atp6ap2*)	Adipocyte-specific KO	High-fat diet	↑BP (basal and high fat diet-induced), ↑glucose tolerance, ↓diet-induced obesity	Wu et al. [86]
Adrenal gland	TASK-1/3 (potassium channel, subfamily K, member 3/9, *Kcnk3/9*)	Zona glomerulosa cells-specific KO	No treatment	↑BP	Guagliardo et al. [88]
	TRPM4 (transient receptor potential cation channel, subfamily M, member 4, *Trpm4*)	Conventional KO	No treatment	↑BP, ↑plasma epinephrine, ↑urinary catecholamine metabolites	Mathar et al. [89]
	Cry1/2 (cryptochrome-1/2, *Cry1/2*)	Conventional KO	High-salt diet	↑BP, ↑increased expression and activity of 3β-HSD	Doi et al. [90]

3. Rat Models

Rats are the generally used experimental animal the same as mice and have some advantages compared with mice such as large body and tissue size and physiological properties similar to those in humans. Despite the advantages, mice have been more frequently used than rats; this is probably due to, except for higher experimental costs than mice, the technical difficulty of creating KO/KI rats. However, genome-editing technologies, i.e., zinc finger nuclease (ZFN), transcription activator-like effector nuclease (TALEN), and clustered regularly interspaced short palindromic repeats (CRISPRs)-associated proteins 9 (CRISPR-Cas9), made it possible to also create KO/KI rats easily. In 2009, Geurts et al. first reported the creation of KO rats by ZFN [100]. Thereafter, a growing body of literature has emerged in the last decade reporting phenotypes of KO/KI rats including genetic hypertensive models as below [101,102].

3.1. KO Models of SHR and SHRSP

SHR (spontaneously hypertensive rat) is a representative genetically hypertensive model that was established by selective breeding of rats with relatively high blood pressure in an outbred colony of Wistar rats that had been maintained in Kyoto University. SHRSP (stroke-prone SHR) is a substrain of SHR that genetically develops more severe hypertension and stroke. Despite the fact that both strains have been widely used for clarifying the responsible genes and the underlying mechanisms of hypertension and its complications [103–105], the literature evaluating cardiovascular phenotypes by using KO/KI models are still scant (Table 5).

SHR is a useful model for hypertensive cardiac hypertrophy [106]. A quantitative trait locus (QTL) related to the left ventricular hypertrophy was previously mapped on chromosome (Chr) 8 by phenotyping of congenic strains between SHR and normotensive Brown Norway (BN) rats [107]. Liška et al. identified promyelocytic leukemia zinc finger (*Plzf*) as a candidate gene on the cardiac QTL and showed that the deletion in SHR did not alter the BP but ameliorated cardiac hypertrophy and fibrosis [108].

Complement 3 (C3) that is overexpressed in aortic smooth muscle cells of SHR has been proposed as a candidate gene responsible for the development of hypertension in this model [109]. Mechanistically, C3-C3a receptor signaling accelerates a change in the characteristic of VSMC and glomerular mesangial cells from contractile to synthetic phenotype via activation of Krüppel-like factor 5 (KLF5) that is a transcription factor to induce the synthetic phenotype of mesenchymal cells [110,111]. Negishi et al. revealed that the C3 deficiency mitigated a salt-sensitive BP elevation and renal injury with decreased renal Ang II level and urinary catecholamine excretion [112].

Rubattu et al. previously identified a QTL on Chr 1 responsible for the susceptibility to salt-induced stroke by a linkage analysis F2 cross between SHR and SHRSP [113]. They identified NADH dehydrogenase (ubiquinone) 1 subunit C2 (*Ndufc2*), encoding a component of the electron transport chain, as a plausible candidate gene in the stroke QTL, then proved that the heterozygous deletion in SHR by ZFN strongly exacerbated the stroke susceptibility with increased oxidative stress and inflammation both in vitro and in vivo [114].

Besides the above, we recently created peroxiredoxin 2 (*Prdx2*) KO SHR to investigate whether the deletion of an antioxidant gene exacerbates cerebro-cardiovascular phenotypes of SHR [115]. Consequently, *Prdx2* KO SHR had greater basal BP compared with WT SHR. Furthermore, the lifespan of *Prdx2* KO SHR under a salt loading condition was shorter than that of WT SHR despite no difference in BP after salt loading between the KO and the WT. No apparent inter-strain differences were found in histopathological evaluation for brain, heart, and kidney lesions, and therefore, the reason for the short life span of *Prdx2* KO SHR under the salt loading condition remains fully unknown.

Lectin-like oxidized low-density lipoprotein receptor-1 (LOX-1) is an endothelial scavenger receptor that is closely involved in the pathogenesis of atherosclerosis [116].

Recently, Liang et al. reported that LOX-1 deficiency had a protective role in spontaneous brain damage in SHRSP with no significant change of BP [117].

We previously found a QTL on Chr1 that affected exaggerated sympathetic responses to the stress of SHRSP by genetic analysis of congenic lines between SHRSP and normotensive Wistar-Kyoto rat (WKY) [118]. Among the genes in the QTL region, stromal interaction molecule 1 (*Stim1*) with a nonsense mutation in SHRSP was identified as a promising candidate ([47], see also Section 2.2 Vasculatures). As STIM1 plays a key role in Ca^{2+} homeostasis in the body, we expected that the *Stim1* mutation was a genetic determinant responsible for cerebro-cardiovascular traits; however, no significant differences were observed in the sympathetic stress responses as well as age-dependent changes in BP between *Stim1* KI SHRSP and SHRSP, i.e., with WT and mutant allele for *Stim1*, respectively [49]. Phenotyping of the *Stim1* KI SHRSP is currently in progress, and the results will be described elsewhere.

3.2. KO Models of Dahl SS

Dahl salt-sensitive (SS) rats originate from a closed colony of Sprague-Dawley (SD) rats and are widely used as a salt-sensitive hypertension model that develop severe hypertension (>200 mmHg) and the complications such as hypertensive kidney injury and heart failure when fed high-salt diets [119]. SS/Jr and DSS/N strains have been separately established by Rapp and Iwai, respectively. Compared with SHR and SHRSP, multiple KO/KI models with SS/Jr genetic backgrounds have been actively created (Table 5).

In 2011, Moreno et al. first reported the phenotype of renin KO SS/Jr, in which a severe decrease in basal BP and abnormal kidney morphologies were observed [120]. Thereafter, a growing literature has shown pathophysiological implications of multiple genes on cardiorenal disease traits in SS/Jr [121–133]. Among them, pleckstrin homology domain containing family A member 7 (*Plekha7*) is a plausible candidate gene for essential hypertension identified by GWAS. A risk variation on *Plekha7*, encoding an adherence junction protein [134], for elevated systolic BP has been found in multiple human populations [135–139]. In this context, Endres et al. created SS/Jr lacking the functional domain of *Plekha7* by ZFN and revealed that the *Plekha7* functional KO SS/Jr had significantly lower BP and renal and cardiac damage under the 8% high-salt diet condition [123]. This is a meaningful study that verified a direct effect of the GWAS gene on the hypertensive phenotype in a genetic rat model with salt sensitivity.

Table 5. Target molecules in rat models.

Strains	Targets (*Official Symbols*)	Methods	Phenotypes	References
SHR/OlaIpcv	Plzf (promyelocytic leukemia zinc finger, *Zbtb16*)	TALEN	↑cardiomyocyte hypertrophy and fibrosis	Liška et al. [108]
SHR/NCrl	Ndufc2 (NADH dehydrogenase (ubiquinone) 1 subunit C2, *Ndufc2*)	ZFN	No effect on BP, ↑salt-induced stroke susceptibility, oxidative stress, and inflammatory signaling	Rubattu et al. [113]
SHR/Izm	C3 (complement 3, *C3*)	ZFN	↓salt-induced BP, ↓renal Ang II level, ↓urinary catecholamine excretion	Negishi et al. [112]
	Prdx2 (peroxiredoxin 2, *Prdx2*)	CRISPR-Cas9	↓basal BP, ↑life span under salt loading condition	Mahal et al. [115]
SHRSP/Izm	LOX-1 (lectin-like oxidized low-density lipoprotein receptor-1, *Olr1*)	ZFN	↓stroke susceptibility independently of BP	Liang et al. [117]
SS/JrHsdMcwi	Renin (*Ren*)	ZFN	↓BP, abnormal kidney morphology	Moreno et al. [120]
	Rag1 (recombination activating 1, *Rag1*)	ZFN	↓BP, ↓renal injury	Mattson et al. [121]
	ROMK (renal outer medullary potassium channel, *Kcnj1*)	ZFN	↓BP	Zhou et al. [122]
	Plekha7 (pleckstrin homology domain containing family A member 7, *Plekha7*)	ZFN	↓BP, ↓renal injury, ↓cardiac fibrosis	Endres et al. [123]
	HV1 (voltage-gated H^+ channel, *Hvcn1*)	ZFN	↓BP, ↓renal injury, ↓oxidative stress	Jin et al. [124]
	CD247 (*Cd247*)	ZFN	↓BP, ↓$CD3^+$ T cells, ↓renal injury	Rudemiller et al. [125]
	BNP (B-type natriuretic peptide, *Nppb*)	ZFN	↑BP, ↑cardiac hypertrophy and fibrosis,↑renal injury	Holditch et al. [126]
	Nr2f2 (nuclear receptor subfamily 2 group F member 2, *Nr2f2*)	ZFN	↓BP, ↑left ventricular/vascular functions, ↑urinary protein	Kumarasamy et al. [127]
	Adora2b (A_{2B} adenosine receptor, *Adora2b*)	ZFN	↑BP, ↑body weight, ↓glucose clearance	Nayak et al. [128]
	Nox4 (NADPH oxidase 4, *Nox4*)	ZFN	↓BP, ↓renal injury, ↓oxidative stress	Cowley et al. [129]
	Rffl-lnc1 (a novel long-noncoding RNA)	CRISPR-Cas9	↑BP, shorter QT intervals	Cheng et al. [130]
	Resp18 (regulated endocrine-specific protein 18, *Resp18*)	ZFN	↑BP, ↑renal injury, ↓survival time	Kumarasamy et al. [131]
	Gper1 (G protein-coupled estrogen receptor 1, *Gper1*)	CRISPR-Cas9	↓BP, ↑vascular relaxation, ↓microbiotal dysbiosis	Waghulde et al. [132]
	$p67^{phox}$ (neutrophil cytosolic factor 2, *Ncf2*)	ZFN	↓BP, ↓renal injury, ↓renal immune cell infiltration	Abais-Battad et al. [133]

4. Conclusions

Hypertension is a multifactorial disease; nevertheless, the majority of previous research has focused on monogenic effects under inducible hypertensive conditions such as Ang II infusion and DOCA-salt in mice. Recent advances in genome-editing techniques, however, have made it possible to create knock-out and knock-in animals more easily, efficiently, and rapidly in rats as well as in mice [140–143]. Accordingly, it is necessary to create knock-out and knock-in models with multiple mutations in different loci to mimic complex genetic backgrounds of hypertensive patients and to uncover how the genetic interactions cause hypertension. As in the case of Dahl SS [144], translation of the findings in the experimental model into human hypertension remains highly challenging. However, translational approaches to bridge the gap between humans and rodent models would be necessary for understanding genetic and molecular mechanisms of essential hypertension in the post-GWAS era. A goal of basic hypertension research using experimental models may reconstruct 'genetically hypertensive mice/rats' from normotensive strains, and vice versa.

Author Contributions: H.O.: conceptualized and drafted the manuscript; H.O. and T.N.: revised the manuscript. All authors have read and agreed to the published version of the manuscript.

Funding: This work was partly supported by JSPS KAKENHI (Grant Number 21H03374) to T.N.

Institutional Review Board Statement: Not applicable.

Informed Consent Statement: Not applicable.

Data Availability Statement: Not applicable.

Conflicts of Interest: The authors declare no conflict of interest.

References

1. Zhou, B.; Carrillo-Larco, R.M.; Danaei, G.; Riley, L.M.; Paciorek, C.J.; Stevens, G.A.; Gregg, E.W.; Bennett, J.E.; Solomon, B.; Singleton, R.K.; et al. Worldwide trends in hypertension prevalence and progress in treatment and control from 1990 to 2019: A pooled analysis of 1201 population-representative studies with 104 million participants. *Lancet* **2021**, *398*, 957–980. [CrossRef]
2. Mashimo, T. Gene targeting technologies in rats: Zinc finger nucleases, transcription activator-like effector nucleases, and clustered regularly interspaced short palindromic repeats. *Dev. Growth Differ.* **2014**, *56*, 46–52. [CrossRef] [PubMed]
3. Kaneko, T.; Mashimo, T. Simple Genome Editing of Rodent Intact Embryos by Electroporation. *PLoS ONE* **2015**, *10*, e0142755. [CrossRef] [PubMed]
4. Stegbauer, J.; Chen, D.; Herrera, M.; Sparks, M.A.; Yang, T.; Konigshausen, E.; Gurley, S.B.; Coffman, T.M. Resistance to hypertension mediated by intercalated cells of the collecting duct. *JCI Insight* **2017**, *2*, e92720. [CrossRef]
5. Li, X.C.; Leite, A.P.O.; Zheng, X.; Zhao, C.; Chen, X.; Zhang, L.; Zhou, X.; Rubera, I.; Tauc, M.; Zhuo, J.L. Proximal Tubule-Specific Deletion of Angiotensin II Type 1a Receptors in the Kidney Attenuates Circulating and Intratubular Angiotensin II-Induced Hypertension in PT-Agtr1a$^{-/-}$ Mice. *Hypertension* **2021**, *77*, 1285–1298. [CrossRef]
6. Cui, T.; Nakagami, H.; Iwai, M.; Takeda, Y.; Shiuchi, T.; Tamura, K.; Daviet, L.; Horiuchi, M. ATRAP, novel AT1 receptor associated protein, enhances internalization of AT1 receptor and inhibits vascular smooth muscle cell growth. *Biochem. Biophys. Res. Commun.* **2000**, *279*, 938–941. [CrossRef]
7. Tsurumi, Y.; Tamura, K.; Tanaka, Y.; Koide, Y.; Sakai, M.; Yabana, M.; Noda, Y.; Hashimoto, T.; Kihara, M.; Hirawa, N.; et al. Interacting molecule of AT1 receptor, ATRAP, is colocalized with AT1 receptor in the mouse renal tubules. *Kidney Int.* **2006**, *69*, 488–494. [CrossRef]
8. Wakui, H.; Tamura, K.; Masuda, S.-I.; Tsurumi-Ikeya, Y.; Fujita, M.; Maeda, A.; Ohsawa, M.; Azushima, K.; Uneda, K.; Matsuda, M.; et al. Enhanced Angiotensin Receptor-Associated Protein in Renal Tubule Suppresses Angiotensin-Dependent Hypertension. *Hypertension* **2013**, *61*, 1203–1210. [CrossRef]
9. Ohsawa, M.; Tamura, K.; Wakui, H.; Maeda, A.; Dejima, T.; Kanaoka, T.; Azushima, K.; Uneda, K.; Tsurumi-Ikeya, Y.; Kobayashi, R.; et al. Deletion of the angiotensin II type 1 receptor–associated protein enhances renal sodium reabsorption and exacerbates angiotensin II–mediated hypertension. *Kidney Int.* **2014**, *86*, 570–581. [CrossRef]
10. Kobayashi, R.; Wakui, H.; Azushima, K.; Uneda, K.; Haku, S.; Ohki, K.; Haruhara, K.; Kinguchi, S.; Matsuda, M.; Ohsawa, M.; et al. An angiotensin II type 1 receptor binding molecule has a critical role in hypertension in a chronic kidney disease model. *Kidney Int.* **2017**, *91*, 1115–1125. [CrossRef]
11. Kinguchi, S.; Wakui, H.; Azushima, K.; Haruhara, K.; Koguchi, T.; Ohki, K.; Uneda, K.; Matsuda, M.; Haku, S.; Yamaji, T.; et al. Effects of ATRAP in Renal Proximal Tubules on Angiotensin-Dependent Hypertension. *J. Am. Heart Assoc.* **2019**, *8*, e012395. [CrossRef]

12. Flister, M.J.; Tsaih, S.W.; O'Meara, C.C.; Endres, B.; Hoffman, M.J.; Geurts, A.M.; Dwinell, M.R.; Lazar, J.; Jacob, H.J.; Moreno, C. Identifying multiple causative genes at a single GWAS locus. *Genome Res.* **2013**, *23*, 1996–2002. [CrossRef]
13. Nguyen, G.; Delarue, F.; Burcklé, C.; Bouzhir, L.; Giller, T.; Sraer, J.D. Pivotal role of the renin/prorenin receptor in angiotensin II production and cellular responses to renin. *J. Clin. Investig.* **2002**, *109*, 1417–1427. [CrossRef]
14. Ramkumar, N.; Kohan, D.E. The nephron (pro)renin receptor: Function and significance. *Am. J. Physiol. Renal Physiol.* **2016**, *311*, F1145–F1148. [CrossRef]
15. Ramkumar, N.; Stuart, D.; Mironova, E.; Bugay, V.; Wang, S.; Abraham, N.; Ichihara, A.; Stockand, J.D.; Kohan, D.E. Renal tubular epithelial cell prorenin receptor regulates blood pressure and sodium transport. *Am. J. Physiol. Renal Physiol.* **2016**, *311*, F186–F194. [CrossRef]
16. Peng, K.; Lu, X.; Wang, F.; Nau, A.; Chen, R.; Zhou, S.F.; Yang, T. Collecting duct (pro)renin receptor targets ENaC to mediate angiotensin II-induced hypertension. *Am. J. Physiol. Renal Physiol.* **2017**, *312*, F245–F253. [CrossRef]
17. Prieto, M.C.; Reverte, V.; Mamenko, M.; Kuczeriszka, M.; Veiras, L.C.; Rosales, C.B.; McLellan, M.; Gentile, O.; Jensen, V.B.; Ichihara, A.; et al. Collecting duct prorenin receptor knockout reduces renal function, increases sodium excretion, and mitigates renal responses in ANG II-induced hypertensive mice. *Am. J. Physiol. Renal Physiol.* **2017**, *313*, F1243–F1253. [CrossRef]
18. Mendoza-Torres, E.; Oyarzún, A.; Mondaca-Ruff, D.; Azocar, A.; Castro, P.F.; Jalil, J.E.; Chiong, M.; Lavandero, S.; Ocaranza, M.P. ACE2 and vasoactive peptides: Novel players in cardiovascular/renal remodeling and hypertension. *Ther. Adv. Cardiovasc. Dis.* **2015**, *9*, 217–237. [CrossRef]
19. Ni, J.; Yang, F.; Huang, X.R.; Meng, J.; Chen, J.; Bader, M.; Penninger, J.M.; Fung, E.; Yu, X.Q.; Lan, H.Y. Dual deficiency of angiotensin-converting enzyme-2 and Mas receptor enhances angiotensin II-induced hypertension and hypertensive nephropathy. *J. Cell. Mol. Med.* **2020**, *24*, 13093–13103. [CrossRef]
20. Ushiogi, Y.; Takabatake, T.; Häberle, D.A. Blood pressure and tubuloglomerular feedback mechanism in chronically salt-loaded spontaneously hypertensive rats. *Kidney Int.* **1991**, *39*, 1184–1192. [CrossRef]
21. Lu, Y.; Wei, J.; Stec, D.E.; Roman, R.J.; Ge, Y.; Cheng, L.; Liu, E.Y.; Zhang, J.; Hansen, P.B.L.; Fan, F.; et al. Macula Densa Nitric Oxide Synthase 1β Protects against Salt-Sensitive Hypertension. *J. Am. Soc. Nephrol.* **2016**, *27*, 2346–2356. [CrossRef]
22. Zhang, J.; Qu, L.; Wei, J.; Jiang, S.; Xu, L.; Wang, L.; Cheng, F.; Jiang, K.; Buggs, J.; Liu, R. A new mechanism for the sex differences in angiotensin II-induced hypertension: The role of macula densa NOS1beta-mediated tubuloglomerular feedback. *Am. J. Physiol. Renal Physiol.* **2020**, *319*, F908–F919. [CrossRef]
23. Hyndman, K.A.; Boesen, E.I.; Elmarakby, A.A.A.; Brands, M.W.; Huang, P.; Kohan, D.E.; Pollock, D.M.; Pollock, J.S. Renal collecting duct NOS1 maintains fluid-electrolyte homeostasis and blood pressure. *Hypertension* **2013**, *62*, 91–98. [CrossRef]
24. Gao, Y.; Stuart, D.; Takahishi, T.; Kohan, D.E. Nephron-Specific Disruption of Nitric Oxide Synthase 3 Causes Hypertension and Impaired Salt Excretion. *J. Am. Heart Assoc.* **2018**, *7*, e009236. [CrossRef]
25. Hyndman, K.A.; Bugaj, V.; Mironova, E.; Stockand, J.D.; Pollock, J.S. NOS1-dependent negative feedback regulation of the epithelial sodium channel in the collecting duct. *Am. J. Physiol. Renal Physiol.* **2015**, *308*, F244–F251. [CrossRef]
26. Raikwar, N.S.; Thomas, C.P. Nedd4-2 isoforms ubiquitinate individual epithelial sodium channel subunits and reduce surface expression and function of the epithelial sodium channel. *Am. J. Physiol. Renal Physiol.* **2008**, *294*, F1157–F1165. [CrossRef]
27. Ronzaud, C.; Loffing-Cueni, D.; Hausel, P.; Debonneville, A.; Malsure, S.R.; Fowler-Jaeger, N.; Boase, N.A.; Perrier, R.; Maillard, M.; Yang, B.; et al. Renal tubular NEDD4-2 deficiency causes NCC-mediated salt-dependent hypertension. *J. Clin. Investig.* **2013**, *123*, 657–665. [CrossRef]
28. Furusho, T.; Uchida, S.; Sohara, E. The WNK signaling pathway and salt-sensitive hypertension. *Hypertens. Res.* **2020**, *43*, 733–743. [CrossRef]
29. Sohara, E.; Uchida, S. Kelch-like 3/Cullin 3 ubiquitin ligase complex and WNK signaling in salt-sensitive hypertension and electrolyte disorder. *Nephrol. Dial. Transplant.* **2016**, *31*, 1417–1424. [CrossRef]
30. Liu, Z.; Xie, J.; Wu, T.; Truong, T.; Auchus, R.J.; Huang, C.-L. Downregulation of NCC and NKCC2 cotransporters by kidney-specific WNK1 revealed by gene disruption and transgenic mouse models. *Hum. Mol. Genet.* **2011**, *20*, 855–866. [CrossRef]
31. Mu, S.; Shimosawa, T.; Ogura, S.; Wang, H.; Uetake, Y.; Kawakami-Mori, F.; Marumo, T.; Yatomi, Y.; Geller, D.S.; Tanaka, H.; et al. Epigenetic modulation of the renal β-adrenergic–WNK4 pathway in salt-sensitive hypertension. *Nat. Med.* **2011**, *17*, 573–580. [CrossRef] [PubMed]
32. Pan, X.; Shao, Y.; Wu, F.; Wang, Y.; Xiong, R.; Zheng, J.; Tian, H.; Wang, B.; Wang, Y.; Zhang, Y.; et al. FGF21 Prevents Angiotensin II-Induced Hypertension and Vascular Dysfunction by Activation of ACE2/Angiotensin-(1-7) Axis in Mice. *Cell Metab.* **2018**, *27*, 1323–1337.e1325. [CrossRef] [PubMed]
33. Crislip, G.R.; Douma, L.G.; Masten, S.H.; Cheng, K.Y.; Lynch, I.J.; Johnston, J.G.; Barral, D.; Glasford, K.B.; Holzworth, M.R.; Verlander, J.W.; et al. Differences in renal BMAL1 contribution to Na^+ homeostasis and blood pressure control in male and female mice. *Am. J. Physiol. Renal Physiol.* **2020**, *318*, F1463–F1477. [CrossRef] [PubMed]
34. Douma, L.G.; Costello, H.M.; Crislip, G.R.; Cheng, K.Y.; Lynch, I.J.; Juffre, A.; Barral, D.; Masten, S.; Roig, E.; Beguiristain, K.; et al. Kidney-specific KO of the circadian clock protein PER1 alters renal Na^+ handling, aldosterone levels, and kidney/adrenal gene expression. *Am. J. Physiol. Renal Physiol.* **2022**, *322*, F449–F459. [CrossRef] [PubMed]
35. Hiramatsu, A.; Izumi, Y.; Eguchi, K.; Matsuo, N.; Deng, Q.; Inoue, H.; Nakayama, Y.; Nonoguchi, H.; Aramburu, J.; Lopez-Rodriguez, C.; et al. Salt-Sensitive Hypertension of the Renal Tubular Cell-Specific NFAT5 (Nuclear Factor of Activated T-Cells 5) Knockout Mice. *Hypertension* **2021**, *78*, 1335–1346. [CrossRef] [PubMed]

36. Ueda, K.; Nishimoto, M.; Hirohama, D.; Ayuzawa, N.; Kawarazaki, W.; Watanabe, A.; Shimosawa, T.; Loffing, J.; Zhang, M.Z.; Marumo, T.; et al. Renal Dysfunction Induced by Kidney-Specific Gene Deletion of Hsd11b2 as a Primary Cause of Salt-Dependent Hypertension. *Hypertension* **2017**, *70*, 111–118. [CrossRef]
37. Shao, S.; Li, X.D.; Lu, Y.Y.; Li, S.J.; Chen, X.H.; Zhou, H.D.; He, S.; Guo, Y.T.; Lu, X.; Gao, P.J.; et al. Renal Natriuretic Peptide Receptor-C Deficiency Attenuates NaCl Cotransporter Activity in Angiotensin II-Induced Hypertension. *Hypertension* **2021**, *77*, 868–881. [CrossRef]
38. Groneberg, D.; KöNig, P.; Wirth, A.; Offermanns, S.; Koesling, D.; Friebe, A. Smooth Muscle–Specific Deletion of Nitric Oxide–Sensitive Guanylyl Cyclase Is Sufficient to Induce Hypertension in Mice. *Circulation* **2010**, *121*, 401–409. [CrossRef]
39. Kuhn, M. Molecular Physiology of Membrane Guanylyl Cyclase Receptors. *Physiol. Rev.* **2016**, *96*, 751–804. [CrossRef]
40. Moyes, A.J.; Hobbs, A.J. C-type Natriuretic Peptide: A Multifaceted Paracrine Regulator in the Heart and Vasculature. *Int. J. Mol. Sci.* **2019**, *20*, 2281. [CrossRef]
41. Nakao, K.; Kuwahara, K.; Nishikimi, T.; Nakagawa, Y.; Kinoshita, H.; Minami, T.; Kuwabara, Y.; Yamada, C.; Yamada, Y.; Tokudome, T.; et al. Endothelium-Derived C-Type Natriuretic Peptide Contributes to Blood Pressure Regulation by Maintaining Endothelial Integrity. *Hypertension* **2017**, *69*, 286–296. [CrossRef]
42. Špiranec, K.; Chen, W.; Werner, F.; Nikolaev, V.O.; Naruke, T.; Koch, F.; Werner, A.; Eder-Negrin, P.; Diéguez-Hurtado, R.; Adams, R.H.; et al. Endothelial C-Type Natriuretic Peptide Acts on Pericytes to Regulate Microcirculatory Flow and Blood Pressure. *Circulation* **2018**, *138*, 494–508. [CrossRef]
43. Tokudome, T.; Otani, K.; Mao, Y.; Jensen, L.J.; Arai, Y.; Miyazaki, T.; Sonobe, T.; Pearson, J.T.; Osaki, T.; Minamino, N.; et al. Endothelial Natriuretic Peptide Receptor 1 Play Crucial Role for Acute and Chronic Blood Pressure Regulation by Atrial Natriuretic Peptide. *Hypertension* **2022**, *79*, 1409–1422. [CrossRef]
44. Hooper, R.; Samakai, E.; Kedra, J.; Soboloff, J. Multifaceted roles of STIM proteins. *Pflügers Arch.-Eur. J. Physiol.* **2013**, *465*, 1383–1396. [CrossRef]
45. Kassan, M.; Ait-Aissa, K.; Radwan, E.; Mali, V.; Haddox, S.; Gabani, M.; Zhang, W.; Belmadani, S.; Irani, K.; Trebak, M.; et al. Essential Role of Smooth Muscle STIM1 in Hypertension and Cardiovascular Dysfunction. *Arterioscler. Thromb. Vasc. Biol.* **2016**, *36*, 1900–1909. [CrossRef]
46. Nishimoto, M.; Mizuno, R.; Fujita, T.; Isshiki, M. Stromal interaction molecule 1 modulates blood pressure via NO production in vascular endothelial cells. *Hypertens. Res.* **2018**, *41*, 506–514. [CrossRef]
47. Ferdaus, M.Z.; Xiao, B.; Ohara, H.; Nemoto, K.; Harada, Y.; Saar, K.; Hübner, N.; Isomura, M.; Nabika, T. Identification of Stim1 as a candidate gene for exaggerated sympathetic response to stress in the stroke-prone spontaneously hypertensive rat. *PLoS ONE* **2014**, *9*, e95091. [CrossRef]
48. Ohara, H.; Nabika, T. A nonsense mutation of Stim1 identified in stroke-prone spontaneously hypertensive rats decreased the store-operated calcium entry in astrocytes. *Biochem. Biophys. Res. Commun.* **2016**, *476*, 406–411. [CrossRef]
49. Odongoo, B.; Ohara, H.; Ngarashi, D.; Kaneko, T.; Kunihiro, Y.; Mashimo, T.; Nabika, T. Pathophysiological significance of Stim1 mutation in sympathetic response to stress and cardiovascular phenotypes in SHRSP/Izm: In vivo evaluation by creation of a novel gene knock-in rat using CRISPR/Cas9. *Clin. Exp. Hypertens.* **2021**, *43*, 34–41. [CrossRef]
50. Fallah, H.P.; Ahuja, E.; Lin, H.; Qi, J.; He, Q.; Gao, S.; An, H.; Zhang, J.; Xie, Y.; Liang, D. A Review on the Role of TRP Channels and Their Potential as Drug Targets_An Insight Into the TRP Channel Drug Discovery Methodologies. *Front. Pharmacol.* **2022**, *13*, 914499. [CrossRef]
51. Iring, A.; Jin, Y.-J.; Albarrán-Juárez, J.; Siragusa, M.; Wang, S.; Dancs, P.T.; Nakayama, A.; Tonack, S.; Chen, M.; Künne, C.; et al. Shear stress–induced endothelial adrenomedullin signaling regulates vascular tone and blood pressure. *J. Clin. Investig.* **2019**, *129*, 2775–2791. [CrossRef]
52. Kita, T.; Kitamura, K. Translational studies of adrenomedullin and related peptides regarding cardiovascular diseases. *Hypertens. Res.* **2022**, *45*, 389–400. [CrossRef]
53. Smyth, E.M.; Grosser, T.; Wang, M.; Yu, Y.; FitzGerald, G.A. Prostanoids in health and disease. *J. Lipid Res.* **2009**, *50*, S423–S428. [CrossRef]
54. Yang, T.; Du, Y. Distinct roles of central and peripheral prostaglandin E2 and EP subtypes in blood pressure regulation. *Am. J. Hypertens.* **2012**, *25*, 1042–1049. [CrossRef]
55. Bryson, T.D.; Harding, P. Prostaglandin E2 EP receptors in cardiovascular disease: An update. *Biochem. Pharmacol.* **2022**, *195*, 114858. [CrossRef]
56. Xu, H.; Fang, B.; Du, S.; Wang, S.; Li, Q.; Jia, X.; Bao, C.; Ye, L.; Sui, X.; Qian, L.; et al. Endothelial cell prostaglandin E2 receptor EP4 is essential for blood pressure homeostasis. *JCI Insight* **2020**, *5*, e138505. [CrossRef]
57. Caillon, A.; Paradis, P.; Schiffrin, E.L. Role of immune cells in hypertension. *Br. J. Pharmacol* **2019**, *176*, 1818–1828. [CrossRef]
58. Lu, X.; Crowley, S.D. The Immune System in Hypertension: A Lost Shaker of Salt 2021 Lewis K. Dahl Memorial Lecture. *Hypertension* **2022**, *79*, 1339–1347. [CrossRef]
59. Rizzoni, D.; De Ciuceis, C.; Szczepaniak, P.; Paradis, P.; Schiffrin, E.L.; Guzik, T.J. Immune System and Microvascular Remodeling in Humans. *Hypertension* **2022**, *79*, 691–705. [CrossRef]
60. Shen, Y.; Shen, Z.; Luo, S.; Guo, W.; Zhu, Y.Z. The Cardioprotective Effects of Hydrogen Sulfide in Heart Diseases: From Molecular Mechanisms to Therapeutic Potential. *Oxid. Med. Cell. Longev.* **2015**, *2015*, 925167. [CrossRef]

61. Yang, G.; Wu, L.; Jiang, B.; Yang, W.; Qi, J.; Cao, K.; Meng, Q.; Mustafa, A.K.; Mu, W.; Zhang, S.; et al. H2S as a physiologic vasorelaxant: Hypertension in mice with deletion of cystathionine gamma-lyase. *Science* **2008**, *322*, 587–590. [CrossRef] [PubMed]
62. Cui, C.; Fan, J.; Zeng, Q.; Cai, J.; Chen, Y.; Chen, Z.; Wang, W.; Li, S.Y.; Cui, Q.; Yang, J.; et al. CD4$^+$ T-Cell Endogenous Cystathionine gamma Lyase-Hydrogen Sulfide Attenuates Hypertension by Sulfhydrating Liver Kinase B1 to Promote T Regulatory Cell Differentiation and Proliferation. *Circulation* **2020**, *142*, 1752–1769. [CrossRef] [PubMed]
63. Sun, X.N.; Li, C.; Liu, Y.; Du, L.J.; Zeng, M.R.; Zheng, X.J.; Zhang, W.C.; Liu, Y.; Zhu, M.; Kong, D.; et al. T-Cell Mineralocorticoid Receptor Controls Blood Pressure by Regulating Interferon-Gamma. *Circ. Res.* **2017**, *120*, 1584–1597. [CrossRef] [PubMed]
64. Stepicheva, N.A.; Song, J.L. Function and regulation of microRNA-31 in development and disease. *Mol. Reprod. Dev.* **2016**, *83*, 654–674. [CrossRef] [PubMed]
65. Li, X.; Cai, W.; Xi, W.; Sun, W.; Shen, W.; Wei, T.; Chen, X.; Sun, L.; Zhou, H.; Sun, Y.; et al. MicroRNA-31 Regulates Immunosuppression in Ang II (Angiotensin II)-induced Hypertension by Targeting Ppp6C (Protein Phosphatase 6c). *Hypertension* **2019**, *73*, e14–e24. [CrossRef] [PubMed]
66. Crowley, S.D.; Rudemiller, N.P. Immunologic Effects of the Renin-Angiotensin System. *J. Am. Soc. Nephrol.* **2017**, *28*, 1350–1361. [CrossRef]
67. Zhang, J.-D.; Patel, M.B.; Song, Y.-S.; Griffiths, R.; Burchette, J.; Ruiz, P.; Sparks, M.A.; Yan, M.; Howell, D.N.; Gomez, J.A.; et al. A Novel Role for Type 1 Angiotensin Receptors on T Lymphocytes to Limit Target Organ Damage in Hypertension. *Circ. Res.* **2012**, *110*, 1604–1617. [CrossRef]
68. Zhang, J.D.; Patel, M.B.; Griffiths, R.; Dolber, P.C.; Ruiz, P.; Sparks, M.A.; Stegbauer, J.; Jin, H.; Gomez, J.A.; Buckley, A.F.; et al. Type 1 angiotensin receptors on macrophages ameliorate IL-1 receptor-mediated kidney fibrosis. *J. Clin. Investig.* **2014**, *124*, 2198–2203. [CrossRef]
69. Lu, X.; Zhang, J.; Wen, Y.; Ren, J.; Griffiths, R.; Rudemiller, N.P.; Ide, S.; Souma, T.; Crowley, S.D. Type 1 Angiotensin Receptors on CD11c-Expressing Cells Protect Against Hypertension by Regulating Dendritic Cell-Mediated T Cell Activation. *Hypertension* **2022**, *79*, 1227–1236. [CrossRef]
70. Sag, C.M.; Schnelle, M.; Zhang, J.; Murdoch, C.E.; Kossmann, S.; Protti, A.; Santos, C.X.C.; Sawyer, G.; Zhang, X.; Mongue-Din, H.; et al. Distinct Regulatory Effects of Myeloid Cell and Endothelial Cell NAPDH Oxidase 2 on Blood Pressure. *Circulation* **2017**, *135*, 2163–2177. [CrossRef]
71. Wen, Y.; Rudemiller, N.P.; Zhang, J.; Lu, X.; Ren, J.; Privratsky, J.R.; Griffiths, R.; Zhang, J.J.; Hammer, G.E.; Crowley, S.D. C-C Motif Chemokine Receptor 7 Exacerbates Hypertension through Effects on T Lymphocyte Trafficking. *Hypertension* **2020**, *75*, 869–876. [CrossRef]
72. Singh, M.V.; Cicha, M.Z.; Nunez, S.; Meyerholz, D.K.; Chapleau, M.W.; Abboud, F.M. Angiotensin II-induced hypertension and cardiac hypertrophy are differentially mediated by TLR3- and TLR4-dependent pathways. *Am. J. Physiol. Heart Circ. Physiol.* **2019**, *316*, H1027–H1038. [CrossRef]
73. Perrotta, M.; Lori, A.; Carnevale, L.; Fardella, S.; Cifelli, G.; Iacobucci, R.; Mastroiacovo, F.; Iodice, D.; Pallante, F.; Storto, M.; et al. Deoxycorticosterone acetate-salt hypertension activates placental growth factor in the spleen to couple sympathetic drive and immune system activation. *Cardiovasc. Res.* **2018**, *114*, 456–467. [CrossRef]
74. Chen, X.H.; Ruan, C.C.; Ge, Q.; Ma, Y.; Xu, J.Z.; Zhang, Z.B.; Lin, J.R.; Chen, D.R.; Zhu, D.L.; Gao, P.J. Deficiency of Complement C3a and C5a Receptors Prevents Angiotensin II-Induced Hypertension via Regulatory T Cells. *Circ. Res.* **2018**, *122*, 970–983. [CrossRef]
75. Caillon, A.; Mian, M.O.R.; Fraulob-Aquino, J.C.; Huo, K.G.; Barhoumi, T.; Ouerd, S.; Sinnaeve, P.R.; Paradis, P.; Schiffrin, E.L. gammadelta T Cells Mediate Angiotensin II-Induced Hypertension and Vascular Injury. *Circulation* **2017**, *135*, 2155–2162. [CrossRef]
76. Zhang, J.; Rudemiller, N.P.; Patel, M.B.; Karlovich, N.S.; Wu, M.; McDonough, A.A.; Griffiths, R.; Sparks, M.A.; Jeffs, A.D.; Crowley, S.D. Interleukin-1 Receptor Activation Potentiates Salt Reabsorption in Angiotensin II-Induced Hypertension via the NKCC2 Co-transporter in the Nephron. *Cell Metab.* **2016**, *23*, 360–368. [CrossRef]
77. Van Beusecum, J.P.; Moreno, H.; Harrison, D.G. Innate immunity and clinical hypertension. *J. Hum. Hypertens.* **2021**, *36*, 503–509. [CrossRef]
78. Su, C.; Xue, J.; Ye, C.; Chen, A. Role of the central renin-angiotensin system in hypertension (Review). *Int. J. Mol. Med.* **2021**, *47*, 95. [CrossRef]
79. Morimoto, S.; Cassell, M.D.; Beltz, T.G.; Johnson, A.K.; Davisson, R.L.; Sigmund, C.D. Elevated Blood Pressure in Transgenic Mice With Brain-Specific Expression of Human Angiotensinogen Driven by the Glial Fibrillary Acidic Protein Promoter. *Circ. Res.* **2001**, *89*, 365–372. [CrossRef]
80. Morimoto, S.; Cassell, M.D.; Sigmund, C.D. Glia- and Neuron-specific Expression of the Renin-Angiotensin System in Brain Alters Blood Pressure, Water Intake, and Salt Preference. *J. Biol. Chem.* **2002**, *277*, 33235–33241. [CrossRef]
81. Sherrod, M.; Davis, D.R.; Zhou, X.; Cassell, M.D.; Sigmund, C.D. Glial-specific ablation of angiotensinogen lowers arterial pressure in renin and angiotensinogen transgenic mice. *Am. J. Physiol. Regul. Integr. Comp. Physiol.* **2005**, *289*, R1763–R1769. [CrossRef]
82. Li, W.; Peng, H.; Mehaffey, E.P.; Kimball, C.D.; Grobe, J.L.; Van Gool, J.M.G.; Sullivan, M.N.; Earley, S.; Danser, A.H.J.; Ichihara, A.; et al. Neuron-Specific (Pro)renin Receptor Knockout Prevents the Development of Salt-Sensitive Hypertension. *Hypertension* **2014**, *63*, 316–323. [CrossRef]

83. Nomura, K.; Hiyama, T.Y.; Sakuta, H.; Matsuda, T.; Lin, C.-H.; Kobayashi, K.; Kobayashi, K.; Kuwaki, T.; Takahashi, K.; Matsui, S.; et al. [Na+] Increases in Body Fluids Sensed by Central Nax Induce Sympathetically Mediated Blood Pressure Elevations via H+-Dependent Activation of ASIC1a. *Neuron* **2019**, *101*, 60–75.e66. [CrossRef]
84. Evans, L.C.; Ivy, J.R.; Wyrwoll, C.; McNairn, J.A.; Menzies, R.I.; Christensen, T.H.; Al-Dujaili, E.A.; Kenyon, C.J.; Mullins, J.J.; Seckl, J.R.; et al. Conditional Deletion of Hsd11b2 in the Brain Causes Salt Appetite and Hypertension. *Circulation* **2016**, *133*, 1360–1370. [CrossRef]
85. Trebak, F.; Li, W.; Feng, Y. Neuronal (pro)renin receptor regulates deoxycorticosterone-induced sodium intake. *Physiological Genomics* **2018**, *50*, 904–912. [CrossRef]
86. Wu, C.H.; Mohammadmoradi, S.; Thompson, J.; Su, W.; Gong, M.; Nguyen, G.; Yiannikouris, F. Adipocyte (Pro)Renin-Receptor Deficiency Induces Lipodystrophy, Liver Steatosis and Increases Blood Pressure in Male Mice. *Hypertension* **2016**, *68*, 213–219. [CrossRef]
87. Davies, L.A.; Hu, C.; Guagliardo, N.A.; Sen, N.; Chen, X.; Talley, E.M.; Carey, R.M.; Bayliss, D.A.; Barrett, P.Q. TASK channel deletion in mice causes primary hyperaldosteronism. *Proc. Natl. Acad. Sci. USA* **2008**, *105*, 2203–2208. [CrossRef] [PubMed]
88. Guagliardo, N.A.; Yao, J.; Stipes, E.J.; Cechova, S.; Le, T.H.; Bayliss, D.A.; Breault, D.T.; Barrett, P.Q. Adrenal Tissue-Specific Deletion of TASK Channels Causes Aldosterone-Driven Angiotensin II–Independent Hypertension. *Hypertension* **2019**, *73*, 407–414. [CrossRef] [PubMed]
89. Mathar, I.; Vennekens, R.; Meissner, M.; Kees, F.; Van Der Mieren, G.; Camacho Londoño, J.E.; Uhl, S.; Voets, T.; Hummel, B.; Van Den Bergh, A.; et al. Increased catecholamine secretion contributes to hypertension in TRPM4-deficient mice. *J. Clin. Investig.* **2010**, *120*, 3267–3279. [CrossRef] [PubMed]
90. Doi, M.; Takahashi, Y.; Komatsu, R.; Yamazaki, F.; Yamada, H.; Haraguchi, S.; Emoto, N.; Okuno, Y.; Tsujimoto, G.; Kanematsu, A.; et al. Salt-sensitive hypertension in circadian clock–deficient Cry-null mice involves dysregulated adrenal Hsd3b6. *Nat. Med.* **2010**, *16*, 67–74. [CrossRef]
91. Wise, I.A.; Charchar, F.J. Epigenetic Modifications in Essential Hypertension. *Int. J. Mol. Sci.* **2016**, *17*, 451. [CrossRef]
92. Arif, M.; Sadayappan, S.; Becker, R.C.; Martin, L.J.; Urbina, E.M. Epigenetic modification: A regulatory mechanism in essential hypertension. *Hypertens. Res.* **2019**, *42*, 1099–1113. [CrossRef]
93. Chaudhary, M. Novel methylation mark and essential hypertension. *J. Genet. Eng. Biotechnol.* **2022**, *20*, 11. [CrossRef]
94. Chakraborty, S.; Mandal, J.; Yang, T.; Cheng, X.; Yeo, J.Y.; McCarthy, C.G.; Wenceslau, C.F.; Koch, L.G.; Hill, J.W.; Vijay-Kumar, M.; et al. Metabolites and Hypertension: Insights into Hypertension as a Metabolic Disorder: 2019 Harriet Dustan Award. *Hypertension* **2020**, *75*, 1386–1396. [CrossRef]
95. Louca, P.; Menni, C.; Padmanabhan, S. Genomic Determinants of Hypertension With a Focus on Metabolomics and the Gut Microbiome. *Am. J. Hypertens.* **2020**, *33*, 473–481. [CrossRef]
96. Mishima, E.; Abe, T. Role of the microbiota in hypertension and antihypertensive drug metabolism. *Hypertens. Res.* **2022**, *45*, 246–253. [CrossRef]
97. Johnson, A.K.; Xue, B. Central nervous system neuroplasticity and the sensitization of hypertension. *Nat. Rev. Nephrol.* **2018**, *14*, 750–766. [CrossRef]
98. Mann, S.J. Neurogenic hypertension: Pathophysiology, diagnosis and management. *Clin. Auton. Res.* **2018**, *28*, 363–374. [CrossRef]
99. Ahmari, N.; Hayward, L.F.; Zubcevic, J. The importance of bone marrow and the immune system in driving increases in blood pressure and sympathetic nerve activity in hypertension. *Exp. Physiol.* **2020**, *105*, 1815–1826. [CrossRef]
100. Geurts, A.M.; Cost, G.J.; Freyvert, Y.; Zeitler, B.; Miller, J.C.; Choi, V.M.; Jenkins, S.S.; Wood, A.; Cui, X.; Meng, X.; et al. Knockout Rats via Embryo Microinjection of Zinc-Finger Nucleases. *Science* **2009**, *325*, 433. [CrossRef]
101. Szpirer, C. Rat models of human diseases and related phenotypes: A systematic inventory of the causative genes. *J. Biomed. Sci.* **2020**, *27*, 84. [CrossRef]
102. Sato, M.; Nakamura, S.; Inada, E.; Takabayashi, S. Recent Advances in the Production of Genome-Edited Rats. *Int. J. Mol. Sci.* **2022**, *23*, 2548. [CrossRef]
103. Nabika, T.; Ohara, H.; Kato, N.; Isomura, M. The stroke-prone spontaneously hypertensive rat: Still a useful model for post-GWAS genetic studies? *Hypertens. Res.* **2012**, *35*, 477–484. [CrossRef]
104. Schulz, A.; Kreutz, R. Mapping genetic determinants of kidney damage in rat models. *Hypertens. Res.* **2012**, *35*, 675–694. [CrossRef]
105. Yang, Y.; Kimura-Ohba, S.; Thompson, J.; Rosenberg, G.A. Rodent Models of Vascular Cognitive Impairment. *Transl. Stroke Res.* **2016**, *7*, 407–414. [CrossRef]
106. Doggrell, S.A.; Brown, L. Rat models of hypertension, cardiac hypertrophy and failure. *Cardiovasc. Res.* **1998**, *39*, 89–105. [CrossRef]
107. Liška, F.; Mancini, M.; Krupková, M.; Chylíková, B.; Křenová, D.; Šeda, O.; Šilhavý, J.; Mlejnek, P.; Landa, V.; Zídek, V.; et al. Plzf as a candidate gene predisposing the spontaneously hypertensive rat to hypertension, left ventricular hypertrophy, and interstitial fibrosis. *Am. J. Hypertens.* **2014**, *27*, 99–106. [CrossRef]
108. Liška, F.; Landa, V.; Zídek, V.; Mlejnek, P.; Šilhavý, J.; Šimáková, M.; Strnad, H.; Trnovská, J.; Škop, V.; Kazdová, L.; et al. Downregulation of Plzf Gene Ameliorates Metabolic and Cardiac Traits in the Spontaneously Hypertensive Rat. *Hypertension* **2017**, *69*, 1084–1091. [CrossRef] [PubMed]

109. Lin, Z.-H.; Fukuda, N.; Jin, X.-Q.; Yao, E.-H.; Ueno, T.; Endo, M.; Saito, S.; Matsumoto, K.; Mugishima, H. Complement 3 Is Involved in the Synthetic Phenotype and Exaggerated Growth of Vascular Smooth Muscle Cells From Spontaneously Hypertensive Rats. *Hypertension* **2004**, *44*, 42–47. [CrossRef] [PubMed]
110. Han, Y.; Fukuda, N.; Ueno, T.; Endo, M.; Ikeda, K.; Xueli, Z.; Matsumoto, T.; Soma, M.; Matsumoto, K. Role of Complement 3a in the Synthetic Phenotype and Angiotensin II-Production in Vascular Smooth Muscle Cells from Spontaneously Hypertensive Rats. *Am. J. Hypertens.* **2012**, *25*, 284–289. [CrossRef] [PubMed]
111. Ikeda, K.; Fukuda, N.; Ueno, T.; Endo, M.; Kobayashi, N.; Soma, M.; Matsumoto, K. Role of complement 3a in the growth of mesangial cells from stroke-prone spontaneously hypertensive rats. *Clin. Exp. Hypertens.* **2014**, *36*, 58–63. [CrossRef]
112. Negishi, E.; Fukuda, N.; Otsuki, T.; Katakawa, M.; Komatsu, K.; Chen, L.; Tanaka, S.; Kobayashi, H.; Hatanaka, Y.; Ueno, T.; et al. Involvement of complement 3 in the salt-sensitive hypertension by activation of renal renin-angiotensin system in spontaneously hypertensive rats. *Am. J. Physiol.-Ren. Physiol.* **2018**, *315*, F1747–F1758. [CrossRef]
113. Rubattu, S.; Volpe, M.; Kreutz, R.; Ganten, U.; Ganten, D.; Lindpaintner, K. Chromosomal mapping of quantitative trait loci contributing to stroke in a rat model of complex human disease. *Nat. Genet.* **1996**, *13*, 429–434. [CrossRef]
114. Rubattu, S.; Di Castro, S.; Schulz, H.; Geurts, A.M.; Cotugno, M.; Bianchi, F.; Maatz, H.; Hummel, O.; Falak, S.; Stanzione, R.; et al. Ndufc2 Gene Inhibition Is Associated with Mitochondrial Dysfunction and Increased Stroke Susceptibility in an Animal Model of Complex Human Disease. *J. Am. Heart Assoc.* **2016**, *5*, e002701. [CrossRef]
115. Mahal, Z.; Fujikawa, K.; Matsuo, H.; Zahid, H.M.; Koike, M.; Misumi, M.; Kaneko, T.; Mashimo, T.; Ohara, H.; Nabika, T. Effects of the Prdx2 depletion on blood pressure and life span in spontaneously hypertensive rats. *Hypertens. Res.* **2019**, *42*, 610–617. [CrossRef]
116. Kattoor, A.J.; Goel, A.; Mehta, J.L. LOX-1: Regulation, Signaling and Its Role in Atherosclerosis. *Antioxidants* **2019**, *8*, 218. [CrossRef]
117. Liang, Y.-Q.; Kakino, A.; Matsuzaka, Y.; Mashimo, T.; Isono, M.; Akamatsu, T.; Shimizu, H.; Tajima, M.; Kaneko, T.; Li, L.; et al. LOX-1 (Lectin-Like Oxidized Low-Density Lipoprotein Receptor-1) Deletion Has Protective Effects on Stroke in the Genetic Background of Stroke-Prone Spontaneously Hypertensive Rat. *Stroke* **2020**, *51*, 1835–1843. [CrossRef]
118. Xiao, B.; Harada, Y.; Kawakami, K.; Nabika, T. A 1.8-Mbp fragment on chromosome 1 affects sympathetic response to stress: Evaluation in reciprocal congenic strains between stroke-prone spontaneously hypertensive rat and Wistar-Kyoto rat. *J. Hypertens.* **2011**, *29*, 257–265. [CrossRef]
119. Rapp, J.P.; Garrett, M.R. Will the real Dahl S rat please stand up? *Am. J. Physiol. Renal Physiol.* **2019**, *317*, F1231–F1240. [CrossRef]
120. Moreno, C.; Hoffman, M.; Stodola, T.J.; Didier, D.N.; Lazar, J.; Geurts, A.M.; North, P.E.; Jacob, H.J.; Greene, A.S. Creation and characterization of a renin knockout rat. *Hypertension* **2011**, *57*, 614–619. [CrossRef] [PubMed]
121. Mattson, D.L.; Lund, H.; Guo, C.; Rudemiller, N.; Geurts, A.M.; Jacob, H. Genetic mutation of recombination activating gene 1 in Dahl salt-sensitive rats attenuates hypertension and renal damage. *Am. J. Physiol. Regul. Integr. Comp. Physiol.* **2013**, *304*, R407–R414. [CrossRef] [PubMed]
122. Zhou, X.; Zhang, Z.; Shin, M.K.; Horwitz, S.B.; Levorse, J.M.; Zhu, L.; Sharif-Rodriguez, W.; Streltsov, D.Y.; Dajee, M.; Hernandez, M.; et al. Heterozygous disruption of renal outer medullary potassium channel in rats is associated with reduced blood pressure. *Hypertension* **2013**, *62*, 288–294. [CrossRef] [PubMed]
123. Endres, B.T.; Priestley, J.R.; Palygin, O.; Flister, M.J.; Hoffman, M.J.; Weinberg, B.D.; Grzybowski, M.; Lombard, J.H.; Staruschenko, A.; Moreno, C.; et al. Mutation of Plekha7 attenuates salt-sensitive hypertension in the rat. *Proc. Natl. Acad. Sci. USA* **2014**, *111*, 12817–12822. [CrossRef] [PubMed]
124. Jin, C.; Sun, J.; Stilphen, C.A.; Smith, S.M.; Ocasio, H.; Bermingham, B.; Darji, S.; Guha, A.; Patel, R.; Geurts, A.M.; et al. HV1 acts as a sodium sensor and promotes superoxide production in medullary thick ascending limb of Dahl salt-sensitive rats. *Hypertension* **2014**, *64*, 541–550. [CrossRef]
125. Rudemiller, N.; Lund, H.; Jacob, H.J.; Geurts, A.A.M.; Mattson, D.L. CD247 modulates blood pressure by altering T-lymphocyte infiltration in the kidney. *Hypertension* **2014**, *63*, 559–564. [CrossRef]
126. Holditch, S.J.; Schreiber, C.A.; Nini, R.; Tonne, J.M.; Peng, K.-W.; Geurts, A.; Jacob, H.J.; Burnett, J.C.; Cataliotti, A.; Ikeda, Y. B-Type Natriuretic Peptide Deletion Leads to Progressive Hypertension, Associated Organ Damage, and Reduced Survival. *Hypertension* **2015**, *66*, 199–210. [CrossRef]
127. Kumarasamy, S.; Waghulde, H.; Gopalakrishnan, K.; Mell, B.; Morgan, E.; Joe, B. Mutation within the hinge region of the transcription factor Nr2f2 attenuates salt-sensitive hypertension. *Nat. Commun.* **2015**, *6*, 6252. [CrossRef]
128. Nayak, S.; Khan, M.A.; Wan, T.C.; Pei, H.; Linden, J.; Dwinell, M.R.; Geurts, A.M.; Imig, J.D.; Auchampach, J.A. Characterization of Dahl salt-sensitive rats with genetic disruption of the A2B adenosine receptor gene: Implications for A2B adenosine receptor signaling during hypertension. *Purinergic Signal.* **2015**, *11*, 519–531. [CrossRef]
129. Cowley, A.W., Jr.; Yang, C.; Zheleznova, N.N.; Staruschenko, A.; Kurth, T.; Rein, L.; Kumar, V.; Sadovnikov, K.; Dayton, A.; Hoffman, M.; et al. Evidence of the Importance of Nox4 in Production of Hypertension in Dahl Salt-Sensitive Rats. *Hypertension* **2016**, *67*, 440–450. [CrossRef]
130. Cheng, X.; Waghulde, H.; Mell, B.; Morgan, E.E.; Pruett-Miller, S.M.; Joe, B. Positional cloning of quantitative trait nucleotides for blood pressure and cardiac QT-interval by targeted CRISPR/Cas9 editing of a novel long non-coding RNA. *PLoS Genet.* **2017**, *13*, e1006961. [CrossRef]

131. Kumarasamy, S.; Waghulde, H.; Cheng, X.; Haller, S.T.; Mell, B.; Abhijith, B.; Ashraf, U.M.; Atari, E.; Joe, B. Targeted disruption of regulated endocrine-specific protein (Resp18) in Dahl SS/Mcw rats aggravates salt-induced hypertension and renal injury. *Physiol. Genom.* **2018**, *50*, 369–375. [CrossRef]
132. Waghulde, H.; Cheng, X.; Galla, S.; Mell, B.; Cai, J.; Pruett-Miller, S.M.; Vazquez, G.; Patterson, A.; Vijay Kumar, M.; Joe, B. Attenuation of Microbiotal Dysbiosis and Hypertension in a CRISPR/Cas9 Gene Ablation Rat Model of GPER1. *Hypertension* **2018**, *72*, 1125–1132. [CrossRef]
133. Abais-Battad, J.M.; Lund, H.; Dasinger, J.H.; Fehrenbach, D.J.; Cowley, A.W., Jr.; Mattson, D.L. NOX2-derived reactive oxygen species in immune cells exacerbates salt-sensitive hypertension. *Free Radic. Biol. Med.* **2020**, *146*, 333–339. [CrossRef]
134. Pulimeno, P.; Bauer, C.; Stutz, J.; Citi, S. PLEKHA7 is an adherens junction protein with a tissue distribution and subcellular localization distinct from ZO-1 and E-cadherin. *PLoS ONE* **2010**, *5*, e12207. [CrossRef]
135. Levy, D.; Ehret, G.B.; Rice, K.; Verwoert, G.C.; Launer, L.J.; Dehghan, A.; Glazer, N.L.; Morrison, A.C.; Johnson, A.D.; Aspelund, T.; et al. Genome-wide association study of blood pressure and hypertension. *Nat. Genet.* **2009**, *41*, 677–687. [CrossRef]
136. Hong, K.W.; Jin, H.S.; Lim, J.E.; Kim, S.; Go, M.J.; Oh, B. Recapitulation of two genomewide association studies on blood pressure and essential hypertension in the Korean population. *J. Hum. Genet.* **2010**, *55*, 336–341. [CrossRef]
137. Fox, E.R.; Young, J.H.; Li, Y.; Dreisbach, A.W.; Keating, B.J.; Musani, S.K.; Liu, K.; Morrison, A.C.; Ganesh, S.; Kutlar, A.; et al. Association of genetic variation with systolic and diastolic blood pressure among African Americans: The Candidate Gene Association Resource study. *Hum. Mol. Genet.* **2011**, *20*, 2273–2284. [CrossRef]
138. Ho, J.E.; Levy, D.; Rose, L.; Johnson, A.D.; Ridker, P.M.; Chasman, D.I. Discovery and replication of novel blood pressure genetic loci in the Women's Genome Health Study. *J. Hypertens.* **2011**, *29*, 62–69. [CrossRef]
139. Lin, Y.; Lai, X.; Chen, B.; Xu, Y.; Huang, B.; Chen, Z.; Zhu, S.; Yao, J.; Jiang, Q.; Huang, H.; et al. Genetic variations in CYP17A1, CACNB2 and PLEKHA7 are associated with blood pressure and/or hypertension in She ethnic minority of China. *Atherosclerosis* **2011**, *219*, 709–714. [CrossRef]
140. Kobayashi, T.; Namba, M.; Koyano, T.; Fukushima, M.; Sato, M.; Ohtsuka, M.; Matsuyama, M. Successful production of genome-edited rats by the rGONAD method. *BMC Biotechnol.* **2018**, *18*, 19. [CrossRef]
141. Miyasaka, Y.; Uno, Y.; Yoshimi, K.; Kunihiro, Y.; Yoshimura, T.; Tanaka, T.; Ishikubo, H.; Hiraoka, Y.; Takemoto, N.; Tanaka, T.; et al. CLICK: One-step generation of conditional knockout mice. *BMC Genom.* **2018**, *19*, 318. [CrossRef]
142. Ohtsuka, M.; Sato, M.; Miura, H.; Takabayashi, S.; Matsuyama, M.; Koyano, T.; Arifin, N.; Nakamura, S.; Wada, K.; Gurumurthy, C.B. i-GONAD: A robust method for in situ germline genome engineering using CRISPR nucleases. *Genome Biol.* **2018**, *19*, 25. [CrossRef]
143. Morisaka, H.; Yoshimi, K.; Okuzaki, Y.; Gee, P.; Kunihiro, Y.; Sonpho, E.; Xu, H.; Sasakawa, N.; Naito, Y.; Nakada, S.; et al. CRISPR-Cas3 induces broad and unidirectional genome editing in human cells. *Nat. Commun.* **2019**, *10*, 5302. [CrossRef]
144. Padmanabhan, S.; Joe, B. Towards Precision Medicine for Hypertension: A Review of Genomic, Epigenomic, and Microbiomic Effects on Blood Pressure in Experimental Rat Models and Humans. *Physiol. Rev.* **2017**, *97*, 1469–1528. [CrossRef]

biomedicines

MDPI

Article

Stroke-Prone SHR as Experimental Models for Cardiovascular Disease Risk Reduction in Humans

Yukio Yamori [1,*], Miki Sagara [1], Hideki Mori [1] and Mari Mori [2] on behalf of the CARDIAC Study Group

1 Institute for World Health Development, Mukogawa Women's University, Nishinomiya 663-8143, Japan
2 Department of Health Management, School of Health Study Tokai University, Hiratsuka 259-1292, Japan
* Correspondence: yamori@cardiacstudy.com

Abstract: Since stroke-prone spontaneously hypertensive rats (SHRSP) develop hypertension and stroke without exception, the prevention or reduction of risk by various nutrients was tested on blood pressure and the mortality caused by stroke and cardiovascular diseases (CVD). In addition to sodium (Na) accelerating hypertension and stroke and potassium (K) counteracting the adverse effect of Na, taurine (Tau), rich in seafood, and magnesium (Mg) contained in soy, nuts, grains, etc., were proven to reduce stroke and CVD and improved survival. Therefore, the Cardiovascular Diseases and Alimentary Comparison Study was started in 1985 to explore the association of biomarkers of diet in 24 hour urine(24U) with CVD risks, and about 100 males and 100 females aged 48–56 in each of 50 populations were studied until 1995. Linear regression analysis indicated that the 24U Tau/creatinine and Mg/creatinine ratios were inversely associated with body mass index, systolic and diastolic blood pressure, and total cholesterol. In comparison with six Euro-Western regions, 24U Tau and Mg collected from six regions, respectively, in Japan and the Mediterranean countries were significantly higher and were significantly associated with lower CVD risks. Diets rich in Tau and Mg were concluded to be contributory to the prevention of CVD in SHRSP and humans.

Keywords: taurine; magnesium; 24 hour urine(24U); stroke-prone spontaneously hypertensive rat

Citation: Yamori, Y.; Sagara, M.; Mori, H.; Mori, M., on behalf of the CARDIAC Study Group. Stroke-Prone SHR as Experimental Models for Cardiovascular Disease Risk Reduction in Humans. *Biomedicines* **2022**, *10*, 2974. https://doi.org/10.3390/biomedicines10112974

Academic Editors: Josef Zicha and Ivana Vaněčková

Received: 31 August 2022
Accepted: 7 November 2022
Published: 18 November 2022

1. Introduction

We successfully established stroke-prone SHR (SHRSP, 1974) [1] by successive breeding from a spontaneously hypertensive rat (SHR, 1963) [2], which died of hemorrhagic and/or ischemic stroke. Since they developed stroke spontaneously [3], they were regarded as an appropriate model for research not only on the pathophysiological mechanisms of stroke in general [4], lacunar stroke [5]' cerebral small vessel diseases [6], and subcortical ischemic stroke [7,8], but also the post-stroke treatments [9,10]. Further, since they develop stroke genetically similar to humans, they are a useful model for the potential prediction of stroke through the analysis of genes related to stroke [11–17]. When stroke can be predicted based on the genes in SHRSP and hopefully in humans, stroke will be prevented by nutrition in humans, as first proven experimentally in SHRSP [3].

Stroke became the leading cause of death in Japan in 1960 after tuberculosis became pharmacologically treated. The nutritional situation in Japan over 60 years ago was supposed to cause stroke among cardiovascular diseases (CVD) because of traditional low protein intake characterized by limited meat supply and owing to habitually low calcium (Ca) and/or magnesium (Mg) intake due to low dairy food consumption. Therefore, we focused on the nutritional prevention of stroke and CVD in the newly established SHRSP by feeding them high-protein diets with Ca and Mg. Since the nutritional intakes were objectively estimated epidemiologically by 24 h urine (24U) analyses of the biomarkers in humans [18], we attempted to show whether experimentally beneficial nutrients for preventing stroke in SHRSP would be epidemiologically associated with cardiovascular risk reduction in humans [19].

Therefore, we proposed to the World Health Organization (WHO) an international cooperative study, "Cardiovascular Diseases and Alimentary Comparison (CARDIAC) Study" [20,21], from our WHO Collaborating Center for Research on Primary Prevention of Cardiovascular Diseases, which was designated by WHO in 1983. In response to our proposal, representative researchers from 61 study sites joined the CARDIAC Study [22].

Since our cooperative studies for over 20 years revealed nutritional biomarkers in worldwide collected 24U samples were significantly associated with CVD risk reduction, we further analyzed these biomarkers in some populations known for their longevity, including Japan (J), which keeps the nearly longest average life expectancy in the world, and Mediterranean (M) countries such as Greece, Sicily of Italy, Spain, and Portugal.

In comparison with Euro-Western countries (EW) (Scotland, Ireland, Sweden, plus Canada, New Zealand, and Australia) where immigrants and descendants from their countries have been living. Among these countries, the popular diets of M countries, rich in polyphenols [23] and related nutrients from fruits, vegetables, olive oil, nuts [24], and fish, have been focused on their basic health effects [25], including cognitive function [26]. However, since no data on M diets compared with EW and J diets has been reported on nutritional biomarkers in 24U samples, we compared them with the CVD risks of M, J, and EW populations in the epidemiological study of this article.

2. Materials and Methods

(1) SHRSP had been used for various nutrition experiments [3,19,27,28], and in the present study, SHRSP from 6 groups (Table 1 and Figure 1) were given control and soy diets (CD, SD) with Mg or Ca at the age of 7 weeks, thereafter until their natural death, and autopsied for macroscopical and microscopical pathological observation [19].

Table 1. Contents of dietary factors in each group of SHRSP.

Group	N	Dietary Contents	Ca, Mg Contents
Control Diet (CD)	10	Control Diet (Crude Protein: 24.6%)	Ca 0.7%, Mg 0.2%
CD + Ca	15	CD (Ca 0.7%, Mg 0.2%) + Ca (0.9%)	Ca 1.6%, Mg 0.2%
CD + Mg	15	CD (Ca 0.7%, Mg 0.2%) + Mg (0.6%)	Ca 0.7%, Mg 0.8%
Soy Protein Diet (SD)	15	Soy Protein Diet (Soy Protein: 24.6%)	Ca 0.7%, Mg 0.2%
SD + Ca	15	SD (Ca 0.7%, Mg 0.2%) + Ca (0.9%)	Ca 1.6%, Mg 0.2%
SD + Ca + Mg	15	SD (Ca0.7%, Mg 0.2%) + Ca (0.9%) + Mg (0.6%)	Ca 1.6%, Mg 0.8%

(2) A health examination was carried out for males and females according to the protocol of the WHO-coordinated Cardiovascular Diseases and Alimentary Comparison (CARDIAC) Study [20,21], and fasting blood and 24U samples were analyzed after anthropological and blood pressure (BP) measurements [20–22,29,30]. About 100 males and 100 females in the age range of 48–56 were randomly invited to the CARDIAC Study health examination after informed consent was obtained from the participants. The study design described in detail [20] was approved at the international committee meeting before starting the CARDIAC Study in 1985.

Informed consent was obtained at the reception of the CARDIAC Study Health Examination from volunteer participants, who were asked to sign the first page of the CARDIAC study questionnaire, and the study was conducted according to the guidelines of the Declaration of Helsinki. Urinary biomarkers such as sodium (Na) for salt intake, potassium (K) for vegetable intake, magnesium (Mg) for grains, nuts, soy, and dietary fiber intake, isoflavones for soy intakes, taurine (Tau) for seafood intakes, urea nitrogen for protein intakes, and creatinine (Cre) for checking the completeness of the collection of 24U samples were analyzed in 50 populations in the world, in total 4211 participants (49.7% females F) in 22 countries worldwide, and also 6 J (864, 53.7% F), 6 M (574, 50.2% F), and 6 EW populations (549, 45.9% F) [18,20–22].

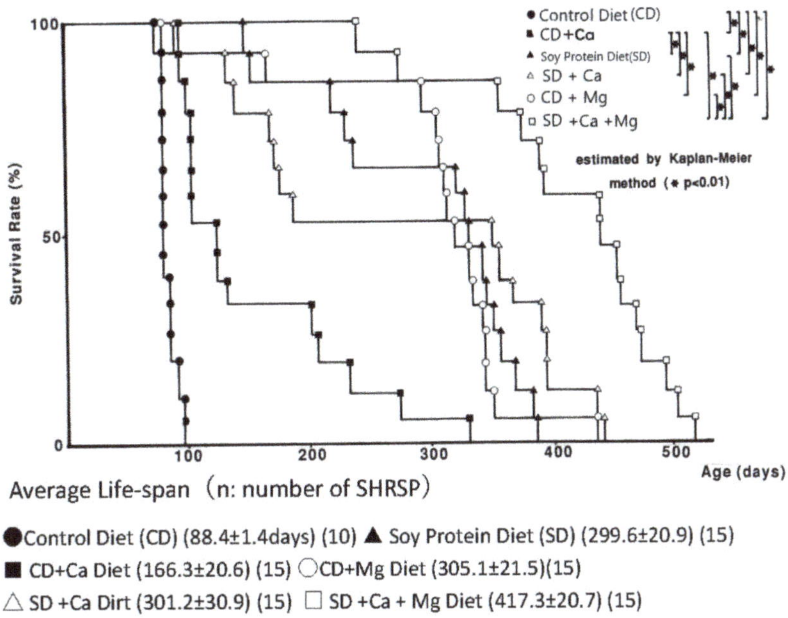

Figure 1. Effect of soy protein, Ca, Mg, and combined diets on the survival rate of salt-loaded SHRSP.

Obese subjects were defined as those with body mass index (BMI) $\geq$ 30 kg/m^2. Participants with hypertension were defined as those with systolic BP (SBP) $\geq$ 140 mmHg or diastolic BP (DBP) $\geq$ 90 mm Hg or those who were receiving anti-hypertensive drug therapy. Hypercholesterolemic subjects were defined as those with serum total cholesterol (TC) $\geq$ 220 mg/dL. General linear models were used to estimate adjusted mean values of BMI, SBP, DBP, and TC across quintiles of the 24 h urinary Tau/Cre (Mg/Cre) ratio after adjustment for age, sex, and use of anti-hypertensive drugs. To evaluate the association of Tau/Cre (Mg/Cre) ratio with cardiovascular disease risk factors, we estimated adjusted odds ratios for obesity, hypercholesterolemia and hypertension in relation to quintiles of Tau/Cre (Mg/Cre) using logistic regression models, adjusting for age and sex as to hypertension and additionally for anti-hypertensive drugs as to obesity and hypercholesterolemia.

ANOVA was used for the comparisons of 24U biomarkers, BMI, SBP, DBP, and TC among the J, M, and EW diets.

3. Results

(1) Experimental Prevention of Stroke in SHRSP.

Since SHRSP developed stroke genetically, they were used to observe the effect of various diets on stroke. For example, SHRSP given 1% salt in drinking water developed severe hypertension and stroke within a short period. However, hypertension was attenuated by increasing K intakes, and even a small reduction of the dietary Na/K ratio significantly improved the survival rate [3,29,30]. The adverse effect of salt was attenuated by alginic acid rich in dietary fibers of the seaweed, which absorbed Na to decrease Na intake via the intestine [27].

The effect of a protein-rich diet was proven in SHRSP fed on a high-fish protein diet with excess salt intake from 1% salt in drinking water. SHRSP fed on low or normal protein diet with excess salt all developed severe hypertension and died from stroke within a shorter period [26]. But the incidence of stroke in SHRSP fed on soy or fish protein-rich diet with excess salt was only 10%. We further analyzed the effect of amino-acids rich in fish and noted that Tau attenuated the development of severe hypertension [28].

Extensive life-long studies on the effect of soy protein with Ca and/or Mg on BP and stroke prevention were designed as shown in Table 1 and Figure 1 in SHRSP, given 1% salt in drinking water. In the present study, soy protein diet (SD) and Mg-fortified control diet (CD + Mg) groups were added to the review of our previous long-term studies [19]. In comparison to SHRSP fed on the control diet (CD) and 1% salt in drinking water, SHRSP fed on a soy protein diet (SP) or a Mg-rich diet (CD + Mg) could survive significantly longer. The average lifespans of these 2 groups (299.6, 305.1 days) were over 200 days longer than the CD group. The effect on lifespan of the 0.6% Mg fortification of a CD diet containing 0.2% Mg (CD + Mg) was similar to the survival of the SD group.

Since the lifespan of 0.9% Ca fortification in CD diets containing 0.7% Ca (CD + Ca) was 166.3 days on average, significantly lower than CD + Mg (305.1 days) or SD (299.6 days). The effect of Mg (0.6%) fortification (CD + Mg) was significantly greater than Ca (0.9%) fortification (CD + Ca).

The average lifespan of salt-loaded SHRSP fed on a soy protein, Mg, and Ca diet was the longest, 417.3 ± 20.7 days.

Their lifespans were significantly longer than salt-loaded SHRSP fed on a control protein diet (88.4 ± days), indicating the intakes of soy protein, Mg, and Ca rich diets were preventive against stroke.

(2) The Association of Urinary Biomarkers with Cardiovascular Risks in the WHO-CARDIAC Study

Since hypertension and stroke were accelerated by Na intake and attenuated by Tau and Mg, 24U samples were collected by the WHO-CARDIAC Study to check the associations of these urinary biomarkers with cardiovascular risks. In addition to the well-known association of 24U Na with BP, stroke mortality rates were significantly positively associated with Na/K ratios [29].

Because of experimental evidence for Tau and Mg attenuating severe hypertension and preventing stroke in SHRSP [28,31], all 24U Tau and Mg data from 50 population samples were divided into five groups, and the adjusted mean values of Tau/Cre and Mg/Cre ratios of the quintiles were inversely associated significantly with BMI, SBP, DBP, and TC in linear regression analyses ($p < 0.001$ for the linear trend and <0.001 for each) [32,33].

The Tau/Cre (Mg/Cre) ratio was significantly inversely associated with obesity, hypercholesterolemia, and hypertension (P for linear trend < 0.001 for the association of the Mg/Cre ratio with obesity, hypercholesterolemia, and hypertension and for the association of the Tau/Cre ratio with obesity and hypercholesterolemia, and <0.05 for the association of the Tau/Cre ratio with hypertension). The odds ratios of obesity, hypercholesterolemia, and hypertension among the subjects in the lowest quintile of Tau/Cre (Mg/Cre) were 2.84 (2.49), 2.20 (2.39), and 1.22 (1.49), compared with the highest quintile (Figures 2 and 3).

These CARDIAC study data indicated that higher intakes of Tau and Mg reduced CVD risk and extended lifespan in humans and suggested that nutritional prevention of stroke for extending lifespan in SHRSP might be applicable to humans. Therefore, CARDIAC study data obtained in J, M, and EW countries were reanalyzed for possible associations with CVD risks.

When Tau/Cre and Mg/Cre of both males and females of J and M were compared with EW, Tau and Mg/Cre in 24U were significantly higher in J and M than in EW (Figure 4). Correspondingly to these 24U data, both systolic and diastolic BP (SBP, DBP*) (*data not shown), TC, and non-HDL cholesterol * were significantly lower in J and M than in EW (Figure 5), although J and M should have significantly higher 24U salt (Figure 6). The significantly higher K excretion in 24 U plus the higher Tau/Cre and Mg/Cre ratios were regarded as the biomarkers contributing to the lower BP in M compared with EW (Figure 6 right).

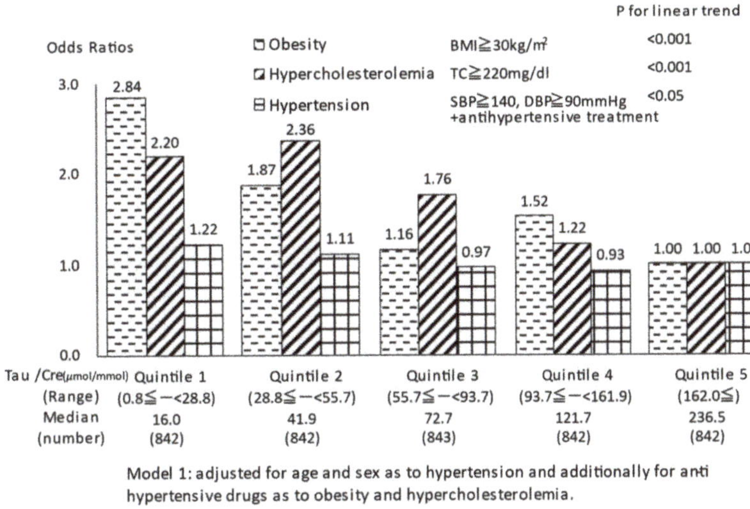

Figure 2. Quintile of Tau/Cre and odds ratios for CVD risks.

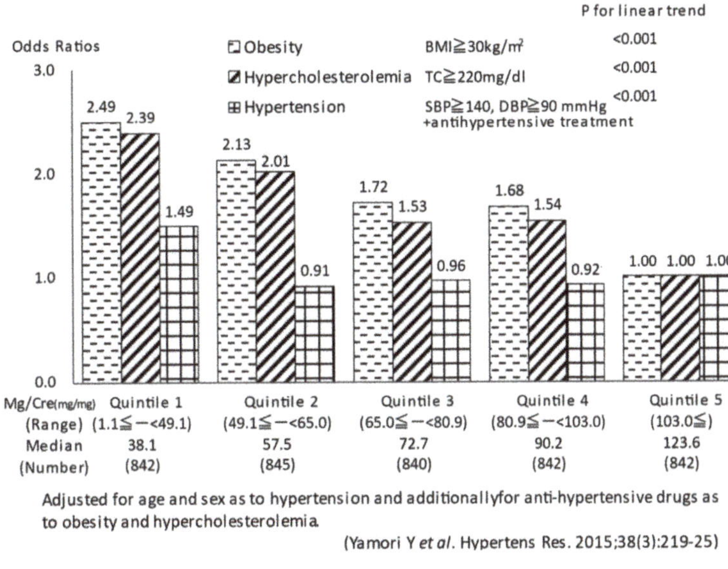

Figure 3. Quintile of Mg/Cre and odds ratios for CVD risks [33].

Despite higher salt intake in J than in EW, the merit of J was the significantly lower BMI compared with M and EW with a higher BMI, which was related to high BP (Figure 7).

One of the factors related to lower BMI in J was supposed to be related to significantly higher 24U isoflavone excretion due to higher intake of soybeans (Figure 7 right), because BMI was inversely associated significantly with 24U isoflavone excretions in CARDIAC Study populations [34].

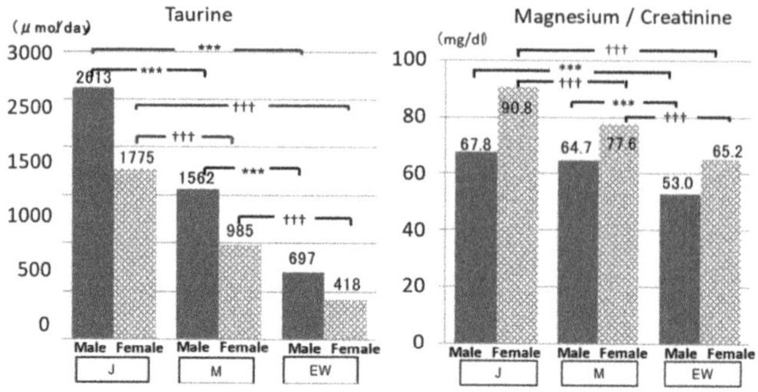

Figure 4. Japanese (J), Mediterranean (M), and Euro-Western (EW) diet populations compared by 24 h urine: Common merits of J and M. J: Aomori, Toyama, Shimane, Chiba, Saga, and Okinawa Prefectures are in Japan. M: Greece, Italy (2), Spain (2), and Portugal. EW: Scotland, Ireland, Sweden, Canada, New Zealand, and Australia. Significant difference: *** $p < 0.001$, ††† $p < 0.001$.

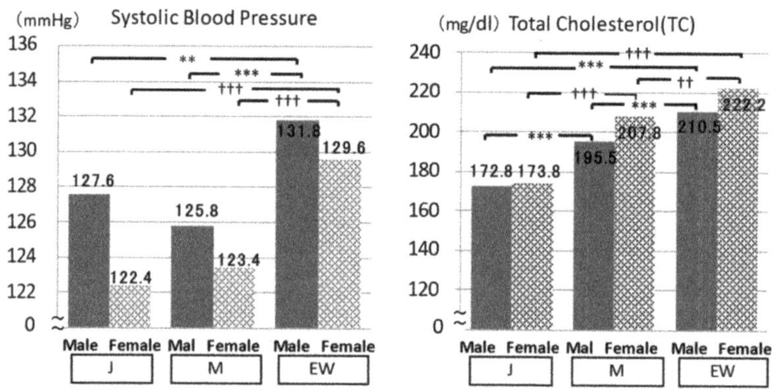

Figure 5. Cardiovascular risks of Japanese (J), Mediterranean (M), and Euro-Western (EW) diet populations: Common merits of J and M compared with EW. Significant difference: ** $p < 0.01$, *** $p < 0.001$, †† $p < 0.01$, ††† $p < 0.001$.

Since J and M showed higher 24U Tau and Mg/Cre and lower SBP, DBP, and TC, CVD risks were compared between individuals with both Tau/Cre and Mg/Cre ratios equal to or higher than their world average (Tau/Cre $\geq$ 639.4 mmol/g and Mg/Cre $\geq$ 82.8 mg/g) and those individuals with both ratios lower than the world average.

BMI, SBP, DBP, TC, and the atherogenic index (AI) calculated from non-HDL/HDL were all significantly lower in the individuals with higher Tau and Mg/Cre ratios than in those with lower ratios of Tau and Mg/Cre, indicating the association of these nutrients with cardiovascular risk reduction in humans (Figure 8).

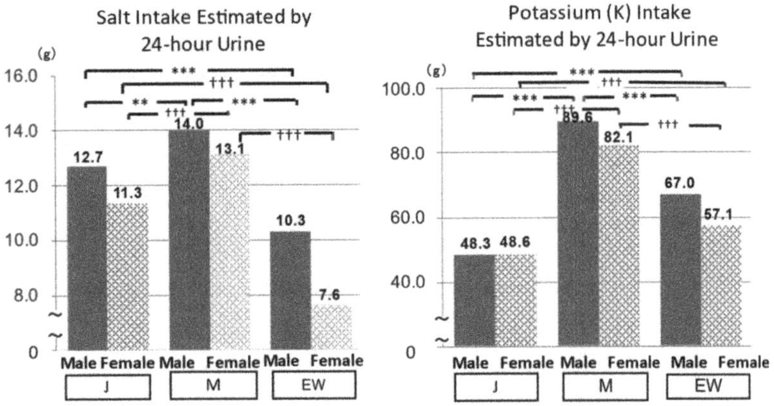

Figure 6. Common demerits of Japanese (J) and Mediterranean (M) diet populations and merits of the Mediterranean (M) diet population (right). Significant difference: ** $p < 0.01$, *** $p < 0.001$, ††† $p < 0.001$.

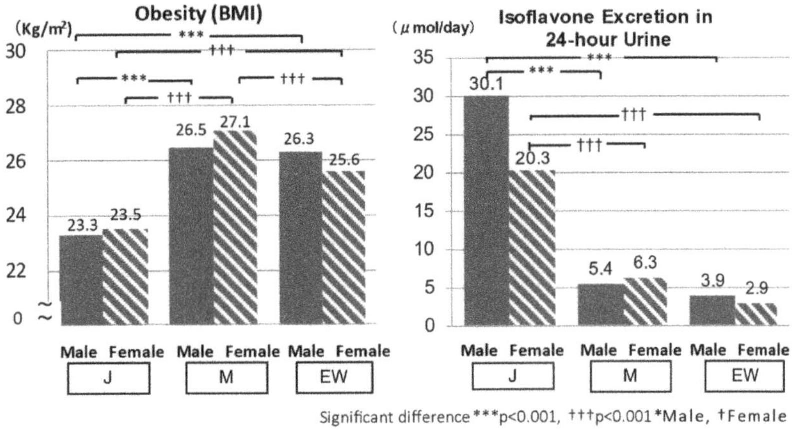

Figure 7. Merit of the Japanese (J) diet population: Low BMI and high isoflavones in 24 h urine excretion in J.

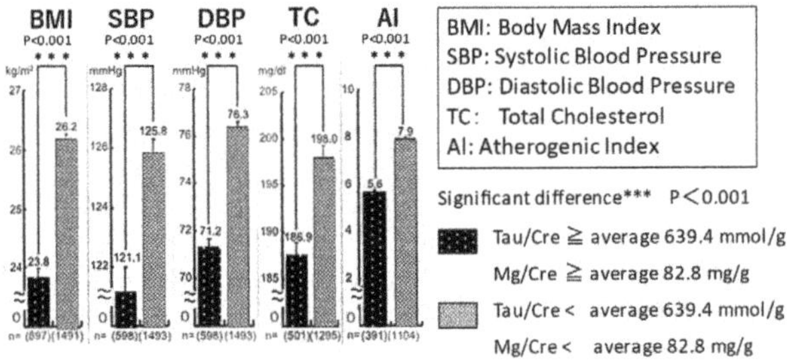

Figure 8. Lower cardiovascular risks were significantly associated with higher Tau (Tau/Cre) and Mg (Mg/Cre) excretions in 24 h urine.

4. Discussion

Since Mg- and Tau-rich diets attenuated the development of hypertension and prevented stroke in SHRSP, 24U of Mg and Tau were examined epidemiologically worldwide in 50 CARDIAC Study populations and were further analyzed in the present study in J and M in comparison with EW populations.

As for Ca, which was proven to prevent stroke and extend lifespan in SHRSP, low Ca intake decreases plasmatic Ca concentration, which stimulates parathyroid hormone (PTH) and renin, angiotensin, and aldosterone secretion to raise BP [35]. Therefore, increased Ca intake attenuates the development of hypertension and stroke. However, since Ca in 24U does not reflect Ca intake in humans [36,37] and is influenced by various factors [38], 24U Ca was not analyzed for its association with the risks of CVD in this study.

Mg and Tau were shown to be related to CVD risks such as hypertension, obesity, and cholesterol-related atherosclerosis.

As for hypertension, Mg activates Na-K ATPase to control electrolyte balance in the cell [39], and therefore the supplementation in the diet reduced intracellular Ca and Na and lowered BP experimentally in SHRSP [31] and clinically in patients with mild hypertension [40]. Tau was observed to decrease BP in SHR and SHRSP [28], and its antihypertensive effect was ascribed to sympathetic modulation [41,42].

In relation to obesity, low Mg status was observed more often in obese individuals [43], and Mg intake was inversely associated with waist size in young Americans [44]. Tau/Cre in 24U was inversely related to obesity in the present study, and the supplementation of Tau decreased body weight in obese mice [45] and clinically in overweight subjects [46].

Mg intake was correlated with the intake of dietary fibers [47], which lowered serum TC [48]. Tau supplementation decreased the effect of a high-fat diet inducing hyperlipidemia in SHRSP and other experimental models [49] by the mechanism of Tau acceleration of bile acid conjugation with cholesterol [50].

Mg and Tau are richly contained in the natural diets, which were obtained commonly from the sea and the mountains even in the Paleolithic period [51]. The recent investigation of a prehistoric kitchen midden in Japan indicated that Mg-rich nuts and seeds, as well as Tau-rich fish and shellfish, were commonly consumed between 5000 and 12,000 years ago [52], and therefore it was speculated that there was no current health problem related to hypertension, obesity, and atherosclerosis. According to the evolutionary concept of human nutrition, such cardiovascular risks as hypertension, obesity, and lipidemia were supposed to be less prevalent in the prehistoric era, and these risks were demonstrated by the present study to be inversely associated with Tau and Mg, the biomarkers of seafood and nuts or seeds, indicating these nutrients commonly taken in the past may potentially reduce current cardiovascular risks. The Japanese are well-known for having the world's longest average life expectancy, which they have maintained as the top-ranking country for the last 30 years (WHO 2016) [53]. The recent evaluation of Japanese dietary intakes by worldwide urinary biomarker analyses revealed the common consumption of soy isoflavones and seafood Tau was also associated with higher Mg intake [54]. Therefore, a Japanese diet containing commonly Mg and Tau may potentially be related to lower cardiovascular risks, which contribute to their longevity.

In conclusion, Tau and Mg, which were effective for reducing cardiovascular diseases in SHRSPs developing genetically transmitted stroke, were epidemiologically associated with lower CVD risks of obesity, hypertension, and hypercholesterolemia worldwide and were proven in the present study to be the nutritional merits of the diet of the J and M populations, which were known for their relatively longer average life expectancy. However, J and M diets contained more salt; therefore, Tau and Mg rich diets with less salt should be recommended for health promotion with fewer cardiovascular diseases.

Author Contributions: Conceptualization, project administration, and funding acquisition, Y.Y. investigation and writing—review and editing, M.S. original draft preparation M.M. data analyses, H.M. All authors have read and agreed to the published version of the manuscript.

Funding: This research was funded by the donation to WHO from the Japan Heart Foundation, which was supported by nation-wide donation in Japan.

Institutional Review Board Statement: The protocol of WHO-coordinated CARDIAC Study was approved by the international committee in 1985 and its excerpts were published for the international approval ([20,21]).

Informed Consent Statement: Participants signed on the record of the CARDIAC Study's health examination after informed consent.

Data Availability Statement: The data that support the findings of this study are available from the corresponding author upon reasonable request.

Conflicts of Interest: The authors declare no conflict of interest.

References

1. Okamoto, K.; Yamori, Y.; Nagaoka, A. Establishment of the stroke-prone spontaneously hypertensive rat (SHR). *Circ. Res.* **1974**, *34* (Suppl. I), 143–153.
2. Okamoto, K.; Aoki, K. Development of a Strain of Spontaneously Hypertensive Rats. *Jpn. Circ. J.* **1963**, *27*, 282–293. [CrossRef] [PubMed]
3. Yamori, Y. The stroke-prone spontaneously hypertensive rat: Contribution to risk factor analysis and prevention of hypertensive diseases. In *Handbook of Hypertension*; De Jong, W., Ed.; Elsevier: Amsterdam, The Netherlands, 1984; pp. 240–255.
4. Fredriksson, K.; Nordborg, C.; Kalimo, H.; Olsson, Y.; Johansson, B.B. Cerebral microangiopathy in stroke-prone spontaneously hypertensive rats. An immunohistochemical and ultrastructural study. *Acta Neuropathol. Vol.* **1988**, *75*, 241–252. [CrossRef]
5. Bailey, E.L.; McCulloch, J.; Sudlow, C.; Wardlaw, J.M. Potential animal models of lacunar stroke: A systematic review. *Stroke* **2009**, *40*, e451–e458. [CrossRef] [PubMed]
6. Wardlaw, J.M.; Smith, C.; Dichgans, M. Mechanisms underlying sporadic cerebral small vessel disease: Insights from neuroimaging. *Lancet Neuurol* **2013**, *12*, 483–497. [CrossRef]
7. Rajani, R.M.; Quick, S.; Ruigrok, S.R.; Graham, D.; Harris, S.E.; Verhaaren, B.F.J.; Fornage, M.; Seshadri, S.; Atanur, S.S.; Dominiczak, A.F.; et al. Reversal of endothelial dysfunction reduces white matter vulnerability in cerebral small vessel disease in rats. *Sci. Transl. Med.* **2018**, *10*, eaam9507. [CrossRef]
8. Bailey, E.L.; Smith, C.; Sudlow, C.L.M.; Wardlaw, J.M. Is the spontaneously hypertensive stroke prone rat a pertinent model of sub cortical ischemic stroke? A systematic review. *Int. J. Stroke* **2011**, *6*, 434–444. [CrossRef]
9. Nagaoka, A.; Kakihana, M.; Fujiwara, K. Effects of idebenone on neurological deficits following cerebrovascular lesions in stroke-prone spontaneously hypertensive rats. *Arch. Gerontol. Geriatr.* **1989**, *8*, 203–212. [CrossRef]
10. Yoshida, H.; Itoh, S.; Ferdousi, F.; Isoda, H. Post-stroke treatment with K-134, a phosphodiesterase 3 inhibitor, improves stroke outcomes in the stroke-prone spontaneously hypertensive rat model–A comparative evaluation of antiplatelet drugs. *J. Pharmacol. Sci.* **2022**, *148*, 229–237. [CrossRef]
11. Rubattu, S.; Volpe, M.; Kreutz, R.; Ganten, U.; Ganten, D.; Lindpaintner, K. Chromosomal mapping of quantitative trait loci contributing to stroke in a rat model of complex human disease. *Nat. Genet.* **1996**, *13*, 429–434. [CrossRef]
12. Jeffs, B.; Clark, J.S.; Anderson, N.H.; Gratton, J.; Brosnan, M.J.; Gauguier, D.; Reid, J.L.; Macrae, I.M.; Dominiczak, A.F. Sensitivity to cerebral ischaemic insult in a rat model of stroke is determined by a single genetic locus. *Nat. Genet.* **1997**, *16*, 364–367. [CrossRef]
13. Rubattu, S.D.; Hubner, N.; Ganten, U.; Evangelista, A.; Stanzione, R.; Di Angelantonio, E.; Plehm, R.; Langanki, R.; Gianazza, E.; Sironi, L.; et al. Reciprocal congenic lines for a major stroke QTL on rat chromosome 1. *Physiol. Genom.* **2006**, *27*, 108–113. [CrossRef]
14. Nabika, T.; Ohara, H.; Kato, N.; Isomura, M. The stroke-prone spontaneously hypertensive rat: Still a useful model for post-GWAS genetic studies? *Hypertens. Res.* **2012**, *35*, 477–484. [CrossRef]
15. Ferdaus, M.Z.; Xiao, B.; Ohara, H.; Nemoto, K.; Harada, Y.; Saar, K.; Hübner, N.; Isomura, M.; Nabika, T. Identification of Stim1 as a candidate gene for exaggerated sympathetic response to stress in the stroke-prone spontaneously hypertensive rat. *PLoS ONE* **2014**, *9*, e95091. [CrossRef]
16. Gandolgor, T.A.; Ohara, H.; Zong, H.C.; Hirashima, T.; Ogawa, T.; Saar, K.; Hübner, N.; Watanabe, T.; Isomura, M.; Nabika, T. Two genomic regions of crhromosomes 1 and 18 rexplain most of the stroke susceptibility under salt loading in stroke-prone spontaneously hypertensive rat/izm. *Hypertension* **2013**, *62*, 55–61. [CrossRef]
17. Takeuchi, F.; Liang, Y.-Q.; Isono, M.; Tajima, M.; Cui, Z.H.; Iizuka, Y.; Gotoda, T.; Nabika, T.; Kato, N. Integrative genomic analysis of blood pressure and related phenotypes in rats. *Dis. Model. Mech.* **2021**, *14*, dmm048090. [CrossRef]
18. Yamori, Y.; Nara, Y.; Kihara, M.; Mano, M.; Horie, R. Simple method for sampling consecutive 24-h urine for epidemiological and clinical studies. *Clin. Exp. Hypertens.* **1984**, *6*, 1161–1167. [CrossRef]
19. Yamori, Y. Predictive and preventive pathology of cardiovascular diseases. *Acta Patho. Jpn.* **1989**, *39*, 683–705. [CrossRef]
20. CARDIAC Study Group (Ed.) *Cardiovascular Diseases and Alimentary Comparison Study Protocol and Manual of Operations. WHO-Collaborating Center for Research on Primary Prevention of Cardiovascular Diseases, and Cardiovascular Diseases Unit*; WHO: Shimane, Japan; Geneva, Switzerland, July 1986.
21. CARDIAC Study Group. Excerpts from the WHO CARDIAC Study Protocol. *J. Cardiovasc. Pharm.* **1990**, (Suppl. 8), S75–S77.

22. Yamori, Y. Chapter 2, History of CARDIAC Study and food culture. Chapter 3, Nutrition for healthier ageing-Two decades of world-wide surveys on diet and human life. In *Proceedings of the International Symposium, Food Culture: Development and Education 20-Year CARDIAC Study Symposium Supported by UNESCO*; Paris, France, 4 October 2005, Nonprofit Organization World Health Frontier Institute: Nishinomiya, Japan, 2008; pp. 11–18, 23–41.
23. Pounis, G.; Costanzo, S.; Bonaccio, M.; Di Castelnuovo, A.; de Curtis, A.; Ruggiero, E.; Persichillo, M.; Cerletti, C.; Donati, M.B.; de Gaetano, G.; et al. Reduced mortality risk by a polyphenol-rich diet: An analysis from the Moli-sani study. *Nutrition* **2018**, *48*, 87–95. [CrossRef]
24. Estruch, R.; Ros, E.; Salas-Salvadó, J.; Covas, M.-I.; Corella, D.; Arós, F.; Gómez-Gracia, E.; Ruiz-Gutiérrez, V.; Fiol, M.; Lapetra, J.; et al. Primary Prevention of Cardiovascular Disease with a Mediterranean Diet Supplemented with Extra-Virgin Olive Oil or Nuts. *N. Engl. J. Med.* **2018**, *378*, e34. [CrossRef] [PubMed]
25. Cesari, F.; Dinu, M.; Pagliai, G.; Rogolino, A.; Giusti, B.; Gori, A.; Casini, A.; Marcucci, R.; Sofi, F. Mediterranean, but not lacto-ovo-vegetarian, diet positively influence circulating progenitor cells for cardiovascular prevention: The CARDIVEG study. *Nutr. Metab. Cardiovasc. Dis.* **2019**, *29*, 604–610. [CrossRef] [PubMed]
26. Keenan, T.D.; Agrón, E.; Mares, J.A.; Clemons, T.E.; van Asten, F.; Swaroop, A.; Chew, E.Y. For the AREDS and AREDS2 Research Groups Adherence to a Mediterranean diet and cognitive function in the Age-Related Eye Disease Studies 1 & 2. *Alzheimer's Dement.* **2020**, *16*, 831–842.
27. Yamori, Y.; Nara, Y.; Tsubouchi, T.; Sogawa, Y.; Ikeda, K.; Horie, R. Dietary prevention of stroke and its mechanisms in stroke-prone spontaneously hypertensive rats—Preventive effect of dietary fibre and palmitoleic acid. *J. Hypertens.* 1986; 4, (Suppl. 3), S449–S452.
28. Nara, Y.; Yamori, Y.; Lovenberg, W. Effect of dietary taurine on blood pressure in spontaneously hypertensive rats. *Biochem. Pharm.* **1978**, *27*, 2689–2692. [PubMed]
29. Yamori, Y.; Liu, L.; Mizushima, S.; Ikeda, K.; Nara, Y. Male cardiovascular mortality and dietary markers in 25 population samples of 16 countries. *J. Hypertens.* **2006**, *24*, 1499–1505. [CrossRef]
30. Yamori, Y. Food factors for atherosclerosis prevention: Asian perspective derived from analyses of worldwide dietary biomarkers. *Exp. Clin. Cardiol.* **2006**, *11*, 94–98.
31. Adachi, M.; Nara, Y.; Mano, M.; Yamori, Y. Effect of dietary magnesium supplementation on intralymphocytic free calcium and magnesium in stroke-prone spontaneously hypertensive rats. *Clin. Exp. Hypertens.* **1994**, *16*, 317–326. [CrossRef]
32. Sagara, M.; Murakami, S.; Mizushima, S.; Liu, L.; Mori, M.; Ikeda, K.; Nara, Y.; Yamori, Y. Taurine in 24-h urine samples is inversely related to cardiovascular risks of middle aged subjects in 50 populations of the world. *Med. Biol.* **2015**, *803*, 623–635.
33. Yamori, Y.; Sagara, M.; Mizushima, S.; Liu, L.; Ikeda, K.; Nara, Y. An inverse association between magnesium in 24-h urine and cardiovascular risk factors in middle-aged subjects in 50 CARDIAC Study populations. *Hypertens Res.* **2015**, *38*, 219–225. [CrossRef]
34. Yamori, Y. Worldwide epidemic of obesity: Hope for Japanese diets. *Clin. Exp. Pharm. Physiol.* **2004**, *31* (Suppl. 2), 2–4. [CrossRef]
35. Villa-Etchegoyen, C.; Lombarte, M.; Matamoros, N.; Belizán, J.M.; Cormick, G. Mechanisms involved in the relationship between low calcium intake and high blood pressure. *Nutrients* **2019**, *11*, 1112. [CrossRef] [PubMed]
36. Ghazali, S.; Barratt, T.M. Urinary excretion of calcium and magnesium in children. *Arch. Dis. Child.* **1974**, *49*, 97–101. [CrossRef] [PubMed]
37. Stapleton, F.B.; Noe, H.N.; Jerkins, G.; Roy, S., III. Urinary excretion of calcium following an oral calcium loading test in healthy children. *Pediatrics* **1982**, *69*, 594–597. [CrossRef] [PubMed]
38. Knapp, E.L. Factors influencing the urinary excretion of calcium. I. In normal persons. *J. Clin. Invest.* **1947**, *26*, 182–202. [CrossRef] [PubMed]
39. Sontia, B.; Touyz, R.M. Role of magnesium in hypertension. *Arch. Biochem. Biophys.* **2007**, *458*, 33–39. [CrossRef] [PubMed]
40. Hatzistavri, L.S.; Sarafidis, P.A.; Georgianos, P.I.; Tziolas, I.M.; Aroditis, C.P.; Zebekakis, P.E.; Pikilidou, M.I.; Lasaridis, A.N. Oral magnesium supplementation reduces ambulatory blood pressure in patients with mild hypertension. *Am. J. Hypertens.* **2009**, *22*, 1070–1075. [CrossRef]
41. Li, N.; Sawamura, M.; Nara, Y.; Ikeda, K.; Yamori, Y. Direct inhibitory effects of taurine on norepinephrine-induced contraction in mesenteric artery of stroke-prone spontaneously hypertensive rats. *Adv. Exp. Med. Biol.* **1996**, *403*, 257–262.
42. Mizushima, S.; Nara, Y.; Sawamura, M.; Yamori, Y. Effects of oral taurine supplementation on lipids and sympathetic nerve tone. *Amino. Acids* **1996**, *46*, 73–80.
43. Nielsen, F.H. Magnesium, inflammation, and obesity in chronic disease. *Nutr. Rev.* **2010**, *68*, 333–340. [CrossRef]
44. He, K.; Liu, K.; Daviglus, M.L.; Morris, S.J.; Loria, C.M.; Van Horn, L.; Jacobs, D.R.; Savage, P.J. Magnesium intake and incidence of metabolic syndrome among young adults. *Circulation* **2006**, *113*, 1675–1682. [CrossRef]
45. Fujihira, E.; Takahashi, H.; Nakazawa, M. Effect of long-term feeding of taurine in hereditary hyperglycemic obese mice. *Chem. Pharm. Bull.* **1970**, *18*, 1636–1642. [CrossRef]
46. Zhang, M.; Bi, L.F.; Fang, J.H.; Su, X.L.; Da, G.L.; Kuwamori, T.; Kagamimori, S. Beneficial effects of taurine on serum lipids in overweight or obese non-diabetic subjects. *Amino Acids* **2003**, *26*, 267–271. [CrossRef]
47. Truswell, A.S. Dietary fibre and blood lipids. *Curr. Opin. Lipidol.* **1995**, *6*, 14–19. [CrossRef]
48. Brown, L.; Rosner, B.; Willett, W.W.; Sacks, F.M. Cholesterol-lowering effects of dietary fiber: A meta-analysis. *Am. J. Clin. Nutr.* **1999**, *69*, 30–42. [CrossRef]

49. Murakami, S.; Kondo, Y.; Toda, Y.; Kitajima, H.; Kameo, K.; Sakono, M.; Fukuda, N. Effect of taurine on cholesterol metabolism in hamsters: Up-regulation of low density lipoprotein (LDL) receptor by taurine. *Life Sci.* **2002**, *70*, 2355–2366. [CrossRef]
50. Murakami, S.; Sakurai, T.; Tomoike, H.; Sakono, M.; Nasu, T.; Fukuda, N. Prevention of hypercholesterolemia and atherosclerosis in the hyperlipidemia- and atherosclerosis-prone Japanese (LAP) quail by taurine supplementation. *Amino Acids* **2010**, *38*, 271–278. [CrossRef]
51. Richards, M. A brief review of the archaeological evidence for Palaeolithic and Neolithic subsistence. *Eur. J. Clin. Nutr.* **2002**, *56*, 1270–1278. [CrossRef]
52. TSMD, Torihama Shell Midden Research Group. The Torihama shell midden—A preliminary report of the excavation in 1984. Fukui Prefecture Education Board and Wakasa History Art Folklore Museum Fukui, Japan, 1985. 2-104, Onyu, Obama, Fukui 917-0241, Japan. Available online: https://wakahaku.pref.fukui.lg.jp/ (accessed on 30 August 2016).
53. World Health Statistics 2016: Monitoring Health for the SDGs. Available online: https://www.who.int/publications/i/item/9789241565264 (accessed on 30 August 2016).
54. Mori, M.; Sagara, M.; Mori, H.; Yamori, Y. Grading of Japanese diet intakes by 24-h urine analysis of taurine and soy isoflavones in relation to cardiovascular risks. *Adv. Exp. Med. Biol.* **2022**, *1370*, 173–184. [CrossRef]

biomedicines MDPI

Review

Chloride Ions, Vascular Function and Hypertension

Kenichi Goto [1,*] and Takanari Kitazono [2]

[1] Department of Health Sciences, Graduate School of Medical Sciences, Kyushu University, Fukuoka 812-8582, Japan
[2] Department of Medicine and Clinical Science, Graduate School of Medical Sciences, Kyushu University, Fukuoka 812-8582, Japan
* Correspondence: goto.kenichi.638@m.kyushu-u.ac.jp

Abstract: Blood pressure is determined by cardiac output and systemic vascular resistance, and mediators that induce vasoconstriction will increase systemic vascular resistance and thus elevate blood pressure. While peripheral vascular resistance reflects a complex interaction of multiple factors, vascular ion channels and transporters play important roles in the regulation of vascular tone by modulating the membrane potential of vascular cells. In vascular smooth muscle cells, chloride ions (Cl^-) are a type of anions accumulated by anion exchangers and the anion–proton cotransporter system, and efflux of Cl^- through Cl^- channels depolarizes the membrane and thereby triggers vasoconstriction. Among these Cl^- regulatory pathways, emerging evidence suggests that upregulation of the Ca^{2+}-activated Cl^- channel TMEM16A in the vasculature contributes to the increased vascular contractility and elevated blood pressure in hypertension. A robust accumulation of intracellular Cl^- in vascular smooth muscle cells through the increased activity of Na^+–K^+–$2Cl^-$ cotransporter 1 (NKCC1) during hypertension has also been reported. Thus, the enhanced activity of both TMEM16A and NKCC1 could act additively and sequentially to increase vascular contractility and hence blood pressure in hypertension. In this review, we discuss recent findings regarding the role of Cl^- in the regulation of vascular tone and arterial blood pressure and its association with hypertension, with a particular focus on TMEM16A and NKCC1.

Keywords: chloride; calcium-activated chloride channel; hypertension; Na^+–K^+–$2Cl^-$ cotransporter 1; TMEM16A; smooth muscle

Citation: Goto, K.; Kitazono, T. Chloride Ions, Vascular Function and Hypertension. *Biomedicines* **2022**, *10*, 2316. https://doi.org/10.3390/biomedicines10092316

Academic Editors: Josef Zicha and Ivana Vaněčková

Received: 17 August 2022
Accepted: 15 September 2022
Published: 18 September 2022

1. Introduction

Hypertension is the most prevalent and important risk factor for cardiovascular disease around the world [1], and cardiovascular complications associated with hypertension accounted for 8.5 million deaths worldwide in 2015 [2]. Nevertheless, global control (<140/90 mmHg) rates among subjects with hypertension in 2019 were only 23% for women and 18% for men [3], and thus more effective treatment strategies for hypertension control are urgently needed.

Lifestyle modifications are recommended for the treatment and prevention of hypertension and hypertension-associated cardiovascular diseases for all subjects, including subjects with high normal blood pressure and patients who are taking antihypertensive agents [4]. In particular, the restriction of dietary sodium chloride (NaCl) has been one of the major focus points among lifestyle modifications for the treatment and prevention of hypertension [5,6]. Indeed, numerous animal and human studies have established a causal relationship between dietary NaCl intake and hypertension as well as hypertension-associated cardiovascular diseases [7–9].

While it is generally assumed that sodium ions (Na^+) but not chloride ions (Cl^-) play a critical role in NaCl-induced hypertension [10,11], the copresence of Na^+ and Cl^- has been reported to be requisite for the development or progression of hypertension in some animal models of hypertension, including desoxycorticosterone-induced hypertensive rats [12],

Dahl salt-sensitive hypertensive rats [13,14] and stroke-prone spontaneously hypertensive rats [15]. Likewise, several studies have suggested the importance of Cl^- in NaCl-induced hypertension in humans [16–18]. These animal and human studies suggest that Na^+ alone may not be sufficient, and that Cl^- may be indispensable or may act cooperatively with Na^+ to give rise to NaCl-induced hypertension. A detailed description of the role of Cl^- in NaCl-induced hypertension in animals and humans can be found in an excellent review by McCallum et al. [19].

The precise mechanisms by which Cl^- contributes to the blood pressure rise in the above studies are yet to be determined, but the ability of Cl^- to modify vascular contractility may play a role. In vascular smooth muscle cells, the intracellular concentration of Cl^- is accumulated by anion exchangers and the anion–proton cotransporter system [20,21]. As the resting membrane potential of smooth muscle in vivo (e.g., −38 mV in the rat caudal artery [22]) is more negative than the reversal potential for Cl^- (e.g., −18 mV in the guinea pig vas deferens [23]), the opening of Cl^- channels leads to an efflux of Cl^- and depolarizes the membrane potential, which would then increase the open probability of L-type Ca^{2+} channels to trigger smooth muscle constriction [20,24].

Thus, in situations with increased intracellular Cl^- concentration or increased Cl^- channel activity in vascular smooth muscle cells, the driving force for the efflux of Cl^- is expected to increase, which in turn could facilitate membrane depolarization and vasoconstriction, and emerging evidence suggests that this scenario is indeed the case in some animal models of hypertension. In this review, we will discuss the possible involvement of Cl^- in the pathogenesis of hypertension. Particular emphasis is given to the roles of Ca^{2+}-activated Cl^- channel transmembrane membrane 16A (TMEM16A; also known as Ano1) and Na^+–K^+–$2Cl^-$ cotransporter 1 (NKCC1) in the increased vascular contractility during hypertension.

2. Role of Chloride Ions in Regulation of Vascular Tone and Blood Pressure

The vascular tone in vivo is regulated by perivascular nerves, including sympathetic, parasympathetic and non-adrenergic non-cholinergic nerves, and the corelease of norepinephrine and ATP from the sympathetic nerve terminals causes vascular smooth muscle membrane depolarization and subsequent constriction [25–29]. Although multiple ionic mechanisms would underpin the nerve-mediated vascular smooth muscle depolarization, several previous studies have suggested that nerve-mediated and exogenously applied norepinephrine-evoked smooth muscle depolarization could be at least partly due to the generation of Ca^{2+}-activated Cl^- currents triggered by the Ca^{2+} release from the intracellular Ca^{2+} stores [24,30–32].

In addition to perivascular nerve-mediated regulation, myogenic response-mediated vascular smooth muscle depolarization and constriction in response to intravascular pressure change also contribute to the regulation of vascular tone [33]: in rat cerebral arteries, intravascular pressure-induced depolarization and constriction have been shown to be inhibited by two distinct Cl^- channel blockers, indanyloxyacetic acid (IAA-94) and 4,4′-diisothiocyanatostilbene-2,2′-disulphonic acid (DIDS), suggesting that the efflux of Cl^- ions through Cl^- channels could contribute to the myogenic response-mediated vasoconstriction [34]. Indeed, in support of this observation, efflux of Cl^- ions was associated with the myogenic constriction in the rat cerebral vascular bed [35]. Nevertheless, because subsequent studies performed in the rat cerebral arteries revealed that IAA-94 depresses L-type calcium current [36], and both IAA-94 and DIDS depress non-selective cationic current [37], the validity of the contribution of Cl^- currents to the myogenic response was called into question.

As such, despite a significant amount of physiological and pharmacological evidence showing that vascular Cl^- channels play a crucial role in regulating vascular tone, the absence of specific inhibitors and the lack of the molecular identities of the channels make it difficult to reach indisputable conclusions. Among other things, there has been a debate regarding the molecular identity of CaCCs ever since the initial report by Byrne and Large in 1987 [38]. Indeed, several proteins have been proposed as the molecular counterpart of CaCCs, and these include CLCA, CLC-3, TWEENTY and bestrophins [39]. However, three independent groups revealed in 2008 that the TMEM16A protein is a molecular counterpart for CaCCs [40–42].

Since these 2008 reports, many studies have confirmed that TMEM16A generates functional CaCC currents in a number of vascular smooth muscle cells and thereby regulates agonist-induced vasoconstriction [21,43–45]. Moreover, it has been revealed that TMEM16A also contributes to intravascular pressure-induced myogenic depolarization and vasoconstriction in the cerebral arteries and renal arterioles of rats [46,47]. Thus, it appears likely that the TMEM16A in vascular smooth muscle cells plays a critical role in regulating vascular tone and blood pressure. Support for this notion comes from the fact that conditional knockout mice of TMEM16A in vascular smooth muscle cells shows a complete deficiency of CaCC currents, decreased responsiveness to vasoconstrictor stimuli and reduced systemic blood pressure [48].

3. Alterations in Vascular Chloride Channels and Transporters in Hypertension

3.1. Ca^{2+}-Activated Chloride Channels (CaCCs) in Vascular Smooth Muscle Cells

It is generally accepted that essential hypertension is characterized by an increased peripheral resistance [49,50]. The increased peripheral resistance in hypertension is determined by an integral and complex interplay between various pathogenic factors, including increased sympathetic nervous activity, enhanced calcium ion mobilization in vascular smooth muscle cells, increased calcium sensitivity of vascular smooth muscle cells and reduced production of endothelium-derived relaxing factors, to name a few [50,51]. Among these factors, alterations in the function of vascular ion channels during hypertension contribute to the increased peripheral resistance by shifting the membrane potential to depolarized levels [22,50,52].

While many studies have demonstrated downregulation of the expression and/or function of vascular potassium (K^+) channels in hypertension [50,51,53,54], emerging evidence reveals an upregulation of expression and/or function of CaCCs in vascular smooth muscle cells of spontaneous hypertensive rats (SHRs), a genetic model of human essential hypertension. Although a previous study suggested an increased activity of CaCCs in vascular smooth muscle cells of SHRs [55], the molecular identity of the CaCCs observed in that study was unclear at the time. A subsequent study by Wang et al. for the first time revealed that TMEM16A is the molecular counterpart for the increased activity of CaCCs in vascular smooth muscle cells of SHRs, and that TMEM16A protein expression is significantly upregulated in the aorta, the carotid arteries, the hindlimb arteries and the mesenteric arteries of SHRs compared to those of normotensive Wistar Kyoto (WKY) rats [56] (Table 1). Consistent with the seminal findings of Wang and colleagues [56], the increased TMEM16A expression levels and the resultant potentiation of vasoconstrictions have also been reported in smooth muscle cells of the coronary arteries [57] and the renal arterioles [47] of SHRs (Table 1).

Table 1. Alterations in vascular smooth muscle Ca^{2+}-activated Cl^- channels during hypertension.

Animals	Alterations in Vascular Smooth Muscle CaCCs during Hypertension	Ref.
SHRs	Increased TMEM16A expression and function in aorta, carotid arteries, hindlimb arteries and mesenteric arteries	[56]
	Increased TMEM16A expression and function in coronary arteries	[57]
	Increased TMEM16A expression and function in renal arterioles	[47]
	Knockdown of TMEM16A by siRNA transfection lowered blood pressure	[56]
	Inhibition of TMEM16A activity by $T16A_{inh}$-A01 lowered blood pressure	[56]
	Treatment of mesenteric resistance arteries with TM_{inh}-23 blocked vasoconstriction	[58]
	Inhibition of TMEM16A activity by TM_{inh}-23 lowered blood pressure	[58]
2K2C renal hypertensive rats	Reduced TMEM16A expression and function in basilar arteries during the development of hypertension	[59,60]

CaCCs, Ca^{2+}-activated Cl^- channels; SHRs, spontaneously hypertensive rats; 2K2C, 2-kidney, 2-clip.

Importantly, the increased expression and function of TMEM16A appear to be associated with blood pressure elevation in SHRs: the in vivo knockdown of TMEM16A by small interfering RNA (siRNA) transfection prevented blood pressure rise, and the in vivo inhibition of TMEM16A activity by $T16A_{inh}$-A01, a TMEM16A inhibitor, reduced blood pressure in SHRs [56] (Table 1). Similarly, a recent study in SHRs showed that in vitro treatment of mesenteric resistance arteries with TM_{inh}-23, a small molecule inhibitor of vascular smooth muscle TMEM16A, blocked vascular smooth muscle constriction in response to vasoconstrictor stimuli, and in vivo treatment with TM_{inh}-23 reduced blood pressure in SHRs with minimal blood pressure change in normotensive rats and mice [58] (Table 1). Although the greater blood pressure lowering effect of TM_{inh}-23 in SHRs appears to be due to an increased sensitivity of TMEM16A to TM_{inh}-23 [58], the mechanisms underlying the increased sensitivity of TMEM16A are unclear and warrant further investigations. Together, these findings implicate vascular smooth muscle CaCC TMEM16A as a possible contributor in the pathogenesis of hypertension in SHRs.

In rat basilar arteries of 2-kidney, 2-clip (2K2C) renal hypertensive rats, exogenously applied angiotensin II (Ang II) induced vasoconstriction that was sensitive to $T16A_{inh}$-A01, and Ang II evoked TMEM16A-mediated CaCC currents in rat basilar smooth muscle cells [59]. These findings suggest that CaCC TMEM16A modulates the vasocontractility of basilar arteries of 2K2C renal hypertensive rats; however, in sharp contrast with SHRs, the activity of CaCCs was decreased gradually during the development of hypertension, and the CaCCs' current density was negatively correlated with blood pressure levels, in basilar arteries of 2K2C renal hypertensive rats [60] (Table 1). Moreover, the TMEM16A protein expression in the smooth muscle layer of the basilar artery decreased during the development of hypertension in 2K2C renal hypertensive rats [59,60] (Table 1).

It is not clear why the activity and the expression of CaCC TMEM16A changed in the opposite direction between SHRs and 2K2C renal hypertensive rats, but the difference might be explained by the different levels of activity of the renin–angiotensin system (RAS) in the vasculature: while the plasma and tissue RASs are suppressed in SHRs [61], the RAS components—particularly the vascular Ang II concentration—are increased in 2K2C renal hypertensive rats [62]. As Ang II decreased TMEM16A expression in some vascular smooth muscle cells, including those from rat basilar arteries [59,60,63], an increase in vascular Ang II concentration in the basilar arteries of 2K2C renal hypertensive rats might downregulate TMEM16A expression and hence reduce the CaCCs' current in this model.

It has been reported that the perivascular sympathetic nerves exert an abnormal trophic influence on the vascular smooth muscle membrane properties of SHRs [64], and a recent report showed that the expression and contractile function of the CaCC TMEM16A in rat arteries were reduced due to the trophic influence of sympathetic nerves during postnatal maturation [65]. Therefore, we speculate that the expression and function of CaCC TMEM16A might also be decreased along with the longer duration of hypertension in SHRs because of the persistent abnormal trophic influence of the sympathetic nerves. This hypothesis might be supported by the observation that the contribution of CaCCs to norepinephrine-induced vasoconstriction in the femoral arteries was decreased in 12-month-old SHRs compared to that of 6-month-old SHRs [66].

TMEM16A may modulate vascular contractility in cooperation with other ion channels in certain vascular beds. Thus, in rat mesenteric and tail arteries, TMEM16A modulates vascular contractility, at least in part, by positively regulating the expression and function of vascular L-type Ca^{2+} channels [67,68]. In another study in rat cerebral arterial smooth muscle cells, transient receptor potential canonical 6 channel (TRPC6) and TMEM16A were found to be spatially localized, and TRPC6 activation led to a local elevation of Ca^{2+}, which in turn activated nearby TMEM16A, leading to vasoconstriction [69]. As the function and expression of both L-type Ca^{2+} channels [70,71] and TRPC6 [72] have been reported to be upregulated in hypertensive rats, it is intriguing to speculate that these mutual interactions of TMEM16A with other vascular ion channels function cooperatively to augment vasoconstriction and hence increase blood pressure in hypertension.

It has been reported that phosphatidylinositol 4,5-bisphosphate (PIP_2), a phospholipid of the plasma membrane, regulates ion channel activity in various cell types [73], and several studies reported that PIP_2 acts as a positive modifier of TMEM16A [74–76]. By contrast, the TMEM16A-mediated CaCC current was not augmented, but rather inhibited by PIP_2 in rat pulmonary artery smooth muscle cells [77]. The reason for the discrepancy is not clear. Nevertheless, a previous report suggests that a significant difference exists between WKY and SHR aortas regarding the PIP_2 hydrolysis response following stimulation with norepinephrine [78], indicating the need for further research to understand the possible regulation of TMEM16A by PIP_2 in blood vessels in hypertension.

Recent evidence suggests that inositol 1,4,5-trisphosphate receptors (IP_3Rs) are spatially colocalized with TMEM16A proteins in nociceptive sensory neurons [79]. If the same holds true in vascular smooth muscle cells, IP_3-induced Ca^{2+} release from intracellular Ca^{2+} stores would activate nearby TMEM16A, and alterations in this signaling pathway might contribute to the TMEM16A-mediated vasoconstriction in SHRs. Indeed, it has been reported that IP_3R channels are upregulated in vascular smooth muscle in hypertension, resulting in enhanced IP_3-induced Ca^{2+} release and increased vasoconstriction [80].

To sum up, while there is a growing body of evidence that CaCC TMEM16A contributes to the increased vascular contractility and elevated blood pressure in SHRs, it is currently unclear whether the upregulation of TMEM16A is specific to SHRs or is present in other hypertensive animal models, and further studies will be needed to clarify the molecular mechanisms that regulate TMEM16A activity during hypertension.

3.2. Ca^{2+}-Activated Chloride Channels (CaCCs) in Vascular Endothelial Cells

In addition to their expression in vascular smooth muscle cells, CaCCs have been reported to be present in some vascular endothelial cells [81–84]. Although the physiological role of endothelial CaCCs is still not well understood, the endothelial CaCCs may contribute to the regulation of the resting membrane potential of the endothelial cells. Indeed, in mouse brain capillary endothelial cells, pharmacological blockade or knockdown of TMEM16A with siRNA induced membrane hyperpolarization, suggesting that the activation of endothelial CaCCs acts to depolarize the membrane potential of the endothelial cells [83]. Further support for this notion comes from the study by Yamamoto et al. [85]. They found that, in the isolated endothelium of guinea pig mesenteric arteries, ACh increased the intracellular concentration of Ca^{2+}, which subsequently activated endothelial small conductance Ca^{2+}-activated K^+ channels ($SK_{Ca}s$), intermediate conductance K_{Ca} (IK_{Ca}) and CaCC simultaneously, and the endothelium-dependent hyperpolarization (EDH) through the activation of both SK_{Ca} and IK_{Ca} was counteracted by the opposing membrane depolarization evoked by the activation of CaCCs [85].

With respect to the alteration of endothelial CaCCs in hypertension, we have previously shown a functional upregulation of endothelial CaCCs in mesenteric resistance arteries of SHRs [86]. In that study, after blockade of EDH with K_{Ca} channel inhibitors, iontophoresed acetylcholine (ACh) evoked a rapid and substantial membrane depolarization in mesenteric resistance arteries of SHRs, but only negligible slow depolarization was detected in those of WKY rats [86,87] (Figure 1).

As the estimated reversal potential of the ACh-evoked depolarization in that study was −18 mV [86], which agrees closely with that reported for Cl^- ions [23,88], and the ACh-evoked depolarization was abolished by endothelium denudation, or reduced either by replacement of external Cl^- ions with impermeant anions or by treatment with the CaCC inhibitors niflumic acid or flufenamic acid, the ACh-evoked depolarization appears to be, at least in part, generated through the activation of endothelial CaCCs in the mesenteric resistance arteries of SHRs [86]. Moreover, the inhibition of the ACh-evoked depolarization by CaCC inhibitors improved the impaired ACh-induced EDH in mesenteric arteries of SHRs, suggesting that an increased activity of endothelial CaCCs may be responsible for the impairment of EDH (Figure 2) (Table 2).

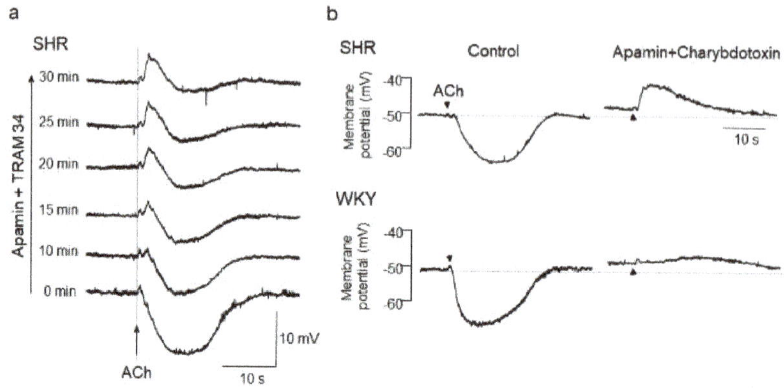

Figure 1. Acetylcholine (ACh)-evoked depolarization in mesenteric arteries of spontaneously hypertensive rats (SHRs). (**a**) A hidden depolarization emerged after blockade of endothelium-dependent hyperpolarization (EDH) with apamin (0.5 μmol/L, a small-conductance Ca^{2+}-sensitive K^+ channel (K_{Ca}) inhibitor) plus TRAM34 (100 nmol/L, an intermediate-conductance K_{Ca} inhibitor) in mesenteric arteries of SHRs. All recordings were from the same cell. (**b**) ACh-evoked depolarization in the presence of apamin (0.5 μmol/L) plus charybdotoxin (60 nmol/L, a large and intermediate-conductance K_{Ca} inhibitor) was larger in amplitude and faster in time course in SHRs than in Wistar Kyoto (WKY) rats. Each paired recording was from the same preparation. Indomethacin (10 μmol/L) and N^{ω}-nitro-L-arginine methyl ester (100 μmol/L) were present throughout. Arrows, application of ACh. Modified from Goto et al. [87].

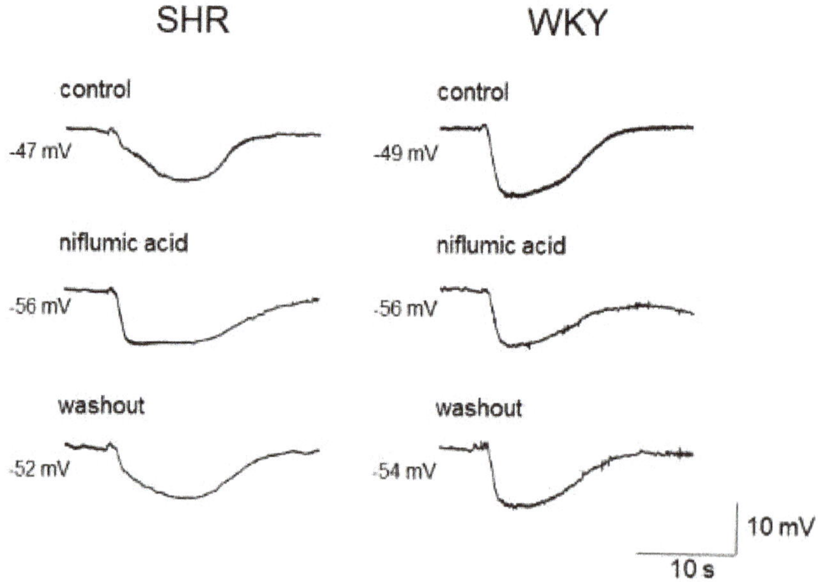

Figure 2. Effects of niflumic acid on acetylcholine-induced, endothelium-dependent hyperpolarization (EDH) in mesenteric arteries of spontaneously hypertensive rats (SHRs) and Wistar Kyoto (WKY) rats. Niflumic acid (50 μmol/L) improved EDH in SHRs but not in WKY rats. Each paired recording was from the same preparation. Indomethacin (10 μmol/L) and N^{ω}-nitro-L-arginine methyl ester (100 μmol/L) were present throughout.

Table 2. Alterations in vascular endothelial Ca^{2+}-activated Cl^- channels during hypertension.

Animals	Alterations in Endothelial CaCCs during Hypertension	Ref.
SHRs	Increased CaCC function in endothelium of mesenteric arteries	[86]
	Increased CaCC function, reduced EDH in mesenteric arteries	[86]
Ang II-induced hypertensive mice	Increased TMEM16A expression in endothelium of aorta	[82]
	Endothelial-specific TMEM16A knockout ameliorated endothelial function and lowered blood pressure	[82]
	Endothelial-specific TMEM16A overexpression deteriorated endothelial function and elevated blood pressure	[82]

CaCCs, Ca^{2+}-activated Cl^-channels; SHRs, spontaneously hypertensive rats; EDH, endothelium-dependent hyperpolarization; Ang II, angiotensin II.

As endothelial cells and adjacent smooth muscle cells are electrically coupled via myoendothelial gap junctions in rat mesenteric arteries [89–91], the impaired EDH leads to attenuated EDH-mediated relaxation and hence to endothelial dysfunction in SHRs [86]. Although some studies have reported that the inhibition of volume-activated Cl^- channels potentiates K^+-induced, EDH-mediated relaxation in rat mesenteric arteries [92,93], the involvement of the volume-activated Cl^- channels in the ACh-evoked depolarization in mesenteric arteries of SHRs is not likely because the volume-regulated Cl^- channel inhibitor NPPB had no effect on the ACh-evoked depolarization in this vascular bed [86].

A negative causal link between the activity of endothelial CaCCs, specifically TMEM16A, and endothelial function has also been reported in other studies [82,84]. Thus, in Ang II-induced hypertensive mice, in which the expression of vascular endothelial TMEM16A is increased, the endothelial-specific TMEM16A knockout ameliorated endothelial function and lowered the systolic blood pressure, whereas the endothelial-specific TMEM16A overexpression deteriorated endothelial function and further elevated the systolic blood pressure [82], and these interactions appear to be related to the facilitating effects of TMEM16A on reactive oxygen species generation via Nox2-containing NADPH oxidase [82] (Table 2).

Another study showed that overexpression of TMEM16A in human pulmonary endothelial cells led to a decrease in ACh-induced NO production [84]. Taken together, these findings suggest that upregulation of endothelial CaCC TMEM16A may contribute to the impaired endothelial function, and if so, that it likely does so via a reduction in the activity of EDH and/or NO; finally, the results suggest that such a reduction in EDH and/or NO activity may be at least partly responsible for the elevated blood pressure in hypertension (Figure 3).

3.3. Na^+–K^+–$2Cl^-$ Cotransporter1 (NKCC1)

NKCC1 located on vascular smooth muscle cells functions to accumulate intracellular Cl^- [20,21]. The most compelling evidence of the functional role of NKCC1 in the regulation of vascular tone and arterial blood pressure comes from studies on NKCC1 knockout mice: the systolic blood pressure was significantly reduced in NKCC1 knockout mice compared to wild-type mice [94], and treatment with bumetanide, an inhibitor of NKCC1 [95], inhibited the vascular contractile activity and lowered mean arterial blood pressure in wild-type mice, with the effects being lost in NKCC1 knockout mice [94,96]. Thus, theoretically, an increase in the activity of the vascular smooth muscle NKCC1 could augment vascular contractility and subsequently lead to enhanced blood pressure, and this is indeed the case in several types of hypertensive rats.

In some experimental models of hypertensive rats, including SHRs [97–99], Milan hypertensive rats [100] and deoxycorticosterone acetate (DOCA) salt hypertensive rats [101], increase in the activity of NKCC in vascular smooth muscle cells has been reported (Table 3). Interestingly, Lee et al. reported that the mRNA and protein expression levels of NKCC1 were epigenetically upregulated in the aorta of SHRs due to *Nkcc1* gene promoter hypomethylation [102], and the *Nkcc1* gene promoter hypomethylation resulted from the decreased activity of DNA methyltransferase 3B [103] (Table 3). Likewise, an epigenetic upregulation of NKCC1 via histone modifications was reported in the aorta of Ang II-induced hypertensive rats [104] (Table 3).

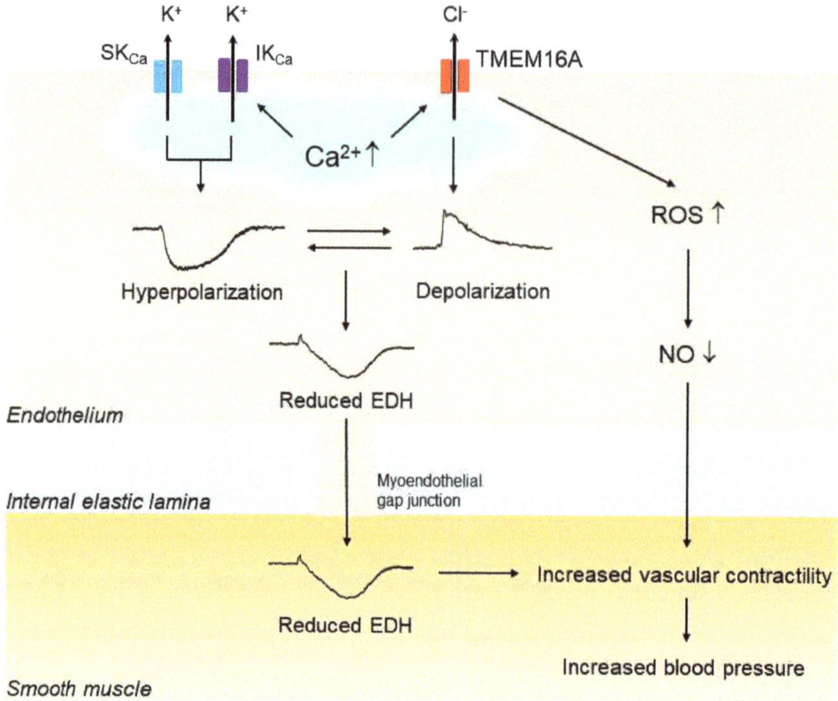

Figure 3. Upregulation of endothelial TMEM16A impairs endothelial function in hypertension. In hypertension, the expression and function of vascular endothelial Ca^{2+}-activated Cl^- channel TMEM16A are increased. Endothelial stimulation with agonists and shear stress increases the intracellular Ca^{2+}concentration, which subsequently activates endothelial small-conductance Ca^{2+}-activated K^+ channels (SK_{Ca}s), intermediate-conductance K_{Ca} (IK_{Ca}) and TMEM16A simultaneously. The endothelium-dependent hyperpolarization (EDH) through the activation of both SK_{Ca} and IK_{Ca} is reduced by the opposing membrane depolarization evoked by the activation of TMEM16A. In addition, activation of TMEM16A may facilitate the generation of reactive oxygen species (ROS) through Nox2-containing NADPH oxidase, leading to reduced bioavailability of nitic oxide (NO). Impaired EDH and/or NO could be at least partly responsible for the blood pressure rise in hypertension.

Table 3. Alterations in vascular smooth muscle NKCC1 during hypertension.

Animals	Alterations in Vascular Smooth Muscle NKCC1 during Hypertension	Ref.
SHRs	Increased NKCC1 function in aorta and carotid arteries	[97–99]
	Epigenetic upregulation of aorta NKCC1 due to *Nkcc1* gene promoter hypomethylation	[102]
	Nkcc1 gene promoter hypomethylation resulted from the decreased activity of DNA methyltransferase 3B	[103]
Milan hypertensive rats	Increased NKCC1 function in thoracic aorta	[100]
DOCA salt hypertensive rats	Increased NKCC1 function in saphenous branch of femoral arteries	[101]
Ang II-induced hypertensive rats	Epigenetic upregulation of aorta NKCC1 due to histone modifications	[104]

NKCC1, Na^+–K^+–$2Cl^-$ cotransporter 1; SHRs, spontaneously hypertensive rats; DOCA, deoxycorticosterone acetate; Ang II, angiotensin II.

In addition to the epigenetic upregulation of NKCC1, another factor may also contribute to the increase in the activity of NKCC1 during hypertension. In fact, some studies have suggested the possible positive regulation of vascular NKCC1 by with-no-lysine kinase (WNK) and sterile-20-related praline–alanine-rich kinase (SPAK): heterozygous WNK1 knockout mouse aorta exhibited reduced phosphorylation of downstream SPAK and NKCC1, leading to decreased responses to vasoconstrictive stimuli [105]. Similarly, the aorta of SPAK knockout mice exhibited reduced phosphorylation of NKCC1 and decreased NKCC1-mediated vascular constriction, and the SPAK knockout mice had low blood pressure [106]. Moreover, activation of the WNK3/SPAK/NKCC1 pathway has been shown to be involved in both the Ang II-induced aortic constriction and Ang II-induced blood pressure rise in mice [107].

These observations suggest that the WNK/SPAK signaling pathway positively regulates the vascular NKCC1 toward vasoconstriction and hypertension. Interestingly, mutations of WNK have been found in patients with familial hyperkalemic hypertension, a form of monogenic hypertension [108]. Nevertheless, there is no evidence to date that demonstrates changes in the WNK/SPAK pathway in animal models of polygenic hypertension such as SHRs or in human essential hypertension.

The studies mentioned above have demonstrated that the expression and/or the function of NKCC1 are upregulated in vascular smooth muscle cells of hypertensive rats. Then, the question arises how the upregulation of NKCC in hypertension contributes to the augmented vascular contractility and increased blood pressure. It has been reported that the intracellular concentration of Cl^- is increased in vascular smooth muscle cells of DOCA salt hypertensive rats because of the increase in the activity of NKCC [101].

The increase in the intracellular Cl^- concentration would increase the driving force for Cl^- efflux via Cl^- channels such as CaCCs upon vasoconstrictor stimulation, and the increase in Cl^- efflux would make the membrane potential more depolarized [20], which in turn would enhance the open probability of voltage-gated L-type Ca^{2+} channels, leading to an increase in vascular tone. In support of this notion, we have shown that the inhibition of the NKCC with bumetanide, an inhibitor of NKCC1 [95], significantly reduced the CaCC-mediated membrane depolarization and constriction in vascular smooth muscle of SHRs [86]. Since, as discussed in the previous section, CaCCs are also functionally upregulated in the vasculature of hypertensive rats, we propose that the enhanced activities of NKCC1 and CaCCs act additively and sequentially to increase vascular contractility and hence blood pressure in hypertension (Figure 4).

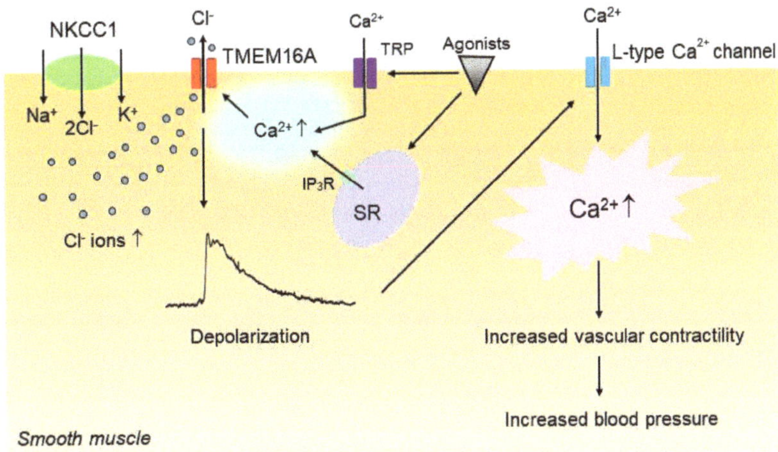

Figure 4. Upregulation of smooth muscle Na^+–K^+–$2Cl^-$ cotransporter 1 (NKCC1) and TMEM16A additively

and sequentially increases vascular contractility in hypertension. In hypertension, the intracellular concentration of Cl^- is increased in vascular smooth muscle cells because of the increased activity of NKCC1. The increase in the intracellular Cl^- concentration then increases the driving force for Cl^- efflux via the Ca^{2+}-activated Cl^- channel TMEM16A when TMEM16A is activated by intracellular Ca^{2+} rise upon stimulation with vasoconstricting agonists, which in turn induces membrane depolarization. TMEM16A might be regulated by a local Ca^{2+} increase that could be generated by IP_3R channels on the sarcoplasmic reticulum (SR) and/or transient receptor potential (TRP) channels. The membrane depolarization would then enhance the open probability of voltage-gated L-type Ca^{2+} channels, leading to an increase in vascular contractility and blood pressure.

4. Clinical Perspectives

While many animal studies suggest that the upregulation of TMEM16A and NKCC1 could contribute to the increased vascular contractility and elevated blood pressure in hypertension as mentioned in the preceding sections, there is very little information concerning their possible involvement in the pathogenesis of hypertension in humans. Interestingly, however, two independent population-based studies reported that some genetic variants of TMEM16A were significantly associated with hypertension in humans [57,109]. Further exploration of the functional impact of the SNP in the TMEM16A coding region could provide a clue to understand the pathophysiological role of TMEM16A in human hypertension.

In addition, it has been reported that there was a positive association between hyperchloremia and in-hospital mortality in hospitalized patients [110]. Moreover, in a recent study in patients with chronic kidney disease, hyperchloremia was an independent predictor of hypertension and proteinuria [111]. Taking these observational studies together, it might be possible to speculate that hyperchloremia might lead to blood pressure elevation and hence to poor prognosis. By contrast, in outpatients with hypertension [112] or chronic heart failure [113], hypochloremia was a predictor of mortality [112,113] but the level of Cl^- was not associated with the level of blood pressure [112]. Thus, while these findings indicate that serum Cl^- alterations are associated with poor prognosis in patients with elevated cardiovascular risk, the ability of changes in serum Cl^- concentration to affect blood pressure is not clear. Further studies are needed to clarify the role of serum Cl^- concentrations on blood pressure regulation and its association with long-term prognosis in patients with elevated cardiovascular risk.

5. Conclusions

Accumulating experimental evidence suggests that Cl^- plays an important role in the regulation of vascular tone through its ability to depolarize vascular smooth muscle cells, and the increased contribution of Cl^- to arterial constriction appears to be associated with the development and progression of hypertension. Of note, there is a growing body of evidence that the upregulation of CaCC TMEM16A in the vasculature contributes to the increased vascular contractility and elevated blood pressure in genetically hypertensive rats. In addition, the increased activity of NKCC1 may also promote hypertension as the result of a robust accumulation of intracellular Cl^- in vascular cells.

Nevertheless, much remains to be determined about the precise molecular mechanisms underlying the increased activity of TMEM16A and NKCC1 as well as their interactions with other signaling pathways during hypertension, and most importantly the pathophysiological roles of these molecules in human hypertension. Further exploration of the arterial tone regulation by Cl^- may facilitate a better understanding of the pathogenesis of hypertension, which may help to develop a novel therapeutic strategy to tackle hypertension and hypertension-associated cardiovascular diseases.

Author Contributions: K.G. was responsible for the conceptualization, literature research, and writing of the manuscript. T.K. assisted in the writing and editing of the manuscript. All authors have read and agreed to the published version of the manuscript.

Funding: This research was funded by JSPS KAKENHI grant number JP21K11023. The APC was funded by JSPS KAKENHI grant number JP21K11023.

Acknowledgments: This work was supported by JSPS KAKENHI Grant Number JP21K11023.

Conflicts of Interest: The authors declare no conflict of interest.

References

1. Lawes, C.M.; Hoorn, S.V.; Rodgers, A. Global Burden of Blood-Pressure-Related Disease, 2001. *Lancet* **2008**, *371*, 1513–1518. [CrossRef]
2. Zhou, B.; Perel, P.; Mensah, G.A.; Ezzati, M. Global Epidemiology, Health Burden and Effective Interventions for Elevated Blood Pressure and Hypertension. *Nat. Rev. Cardiol.* **2021**, *18*, 785–802. [CrossRef]
3. Zhou, B.; Carrillo-Larco, R.M.; Danaei, G.; Riley, L.M.; Paciorek, C.J.; Stevens, G.A.; Gregg, E.W.; Bennett, J.E.; Solomon, B.; Singleton, R.K.; et al. Worldwide Trends in Hypertension Prevalence and Progress in Treatment and Control from 1990 to 2019: A Pooled Analysis of 1201 Population-Representative Studies with 104 Million Participants. *Lancet* **2021**, *398*, 957–980. [CrossRef]
4. Umemura, S.; Arima, H.; Arima, S.; Asayama, K.; Dohi, Y.; Hirooka, Y.; Horio, T.; Hoshide, S.; Ikeda, S.; Ishimitsu, T.; et al. The Japanese Society of Hypertension Guidelines for the Management of Hypertension (JSH 2019). *Hypertens. Res.* **2019**, *42*, 1235–1481. [CrossRef] [PubMed]
5. Tsuchihashi, T. Dietary Salt Intake in Japan-Past, Present, and Future. *Hypertens. Res.* **2022**, *45*, 748–757. [CrossRef] [PubMed]
6. Cicero, A.F.G.; Veronesi, M.; Fogacci, F. Dietary Intervention to Improve Blood Pressure Control: Beyond Salt Restriction. *High Blood Press. Cardiovasc. Prev.* **2021**, *28*, 547–553. [CrossRef] [PubMed]
7. Intersalt. Intersalt: An International Study of Electrolyte Excretion and Blood. *BMJ* **1988**, *297*, 319–328.
8. Mozaffarian, D.; Fahimi, S.; Singh, G.M.; Micha, R.; Khatibzadeh, S.; Engell, R.E.; Lim, S.; Danaei, G.; Ezzati, M.; Powles, J. Global Sodium Consumption and Death from Cardiovascular Causes. *N. Engl. J. Med.* **2014**, *371*, 624–634. [CrossRef]
9. Meneton, P.; Jeunemaitre, X.; De Wardener, H.E.; Macgregor, G.A. Links between Dietary Salt Intake, Renal Salt Handling, Blood Pressure, and Cardiovascular Diseases. *Physiol. Rev.* **2005**, *85*, 679–715. [CrossRef]
10. Dahl, L.K.; Love, R.A. Evidence for Relationship between Sodium (Chloride) Intake and Human Essential Hypertension. *AMA Arch. Intern. Med.* **1954**, *94*, 525–531. [CrossRef] [PubMed]
11. Kempner, W. Treatment of Hypertensive Vascular Disease with Rice Diet. *Am. J. Med.* **1948**, *4*, 545–577. [CrossRef]
12. Kurtz, T.W.; Morris, R.C. Dietary Chloride as a Determinant of "Sodium-Dependent" Hypertension. *Science* **1983**, *222*, 1139–1141. [CrossRef] [PubMed]
13. Kotchen, T.A.; Luke, R.G.; Ott, C.E.; Galla, J.H.; Whitescarver, S. Effect of Chloride on Renin and Blood Pressure Responses to Sodium Chloride. *Ann. Intern. Med.* **1983**, *98 Pt 2*, 817–822. [CrossRef] [PubMed]
14. Whitescarver, S.A.; Holtzclaw, B.J.; Downs, J.H.; Ott, C.E.; Sowers, J.R.; Kotchen, T.A. Effect of Dietary Chloride on Salt-Sensitive and Renin-Dependent Hypertension. *Hypertension* **1986**, *8*, 56–61. [CrossRef] [PubMed]
15. Tanaka, M.; Schmidlin, O.; Yi, S.L.; Bollen, A.W.; Morris, R.C. Genetically Determined Chloride-Sensitive Hypertension and Stroke. *Proc. Natl. Acad. Sci. USA* **1997**, *94*, 14748–14752. [CrossRef] [PubMed]
16. Kurtz, T.W.; Al-Bander, H.A.; Morris, R.C. "Salt-Sensitive" Essential Hypertension in Men. Is the Sodium Ion Alone Important? *N. Engl. J. Med.* **1987**, *317*, 1043–1048. [CrossRef] [PubMed]
17. Luft, F.C.; Zemel, M.B.; Sowers, J.A.; Fine Berg, N.S.; Weinberger, M.H. Sodium Bicarbonate and Sodium Chloride: Effects on Blood Pressure and Electrolyte Homeostasis in Normal and Hypertensive Man. *J. Hypertens.* **1990**, *8*, 663–670. [CrossRef]
18. Shore, A.C.; Markandu, N.D.; MacGregor, G.A. A Randomized Crossover Study to Compare the Blood Pressure Response to Sodium Loading with and without Chloride in Patients with Essential Hypertension. *J. Hypertens.* **1988**, *6*, 613–617. [CrossRef] [PubMed]
19. McCallum, L.; Lip, S.; Padmanabhan, S. The Hidden Hand of Chloride in Hypertension. *Pflug. Arch.* **2015**, *467*, 595–603. [CrossRef] [PubMed]
20. Chipperfield, A.R.; Harper, A.A. Chloride in Smooth Muscle. *Prog. Biophys. Mol. Biol.* **2000**, *74*, 175–221. [CrossRef]
21. Bulley, S.; Jaggar, J.H. Cl^- Channels in Smooth Muscle Cells. *Pflug. Arch.* **2014**, *466*, 861–872. [CrossRef] [PubMed]
22. Bryant, H.J.; Harder, D.R.; Pamnani, M.B.; Haddy, F.J. In Vivo Membrane Potentials of Smooth Muscle Cells in the Caudal Artery of the Rat. *Am. J. Physiol.* **1985**, *249 Pt 1*, C78–C83. [CrossRef]
23. Aickin, C.C.; Brading, A.F. Measurement of Intracellular Chloride in Guinea-pig Vas Deferens by Ion Analysis, 36chloride and Micro-electrodes. *J. Physiol.* **1982**, *326*, 139–154. [CrossRef] [PubMed]
24. Large, W.A.; Wang, Q. Characteristics and Physiological Role of the Ca(2+)-Activated Cl^- Conductance in Smooth Muscle. *Am. J. Physiol.* **1996**, *271 Pt 1*, C435–C454. [CrossRef]
25. Goto, K.; Fujii, K.; Onaka, U.; Abe, I.; Fujishima, M. Effects of Adrenomedullin and PAMP on Membrane Potential and Neurotransmission. *Peptides* **2000**, *21*, 257–263. [CrossRef]
26. Brock, M.A.; Cunnane, T.C. Effects of Ca^{2+} Concentration and Ca^{2+} Channel Blockers on Noradrenaline Release and Purinergic Neuroeffector Transmission in Rat Tail Artery. *Br. J. Pharmacol.* **1999**, *126*, 11–18. [CrossRef]
27. Burnstock, G.; Ralevic, V. New Insights into the Local Regulation of Blood Flow by Perivascular Nerves and Endothelium. *Br. J. Plast. Surg.* **1994**, *47*, 527–543. [CrossRef]

28. Hill, C.E.; Phillips, J.K.; Sandow, S.L. Heterogeneous Control of Blood Flow amongst Different Vascular Beds. *Med. Res. Rev.* **2001**, *21*, 1–60. [CrossRef]
29. Plane, F.; Garland, C.J. Electrophysiology of Cerebral Blood Vessels. *Pharmacol. Ther.* **1992**, *56*, 341–358. [CrossRef]
30. Gould, D.J.; Hill, C.E. Alpha-Adrenoceptor Activation of a Chloride Conductance in Rat Iris Arterioles. *Am. J. Physiol.* **1996**, *271 Pt 2*, H2469–H2476. [CrossRef]
31. Kitamura, K.; Yamazaki, J. Chloride Channels and Their Functional Roles in Smooth Muscle Tone in the Vasculature. *Jpn. J. Pharmacol.* **2001**, *85*, 351–357. [CrossRef]
32. Leblanc, N.; Ledoux, J.; Saleh, S.; Sanguinetti, A.; Angermann, J.; O'Driscoll, K.; Britton, F.; Perrino, B.A.; Greenwood, I.A. Regulation of Calcium-Activated Chloride Channels in Smooth Muscle Cells: A Complex Picture Is Emerging. *Can. J. Physiol. Pharmacol.* **2005**, *83*, 541–556. [CrossRef] [PubMed]
33. Davis, M.J.; Hill, M.A. Signaling Mechanisms Underlying the Vascular Myogenic Response. *Physiol. Rev.* **1999**, *79*, 387–423. [CrossRef] [PubMed]
34. Nelson, M.T.; Conway, M.A.; Knot, H.J.; Brayden, J.E. Chloride Channel Blockers Inhibit Myogenic Tone in Rat Cerebral Arteries. *J. Physiol.* **1997**, *502 Pt 2*, 259–264. [CrossRef]
35. Doughty, J.M.; Langton, P.D. Measurement of Chloride Flux Associated with the Myogenic Response in Rat Cerebral Arteries. *J. Physiol.* **2001**, *534 Pt 3*, 753–761. [CrossRef]
36. Doughty, J.M.; Miller, A.L.; Langton, P.D. Non-Specificity of Chloride Channel Blockers in Rat Cerebral Arteries: Block of the L-Type Calcium Channel. *J. Physiol.* **1998**, *507 Pt 2*, 433–439. [CrossRef] [PubMed]
37. Welsh, D.G.; Nelson, M.T.; Eckman, D.M.; Brayden, J.E. Swelling-Activated Cation Channels Mediate Depolarization of Rat Cerebrovascular Smooth Muscle by Hyposmolarity and Intravascular Pressure. *J. Physiol.* **2000**, *527 Pt 1*, 139–148. [CrossRef]
38. Byrne, N.G.; Large, W.A. Membrane Mechanism Associated with Muscarinic Receptor Activation in Single Cells Freshly Dispersed from the Rat Anococcygeus Muscle. *Br. J. Pharmacol.* **1987**, *92*, 371–379. [CrossRef]
39. Matchkov, V.V. Mechanisms of Cellular Synchronization in the Vascular Wall. Mechanisms of Vasomotion. *Dan. Med. Bull.* **2010**, *57*, B4191.
40. Yang, Y.D.; Cho, H.; Koo, J.Y.; Tak, M.H.; Cho, Y.; Shim, W.S.; Park, S.P.; Lee, J.; Lee, B.; Kim, B.M.; et al. TMEM16A Confers Receptor-Activated Calcium-Dependent Chloride Conductance. *Nature* **2008**, *455*, 1210–1215. [CrossRef]
41. Caputo, A.; Caci, E.; Ferrera, L.; Pedemonte, N.; Barsanti, C.; Sondo, E.; Pfeffer, U.; Ravazzolo, R.; Zegarra-Moran, O.; Galietta, L.J.V. TMEM16A, a Membrane Protein Associated with Calcium-Dependent Chloride Channel Activity. *Science* **2008**, *322*, 590–594. [CrossRef] [PubMed]
42. Schroeder, B.C.; Cheng, T.; Jan, Y.N.; Jan, L.Y. Expression Cloning of TMEM16A as a Calcium-Activated Chloride Channel Subunit. *Cell* **2008**, *134*, 1019–1029. [CrossRef] [PubMed]
43. Dam, V.S.; Boedtkjer, D.M.B.; Aalkjaer, C.; Matchkov, V. The Bestrophin- and TMEM16A-Associated Ca(2+)- Activated Cl(−) Channels in Vascular Smooth Muscles. *Channels* **2014**, *8*, 361–369. [CrossRef] [PubMed]
44. Hawn, M.B.; Akin, E.; Hartzell, H.C.; Greenwood, I.A.; Leblanc, N. Molecular Mechanisms of Activation and Regulation of ANO1-Encoded Ca^{2+}-Activated Cl^- Channels. *Channels* **2021**, *15*, 569–603. [CrossRef]
45. Wray, S.; Prendergast, C.; Arrowsmith, S. Calcium-Activated Chloride Channels in Myometrial and Vascular Smooth Muscle. *Front. Physiol.* **2021**, *12*, 1805. [CrossRef]
46. Bulley, S.; Neeb, Z.P.; Burris, S.K.; Bannister, J.P.; Thomas-Gatewood, C.M.; Jangsangthong, W.; Jaggar, J.H. TMEM16A/ANO1 Channels Contribute to the Myogenic Response in Cerebral Arteries. *Circ. Res.* **2012**, *111*, 1027–1036. [CrossRef]
47. Yip, K.P.; Balasubramanian, L.; Kan, C.; Wang, L.; Liu, R.; Ribeiro-Silva, L.; Sham, J.S.K. Intraluminal Pressure Triggers Myogenic Response via Activation of Calcium Spark and Calcium-Activated Chloride Channel in Rat Renal Afferent Arteriole. *Am. J. Physiol. Renal Physiol.* **2018**, *315*, F1592–F1600. [CrossRef]
48. Heinze, C.; Seniuk, A.; Sokolov, M.V.; Huebner, A.K.; Klementowicz, A.E.; Szijártó, I.A.; Schleifenbaum, J.; Vitzthum, H.; Gollasch, M.; Ehmke, H.; et al. Disruption of Vascular Ca^{2+}-Activated Chloride Currents Lowers Blood Pressure. *J. Clin. Investig.* **2014**, *124*, 675–686. [CrossRef]
49. Mulvany, M.J.; Halpern, W. Contractile Properties of Small Arterial Resistance Vessels in Spontaneously Hypertensive and Normotensive Rats. *Circ. Res.* **1977**, *41*, 19–26. [CrossRef]
50. Pintérová, M.; Kuneš, J.; Zicha, J. Altered Neural and Vascular Mechanisms in Hypertension. *Physiol. Res.* **2011**, *60*, 381–402. [CrossRef]
51. Goto, K.; Ohtsubo, T.; Kitazono, T. Endothelium-Dependent Hyperpolarization (EDH) in Hypertension: The Role of Endothelial Ion Channels. *Int. J. Mol. Sci.* **2018**, *19*, 315. [CrossRef] [PubMed]
52. Stekiel, W.J.; Contney, S.J.; Lombard, J.H. Small Vessel Membrane Potential, Sympathetic Input, and Electrogenic Pump Rate in SHR. *Am. J. Physiol.* **1986**, *250 Pt 1*, C547–C556. [CrossRef] [PubMed]
53. Fujii, K.; Tominaga, M.; Ohmori, S.; Kobayashi, K.; Koga, T.; Takata, Y.; Fujishima, M. Decreased Endothelium-Dependent Hyperpolarization to Acetylcholine in Smooth Muscle of the Mesenteric Artery of Spontaneously Hypertensive Rats. *Circ. Res.* **1992**, *70*, 660–669. [CrossRef] [PubMed]
54. Goto, K.; Fujii, K.; Abe, I. Impaired β-Adrenergic Hyperpolarization in Arteries from Prehypertensive Spontaneously Hypertensive Rats. *Hypertension* **2001**, *37*, 609–613. [CrossRef]

55. Wang, Z.; Chai, Q.; Liu, Z.; Liu, D.; Chen, L. Chloride Channel Activity of Vascular Smooth Muscle in the Spontaneous Hypertensive Rats. *Chin. J. Physiol.* **2004**, *47*, 129–135. [PubMed]
56. Wang, B.; Li, C.; Huai, R.; Qu, Z. Overexpression of ANO1/TMEM16A, an Arterial Ca^{2+}-Activated Cl^- Channel, Contributes to Spontaneous Hypertension. *J. Mol. Cell. Cardiol.* **2015**, *82*, 22–32. [CrossRef] [PubMed]
57. Askew Page, H.R.; Dalsgaard, T.; Baldwin, S.N.; Jepps, T.A.; Povstyan, O.; Olesen, S.P.; Greenwood, I.A. TMEM16A Is Implicated in the Regulation of Coronary Flow and Is Altered in Hypertension. *Br. J. Pharmacol.* **2019**, *176*, 1635–1648. [CrossRef] [PubMed]
58. Cil, O.; Chen, X.; Askew Page, H.R.; Baldwin, S.N.; Jordan, M.C.; Myat Thwe, P.; Anderson, M.O.; Haggie, P.M.; Greenwood, I.A.; Roos, K.P.; et al. A Small Molecule Inhibitor of the Chloride Channel TMEM16A Blocks Vascular Smooth Muscle Contraction and Lowers Blood Pressure in Spontaneously Hypertensive Rats. *Kidney Int.* **2021**, *100*, 311–320. [CrossRef]
59. Li, R.S.; Wang, Y.; Chen, H.S.; Jiang, F.Y.; Tu, Q.; Li, W.J.; Yin, R.X. TMEM16A Contributes to Angiotensin II-Induced Cerebral Vasoconstriction via the RhoA/ROCK Signaling Pathway. *Mol. Med. Rep.* **2016**, *13*, 3691–3699. [CrossRef]
60. Wang, M.; Yang, H.; Zheng, L.Y.; Zhang, Z.; Tang, Y.B.; Wang, G.L.; Du, Y.H.; Lv, X.F.; Liu, J.; Zhou, J.G.; et al. Downregulation of TMEM16A Calcium-Activated Chloride Channel Contributes to Cerebrovascular Remodeling during Hypertension by Promoting Basilar Smooth Muscle Cell Proliferation. *Circulation* **2012**, *125*, 697–707. [CrossRef]
61. Shiono, K.; Sokabe, H. Renin-Angiotensin System in Spontaneously Hypertensive Rats. *Am. J. Physiol.* **1976**, *231*, 1295–1299. [CrossRef]
62. Okamura, T.; Miyazaki, M.; Inagami, T.; Toda, N. Vascular Renin-Angiotensin System in Two-Kidney, One Clip Hypertensive Rats. *Hypertension* **1986**, *8*, 560–565. [CrossRef] [PubMed]
63. Zhang, X.H.; Zheng, B.; Yang, Z.; He, M.; Yue, L.Y.; Zhang, R.N.; Zhang, M.; Zhang, W.; Zhang, X.; Wen, J.K. TMEM16A and Myocardin Form a Positive Feedback Loop That Is Disrupted by KLF5 during Ang II-Induced Vascular Remodeling. *Hypertension* **2015**, *66*, 412–421. [CrossRef] [PubMed]
64. Abel, P.W.; Hermsmeyer, K. Sympathetic Cross-Innervation of SHR and Genetic Controls Suggests a Trophic Influence on Vascular Muscle Membranes. *Circ. Res.* **1981**, *49*, 1311–1318. [CrossRef] [PubMed]
65. Kostyunina, D.S.; Zhang, L.; Shvetsova, A.A.; Selivanova, E.K.; Tarasova, O.S.; Matchkov, V.V.; Gaynullina, D.K. Trophic Sympathetic Influence Weakens Pro-Contractile Role of Cl^- Channels in Rat Arteries during Postnatal Maturation. *Sci. Rep.* **2020**, *10*, 20002. [CrossRef] [PubMed]
66. Liskova, S.; Petrova, M.; Karen, P.; Behuliak, M.; Zicha, J. Contribution of Ca^{2+}-Dependent Cl^- Channels to Norepinephrine-Induced Contraction of Femoral Artery Is Replaced by Increasing EDCF Contribution during Ageing. *Biomed Res. Int.* **2014**, *2014*, 289361. [CrossRef]
67. Jensen, A.B.; Joergensen, H.B.; Dam, V.S.; Kamaev, D.; Boedtkjer, D.; Füchtbauer, E.M.; Aalkjaer, C.; Matchkov, V.V. Variable Contribution of TMEM16A to Tone in Murine Arterial Vasculature. *Basic Clin. Pharmacol. Toxicol.* **2018**, *123*, 30–41. [CrossRef]
68. Dam, V.S.; Boedtkjer, D.M.B.; Nyvad, J.; Aalkjaer, C.; Matchkov, V. TMEM16A Knockdown Abrogates Two Different Ca(2+)-Activated Cl(−) Currents and Contractility of Smooth Muscle in Rat Mesenteric Small Arteries. *Pflug. Arch.* **2014**, *466*, 1391–1409. [CrossRef]
69. Wang, Q.; Dennis Leo, M.; Narayanan, D.; Kuruvilla, K.P.; Jaggar, J.H. Local Coupling of TRPC6 to ANO1/TMEM16A Channels in Smooth Muscle Cells Amplifies Vasoconstriction in Cerebral Arteries. *Am. J. Physiol. Cell Physiol.* **2016**, *310*, C1001–C1009. [CrossRef]
70. Ohya, Y.; Abe, I.; Fujii, K.; Takata, Y.; Fujishima, M. Voltage-Dependent Ca^{2+} Channels in Resistance Arteries from Spontaneously Hypertensive Rats. *Circ. Res.* **1993**, *73*, 1090–1099. [CrossRef]
71. Pesic, A.; Madden, J.A.; Pesic, M.; Rusch, N.J. High Blood Pressure Upregulates Arterial L-Type Ca^{2+} Channels: Is Membrane Depolarization the Signal? *Circ. Res.* **2004**, *94*, e97–e104. [CrossRef] [PubMed]
72. Bae, Y.M.; Kim, A.; Lee, Y.J.; Lim, W.; Noh, Y.H.; Kim, E.J.; Kim, J.; Kim, T.K.; Park, S.W.; Kim, B.; et al. Enhancement of Receptor-Operated Cation Current and TRPC6 Expression in Arterial Smooth Muscle Cells of Deoxycorticosterone Acetate-Salt Hypertensive Rats. *J. Hypertens.* **2007**, *25*, 809–817. [CrossRef]
73. Suh, B.C.; Hille, B. PIP2 Is a Necessary Cofactor for Ion Channel Function: How and Why? *Annu. Rev. Biophys.* **2008**, *37*, 175–195. [CrossRef]
74. Ta, C.M.; Acheson, K.E.; Rorsman, N.J.G.; Jongkind, R.C.; Tammaro, P. Contrasting Effects of Phosphatidylinositol 4,5-Bisphosphate on Cloned TMEM16A and TMEM16B Channels. *Br. J. Pharmacol.* **2017**, *174*, 2984–2999. [CrossRef] [PubMed]
75. Tembo, M.; Wozniak, K.L.; Bainbridge, R.E.; Carlson, A.E. Phosphatidylinositol 4,5-Bisphosphate (PIP 2) and Ca^{2+} Are Both Required to Open the Cl^- Channel TMEM16A. *J. Biol. Chem.* **2019**, *294*, 12556–12564. [CrossRef] [PubMed]
76. Jia, Z.; Chen, J. Specific PIP 2 Binding Promotes Calcium Activation of TMEM16A Chloride Channels. *Commun. Biol.* **2021**, *4*, 259. [CrossRef]
77. Pritchard, H.A.T.; Leblanc, N.; Albert, A.P.; Greenwood, I.A. Inhibitory Role of Phosphatidylinositol 4,5-Bisphosphate on TMEM16A-Encoded Calcium-Activated Chloride Channels in Rat Pulmonary Artery. *Br. J. Pharmacol.* **2014**, *171*, 4311–4321. [CrossRef]
78. Ek, T.P.; Campbell, M.D.; Deth, R.C. Reduction of Norepinephrine-Induced Tonic Contraction and Phosphoinositide Turnover in Arteries of Spontaneously Hypertensive Rats. A Possible Role for Protein Kinase C. *Am. J. Hypertens.* **1989**, *2*, 40–45. [CrossRef] [PubMed]

79. Jin, X.; Shah, S.; Liu, Y.; Zhang, H.; Lees, M.; Fu, Z.; Lippiat, J.D.; Beech, D.J.; Sivaprasadarao, A.; Baldwin, S.A.; et al. Activation of the Cl$^-$ Channel ANO1 by Localized Calcium Signals in Nociceptive Sensory Neurons Requires Coupling with the IP3 Receptor. *Sci. Signal.* **2013**, *6*, ra73. [CrossRef]
80. Abou-Saleh, H.; Pathan, A.R.; Daalis, A.; Hubrack, S.; Abou-Jassoum, H.; Al-Naeimi, H.; Rusch, N.J.; Machaca, K. Inositol 1,4,5-Trisphosphate (IP3) Receptor up-Regulation in Hypertension Is Associated with Sensitization of Ca2+ Release and Vascular Smooth Muscle Contractility. *J. Biol. Chem.* **2013**, *288*, 32941–32951. [CrossRef] [PubMed]
81. Nilius, B.; Droogmans, G. Ion Channels and Their Functional Role in Vascular Endothelium. *Physiol. Rev.* **2001**, *81*, 1415–1459. [CrossRef]
82. Ma, M.M.; Gao, M.; Guo, K.M.; Wang, M.; Li, X.Y.; Zeng, X.L.; Sun, L.; Lv, X.F.; Du, Y.H.; Wang, G.L.; et al. TMEM16A Contributes to Endothelial Dysfunction by Facilitating Nox2 NADPH Oxidase-Derived Reactive Oxygen Species Generation in Hypertension. *Hypertension* **2017**, *69*, 892–901. [CrossRef] [PubMed]
83. Suzuki, T.; Yasumoto, M.; Suzuki, Y.; Asai, K.; Imaizumi, Y.; Yamamura, H. TMEM16A Ca^{2+}-Activated Cl$^-$ Channel Regulates the Proliferation and Migration of Brain Capillary Endothelial Cells. *Mol. Pharmacol.* **2020**, *98*, 61–71. [CrossRef]
84. Skofic Maurer, D.; Zabini, D.; Nagaraj, C.; Sharma, N.; Lengyel, M.; Nagy, B.M.; Frank, S.; Klepetko, W.; Gschwandtner, E.; Enyedi, P.; et al. Endothelial Dysfunction Following Enhanced TMEM16A Activity in Human Pulmonary Arteries. *Cells* **2020**, *9*, 1984. [CrossRef] [PubMed]
85. Yamamoto, Y.; Suzuki, H. Effects of Increased Intracellular Cl$^-$ Concentration on Membrane Responses to Acetylcholine in the Isolated Endothelium of Guinea Pig Mesenteric Arteries. *J. Physiol. Sci.* **2007**, *57*, 31–41. [CrossRef] [PubMed]
86. Goto, K.; Edwards, F.R.; Hill, C.E. Depolarization Evoked by Acetylcholine in Mesenteric Arteries of Hypertensive Rats Attenuates Endothelium-Dependent Hyperpolarizing Factor. *J. Hypertens.* **2007**, *25*, 345–359. [CrossRef]
87. Goto, K.; Rummery, N.M.; Grayson, T.H.; Hill, C.E. Attenuation of Conducted Vasodilation in Rat Mesenteric Arteries during Hypertension: Role of Inwardly Rectifying Potassium Channels. *J. Physiol.* **2004**, *561*, 215–231. [CrossRef] [PubMed]
88. Hirst, G.D.S.; Bramich, N.J.; Teramoto, N.; Suzuki, H.; Edwards, F.R. Regenerative Component of Slow Waves in the Guinea-Pig Gastric Antrum Involves a Delayed Increase in $[Ca^{2+}]_i$ and Cl$^-$ Channels. *J. Physiol.* **2002**, *540 Pt 3*, 907–919. [CrossRef] [PubMed]
89. Sandow, S.L.; Hill, C.E. Incidence of Myoendothelial Gap Junctions in the Proximal and Distal Mesenteric Arteries of the Rat Is Suggestive of a Role in Endothelium-Derived Hyperpolarizing Factor–Mediated Responses. *Circ. Res.* **2000**, *86*, 341–346. [CrossRef] [PubMed]
90. Goto, K.; Fujii, K.; Kansui, Y.; Abe, I.; Iida, M. Critical Role of Gap Junctions in Endothelium-Dependent Hyperpolarization in Rat Mesenteric Arteries. *Clin. Exp. Pharmacol. Physiol.* **2002**, *29*, 595–602. [CrossRef] [PubMed]
91. Mather, S.; Dora, K.A.; Sandow, S.L.; Winter, P.; Garland, C.J. Rapid Endothelial Cell-Selective Loading of Connexin 40 Antibody Blocks Endothelium-Derived Hyperpolarizing Factor Dilation in Rat Small Mesenteric Arteries. *Circ. Res.* **2005**, *97*, 399–407. [CrossRef]
92. Doughty, J.M.; Boyle, J.P.; Langton, P.D. Blockade of Chloride Channels Reveals Relaxations of Rat Small Mesenteric Arteries to Raised Potassium. *Br. J. Pharmacol.* **2001**, *132*, 293–301. [CrossRef] [PubMed]
93. Yang, C.; Kwan, Y.W.; Chan, S.W.; Lee, S.M.Y.; Leung, G.P.H. Potentiation of EDHF-Mediated Relaxation by Chloride Channel Blockers. *Acta Pharmacol. Sin.* **2010**, *31*, 1303–1311. [CrossRef] [PubMed]
94. Meyer, J.W.; Flagella, M.; Sutliff, R.L.; Lorenz, J.N.; Nieman, M.L.; Weber, C.S.; Paul, R.J.; Shull, G.E. Decreased Blood Pressure and Vascular Smooth Muscle Tone in Mice Lacking Basolateral Na^+-K^+-2Cl$^-$ Cotransporter. *Am. J. Physiol. Heart Circ. Physiol.* **2002**, *283*, H1846–H1855. [CrossRef] [PubMed]
95. Zhao, Y.; Roy, K.; Vidossich, P.; Cancedda, L.; De Vivo, M.; Forbush, B.; Cao, E. Structural Basis for Inhibition of the Cation-Chloride Cotransporter NKCC1 by the Diuretic Drug Bumetanide. *Nat. Commun.* **2022**, *13*, 2747. [CrossRef]
96. Garg, P.; Martin, C.F.; Elms, S.C.; Gordon, F.J.; Wall, S.M.; Garland, C.J.; Sutliff, R.L.; O'Neill, W.C. Effect of the Na-K-2Cl Cotransporter NKCC1 on Systemic Blood Pressure and Smooth Muscle Tone. *Am. J. Physiol. Heart Circ. Physiol.* **2007**, *292*. [CrossRef]
97. Orlov, S.N.; Resink, T.J.; Bernhardt, J.; Bühler, F.R. Na(+)-K+ Pump and Na(+)-K+ Co-Transport in Cultured Vascular Smooth Muscle Cells from Spontaneously Hypertensive and Normotensive Rats: Baseline Activity and Regulation. *J. Hypertens.* **1992**, *10*, 733–740. [CrossRef] [PubMed]
98. Kuriyama, S.; Denny, T.N.; Aviv, A. 22Na^+ and 86Rb^+ Transport in Vascular Smooth Muscle of SHR, Wistar Kyoto, and Wistar Rats. *J. Cardiovasc. Pharmacol.* **1988**, *11*, 722–729. [CrossRef]
99. Tokushige, A.; Kino, M.; Tamura, H.; Hopp, L.; Searle, B.M.; Aviv, A. Bumetanide-Sensitive Sodium-22 Transport in Vascular Smooth Muscle Cell of the Spontaneously Hypertensive Rat. *Hypertension* **1986**, *8*, 379–385. [CrossRef] [PubMed]
100. Canessa, M.; Salazar, G.; Werner, E.; Vallega, G.; Gonzalez, A. Cell Growth and Na-K-Cl Cotransport Responses of Vascular Smooth Muscle Cells of Milan Rats. *Hypertension* **1994**, *23 Pt 2*, 1022–1026. [CrossRef] [PubMed]
101. Davis, J.P.L.; Chipperfield, A.R.; Harper, A.A. Accumulation of Intracellular Chloride by (Na-K-Cl) Co-Transport in Rat Arterial Smooth Muscle Is Enhanced in Deoxycorticosterone Acetate (DOCA)/Salt Hypertension. *J. Mol. Cell. Cardiol.* **1993**, *25*, 233–237. [CrossRef]
102. Lee, H.A.; Baek, I.; Seok, Y.M.; Yang, E.; Cho, H.M.; Lee, D.Y.; Hong, S.H.; Kim, I.K. Promoter Hypomethylation Upregulates Na^+-K^+-2Cl$^-$ Cotransporter 1 in Spontaneously Hypertensive Rats. *Biochem. Biophys. Res. Commun.* **2010**, *396*, 252–257. [CrossRef]

103. Cho, H.M.; Lee, H.A.; Kim, H.Y.; Han, H.S.; Kim, I.K. Expression of Na^{+}-K^{+}-2Cl Cotransporter 1 Is Epigenetically Regulated during Postnatal Development of Hypertension. *Am. J. Hypertens.* **2011**, *24*, 1286–1293. [CrossRef]
104. Cho, H.M.; Lee, D.Y.; Kim, H.Y.; Lee, H.A.; Seok, Y.M.; Kim, I.K. Upregulation of the Na(+)-K(+)-2Cl(−) Cotransporter 1 via Histone Modification in the Aortas of Angiotensin II-Induced Hypertensive Rats. *Hypertens. Res.* **2012**, *35*, 819–824. [CrossRef]
105. Bergaya, S.; Faure, S.; Baudrie, V.; Rio, M.; Escoubet, B.; Bonnin, P.; Henrion, D.; Loirand, G.; Achard, J.M.; Jeunemaitre, X.; et al. WNK1 Regulates Vasoconstriction and Blood Pressure Response to α 1-Adrenergic Stimulation in Mice. *Hypertension* **2011**, *58*, 439–445. [CrossRef]
106. Yang, S.S.; Lo, Y.F.; Wu, C.C.; Lin, S.W.; Yeh, C.J.; Chu, P.; Sytwu, H.K.; Uchida, S.; Sasaki, S.; Lin, S.H. SPAK-Knockout Mice Manifest Gitelman Syndrome and Impaired Vasoconstriction. *J. Am. Soc. Nephrol.* **2010**, *21*, 1868–1877. [CrossRef]
107. Zeniya, M.; Sohara, E.; Kita, S.; Iwamoto, T.; Susa, K.; Mori, T.; Oi, K.; Chiga, M.; Takahashi, D.; Yang, S.; et al. Dietary Salt Intake Regulates WNK3-SPAK-NKCC1 Phosphorylation Cascade in Mouse Aorta through Angiotensin II. *Hypertension* **2013**, *62*, 872–878. [CrossRef]
108. Murthy, M.; Kurz, T.; O'Shaughnessy, K.M. WNK Signalling Pathways in Blood Pressure Regulation. *Cell. Mol. Life Sci.* **2017**, *74*, 1261–1280. [CrossRef]
109. Jin, H.-S.; Jung, D. Gender-Specific Association of the ANO1 Genetic Variations with Hypertension. *Biomed. Sci. Lett.* **2015**, *21*, 144–151. [CrossRef]
110. Thongprayoon, C.; Cheungpasitporn, W.; Hansrivijit, P.; Thirunavukkarasu, S.; Chewcharat, A.; Medaura, J.; Mao, M.A.; Kashani, K. Association of Serum Chloride Level Alterations with In-Hospital Mortality. *Postgrad. Med. J.* **2020**, *96*, 731–736. [CrossRef]
111. Takahashi, A.; Maeda, K.; Sasaki, K.; Doi, S.; Nakashima, A.; Doi, T.; Masaki, T. Relationships of Hyperchloremia with Hypertension and Proteinuria in Patients with Chronic Kidney Disease. *Clin. Exp. Nephrol.* **2022**, *26*, 880–885. [CrossRef] [PubMed]
112. McCallum, L.; Jeemon, P.; Hastie, C.E.; Patel, R.K.; Williamson, C.; Redzuan, A.M.; Dawson, J.; Sloan, W.; Muir, S.; Morrison, D.; et al. Serum Chloride Is an Independent Predictor of Mortality in Hypertensive Patients. *Hypertension* **2013**, *62*, 836–843. [CrossRef] [PubMed]
113. Bellino, M.C.; Massari, F.; Albanese, M.; Ursi, R.; Angelini, G.; Lisi, F.; Amato, L.; Scicchitano, P.; Guida, P.; Brunetti, N.D.; et al. Baseline and Incident Hypochloremia in Chronic Heart Failure Outpatients: Clinical Correlates and Prognostic Role. *Eur. J. Intern. Med.* **2021**, *84*, 32–37. [CrossRef] [PubMed]

biomedicines

MDPI

Article

Elevated Vascular Sympathetic Neurotransmission and Remodelling Is a Common Feature in a Rat Model of Foetal Programming of Hypertension and SHR

Maria Sofia Vieira-Rocha [1,2], Joana Beatriz Sousa [1,2,*], Pilar Rodríguez-Rodríguez [3], Silvia Madaglena Arribas [3] and Carmen Diniz [1,2,*]

1 Laboratory of Pharmacology, Department of Drug Science, Faculty of Pharmacy, University of Porto, 4050-313 Porto, Portugal
2 LAQV/REQUIMTE, Faculty of Pharmacy, University of Porto, 4050-313 Porto, Portugal
3 Department of Physiology, Faculty of Medicine, Universidad Autonoma de Madrid, 28029 Madrid, Spain
* Correspondence: jbsousa@ff.up.pt (J.B.S.); cdiniz@ff.up.pt (C.D.)

Abstract: Hypertension is of unknown aetiology, with sympathetic nervous system hyperactivation being one of the possible contributors. Hypertension may have a developmental origin, owing to the exposure to adverse factors during the intrauterine period. Our hypothesis is that sympathetic hyperinnervation may be implicated in hypertension of developmental origins, being this is a common feature with essential hypertension. Two-animal models were used: spontaneously hypertensive rats (SHR-model of essential hypertension) and offspring from dams exposed to undernutrition (MUN-model of developmental hypertension), with their respective controls. In adult males, we assessed systolic blood pressure (SBP), diastolic blood pressure (DBP), heart rate (HR), sympathetic nerve function (^{3}H-tritium release), sympathetic innervation (immunohistochemistry) and vascular remodelling (histology). MUN showed higher SBP/DBP, but not HR, while SHR exhibited higher SBP/DBP/HR. Regarding the mesenteric arteries, MUN and SHR showed reduced lumen, increased media and adventitial thickness and increased wall/lumen and connective tissue compared to respective controls. Regarding sympathetic nerve activation, MUN and SHR showed higher tritium release compared to controls. Total tritium tissue/tyrosine hydroxylase detection was higher in SHR and MUN adventitia arteries compared to respective controls. In conclusion, sympathetic hyperinnervation may be one of the contributors to vascular remodelling and hypertension in rats exposed to undernutrition during intrauterine life, which is a common feature with spontaneous hypertension.

Keywords: foetal programming of hypertension; sympathetic neurotransmission; sympathetic innervation; vascular remodelling; fibrosis; foetal undernutrition

Citation: Vieira-Rocha, M.S.; Sousa, J.B.; Rodríguez-Rodríguez, P.; Arribas, S.M.; Diniz, C. Elevated Vascular Sympathetic Neurotransmission and Remodelling Is a Common Feature in a Rat Model of Foetal Programming of Hypertension and SHR. *Biomedicines* **2022**, *10*, 1902. https://doi.org/10.3390/biomedicines10081902

Academic Editor: Ivana Vaněčková

Received: 1 July 2022
Accepted: 4 August 2022
Published: 5 August 2022

1. Introduction

Hypertension is one of the most important risk factors of cardiovascular disease, and despite the current treatment options, a substantial portion of the population still have uncontrolled or suboptimal controlled blood pressure (BP) [1]. Additionally, the aetiology of hypertension remains unknown in most cases. Sympathetic nervous system (SNS) has an integral role in the regulation of heart rate and contractility, vascular tone and fluid volume. SNS hyperactivation leads to retention of salt and/or water and increases cardiac output and peripheral resistance [2]. In addition, elevated SNS also participates in vasculokar remodelling related to smooth-muscle hypertrophy and fibrosis [3], contributing to the development and/or the maintenance of hypertension through an increase in peripheral resistance [4–6]. Recent advances have led to the understanding that hypertension may have a developmental origin. It is now well accepted that the foetus can adapt to adverse intrauterine conditions promoting physiological alterations in foetal development to ensure survival [7]. Such alterations, later in life, may increase the susceptibility to develop hypertension and cardiometabolic diseases, in a process named as foetal programming [8–10]. Several adverse factors during intrauterine life have been demonstrated to contribute to inadequate foetal development and foetal programming. The most important ones are malnutrition [9,11], oxygen deprivation [12–14], placental insufficiency [15], exposure to excess of glucocorticoids [16–18], toxic substances (alcohol, tobacco) [19–21] and environmental pollutants [22]. The mechanisms underlying an offspring's predisposition to develop hypertension in adulthood have not been completely addressed. Nevertheless, implication of increased oxidative stress [23,24], alterations in the glucocorticoid axis [16,25–27] or activation of the renin-angiotensin system (RAS) [28–30] have been suggested. Some of these alterations might be mediated by epigenetic modulation of genes implicated in cardiovascular control [24,31–34] and/or alterations in renal or vascular autonomic functions [14,26,35–38]. In the context of foetal programming hypertension (FPH), the contribution of the peripheral sympathetic nervous system is still not completely understood [39]. The spontaneously hypertensive rats (SHR), developed by Okamoto and Aoki 1963 [40] is a well-established model, which resembles essential hypertension in humans [41]. Among other common features, the SHR exhibits alterations in RAS [42], sympathetic hyperinnervation [43] and resistance artery remodelling [44], features also present in human essential hypertension. SHR and rat models of FPH also share some similarities regarding blood pressure development, i.e., the sexual dimorphism and the time course of development [45,46].

In the current work, we hypothesize that hypertension of developmental origin may share features with SHR regarding sympathetic hyperinnervation, which may contribute to vascular remodelling and hypertension development. Our aims were to evaluate, in mesenteric arteries from SHR and from a rat model of FPH induced by maternal undernutrition during gestation (MUN), the (i) sympathetic innervation, (ii) sympathetic activation and (iii) vascular remodelling.

2. Materials and Methods

2.1. Animals

Sprague–Dawley, Wistar Kyoto (WKY) and spontaneously hypertensive (SHR) rats from the colony maintained at the animal house facility of the Universidad Autónoma de Madrid were used. All experimental procedures were approved by the Ethics Review Board of Universidad Autónoma de Madrid and Comunidad Autónoma de Madrid (CEI63-1112-A097 and PROEX 04/19) according to the Guidelines for the Care and Use of Laboratory Animals (National Institutes of Health publication no. 85-23, revised in 1996), the Spanish legislation (RD 1201/2005) and the Directive 2010/63/EU on the protection of animals used for scientific purposes. The rats were housed in buckets 36.5/21.5/18.5 cm (length/width/height) on aspen wood bedding, under controlled conditions of 22 °C, 40% relative humidity and 12/12 light/dark photoperiod. The animal health monitoring indicated that they were free from pathogens that may interact with any of the parameters studied. The ARRIVE Guidelines were followed for reporting in vivo experiments [47].

2.1.1. Experimental Model of FPH

FPH model based on global maternal nutrient restriction was induced as previously described [46,48]. Two study groups were established using the Sprague–Dawley strain: a CONTROL group, with ad libitum feeding throughout pregnancy and lactation, and a group with intake restriction during part of gestation (maternal undernutrition model, MUN). This last group of rats had ad libitum diet during the first half of the gestation (from day 1 to 10), and then they were fed with 50% of the intake of a pregnant rat (from day 11 until delivery). The maximum daily intake of rat chow was previously determined in a group of pregnant rats as 24 g/day. After delivery and through the lactation period, the mothers were fed ad libitum. The mothers were fed with a breeding diet (Euro Rodent Diet 22; 5LF5, Labdiet, Madrid, Spain) containing 55% carbohydrates, 22% protein, 4.4% fat,

4.1% fibre and 5.4% mineral at 12.2% humidity. Drinking water was provided ad libitum to all animals. Immediately after birth, the offspring were weighed individually and sexed, and the litter was randomly standardised to 12 rats, 6 males and 6 females, if possible. The rest of the litter was sacrificed with CO_2. At the age of 6 months, male offspring from the two experimental groups (control and MUN) were analysed.

2.1.2. Experimental Model of Spontaneous Hypertension

SHR, a well-known animal model of essential hypertension, was also used to make comparisons with the animal model of FPH induced by foetal undernutrition. Wistar-Kyoto rats (WKY) were chosen as the control model, representing a normotensive state. Both WKY and SHR male rats were bred at the Animal House of Universidad Autónoma de Madrid and used at the age of 6 months.

2.1.3. Experimental Protocol

The animals from the different experimental groups (WKY, SHR, CONTROL and MUN) were first weighed and then anesthetized to measure the haemodynamic parameters (see below). Thereafter, the rats were sacrificed using a guillotine, the method being reported as the sacrificial advisable in studies involving the nervous system. The tissue samples were collected immediately after sacrifice and the mesenteric bed collected from which the main mesenteric artery was dissected. Four segments were obtained from each tissue.

2.2. Chemicals

The following drugs were used: levo-[ring-2,5,6-3H]-noradrenaline, specific activity 44.8 Ci/mmol (DuPont NEN, I.L.C., Lisboa, Portugal); the scintillation mixture used was from OptiPhase 'Hisafe' 3, PerkinElmer, I.L.C. (Lisboa, Portugal); desipramine hydrochloride purchased from Sigma-Aldrich (Sintra, Portugal). Entellan (mounting medium), Orcein, Masson's trichrome and haematoxylin/eosin from Merck (Darmstadt, Germany). The following antibodies were used: mouse monoclonal anti-tyrosine hydroxylase antibody (ab137869), from Abcam, London, UK) and Alexa Fluor 488 goat anti-mouse IgG (H + L) antibody, highly cross-adsorbed (Invitrogen, Life Technologies, SA, Madrid, Spain); vectashield mounting medium with DAPI (Vector Laboratories, London, UK). Stock solutions were made up in ultrapure water and diluted in superfusion medium immediately before use.

2.3. Haemodynamic Parameters Measurement

Haemodynamic parameters were determined as previously described [49]. Briefly, the rats were anesthetized (37.5 mg/kg Ketamine hydrochloride and 0.25 mg/kg Medetomidine hydrochloride i.p.). Thereafter, a cannula was inserted into the right iliac artery and connected to a Pressure transducer (Statham, Harvard Apparatus GmbH, Berlin, Germany). The blood pressure wave was recorded on a PC computer, using the data acquisition with the PowerLab system (ADInstruments) for 60 min and systolic blood pressure (SBP), diastolic blood pressure (DBP) and heart rate (HR) were measured in the final portion of the recorded trace.

2.4. [^{3}H]-Noradrenaline Release Experiments

Evaluation of [^{3}H]-noradrenaline (NA) release experiments was carried out as previously described [50–52]. Arteries were preincubated in 2 mL of Krebs-Henseleit solution containing 0.1 µmol/L [^{3}H]-NA (for 60 min at 37 °C) and transferred into superfusion chambers, superfused with [^{3}H]-NA-free medium (1 mL/min; constant rate: Krebs-Henseleit solution with desipramine 400 nmol/L to inhibit NA's neuronal uptake). Three identical periods of electrical stimulation were applied (Hugo Sachs Elektronik, March-Hugstetten, Germany; constant current mode, rectangular pulses; 1 ms, with current strength 50 mA; 5 Hz, 100 pulses). The first, starting at t = 30 min (S0) was not used for determination of tritium outflow. The subsequent periods (S_1 and S_2) were applied at t = 90 min and t = 120 min, respectively. The superfusate was collected each 5-min period, starting from minute 85 of superfusion onwards. At the end of the experiments (t = 130 min), tritium was measured in superfusate samples and solubilized arteries (sonicated 1 h with 2.5 mL of 0.2 mol/L perchloric acid) by liquid scintillation spectrometry (LS 6500, Beckman Instruments, Fullerton, CA, USA) after adding 6 mL of a scintillation mixture to each sample.

Electrically evoked tritium overflow from artery segments incubated with [^{3}H]-NA was shown to reflect action potential-evoked neuronal NA release and evoked tritium overflow are assumed to reflect changes in neuronal NA release [43,50,53]. Values of [^{3}H]-NA uptake were estimated as previously described [54]: the tissue tritium content was obtained at the end of each [^{3}H]-NA release experiment and was summed to values of [^{3}H]-NA previously collected in the 5-min superfusate samples (from t = 85 min to t = 130 min, control segments). The final value was considered as the total amount of incorporated [^{3}H]-NA in individual mesenteric artery segments (total tissue tritium content).

2.5. *Immunohistochemistry*

Immunohistochemistry procedures were previously described [53]. Briefly, four artery segments were obtained from each artery and immediately placed in cold phosphate buffer solution (PBS; in g/L): NaCl 8.0, $Na_2HPO_4.2H_2O$ 0.77, KCl 0.20 and KH_2PO_4 0.19 (pH 7.2). Each segment was longitudinally opened and fixed (paraformaldehyde 4% PBS; 50 min; room temperature). After two 15-min PBS washing cycles, artery segments were incubated with the primary antibody (mouse monoclonal anti-tyrosine hydroxylase, TH, 1:100 dilution, overnight, 4 °C to stain noradrenergic nerve terminals). Thereafter, tissues were incubated with Alexa 488 anti-mouse fluorescent secondary antibody (1:1000 dilution, 1 h, room temperature). The primary antibody was previously validated by the manufacturer. Negative controls were incubated on adjacent sections using 10% normal horse serum or blocking solution instead of the primary antibody. After three PBS washing cycles, the arteries were mounted intact with antifading agent (headshield mounting medium with DAPI).

Artery segments were visualized with a Leica SP2 laser scanning confocal microscopy (LSCM) system (Leica Microsystems, Metzler, Germany) fitted with an inverted microscope (×63 oil immersion lens). Stacks of 1-μm-thick serial optical images were captured from five randomly chosen regions along the adventitial layer of the mesenteric artery, which was identified by the shape and orientation of the nuclei stained with DAPI [55]. Adventitia was scanned along each mesenteric artery and the resulting images were reconstructed separately for each wavelength. Two stacks of images were sequentially obtained at different wavelengths: the first stack was taken with the Ex 405 nm and Me 412–470 nm wavelength to visualize cell nuclei (DAPI staining). The second was taken with the Ex 488 nm and Em 490–570 nm wavelength to visualize the TH (location of sympathetic terminals). Image acquisition was always performed under the same laser power, brightness, and contrast conditions. The resulting images were reconstructed separately for each wavelength for later quantification.

Quantitative analysis of confocal z-stacks images was performed using image analysis software (PAQI, CEMUP, Porto, Portugal) as previously described [56]. Briefly, a sequential routine was designed and developed to analyse each fluorescent signal used. PAQI software measured the surface area and strength of the fluorescence signal marking the postganglionic sympathetic nerves.

2.6. *Histology*

Serial 2-μm thickness sections of mesenteric arteries, previously fixed in paraformaldehyde 4% PBS, were dewaxed in xylene, then they were hydrated in decreasing concentrations of alcohols and stained with orcein, haematoxylin/eosin and with Masson Trichrome. Each tissue was cut in five levels along the length of the vessel to ensure data represents the putative mesenteric artery heterogeneity rather than only a specific location of the artery. Each batch represents histochemical staining and includes sections from all five levels of mesenteric artery from each animal group. This procedure was repeated three times (3 batches). In total, 150 sections were obtained. Sections were stained and divided according to animal source. Within each of these groups, a random selection of the sections was carried out.

Stained sections were visualized using a high-resolution Zeiss Axiocam 105 colour digital camera mounted on a Zeiss Primo Star microscope, using an ×10 objective, to analyse the arterial lumen, media and adventitia layer. Histomorphometry was performed with ImageJ software [57], and data of lumen diameter and cross sectional area of the media and the adventitia were obtained.

2.7. Statistics

Statistics were performed with GraphPad Prism (version 8.3) software (San Diego, CA, USA). Sample size was calculated assuming a probability error of alpha type of 5% ($p < 0.05$) and potency of 80%. The normality of the variables was evaluated with Kolmogorov-Smirnov test. Results were expressed as mean ± s.e.m. Differences of means were compared using one- or two-way ANOVA, followed by post-hoc Holm-Sidak's multicomparison t test or Student's t test. A p value lower than 0.05 was considered to denote statistically significant differences.

3. Results

Body weight was not different between male MUN (460.18 ± 8.1 g; $n = 6$) and CONTROL rats (486.6 ± 18.6 g; $n = 6$; $p = 0.21$). Moreover, no differences between body weight of SHR (377.4 ± 8.9; $n = 6$) and WKY (367.0 ± 11.9 g; $n = 6$, $p = 0.52$) were found.

MUN exhibited larger SPB and DBP, but not HR, when compared to CONTROL, whereas SHR evidenced larger SBP, DBP and HR when compared to WKY rats (Table 1). MUN showed lower SBP and DBP levels compared to SHR. However, HR was not enhanced in MUN, contrasting with the HR values observed in SHR.

Table 1. Blood Pressure and Heart Rate from CONTROL/MUN and WKY/SHR rats.

Mesenteric Artery	DBP (mmHg)	SBP (mmHg)	HR (Beat/min)	n
CONTROL	68.7 ± 4.2 †	125.6 ± 5.1	258 ± 8	6
MUN	90.2 ± 4.5 $^{*\#}$	151.1 ± 4.1 $^{*\#}$	253 ± 9 $^{\#}$	6
WKY	98.5 ± 5.5	133.1 ± 4.3	250 ± 10	6
SHR	140.3 ± 5.9 *	186.2 ± 8.3 *	318 ± 8 *	6

Four animal groups in the study: MUN, offspring from rats exposed to maternal undernutrition during pregnancy; CONTROL, offspring from mothers fed ad libitum during pregnancy; SHR, spontaneously hypertensive rats; WKY, Wistar Kyoto rats. SBP, systolic blood pressure; DBP, diastolic blood pressure; HR, heart rate. Values are mean ± s.e.m, from n rats from each group in study. Significant differences from the appropriate animal control: * $p < 0.05$; from the WKY model: † $p < 0.05$; from the SHR model: $^{\#}$ $p < 0.05$.

3.1. Influence of Foetal Undernutrition on Vascular Morphology

Histological sections of the mesenteric artery showed a reduced lumen area in MUN compared to CONTROL (Figures 1a and 2b), a reduction of MUN arterial lumen near 0.6-fold relatively to CONTROL values. Such effect was also observed in arteries from SHR and WKY, a reduction of SHR arterial lumen near 0.7-fold relatively to WKY values.

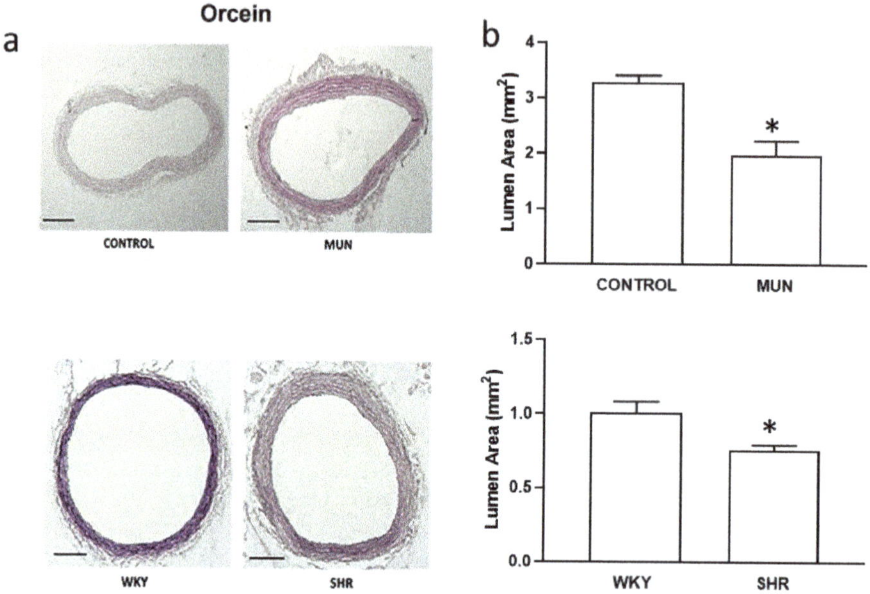

Figure 1. Histomorphometry of the lumen of mesenteric arteries from CONTROL and MUN (upper panel) and WKY and SHR rats (lower panel). (**a**) Images were obtained from orcein stained arteries (scale bar = 500 μm); (**b**) the graphics show the lumen area. MUN, offspring exposed to maternal undernutrition during pregnancy; CONTROL, offspring from mothers fed ad libitum during pregnancy; SHR, spontaneously hypertensive rats; WKY, Wistar Kyoto rats. Values are mean ± s.e.m. from 6 rats of each group. Significant differences from the respective control rat: * $p < 0.05$.

The media and adventitia layers of MUN were thicker compared to CONTROL (with an increase near 1.3-fold and 1.5-fold, respectively) similarly to what occurred in SHR compared to WKY (with an increase near 1.4-fold and 1.5-fold, respectively) (Figure 2a,b). Moreover, ratios of media/lumen and of adventitia/lumen were significantly increased in both MUN and SHR compared to data from respective controls, CONTROL and WKY (Table 2).

Table 2. Vascular wall morphology changes associated with hypertension.

Mesenteric Artery	Ratio Media/Lumen	Ratio Adventitia/Lumen	*n*
CONTROL	0.626 ± 0.016	0.419 ± 0.057	6
MUN	0.989 ± 0.314 *	0.829 ± 0.303 *	6
WKY	0.307 ± 0.092	0.266 ± 0.112	6
SHR	0.519 ± 0.165 *	0.398 ± 0.187 *	6

Images were obtained from haematoxylin/eosin-stained arteries. Four animal groups in the study: MUN, offspring from rats exposed to maternal undernutrition during pregnancy; CONTROL, offspring from mothers fed ad libitum during pregnancy; SHR, spontaneously hypertensive rats; WKY, Wistar Kyoto rats. Values are mean ± s.e.m, from 6 rats from each group in study. Significant differences from the appropriate animal control: * $p < 0.05$.

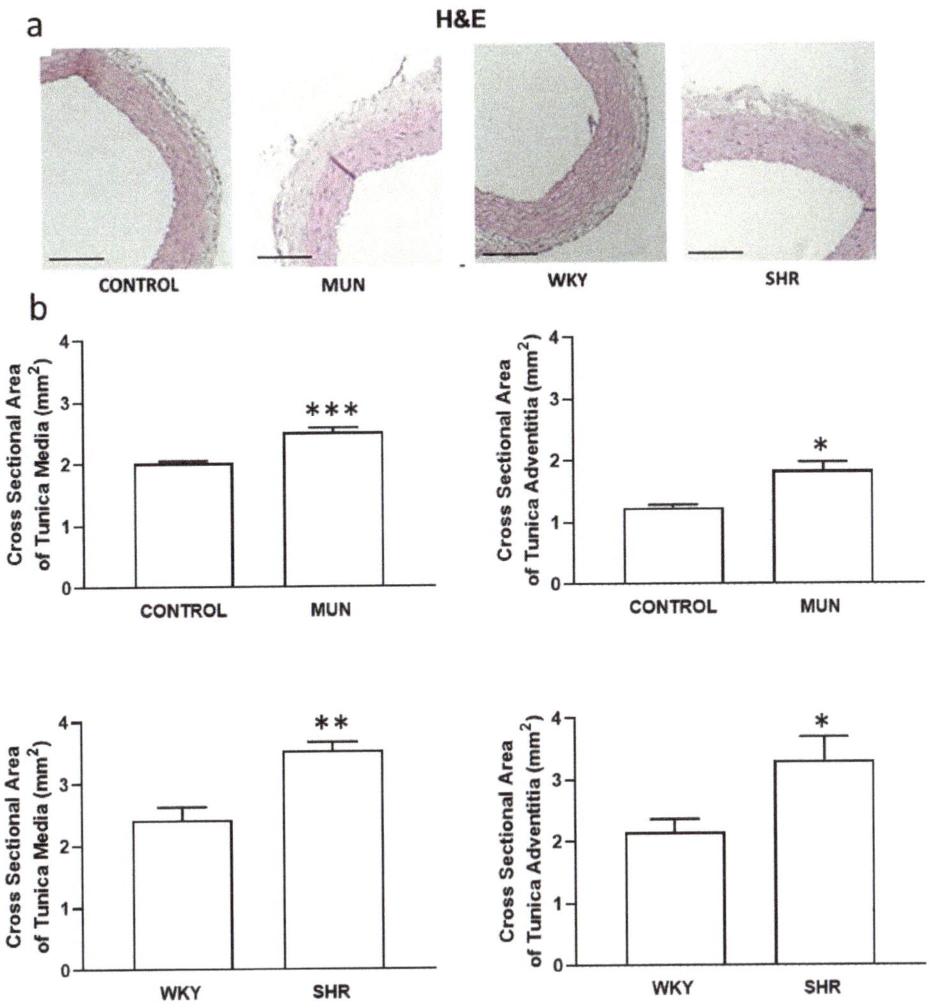

Figure 2. Histomorphometry of hypertensive (MUN and SHR) and normotensive (CONTROL and WKY) mesenteric wall. (**a**) Images were obtained from haematoxylin/eosin-stained arteries (scare bar = 300 µm). (**b**) Cross-sectional area of tunica media (right panel) and cross-sectional area of tunica adventitia (left panel). MUN, offspring exposed to maternal undernutrition during pregnancy; CONTROL, offspring from mothers fed ad libitum during pregnancy; SHR, spontaneously hypertensive rats; WKY, Wistar Kyoto rats. Values are mean ± s.e.m. from 6 rats of each group. Significant differences from the respective control rat: * $p < 0.05$; ** $p < 0.01$ *** $p < 0.001$.

An increase in the connective tissue content was also observed, both in MUN and SHR, compared to their respective controls (CONTROL and WKY, respectively, Figure 3); the connective tissue increased in MUN near 2.1-fold relatively to CONTROL, whereas in SHR, it increased 1.5-fold relatively to WKY).

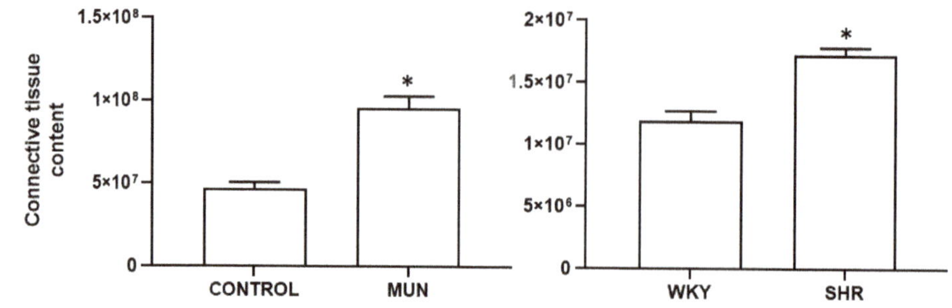

Figure 3. Histomorphometry of hypertensive (MUN and SHR) and normotensive (CONTROL and WKY) mesenteric arteries, stained with Trichrome. MUN, offspring exposed to maternal undernutrition during pregnancy; CONTROL, offspring from mothers fed ad libitum during pregnancy; SHR, spontaneously hypertensive rats; WKY, Wistar Kyoto rats. Values are mean ± s.e.m. from 6 rats of each group. Significant differences from the respective control rat: * $p < 0.05$.

3.2. *Influence of Foetal Undernutrition in Sympathetic Postganglionic Nerves Activation*

Electrical field stimulation (5 Hz, 1 ms, 100 pulses, 50 mA) tritium outflow was higher in mesenteric arteries from MUN compared to CONTROL (Figure 4). Similarly, there was a larger tritium outflow in mesenteric arteries from SHR compared to WKY (Figure 4). The fractional rate of basal tritium outflow (b_1), electrically evoked tritium overflow (S_1) and S_2/S_1 ratios are shown in Table 3. Basal outflow and electrically evoked tritium overflow remained constant throughout the control experiments, with b_n/b_1 and S_n/S_1 values close to unity. Electrically evoked tritium overflow (S_1) was higher in hypertensive arteries (both SHR and MUN) compared to their respective control (WKY or CONTROL) vessels (Table 3). However, the evoked overflow was similar between the hypertensive arteries (MUN *versus* SHR arteries: $p > 0.05$).

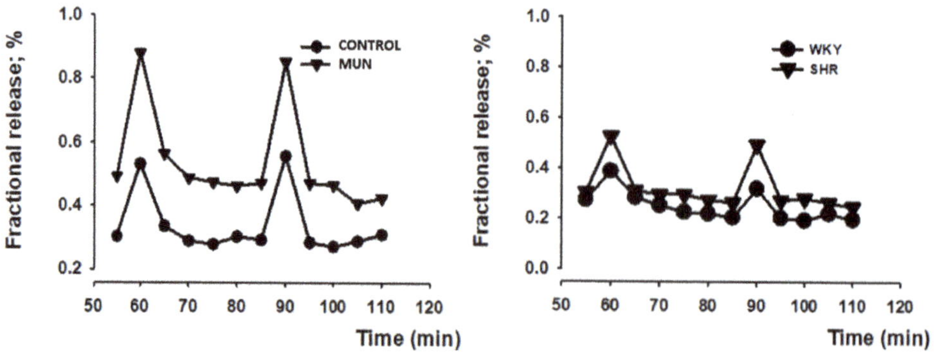

Figure 4. Representative examples of time course tritium outflow from: mesenteric arteries from normotensive animals, CONTROL (circles, left panel) and WKY (circles, right panel), and hypertensive animals, MUN (triangles, left panel) and SHR (triangles, right panel) from typical experiments. After pre-incubation with [^{3}H]-noradrenaline, tissues were superfused with [^{3}H]-noradrenaline-free medium containing desipramine (400 nM). Tritium outflow (ordinates) is expressed as a percentage of the total radioactivity present in the tissue at the beginning of the collection period and was measured in samples collected every 5 min. Artery segments were stimulated twice by 100 pulses/5 Hz, (S_1, S_2). Each line represents the outflow of tritium from a single superfusion chamber; MUN, offspring exposed to maternal undernutrition during pregnancy; CONTROL, offspring from mothers fed ad libitum during pregnancy; SHR, spontaneously hypertensive rats; WKY, Wistar Kyoto rats.

Table 3. Basal tritium outflow (b_1), electrically evoked tritium overflow (S_1) and S_2/S_1 ratios from mesenteric arteries from CONTROL/MUN and WKY/SHR rats.

Mesenteric Artery	Basal Outflow (b_1) (Fractional Rate Outflow; min^{-1})	Evoked Overflow (S_1) (% of Tissue Tritium Content)	n
CONTROL	0.097 ± 0.008	0.220 ± 0.040	6
MUN	0.093 ± 0.006	0.347 ± 0.030 *	6
WKY	0.088 ± 0.006	0.265 ± 0.021	6
SHR	0.073 ± 0.006	0.381 ± 0.032 *	6

Tissues were stimulated twice at 30-min intervals (S_1–S_2; 100 pulses, 5 Hz, 1 ms, 50 mA): b_1 refers to the 5-min period immediately before S_1. The electrically evoked tritium overflow is expressed as a percentage of the tissue tritium content at the onset of stimulation. Basal tritium outflow (b_1), electrically evoked tritium overflow (S_1) from the four animal groups in study. MUN, offspring exposed to maternal undernutrition during pregnancy; CONTROL, offspring from mothers fed ad libitum during pregnancy; SHR, spontaneously hypertensive rats; WKY, Wistar Kyoto rats. Values are mean ± s.e.m, from 6 rats from each group in study. Significant differences from the appropriate animal control: * $p < 0.05$.

3.3. Influence of Foetal Undernutrition on Perivascular Sympathetic Innervation

In our experimental conditions, total tissue tritium content (per mg of tissue) was higher in MUN compared to CONTROL arteries (Figure 5a): total tissue content in MUN increased near 2.1-fold relatively to CONTROL. Moreover, tritium uptake was also larger in SHR mesenteric arteries compared to the values from WKY arteries (Figure 5b): total tissue content in SHR increased 1.7-fold relatively to WKY.

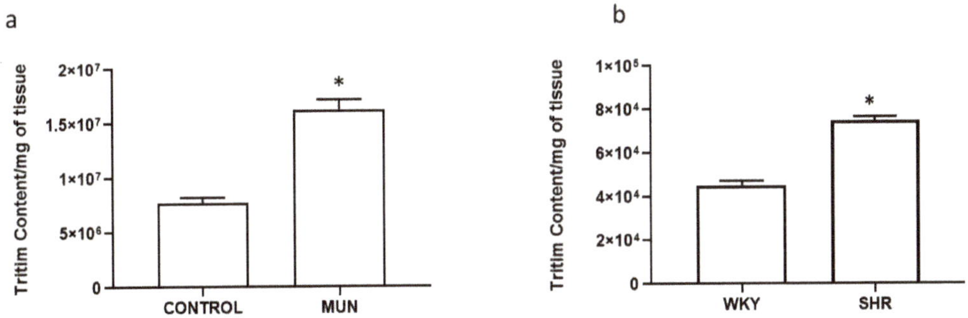

Figure 5. [^{3}H]-Tritium uptake in mesenteric arteries from (**a**) CONTROL and MUN and (**b**) WKY and SHR rats. MUN, offspring exposed to maternal undernutrition during pregnancy; CONTROL, offspring from mothers fed ad libitum during pregnancy; SHR, spontaneously hypertensive rats; WKY, Wistar Kyoto rats. Values are mean ± s.e.m. from 6 animals per group. Significant differences from the respective control rats: * $p < 0.05$.

The influence of foetal undernutrition on perivascular sympathetic innervation in the adventitial layer of mesenteric arteries (identified from LSCM images by the shape and orientation of the nuclei and by exhibiting scattered fibres [56], was also assessed using a sympathetic neuronal marker, thyroxine hydroxylase (TH). Non-significant immunoreactivity was observed when the primary antibody was omitted (negative controls, data not shown). In mesenteric arteries from all experimental groups, immunoreactivity for the sympathetic neuronal marker TH evidenced the presence of sympathetic nervous fibres (Figure 6a, green marker). However, the pattern of TH immunoreactivity in arteries from SHR and from MUN exceeded those observed in WKY and CONTROL arteries, respectively (Figure 6b), revealing a denser and thicker sympathetic innervation: TH immunoreactivity, in MUN, increased near 1.9-fold relatively to CONTROL, whereas, in SHR, it increased 1.5-fold relatively to WKY.

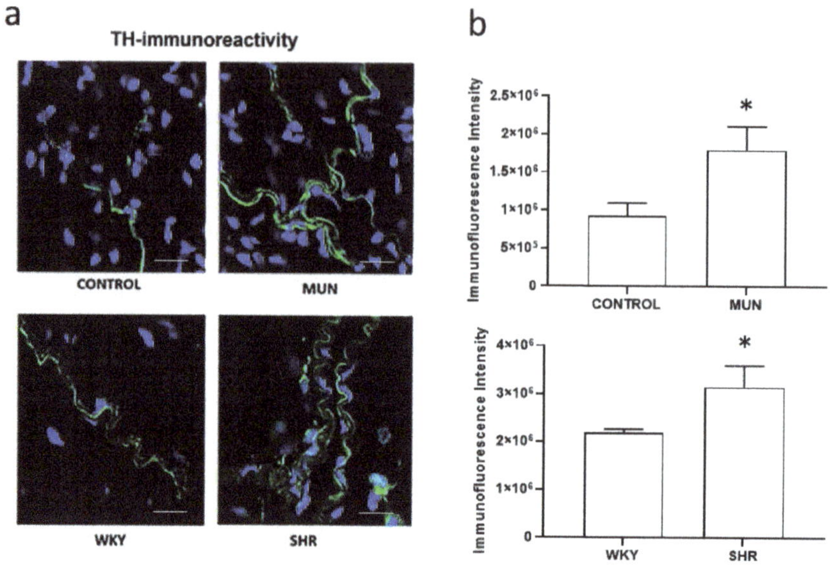

Figure 6. (**a**) Laser scanning confocal microscopy representative images of the adventitia layer of mesenteric arteries from CONTROL and MUN (upper panel) WKY and SHR (lower panel) rats. Images show the immunofluorescence reactivity to TH (green) with DAPI-stained nuclei (blue). Scale bar: 25 μm. (**b**) Quantitative analysis of LSCM images. MUN, offspring exposed to maternal undernutrition during pregnancy; CONTROL, offspring from mothers fed ad libitum during pregnancy; SHR, spontaneously hypertensive rats; WKY, Wistar Kyoto rats. In the graphics, values are mean ± s.e.m. from 6 rats from each group. Significant differences from the respective control rats: * $p < 0.05$.

4. Discussion

The current study shows that foetal undernutrition induces an increase in the arterial sympathetic neurotransmission in rat adulthood (6-months old) which can justify, at least in part, the arterial inward hypertrophic remodelling observed and further highlighted the role of adventitia in the pathophysiology of foetal programming of hypertension. In addition, this study also shows that MUN rats share with SHR, a rat model of essential hypertension, these pathological aspects, which may contribute to the development of hypertension.

Foetal undernutrition was associated with the development of sympathetic hyperinnervation observed in the adventitial layer of the mesenteric artery. LSCM data revealed the presence of nerve fibres positive for TH (a sympathetic nervous fibre marker), spread through the adventitia reaching the medial layer, being thicker in MUN mesenteric arteries. Tyrosine hydroxylase (TH) is an enzyme localized inside sympathetic neurotransmitter storage vesicles [58] that is used as a peripheral sympathetic marker [43,58]. The higher TH immunoreactivity observed (induced by undernutrition during foetal life) indicates the occurrence of a sympathetic hyperinnervation in MUN mesenteric arteries and is in accordance with a previous report in FPH induced by prenatal hypoxia, in which a sympathetic hyperinnervation in tibial arteries was described [59]. A denser innervation and higher NA content was also observed in SHR mesenteric arteries, in accordance with previous studies [54,60]. The larger innervation was associated with higher noradrenaline (NA) content and release, assessed by functional studies using [^{3}H]-NA, which can be taken up into vesicles in sympathetic nerve terminals by a specific NA transporter [61,62]. The total tissue tritium content can be considered an indicator of the sympathetic innervation density [54]. The larger release of tritium upon electrical stimulation can be a consequence of the hyperinnervation observed. However, it may also reflect changes in the presynaptic machinery and regulation of noradrenaline release. We have previous evidence that these mechanisms are altered in both SHR [43,53,54] and MUN rats [63]. Our data also showed that the total tissue content of the MUN mesenteric artery was higher compared to CONTROL, indicating that the MUN nervous terminals can incorporate more NA. Our functional data showed a higher sympathetic output in both SHR and MUN rats, compared to their respective controls. While the difference in innervation was similar for both hypertensive models, the tritium outflow and NA content was relatively higher in MUN rats, suggesting an increased availability of NA in the synaptic cleft of sympathetic MUN nerves. Similar data was found in SHR mesenteric [54,64] and tail [65] arteries. Our data agrees with studies in other models of FPH showing that SNS activation seems to be increased [36,58,66–69]: it was described an increased circulating levels of NA in animal models of FPH [70–72] and in LBW humans [73].

An increased sympathetic neurotransmission can contribute to increase vascular tone and to the observed remodelling, which, in turn, can be one of the mechanisms behind hypertension development in MUN. When compared to SHR, the increased NA release observed was analogous to MUN, although SHR presented a smaller basal outflow. This data is in line with previous studies in SHR, in which a sympathetic hyperactivity was also described in the cerebral artery [74], tail artery [49] and mesenteric bed [52]. Morphometric analysis indicated that foetal undernutrition induces inward hypertrophic remodelling in the mesenteric artery, with the contribution of both media and adventitia layers, and the presence of vascular fibrosis, in agreement with data from other studies, in SHR [75] and MUN [76,77]. The MUN mesenteric artery exhibits a reduced lumen area as well as an increased media and adventitia layers compared to CONTROL, indicative of inward hypertrophic vascular remodelling. A similar profile of vascular wall changes between SHR and WKY was also observed. This type of remodelling is a characteristic of hypertension and can contribute to increase total peripheral vascular resistance [78]. In mesenteric arteries from rats exposed to undernutrition in utero, the same type of remodelling was also reported [79].

The hypertrophy observed in the media layer of MUN mesenteric artery can be due to cellular hypertrophy caused by vascular smooth muscle cell (VSMC) proliferation or growth. Sympathetic hyperinnervation was shown to be related to media hypertrophy in jejunal arteries of SHR, since an increase in the number of nerve fibres occurred before the development of hypertension or an increase in the thickness of the arterial media [79]. Such finding suggests that increased sympathetic activity possibly plays a causal role in the development of hypertension, through vascular remodelling. Our study proves that the remodelling process is relatively similar in MUN and SHR (inward hypertrophic).

MUN had higher SBP and DBP than CONTROL, confirming previous results with the same foetal model [11,45,80–82]. SHR also exhibited higher values of SBP compared to WKY, as shown in many studies [54,83–85]. It was not possible to compare the blood pressure directly between SHR and MUN male rats, since they have a control strain with different background. However, the current study indicates that foetal undernutrition induces a milder form of hypertension when compared to a genetic model of essential hypertension, with lower levels of blood pressure (around 50 mmHg difference for SHR-WKY and 25 mm Hg for MUN-SD control). Additionally, MUN rats did not show an elevated HR, which was detected in the SHR model. The lack of HR alterations of MUN rats is in line with other studies using maternal low protein [35,86] or low micronutrients, such as zinc [87] or vitamin B12 [88] diets during pregnancy. In SHR, the tachycardia probably reflects an alteration in baroreceptor regulation, as previously reported [86], which may contribute to the higher blood pressure in this strain. It should be considered that the blood pressure data was obtained under ketamine/medetomidine anaesthesia, which exerts an influence on the sympathetic nervous system: ketamine has been shown to reduce both exocytosis and NA uptake [89]; medetomidine has been reported to reduce noradrenaline outflow within the central nervous system, dampening the central sympathetic tone [90]. Given our findings of an increased sympathetic activity and innervation in both SHR and MUN rat, it is likely that the blood pressure in both models of hypertension was underestimated. Accordingly, previous studies in adult SHR assessed by tail cuff showed high blood pressure levels over 220 mmHg of SBP in awake animals [91], relatively to those reported in the present study under anaesthesia (SBP = 186 mmHg). In MUN rats, recent reports in awake 6-month-old rats showed SBP values of 163.1 mmHg [92], which are higher than those reported in the present study under anaesthesia (SBP = 151.1 mmHg).

In SHR rats, sex differences have been associated, among other causes, with increased sympathetic outflow due to dysfunctional regulation of presynaptic α-adrenoceptors in males [93]. Regarding the MUN model, we have reported that females counteract better the effects of foetal stress and do not develop hypertension in adult life [46]. It is important to know that these studies were performed under medetomidine/ketamine anaesthesia. However, recent reports have revealed that, in non-anesthetized rats, there is a tendency toward hypertension in 6-month-old MUN female rats, which is established and evident in old age [91]. Such data suggested that MUN females may also have increased blood pressure levels, and this could also explain the similarities in vascular remodelling between males and females [81]. Thus, the analysis of sex differences in MUN models deserves further attention.

5. Conclusions

In conclusion, and as depicted in Figure 7, elevated sympathetic neurotransmission in MUN and SHR supports that SNS plays an important role in the development of hypertension in both foetal programming and essential hypertension, which is in line with the neurogenic hypothesis of hypertension [94]. The similarities regarding sympathetic neurotransmission and remodelling in both models suggest that the increased sympathetic activity observed in FPH plays a causal role in the development of hypertension through vascular remodelling.

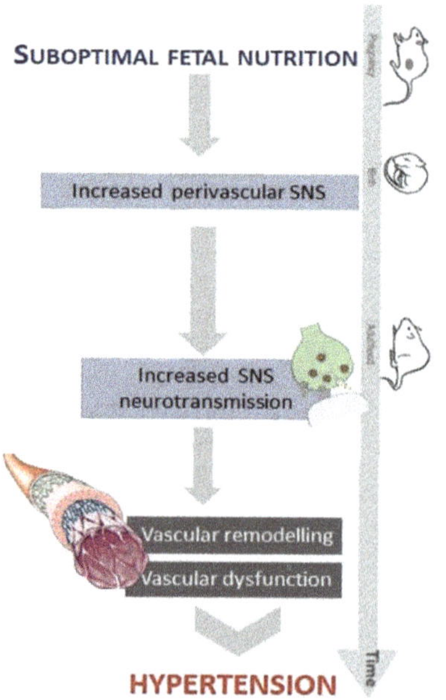

Figure 7. Effect of suboptimal foetal undernutrition on SNS and vascular wall from a rat FPH model.

Author Contributions: Conceptualization, S.M.A. and C.D.; methodology, P.R.-R., J.B.S. and M.S.V.-R.; software, M.S.V.-R., S.M.A., J.B.S. and C.D.; validation M.S.V.-R., S.M.A. and C.D.; formal analysis, M.S.V.-R., S.M.A., J.B.S. and C.D.; investigation, M.S.V.-R., S.M.A. and C.D.; resources, M.S.V.-R., S.M.A. and C.D.; data curation, M.S.V.-R., S.M.A. and C.D.; writing—original draft preparation, M.S.V.-R., S.M.A. and C.D.; writing—review and editing, all authors; visualization, M.S.V.-R.; supervision, S.M.A. and C.D.; project administration, S.M.A. and C.D.; funding acquisition, S.M.A. and C.D. All authors have read and agreed to the published version of the manuscript.

Funding: Portuguese Foundation for Science and Technology (FCT) is acknowledged for UIDB/QUI/50006/2020 and Spanish Ministry of Science and Innovation, COCARDIOLAC project (RTI 2018-097504-B-I00).

Institutional Review Board Statement: The study was conducted according to the guidelines of the Declaration of Helsinki and approved by the Ethics Review Board of Universidad Autónoma de Madrid (CEI-UAM 96-1776-A286), the Regional Environment Committee of the Comunidad Autónoma de Madrid (PROEX 04/19; date of approval 5 July 2019).

Informed Consent Statement: Not applicable.

Data Availability Statement: The data presented in this study are available on request from the author (M.S.V.-R.).

Acknowledgments: We would like to thank the personnel of the Animal Facilities at Universidad Autónoma de Madrid for the care of the animals.

Conflicts of Interest: The authors declare no conflict of interest.

References

1. van der Linden, E.L.; Collard, D.; Beune, E.J.A.J.; Nieuwkerk, P.T.; Galenkamp, H.; Haafkens, J.A.; van Charante, E.P.M.; van den Born, B.-J.H.; Agyemang, C. Determinants of suboptimal blood pressure control in a multi-ethnic population: The Healthy Life in an Urban Setting (HELIUS) study. *J. Clin. Hypertens.* **2021**, *23*, 1068–1076. [CrossRef] [PubMed]
2. Thomas, P.; Dasgupta, I. The role of the kidney and the sympathetic nervous system in hypertension. *Pediatr. Nephrol.* **2015**, *30*, 549–560. [CrossRef] [PubMed]
3. Oparil, S.; Zaman, M.A.; Calhoun, D.A. Pathogenesis of hypertension. *Ann. Inter. Med.* **2003**, *139*, 761–776. [CrossRef] [PubMed]
4. Grassi, G.; Bertoli, S.; Seravalle, G. Sympathetic nervous system: Role in hypertension and in chronic kidney disease. *Curr. Opin. Nephrol. Hypertens.* **2012**, *21*, 46–51. [CrossRef] [PubMed]
5. Grassi, G.; Ram, V.S. Evidence for a critical role of the sympathetic nervous system in hypertension. *J. Am. Soc. Hypertens.* **2016**, *10*, 457–466. [CrossRef]
6. Manolis, A.J.; Poulimenos, L.E.; Kallistratos, M.S.; Gavras, I.; Gavras, H. Sympathetic overactivity in hypertension and cardiovascular disease. *Curr. Vasc. Pharmacol.* **2014**, *12*, 4–15.
7. Alexander, B.T.; Dasinger, J.H.; Intapad, S. Fetal programming and cardiovascular pathology. *Compr. Physiol.* **2015**, *5*, 997–1025.
8. Barker, D.J.; Osmond, C. Low birth weight and hypertension. *BMJ* **1988**, *297*, 134–135. [CrossRef]
9. Barker, D.J.; Clark, P.M. Fetal undernutrition and disease in later life. *Rev. Reprod.* **1997**, *2*, 105–112. [CrossRef]
10. Alexander, B.T. Fetal programming of hypertension. *Am. J. Physiol. Regul. Integr. Comp. Physiol.* **2006**, *290*, R1–R10. [CrossRef]
11. Edwards, L.J.; McMillen, I.C. Maternal undernutrition increases arterial blood pressure in the sheep fetus during late gestation. *J. Physiol.* **2001**, *533*, 561–570. [CrossRef] [PubMed]
12. Tintu, A.N.; Noble, F.A.; Rouwet, E.V. Hypoxia disturbs fetal hemodynamics and growth. *Endothelium* **2007**, *14*, 353–360. [CrossRef] [PubMed]
13. Goyal, R.; Leitzke, A.; Goyal, D.; Gheorghe, C.P.; Longo, L.D. Antenatal maternal hypoxic stress: Adaptations in fetal lung Renin-Angiotensin system. *Reprod. Sci.* **2011**, *18*, 180–189. [CrossRef] [PubMed]
14. Svitok, P.; Molcan, L.; Stebelova, K.; Vesela, A.; Sedlackova, N.; Ujhazy, E.; Mach, M.; Zeman, M. Prenatal hypoxia in rats increased blood pressure and sympathetic drive of the adult offspring. *Hypertens. Res.* **2016**, *39*, 501–505. [CrossRef] [PubMed]
15. Alexander, B.T. Placental insufficiency leads to development of hypertension in growth-restricted offspring. *Hypertension* **2003**, *41*, 457–462. [CrossRef]
16. Dodic, M.; May, C.N.; Wintour, E.M.; Coghlan, J.P. An early prenatal exposure to excess glucocorticoid leads to hypertensive offspring in sheep. *Clin. Sci.* **1998**, *94*, 149–155. [CrossRef]
17. Gwathmey, T.M.; Shaltout, A.H.; Rose, J.C.; Diz, D.I.; Chappell, M.C. Glucocorticoid-induced fetal programming alters the functional complement of angiotensin receptor subtypes within the kidney. *Hypertension* **2011**, *57*, 620–626. [CrossRef]
18. Nguyen, P.; Khurana, S.; Peltsch, H.; Grandbois, J.; Eibl, J.; Crispo, J.; Ansell, D.; Tai, T.C. Prenatal glucocorticoid exposure programs adrenal PNMT expression and adult hypertension. *J. Endocrinol.* **2015**, *227*, 117–127. [CrossRef]

19. Nykjaer, C.; Alwan, N.A.; Greenwood, D.C.; Simpson, N.A.; Hay, A.W.; White, K.L.; Cade, J.E. Maternal alcohol intake prior to and during pregnancy and risk of adverse birth outcomes: Evidence from a British cohort. *J. Epidemiol. Community Health* **2014**, *68*, 542–549. [CrossRef]
20. Nordentoft, M.; Lou, H.C.; Hansen, D.; Nim, J.; Pryds, O.; Rubin, P.; Hemmingsen, R. Intrauterine growth retardation and premature delivery: The influence of maternal smoking and psychosocial factors. *Am. J. Public Health* **1996**, *86*, 347–354. [CrossRef]
21. Banderali, G.; Martelli, A.; Landi, M.; Moretti, F.; Betti, F.; Radaelli, G.; Lassandro, C.; Verduci, E. Short and long term health effects of parental tobacco smoking during pregnancy and lactation: A descriptive review. *J. Transl. Med.* **2015**, *13*, 327. [CrossRef] [PubMed]
22. Sprauve, M.E.; Lindsay, M.K.; Drews-Botsch, C.D.; Graves, W. Racial patterns in the effects of tobacco use on fetal growth. *Am. J. Obstet. Gynecol.* **1999**, *181*, S22–S27. [CrossRef]
23. Poston, L.; Igosheva, N.; Mistry, H.D.; Seed, P.T.; Shennan, A.H.; Rana, S.; Karumanchi, S.A.; Chappell, L.C. Role of oxidative stress and antioxidant supplementation in pregnancy disorders. *Am. J. Clin. Nutr.* **2011**, *94*, 1980S–1985S. [CrossRef] [PubMed]
24. Rodriguez-Rodriguez, P.; Ramiro-Cortijo, D.; Reyes-Hernández, C.G.; López de Pablo, A.L.; González, M.C.; Arribas, S.M. Implication of oxidative stress in fetal programming of cardiovascular disease. *Front. Physiol.* **2018**, *9*, 602. [CrossRef]
25. Nuyt, A.M. Mechanisms underlying developmental programming of elevated blood pressure and vascular dysfunction: Evidence from human studies and experimental animal models. *Clin. Sci.* **2008**, *114*, 1–17. [CrossRef]
26. Lesage, J.; Blondeau, B.; Grino, M.; Bréant, B.; Dupouy, J.P. Maternal undernutrition during late gestation induces fetal overexposure to glucocorticoids and intrauterine growth retardation, and disturbs the hypothalamo-pituitary adrenal axis in the newborn rat. *Endocrinology* **2001**, *142*, 1692–1702. [CrossRef]
27. Chisari, A.N.; Giovambattista, A.; Perelló, M.; Gaillard, R.C.; Spinedi, E.S. Maternal undernutrition induces neuroendocrine immune dysfunction in male pups at weaning. *Neuroimmunomodulation* **2001**, *9*, 41–48. [CrossRef]
28. Vehaskari, V.M.; Stewart, T.; Lafont, D.; Soyez, C.; Seth, D.; Manning, J. Kidney angiotensin and angiotensin receptor expression in prenatally programmed hypertension. *Am. J. Physiol. Ren. Physiol.* **2004**, *287*, F262–F267. [CrossRef]
29. Manning, J.; Vehaskari, V.M. Postnatal modulation of prenatally programmed hypertension by dietary Na and ACE inhibition. *Am. J. Physiol. Regul. Integr. Comp. Physiol.* **2005**, *288*, R80–R84. [CrossRef]
30. Mesquita, F.F.; Gontijo, J.A.; Boer, P.A. Expression of renin-angiotensin system signalling compounds in maternal protein-restricted rats: Effect on renal sodium excretion and blood pressure. *Nephrol. Dial. Transplant.* **2010**, *25*, 380–388. [CrossRef]
31. Bogdarina, I.; Welham, S.; King, P.J.; Burns, S.P.; Clark, A.J. Epigenetic modification of the renin-angiotensin system in the fetal programming of hypertension. *Circ. Res.* **2007**, *100*, 520–526. [CrossRef] [PubMed]
32. Wu, L.; Shi, A.; Zhu, D.; Bo, L.; Zhong, Y.; Wang, J.; Xu, Z.; Mao, C. High sucrose intake during gestation increases angiotensin II type 1 receptor-mediated vascular contractility associated with epigenetic alterations in aged offspring rats. *Peptides* **2016**, *86*, 133–144. [CrossRef]
33. Goyal, R.; Leitzke, A.; Gheorghe, C.P.; Longo, L.D. Brain renin-angiotensin system: Fetal epigenetic programming by maternal protein restriction during pregnancy. *Reprod. Sci.* **2010**, *17*, 227–238. [CrossRef]
34. De Mello, W.C. Intracellular angiotensin II as a regulator of muscle tone in vascular resistance vessels. Pathophysiological implications. *Peptides* **2016**, *78*, 87–90. [CrossRef] [PubMed]
35. Barros, M.A.; De Brito Alves, J.L.; Nogueira, V.O.; Wanderley, A.G.; Costa-Silva, J.H. Maternal low-protein diet induces changes in the cardiovascular autonomic modulation in male rat offspring. *Nutr. Metab. Cardiovasc. Dis.* **2015**, *25*, 123–130. [CrossRef] [PubMed]
36. Mizuno, M.; Siddique, K.; Baum, M.; Smith, S.A. Prenatal programming of hypertension induces sympathetic overactivity in response to physical stress. *Hypertension* **2013**, *61*, 180–186. [CrossRef] [PubMed]
37. Vehaskari, V.M.; Aviles, D.H.; Manning, J. Prenatal programming of adult hypertension in the rat. *Kidney Int.* **2001**, *59*, 238–245. [CrossRef]
38. Silvagni, A.; Barros, V.G.; Mura, C.; Antonelli, M.C.; Carboni, E. Prenatal restraint stress differentially modifies basal and stimulated dopamine and noradrenaline release in the nucleus accumbens shell: An 'in vivo' microdialysis study in adolescent and young adult rats. *Eur. J. Neurosci.* **2008**, *28*, 744–758. [CrossRef]
39. Vieira-Rocha, M.S.; Sousa, J.B.; Rodriguez-Rodriguez, P.; Morato, M.; Arribas, S.M.; Diniz, C. Insights into sympathetic nervous system and GPCR interplay in fetal programming of hypertension: A bridge for new pharmacological strategies. *Drug Discov. Today* **2020**, *25*, 739–747. [CrossRef]
40. Okamoto, K.; Aoki, K. Development of a strain of spontaneously hypertensive rats. *Jpn. Circ. J.* **1963**, *27*, 282–293. [CrossRef]
41. Trippodo, N.C.; Frohlich, E.D. Similarities of genetic (spontaneous) hypertension. Man and rat. *Circ. Res.* **1981**, *48*, 309–319. [CrossRef] [PubMed]
42. Sasamura, H.; Hayashi, K.; Ishiguro, K.; Nakaya, H.; Saruta, T.; Itoh, H. Prevention and regression of hypertension: Role of renal microvascular protection. *Hypertens. Res.* **2009**, *32*, 658–664. [CrossRef] [PubMed]
43. Sousa, J.B.; Vieira-Rocha, M.S.; Sá, C.; Ferreirinha, F.; Correia-de-Sá, P.; Fresco, P.; Diniz, C. Lack of endogenous adenosine tonus on sympathetic neurotransmission in spontaneously hypertensive rat mesenteric artery. *PLoS ONE* **2014**, *9*, e105540. [CrossRef] [PubMed]

44. Arribas, S.M.; Briones, A.M.; Bellingham, C.; González, M.C.; Salaices, M.; Liu, K.; Wang, Y.; Hinek, A. Heightened aberrant deposition of hard-wearing elastin in conduit arteries of prehypertensive SHR is associated with increased stiffness and inward remodeling. *Am. J. Physiol. Heart Circ. Physiol.* **2008**, *295*, H2299–H2307. [CrossRef]
45. González, J.M.; Briones, A.M.; Somoza, B.; Daly, C.J.; Vila, E.; Starcher, B.; McGrath, J.C.; González, M.C.; Arribas, S.M. Postnatal alterations in elastic fiber organization precede resistance artery narrowing in SHR. *Am. J. Physiol. Heart Circ. Physiol.* **2006**, *291*, H804–H812. [CrossRef]
46. Rodríguez-Rodríguez, P.; de Pablo, A.L.; Condezo-Hoyos, L.; Martín-Cabrejas, M.A.; Aguilera, Y.; Ruiz-Hurtado, G.; Gutierrez-Arzapalo, P.Y.; Ramiro-Cortijo, D.; Fernández-Alfonso, M.S.; González, M.C.; et al. Fetal undernutrition is associated with perinatal sex-dependent alterations in oxidative status. *J. Nutr. Biochem.* **2015**, *26*, 1650–1659. [CrossRef]
47. Percie du Sert, N.; Hurst, V.; Ahluwalia, A.; Alam, S.; Avey, M.T.; Baker, M.; Browne, W.J.; Clark, A.; Cuthill, I.C.; Howells, D.W.; et al. The ARRIVE guidelines 2.0: Updated guidelines for reporting animal research. *PLoS Biol.* **2020**, *18*, e3000410.
48. Muñoz-Valverde, D.; Rodríguez, P.R.; Gutierrez-Arzapalo, P.Y.; De Pablo, A.L.L.; González, M.C.; López-Giménez, R.; Somoza, B.; Arribas, S.M. Effect of Fetal Undernutrition and Postnatal Overfeeding on Rat Adipose Tissue and Organ Growth at Early Stages of Postnatal Development. *Physiol. Res.* **2015**, *64*, 547–559. [CrossRef]
49. Rodríguez-Rodríguez, P.; De Pablo, L.L.; García-Prieto, C.F.; Somoza, B.; Quintana-Villamandos, B.; De Diego, J.J.G.; Gutierrez-Arzapalo, P.Y.; Ramiro-Cortijo, D.; Gonzalez-Garcia, M.C.; Arribas, S.M. Long term effects of fetal undernutrition on rat heart. Role of hypertension and oxidative stress. *PLoS ONE* **2017**, *12*, e0171544. [CrossRef]
50. Sousa, J.B.; Vieira-Rocha, M.S.; Arribas, S.M.; González, M.C.; Fresco, P.; Diniz, C. Endothelial and Neuronal Nitric Oxide Activate Distinct Pathways on Sympathetic Neurotransmission in Rat Tail and Mesenteric Arteries. *PLoS ONE* **2015**, *10*, e0129224. [CrossRef]
51. Diniz, C.; Fresco, P.; Leal, S.; Gonçalves, J. Adenosine receptors involved in modulation of noradrenaline release in isolated rat tail artery. *Eur. J. Pharmacol.* **2004**, *504*, 17–25. [CrossRef] [PubMed]
52. Fresco, P.; Diniz, C.; Goncalves, J. Facilitation of noradrenaline release by activation of adenosine A(2A) receptors triggers both phospholipase C and adenylate cyclase pathways in rat tail artery. *Cardiovasc. Res.* **2004**, *63*, 739–746. [CrossRef] [PubMed]
53. Rocha-Pereira, C.; Sousa, J.B.; Vieira-Rocha, M.S.; Fresco, P.; Gonçalves, J.; Diniz, C. Differential inhibition of noradrenaline release mediated by inhibitory A(1)-adenosine receptors in the mesenteric vein and artery from normotensive and hypertensive rats. *Neurochem. Int.* **2013**, *62*, 399–405. [CrossRef] [PubMed]
54. Rocha-Pereira, C.; Arribas, S.M.; Fresco, P.; González, M.C.; Gonçalves, J.; Diniz, C. Impaired inhibitory function of presynaptic A1-adenosine receptors in SHR mesenteric arteries. *J. Pharmacol. Sci.* **2013**, *122*, 59–70. [CrossRef] [PubMed]
55. Arribas, S.M.; Hillier, C.; González, C.; McGrory, S.; Dominiczak, A.; McGrath, J. Cellular Aspects of Vascular Remodeling in Hypertension Revealed by Confocal Microscopy. *Hypertension* **1997**, *30*, 1455–1464. [CrossRef]
56. Sousa, J.B.; Fresco, P.; Diniz, C. Imaging receptors with Laser Scanning Confocal Microscopy: Qualitative and quantitative analysis. In *Microscopy: Advances in Scientific Research and Education*; Méndez-Villas, A., Ed.; Formatex Research Center: Badajoz, Spain, 2014; pp. 201–208. ISBN 978-84-942134-3-4.
57. Schindelin, J.; Arganda-Carreras, I.; Frise, E.; Kaynig, V.; Longair, M.; Pietzsch, T.; Preibish, S.; Rueden, C.; Saalfeld, S.; Schmid, B.; et al. Fiji: An open-source platform for biological-image analysis. *Nat. Methods* **2012**, *9*, 676–682. [CrossRef]
58. De Fontgalland, D.; Wattchow, D.A.; Costa, M.; Brookes, S.J. Immunohistochemical characterization of the innervation of human colonic mesenteric and submucosal blood vessels. *Neurogastroenterol. Motil.* **2008**, *20*, 1212–1226. [CrossRef]
59. Rook, W.; Johnson, C.D.; Coney, A.M.; Marshall, J.M. Prenatal hypoxia leads to increased muscle sympathetic nerve activity, sympathetic hyperinnervation, premature blunting of neuropeptide Y signaling, and hypertension in adult life. *Hypertension* **2014**, *64*, 1321–1327. [CrossRef]
60. Luff, S.E.; Young, S.B.; McLachlan, E.M. Hyperinnervation of mesenteric arteries in spontaneously hypertensive rats by sympathetic but not primary afferent axons. *J. Vasc. Res.* **2005**, *42*, 348–358. [CrossRef]
61. Blakely, R.D.; Bauman, A.L. Biogenic amine transporters: Regulation in flux. *Curr. Opin. Neurobiol.* **2000**, *10*, 328–336. [CrossRef]
62. Zhou, J. Norepinephrine transporter inhibitors and their therapeutic potential. *Drugs Future* **2004**, *29*, 1235–1244. [CrossRef] [PubMed]
63. Vieira-Rocha, M.S.; Rodríguez-Rodríguez, P.; Sousa, J.B.; González, M.C.; Arribas, S.M.; López de Pablo, A.L.; Diniz, C. Vascular angiotensin AT1 receptor neuromodulation in fetal programming of hypertension. *Vascul. Pharmacol.* **2019**, *117*, 27–34. [CrossRef] [PubMed]
64. Whall, C.W., Jr.; Myers, M.M.; Halpern, W. Norepinephrine sensitivity, tension development and neuronal uptake in resistance arteries from spontaneously hypertensive and normotensive rats. *Blood Vessels* **1980**, *17*, 1–15. [PubMed]
65. Zsoter, T.T.; Wolchinsky, C. Norepinephrine uptake in arteries of spontaneously hypertensive rats. *Clin. Investig. Med.* **1983**, *6*, 191–195.
66. IJzerman, R.G.; Stehouwer, C.D.; de Geus, E.J.; van Weissenbruch, M.M.; Delemarre-van de Waal, H.A.; Boomsma, D.I. Low birth weight is associated with increased sympathetic activity: Dependence on genetic factors. *Circulation* **2003**, *108*, 566–571. [CrossRef]
67. Mizuno, M.; Lozano, G.; Siddique, K.; Baum, M.; Smith, S.A. Enalapril attenuates the exaggerated sympathetic response to physical stress in prenatally programmed hypertensive rats. *Hypertension* **2014**, *63*, 324–329. [CrossRef]

68. Marshall, J.M. Interactions between local dilator and sympathetic vasoconstrictor influences in skeletal muscle in acute and chronic hypoxia. *Exp. Physiol.* **2015**, *100*, 1400–1411. [CrossRef]
69. de Brito Alves, J.L.; de Oliveira, J.M.; Ferreira, D.J.; Barros, M.A.; Nogueira, V.O.; Alves, D.S.; Vidal, H.; Leandro, C.G.; Lagranha, C.J.; Pirola, L.; et al. Maternal protein restriction induced-hypertension is associated to oxidative disruption at transcriptional and functional levels in the medulla oblongata. *Clin. Exp. Pharmacol. Physiol.* **2016**, *43*, 1177–1184. [CrossRef]
70. Hiraoka, T.; Kudo, T.; Kishimoto, Y. Catecholamines in experimentally growth-retarded rat fetus. *Asia Ocean. J. Obstet. Gynaecol.* **1991**, *17*, 341–348. [CrossRef]
71. Petry, C.J.; Dorling, M.W.; Wang, C.L.; Pawlak, D.B.; Ozanne, S.E. Catecholamine levels and receptor expression in low protein rat offspring. *Diabet. Med.* **2000**, *17*, 848–853. [CrossRef]
72. Jones, C.T.; Robinson, J.S. Studies on experimental growth retardation in sheep. Plasma catecholamines in fetuses with small placenta. *J. Dev. Physiol.* **1983**, *5*, 77–87. [PubMed]
73. Johansson, S.; Norman, M.; Legnevall, L.; Dalmaz, Y.; Lagercrantz, H.; Vanpée, M. Increased catecholamines and heart rate in children with low birth weight: Perinatal contributions to sympathoadrenal overactivity. *J. Intern. Med.* **2007**, *261*, 480–487. [CrossRef] [PubMed]
74. Butcher, K.S.; Hachinski, V.C.; Wilson, J.X.; Guiraudon, C.; Cechetto, D.F. Cardiac and sympathetic effects of middle cerebral artery occlusion in the spontaneously hypertensive rat. *Brain Res.* **1993**, *621*, 79–86. [CrossRef]
75. Susic, D.; Varagic, J.; Ahn, J.; Matavelli, L.C.; Frohlich, E.D. Beneficial cardiovascular actions of eplerenone in the spontaneously hypertensive rat. *J. Cardiovasc. Pharmacol. Ther.* **2005**, *10*, 197–220. [CrossRef] [PubMed]
76. Darby, J.R.T.; McMillen, I.C.; Morrison, J.L. Maternal undernutrition in late gestation increases IGF2 signalling molecules and collagen deposition in the right ventricle of the fetal sheep heart. *J. Physiol.* **2018**, *596*, 2345–2358. [CrossRef] [PubMed]
77. Vieira-Rocha, M.S.; Rodriguez-Rodriguez, P.; Ferreira-Duarte, M.; Faria, M.; Sousa, J.B.; Morato, M.; Arribas, S.M.; Diniz, C. Fetal Undernutrition Modifies Vascular RAS Balance Enhancing Oxidative Damage and Contributing to Remodeling. *Int. J. Mol. Sci.* **2022**, *23*, 1233. [CrossRef]
78. Renna, N.F.; de Las Heras, N.; Miatello, R.M. Pathophysiology of vascular remodeling in hypertension. *Int. J. Hypertens.* **2013**, *2013*, 808353. [CrossRef]
79. Khorram, O.; Momeni, M.; Ferrini, M.; Desai, M.; Ross, M.G. In utero undernutrition in rats induces increased vascular smooth muscle content in the offspring. *Am. J. Obstet. Gynecol.* **2007**, *196*, 486.e1–486.e8. [CrossRef]
80. Khorram, O.; Khorram, N.; Momeni, M.; Han, G.; Halem, J.; Desai, M.; Ross, M.G. Maternal undernutrition inhibits angiogenesis in the offspring: A potential mechanism of programmed hypertension. *Am. J. Physiol. Regul. Integr. Comp. Physiol.* **2007**, *293*, R745–R753. [CrossRef]
81. Khorram, O.; Ghazi, R.; Chuang, T.D.; Han, G.; Naghi, J.; Ni, Y. Pearce, W.J. Excess maternal glucocorticoids in response to in utero undernutrition inhibit offspring angiogenesis. *Reprod. Sci.* **2014**, *21*, 601–611. [CrossRef]
82. Gutierrez-Arzapalo, P.Y.; Rodríguez-Rodríguez, P.; Ramiro-Cortijo, D.; López de Pablo, Á.L.; López-Giménez, M.R.; Condezo-Hoyos, L.; Greenwald, S.E.; González, M.D.C.; Arribas, S.M. Role of fetal nutrient restriction and postnatal catch-up growth on structural and mechanical alterations of rat aorta. *J. Physiol.* **2018**, *596*, 5791–5806. [CrossRef] [PubMed]
83. Somoza, B.; Abderrahim, F.; González, J.M.; Conde, M.V.; Arribas, S.M.; Starcher, B.; Regadera, J.; Fernández-Alfonso, M.S.; Díaz-Gil, J.J.; González, M.C. Short-term treatment of spontaneously hypertensive rats with liver growth factor reduces carotid artery fibrosis, improves vascular function, and lowers blood pressure. *Cardiovasc. Res.* **2006**, *69*, 764–771. [CrossRef] [PubMed]
84. Scott, T.M.; Pang, S.C. The correlation between the development of sympathetic innervation and the development of medial hypertrophy in jejunal arteries in normotensive and spontaneously hypertensive rats. *J. Auton. Nerv. Syst.* **1983**, *8*, 25–32. [CrossRef]
85. Cai, X.N.; Wang, C.Y.; Cai, Y.; Peng, F. Effects of renal denervation on blood-pressure response to hemorrhagic shock in spontaneously hypertensive rats. *Chin. J. Traumatol.* **2018**, *21*, 293–300. [CrossRef] [PubMed]
86. Pladys, P.; Lahaie, I.; Cambonie, G.; Thibault, G.; Lê, N.L.; Abran, D.; Nuyt, A.M. Role of brain and peripheral angiotensin II in hypertension and altered arterial baroreflex programmed during fetal life in rat. *Pediatr. Res.* **2004**, *55*, 1042–1099. [CrossRef] [PubMed]
87. Spann, M.N.; Smerling, J.; Gustafsson, H.; Foss, S.; Altemus, M.; Monk, C. Deficient maternal zinc intake-but not folate-is associated with lower fetal heart rate variability. *Early Hum. Dev.* **2015**, *91*, 169–172. [CrossRef]
88. Sucharita, S.; Dwarkanath, P.; Thomas, T.; Srinivasan, K.; Kurpad, A.V.; Vaz, M. Low maternal vitamin B12 status during pregnancy is associated with reduced heart rate variability indices in young children. *Matern. Child Nutr.* **2014**, *10*, 226–233. [CrossRef]
89. Kitagawa, H.; Yamazaki, T.; Akiyama, T.; Mori, H.; Sunagawa, K. Effects of ketamine on exocytotic and non-exocytotic noradrenaline release. *Neurochem. Int.* **2003**, *42*, 261–267. [CrossRef]
90. Sinclair, M.D. A review of the physiological effects of alpha2-agonists related to the clinical use of medetomidine in small animal practice. *Can. Vet. J.* **2003**, *44*, 885–897.
91. Briones, A.M.; González, J.M.; Somoza, B.; Giraldo, J.; Daly, C.J.; Vila, E.; González, M.C.; McGrath, J.C.; Arribas, S.M. Role of elastin in spontaneously hypertensive rat small mesenteric artery remodelling. *J. Physiol.* **2003**, *552*, 185–195. [CrossRef]

92. Gutiérrez-Arzapalo, P.Y.; Rodríguez-Rodríguez, P.; Ramiro-Cortijo, D.; Gil-Ortega, M.; Somoza, B.; de Pablo, Á.L.L.; González, M.D.C.; Arribas, S.M. Fetal Undernutrition Induces Resistance Artery Remodeling and Stiffness in Male and Female Rats Independent of Hypertension. *Biomedicines* **2020**, *8*, 424. [CrossRef] [PubMed]
93. Elmarakby, A.A.; Sullivan, J.C. Sex differences in hypertension: Lessons from spontaneously hypertensive rats (SHR). *Clin. Sci.* **2021**, *135*, 1791–1804. [CrossRef] [PubMed]
94. Grassi, G.; Seravalle, G.; Quarti-Trevano, F. The 'neuroadrenergic hypothesis' in hypertension: Current evidence. *Exp. Physiol.* **2010**, *95*, 581–586. [CrossRef] [PubMed]

biomedicines

Article

Slower Growth during Lactation Rescues Early Cardiovascular and Adipose Tissue Hypertrophy Induced by Fetal Undernutrition in Rats

Pilar Rodríguez-Rodríguez [1,2], Ignacio Monedero-Cobeta [1,2], David Ramiro-Cortijo [1,2], Sophida Puthong [3], Begoña Quintana-Villamandos [4], Alicia Gil-Ramírez [2,5], Silvia Cañas [2,5], Santiago Ruvira [1,2] and Silvia M. Arribas [1,2,*]

1 Department of Physiology, Faculty of Medicine, Universidad Autónoma de Madrid, 28029 Madrid, Spain
2 Food, Oxidative Stress and Cardiovascular Health (FOSCH) Research Group, Universidad Autónoma de Madrid, Ciudad Universitaria de Cantoblanco, 28049 Madrid, Spain
3 Department of Physiology, Khon Kaen University, Khon Kaen 40002, Thailand
4 Department of Anesthesiology, Hospital General Universitario Gregorio Marañón, 28007 Madrid, Spain
5 Department of Agricultural Chemistry and Food Science, Faculty of Science, Universidad Autónoma de Madrid, Ciudad Universitaria de Cantoblanco, 28049 Madrid, Spain
* Correspondence: silvia.arribas@uam.es

Citation: Rodríguez-Rodríguez, P.; Monedero-Cobeta, I.; Ramiro-Cortijo, D.; Puthong, S.; Quintana-Villamandos, B.; Gil-Ramírez, A.; Cañas, S.; Ruvira, S.; Arribas, S.M. Slower Growth during Lactation Rescues Early Cardiovascular and Adipose Tissue Hypertrophy Induced by Fetal Undernutrition in Rats. *Biomedicines* **2022**, *10*, 2504. https://doi.org/10.3390/biomedicines10102504

Academic Editors: Josef Zicha and Ivana Vaněčková

Received: 11 August 2022
Accepted: 5 October 2022
Published: 7 October 2022

Abstract: Low birth weight (LBW) and accelerated growth during lactation are associated with cardiometabolic disease development. LBW offspring from rats exposed to undernutrition during gestation (MUN) develops hypertension. In this rat model, we tested if slower postnatal growth improves early cardiometabolic alterations. MUN dams were fed *ad libitum* during gestation days 1–10, with 50% of the daily intake during days 11–21 and *ad libitum* during lactation. Control dams were always fed *ad libitum*. Pups were maintained with their own mother or cross-fostered. Body weight and length were recorded weekly, and breastmilk was obtained. At weaning, the heart was evaluated by echocardiography, and aorta structure and adipocytes in white perivascular fat were studied by confocal microscopy (size, % beige-adipocytes by Mitotracker staining). Breastmilk protein and fat content were not significantly different between groups. Compared to controls, MUN males significantly accelerated body weight gain during the exclusive lactation period (days 1–14) while females accelerated during the last week; length growth was slower in MUN rats from both sexes. By weaning, MUN males, but not females, showed reduced diastolic function and hypertrophy in the heart, aorta, and adipocytes; the percentage of beige-type adipocytes was smaller in MUN males and females. Fostering MUN offspring on control dams significantly reduced weight gain rate, cardiovascular, and fat hypertrophy, increasing beige-adipocyte proportion. Control offspring nursed by MUN mothers reduced body growth gain, without cardiovascular modifications. In conclusion, slower growth during lactation can rescue early cardiovascular alterations induced by fetal undernutrition. Exclusive lactation was a key period, despite no modifications in breastmilk macronutrients, suggesting the role of bioactive components. Our data support that lactation is a key period to counteract cardiometabolic disease programming in LBW and a potential intervention window for the mother.

Keywords: fetal undernutrition programming; cross-fostering; lactation period; cardiovascular hypertrophy; adipose tissue browning

1. Introduction

Adverse environmental conditions during the intrauterine period increase the risk to develop cardiometabolic diseases (CMD), a process known as fetal programming [1,2]. Exposure to the fetus to insufficient nutrients is one of the key programming factors, widely demonstrated in human populations by the association between low birth weight (LBW) and higher rates of hypertension, coronary heart disease, or type 2 diabetes mellitus [3–5]. Fetal programming has also been validated in animal models [6–8]. The postnatal period seems to be a second critical window for programming, particularly after exposure to a fetal stressor. During this period, LBW individuals frequently experience a rapid weight gain, and compelling evidence indicates that this growth pattern is also deleterious. Adolescents born prematurely with a quick postnatal growth have a lower flow-mediated-endothelial dilatation compared to those with slower weight gain and a higher risk of hypertension [8]; quick growth and elevated central adiposity in infancy also contribute to the future development of metabolic disorders [9,10]. A systematic review highlights that rapid postnatal catch-up growth is a more important factor than LBW alone for the development of cardiovascular disease and its risk factors [11].

Nutrition in LBW individuals represents a challenge to ensure a postnatal growth pattern adequate to achieve brain and bone development while preventing CMD [12]. In this context, the lactation period has been proposed as a critical window to counteract alterations initiated during fetal life [13], since human breast milk is recognized as the gold standard for infant nutrition tailored to provide the nutrients and bioactive factors for healthy growth [14]. Breastfeeding provides protection against CMD, is associated with lower blood pressure levels and a lower risk of overweight in children [15,16], and a systematic review evidences slower rates of weight gain in preterm and LBW infants fed with breastmilk compared to the formula [17].

Animal models provide the tools to analyze the influence of both intrauterine and lactation periods in a controlled environment. In a rat model of fetal programming induced by maternal undernutrition during gestation (MUN), we have demonstrated that LBW offspring accelerate growth during lactation, reaching the same body weight as control counterparts by wearing. This catch-up trajectory in MUN rats leads to excessive growth of cardiovascular organs in males [18,19] and white adipose tissue (WAT) deposits in both sexes [20].

In this study, we aim to demonstrate in an animal model of LBW if a slower growth during the lactation period can rescue the cardiometabolic alterations induced by undernutrition. To test this hypothesis, in MUN rats we have modified the lactation environment using a cross-fostering protocol and analyzed the impact on (1) body weight and length growth pattern during the first three weeks of postnatal life, (2) milk macronutrient composition, and (3) heart, aorta and perivascular adipose tissue growth and cardiac function at weaning, assessing the influence of sex in the abovementioned parameters.

2. Materials and Methods

Experiments were performed on Sprague Dawley rats from the colony bred at the Animal House of Universidad Autónoma de Madrid (ES-28079-0000097). This study was performed in accordance with European Union Guidelines for the Care and Use of Laboratory Animals (Directive 2010/63/EU and Spanish RD 1201/2005) and was approved by the Ethics Review Board of Universidad Autónoma de Madrid and by the local government (Comunidad Autónoma de Madrid, Spain; PROEX 004-19, approved on 7 May 2019).

The rats were housed in buckets 36.5 × I21.5 × 18.5 cm (length×width×height) on aspen wood bedding, maintained under controlled conditions of temperature and 12/12 light/dark photoperiod, and the welfare of the animals was monitored daily by staff. The animals were fed with a EuroRodent breeding diet (5LF5; Labdiet, Madrid, Spain) containing 22.0% protein, 4.4% fat, 55.0% carbohydrates, 4.1% fiber, and 5.4% mineral (sodium content 0.26%) and drinking water were provided *ad libitum*.

2.1. Maternal Undernutrition (MUN) Model and Cross-Fostering Protocol

We used a rat model of fetal programming induced by maternal global nutrient restriction during the second part of gestation, as previously described [21]. After observation of sperm in the vaginal smear (gestation = day 0) the dam was allocated to either MUN or the control group. Control dams were fed *ad libitum* throughout pregnancy and lactation; MUN dams were fed *ad libitum* from day 1–10 and with 50% of the averaged control daily intake (previously calculated as 24 g/day) from day 11 to the end of gestation, returning to *ad libitum* diet during lactation. 24 h after birth the pups were sexed and weighed individually and the litter was standardized to 12 individuals, 6 males, and 6 females if possible (smaller litters were not used for this work, being used for other studies). Control and MUN litters were left with their mothers or cross-fostered.

Cross-fostering protocol. We used a protocol similar to [22] with some modifications. Two dams were mated at the same time, ensuring the same gestation day 0, and those with birth on the same day were cross-fostered. On postnatal day 1, the entire litter was exchanged to the foster's dam cage with bedding from the original mother to avoid rejection. The litter was standardized to 12 individuals and the pups were sexed. The rats were not individually marked to help acceptance (no rejection was detected in any of the groups).

The name of the groups was "pup type–on–mother type" (i.e., M-on-C refers to a pup from a MUN dam fostered by a control mother). Four experimental groups were studied: control-on-control (C-on-C; n = 6 mothers), MUN-on-MUN (M-on-M; n = 6 mothers); control-on-MUN (C-on-M, n = 4 mothers), MUN-on-control (M-on-C; n = 4 mothers).

2.2. Experimental Protocols

The entire offspring from the four groups were weighted and head-to-tail length measured at days 1, 14, and 21 of the lactation periods with an analytical balance (Conecta, Barcelona, Spain) and a digital caliper (Conecta, Barcelona, Spain), respectively. The body weight gain from day 1 to 14 was calculated as Formula (1). A similar formula was used to assess body weight gain during the last week and length gain in both lactation periods. Weigh gain was expressed as g/day and length gain as cm/day.

$$Weight\ gain = \frac{\left(Weight_{day\ 14} - Weight_{day\ 1}\right)}{number\ of\ days} \quad (1)$$

Breast milk was collected from MUN and control dams on days 12 and 14 of lactation (see protocol below). On day 21 (weaning), 1–2 males and 1–2 females from each litter were studied and the remaining animals were used for other studies according to the 3Rs rule for experimental animals. Firstly, cardiac structure and function were evaluated by echocardiography. Thereafter, the rats were placed in a CO_2 chamber and killed by exsanguination. Then, the thoracic aorta and mesenteric bed were removed and stored in a cold saline solution. The aorta was cleaned, and the aortic arch and immediate region were fixed in 4% paraformaldehyde solution (PFA). Perivascular adipocytes from the mesenteric bed were also dissected and fixed in PFA. Both tissues were stored for later confocal microscopy analysis. Figure 1 shows a summary of experimental protocols.

2.3. Breast Milk Collection and Measurements

Breast milk was collected on day 12th and day 14th of lactation (the period of maximal milk yield) and was pooled, allowing the mother and pups to recover during day 13. On the day of collection, the pups were separated from the mother for 3 h; the dam was anesthetized with isoflurane inhalation anesthesia (1000 mg/g; Isoflutek, Karizoo Lab., Barcelona, Spain) and injected with 0.1–0.15 mL of oxytocin i.p. depending on rat body weight (10 UI/mL; Syntocinon, Mylan Products Ltd., Hertfordshire, UK). A 1.0–1.5 mL volume of breast milk was obtained by gentle massage of the mammary glands and collection by a glass micropipette in Eppendorf tubes. The dam was milked for a maximum of 1 h. Each breastmilk sample was divided into two aliquots, one to assess total fat and the second one to assess total proteins.

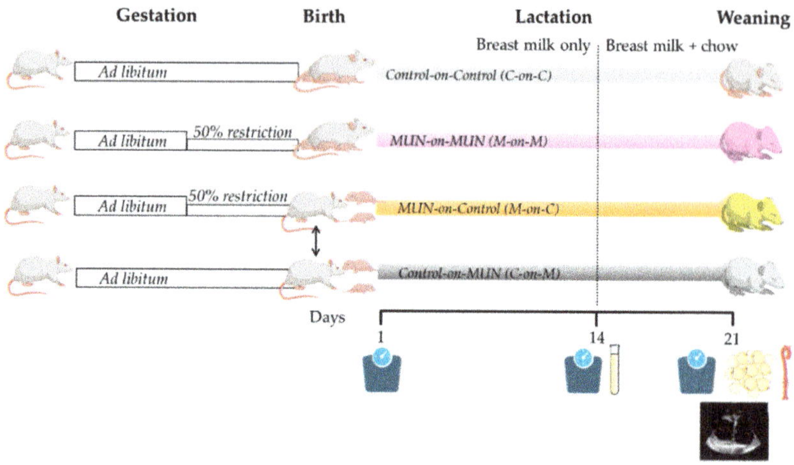

Figure 1. Experimental protocol diagram. MUN, maternal undernutrition during gestation; C-on-C, control pups fostered by a control mother (n = 6 dams); M-on-M, MUN rats fostered by their own mothers (n = 6 dams); C-on-M control rats fostered by a MUN mother (n = 4 dams); MUN-on-C, MUN rats fostered by a control mother (n = 4 dams).

Total fat content. Fat was analyzed by the Mojonnier method [23] with modifications. Briefly, a 0.5 mL volume of breast milk was mixed with 0.5 mL of ammonium hydroxide, followed by the addition of 50 µL of ethanolic phenolphthalein solution (0.5% *w*/*v*) and shaken. Thereafter, 2.5 mL of ethanol, 2.5 mL of petroleum ether, and 2.5 mL of ethyl ether were added and vigorously mixed for 30 s. Thereafter, the mix was centrifuged at 4000× *g* (3 min, RT) and the upper phase containing the fat was stored. This process was repeated three times with the remaining aqueous phase, adding the ethanol, ethyl ether, and petroleum ether. Thereafter, the fat from the 3 centrifugation was pooled, and to evaporate the ether solvent, it was placed overnight in a gravity convection oven at 50 °C, uncapped. Finally, the total fat content in the sample was measured by gravimetry and expressed as mg/mL of BM. A blank reaction was performed by substituting the 0.5 mL breast milk volume with H_2O-Q.

Total protein level. Protein quantification was carried out in the defatted phase of breast milk using the Bradford method [24]. It has to be noted that centrifugated breast milk samples continue to contain proteins such as caseins. Briefly, 10 µL of BM (1:50 *v*/*v* in H_2O-Q) were mixed in a microplate with a 200 µL volume of Coomassie-blue dye (1:4 *v*/*v* in H_2O-Q; Bio-Rad Lab., Hercules, CA, USA). After shaking the mix for 1 min, the absorbance was measured at 595 nm in a microplate reader (Synergy HT Multimode; BioTek Instruments, Winooski, VT, USA). Total protein levels were expressed as mg of BSA/mL. BSA (Sigma-Aldrich, St. Louis, MO, USA) was used for the standard curve (range 0.1–0.5 mg/mL) and Coomassie-blue dye was substituted by H_2O-Q for the blank curve.

2.4. Transthoracic Echocardiography

Transthoracic echocardiography was performed in 21 days-old rats using the VIVID q-system (GE Healthcare, Munich, Germany) equipped with a 13 MHz probe (12S-RS, GE Healthcare, Munich, Germany) as previously described [25]. Briefly, the rats were anesthetized by i.p injection with 80 mg/kg Ketamine hydrochloride (AuroMedics Pharma LLC., Dublin, Ireland) and 10 mg/kg Diazepam (Hospira, Inc., Lake Forest, IL, USA). The images were acquired with the animals in left lateral decubitus. M-mode imaging of the parasternal short-axis (papillary level) view allowed measurement of end-diastolic (mm) and end-systolic (mm) internal diameter, posterior wall thickness (mm), and interventricular septum thickness (mm) at diastole. Additionally, it was calculated the systolic functionality as the ejection fraction (%). The pulsed-wave Doppler early-to-late transmitral peak diastolic flow velocity ratio (E/A ratio; arbitrary units) was measured to assess diastolic function (E, mitral peak early-filling velocity and A, mitral peak flow velocity at atrial contraction). The transmitral flow velocity profile was determined by positioning a sample volume at the tip of the mitral valve on the para-apical long-axis view. The E-wave deceleration time was measured as the time interval between peak E-wave velocity and the point where the descending E-wave (or its extrapolation) intercepted the zero line. Values were determined by averaging the measurements of three consecutive cardiac cycles.

2.5. Aorta Structure by Confocal Microscopy

Aorta structure was assessed in a ring cut from the PFA fixed thoracic aorta proximal to the aortic arch. The ring was mounted intact on a slide provided with a small well, to avoid compression containing Citifluor AF2 mounting medium (Aname, Madrid, Spain), as previously described [19]. The ring was visualized with a ×20 objective with a laser scanning confocal microscope (Leica TCS SP2, Leica Microsystems, Barcelona, Spain) at Excitation = 488 nm/Emission = 500–560 nm, the wavelength at which elastin can be detected by its autofluorescence allowing to detect the medial layer. Single images were captured with a ×20 air objective, ensuring the elastic lamellae were clearly visible. Quantification was performed with FIJI software [26], measuring internal and external perimeters (µm); from these measurements medial cross-sectional areas (CSAs) were calculated in μm^2, assuming the sections were circular.

2.6. Adipocyte Size and Browning

PFA-fixed adipocytes from the perivascular tissue of the mesenteric bed were incubated with Mitotracker red FM (Invitrogen, ThermoFisher, Madrid, Spain) prepared in phosphate buffer saline (PBS) to stain mitochondria-positive cells [27], evaluating the percentage of beige adipocytes, since they are rich in mitochondria. Adipocytes were incubated for 60 min (1:4000 *v*/*v*, RT, darkness) and washed 3 times for 10 min in PBS. Thereafter, they were incubated for 15 min in a solution of DAPI (Invitrogen, ThermoFisher, Madrid, Spain) to stain nuclei (1:500 *v*/*v* from stock solution, RT, darkness,) followed by 3 washes in PBS for 10 min each at RT. The samples were mounted intact on a slide with a small well and were visualized with a Laser Scanning Confocal Microscope (Leica TCS SP2, Leica Microsystems, Barcelona, Spain). Five regions were randomly chosen based on DAPI wavelength and single images were captured with a ×40 objective at three wavelengths: Excitation = 405 nm/Emission = 410–475 nm to visualize nuclei, Excitation = 488 nm/Emission = 500–560 nm to visualize adipocyte size by autofluorescence [20], and Excitation = 581 nm/Emission = 644 nm to visualize mitochondria-positive cells.

The quantification of adipocyte size and proportion of mitochondria-positive cells was performed by FIJI software. The total number and area occupied by the adipocytes were quantified in the Excitation = 488 nm/Emission = 500–560 nm wavelength images (autofluorescence), and the average adipocyte size was calculated in each image. To assess Mitotracker-positive cells in perivascular WAT, the area occupied by adipocytes was measured in autofluorescent images, Mitotracker-positive areas were measured in the same region in the red wavelength images, and the percentage was calculated.

2.7. Statistical Analysis

Statistical analysis was performed with R software (version 3.6.0, 2018, R Core Team, Vienna; Austria) within the R Studio interface using *rio, dplyr, compareGroups, ggpubr, devtools,* and *ggplot2* packages. Data was expressed as the median and interquartile range [Q1; Q3]. The differences in macronutrients of breast milk between control and MUN were performed by Mann-Whitney's U test. The Kruskal–Wallis test by ranks was used to test the independent experimental groups. The significant Kruskal–Wallis test was used with Dunn's test with directed pairwise comparison using C-on-C as a reference group. Second, it was considered the comparison between M-on-M versus M-on-C to determine, if pups with prior fetal programming, the lactation could reverse the modification. A p-value (p) < 0.05 was considered significant.

3. Results

3.1. Proteins and Fats in Breast Milk

No differences between control and MUN were detected in breast milk proteins (Control = 125.0 [113.0; 129.0] mg eq. BSA/mL, n = 12 rats; MUN = 117.0 [111.0; 123.0] mg eq. BSA/mL, n = 10 rats; p-Value=0.539). Fat levels in MUN milk tended to be lower but did not reach statistical significance (Control = 98.5 [89.3; 113.0] mg/mL, n = 12 rats; MUN = 84.0 [66.7; 107.0] mg/mL, n = 10 rats; p-Value = 0.090).

3.2. Body Growth Gain

From day 1 to 14 of lactation (exclusive lactation period), male rats from the M-on-M group exhibited a significantly higher increase in body weight gain compared to the C-on-C group. MUN males fostered by a C mother (M-on-C group) did not show differences compared C-on-C group, being significantly smaller compared to M-on-M males. Males from the C-on-M group had a significantly lower body weight gain compared to C-on-C (Figure 2A). By contrast, during the first lactation period, M-on-M female rats did not show differences in body weight gain compared to C-on-C or compared to M-on-C. As observed in males, control females nourished by a MUN mother (C-on-M) showed a lower body weight gain compared to C-on-C females (Figure 2A).

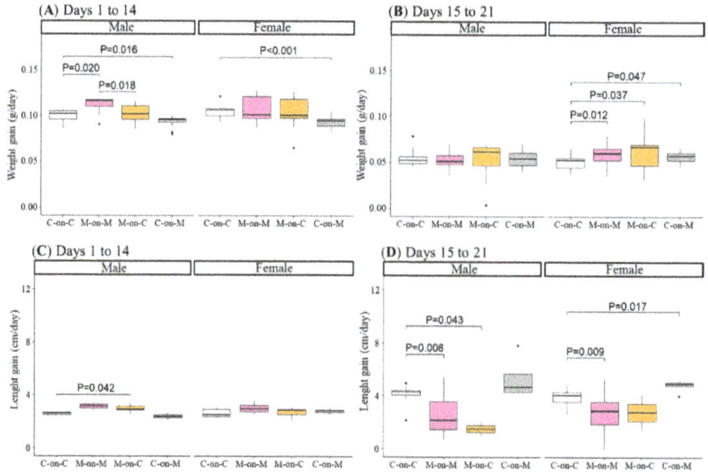

Figure 2. Body weight from day 1 to 14 of lactation (**A**) and from day 15 to 21 of lactation (**B**), and body length gain from day 1 to 14 of lactation (**C**) and from day 15 to 21 of lactation (**D**) in male and female rats from MUN and Control mothers. MUN, maternal undernutrition during gestation; C-on-C, control pups fostered by their mothers (34 males and 38 females); M-on-M, MUN rats fostered by their mothers (32 males and 40 females); C-on-M, control rats fostered by MUN mothers (21 males and 27 females); M-on-C, MUN rats fostered by a control mother (25 males and 23 females). Data show the median and interquartile range [Q1; Q3] and the showed p-Value (P) was extracted from Dunnett's post-hoc pairwise comparison test when the Kruskal-Wallis test was P < 0.05. Dots show outliers.

From day 15 to 21, when the pups suckle from mothers and eat by themselves, male M-on-M did not show statistical differences in body weight gain compared to C-on-C, or M-on-C males. However, M-on-M females accelerated growth during this period, as shown by the higher weight gain compared to C-on-C female rats. M-on-C rats also had higher weight gain compared to C-on-C females (Figure 2B).

From day 1 to 14, males of M-on-M length gain tended to be smaller compared to the C-on-C group but did not reach statistical significance (p-Value = 0.083). Cross-fostered MUN males (M-on-C) had a significantly higher length gain compared to the C-on-C group. In female rats, no differences were detected in any of the experimental groups in length gain during the exclusive lactation period (Figure 2C).

From day 15 to 21 length gain was significantly lower in M-on-M males and females compared to C-on-C control sex-matched counterparts. M-on-C males maintained a lower length gain, while M-on-C females improved the length rate, not significantly different from C-on-C females (Figure 2D).

3.3. *Perivascular Adipocyte Size and Type*

WAT adipocyte size was assessed in perivascular fat from the mesenteric bed obtained from 21-day-old offspring. M-on-M males had significantly larger adipocytes compared to C-on-C rats (Figure 3A). Adipocytes from MUN males fostered by a control mother (M-on-C) were of similar size compared to C-on-C males and significantly smaller compared to M-on-M. Control males fostered by a MUN mother (C-on-M) also showed smaller adipocytes size compared to C-on-C males (Figure 3A).

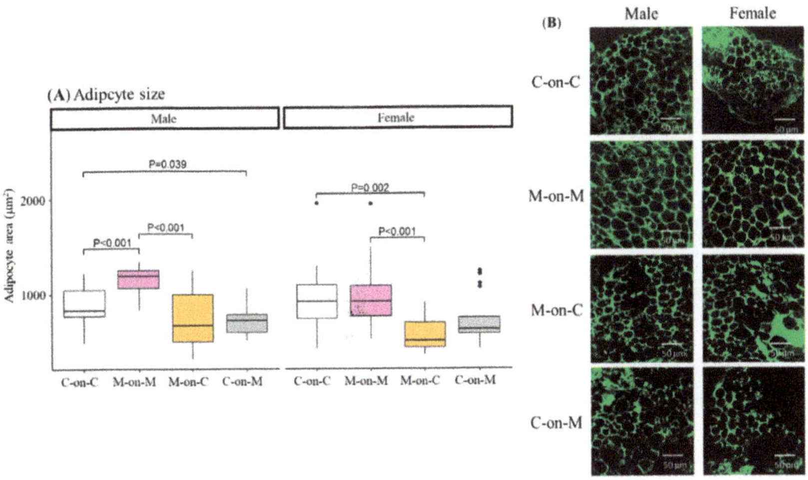

Figure 3. Average size of perivascular adipocytes (**A**) and representative images obtained by confocal microscopy (**B**) of 21-day-old male and female offspring from MUN and Control mothers. MUN, maternal undernutrition during gestation; C-on-C, control pups fostered by their mothers (6 males and 6 females); M-on-M, MUN rats fostered by their mothers (6 males and 6 females); C-on-M, control rats fostered by MUN mothers (6 males and 6 females); M-on-C, MUN rats fostered by a control mother (6 males and 6 females). Data show the median and interquartile range [Q1; Q3], and the showed p-Value (P) was extracted from Dunnett's post-hoc pairwise comparison test when the Kruskal-Wallis test was $P < 0.05$. Images were acquired with a Leica TCS SP2 confocal microscope, ×40 objective at Ex = 488 nm/Em = 500–560 nm wavelength; scale bar = 50 µm. Dots shown outliers.

Regarding females, M-on-M rats had similar adipocytes compared to C-on-C. M-on-C females showed significantly smaller adipocyte size compared to C-on-C females and also compared to M-on-M. No differences were detected between control rats fostered by a MUN mother (C-on-M) and C-on-C females (Figure 3A).

We used Mitotracker, a dye that stains mitochondria, to evaluate beige-type adipocytes within perivascular WAT. Mitotracker-positive cells showed an intense stain. M-on-M males had a smaller percentage of Mitotracker-positive cells compared to C-on-C controls (Figure 4). MUN males fostered by a control dam (M-on-C) had a similar % compared to C-on-C males, being larger compared to M-on-M offspring. Control rats fostered by a MUN mother had lower Mitotracker-positive stained areas compared to C-on-C males.

As observed in males, M-on-M females had significantly lower Mitotracker positive area compared to C-on-C females. MUN rats fostered by a control mother (M-on-C) had a similar % compared to C-on-C offspring, being significantly larger compared to M-on-M rats (Figure 4).

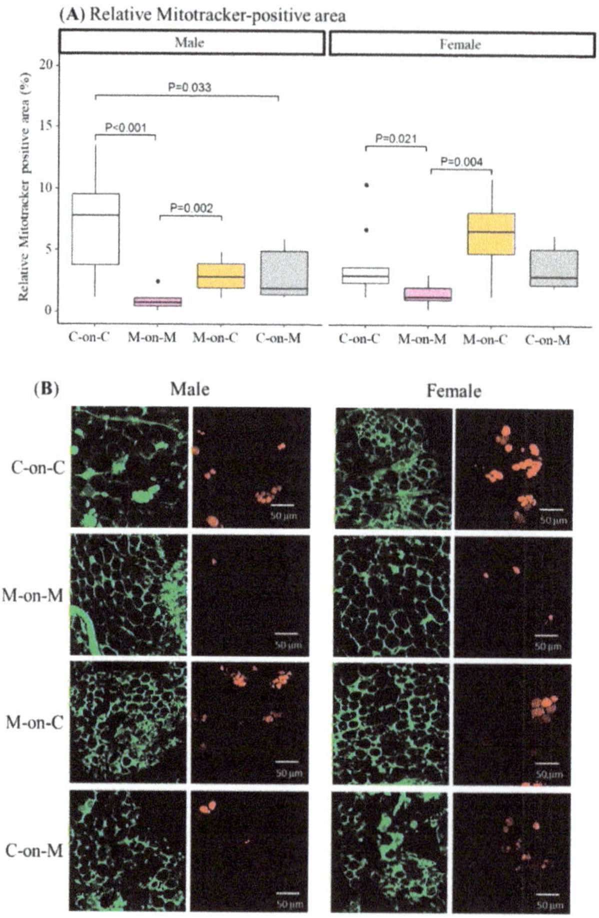

Figure 4. Percentage of Mitotracker-positive adipocytes in perivascular fat (**A**) and representative images obtained by confocal microscopy (**B**) of 21-day-old male and female offspring from MUN and Control mothers. MUN, maternal undernutrition during gestation; C-on-C, control pups fostered by their mothers (6 males and 6 females); M-on-M, MUN rats fostered by their mothers (6 males and 6 females); C-on-M, control rats fostered by MUN mothers (6 males and 6 females); M-on-C, MUN rats fostered by a control mother (6 males and 6 females). Data show the median and interquartile range [Q1; Q3], and the showed *p*-Value (P) was extracted from Dunnett's post-hoc pairwise comparison test when the Kruskal-Wallis test was P < 0.05. Images were acquired with a Leica TCS SP2 confocal microscope, ×40 objective at Ex = 488 nm/Em = 500–560 nm (green images) and Ex = 581 nm/Em = 644 nm (red images); scale bar = 50 µm. Dots show outliers.

3.4. Heart Structure and Function

Heart structure and function were assessed by transthoracic echocardiography at the end of the lactation period. In male M-on-M rats, posterior wall thickness tended to be higher and interventricular septum was significantly higher compared to C-on-C males. MUN males fostered by a control mother (M-on-C) did not show differences with C-on-C males. Female pups did not show significant differences in any of the studied parameters (Figure 5a,b).

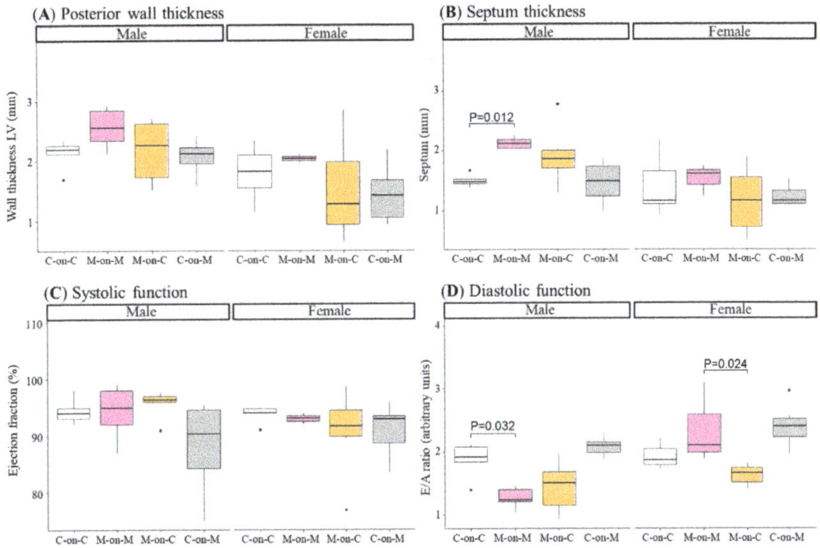

Figure 5. Posterior wall thickness (**A**), septum thickness (**B**), ejection fraction (**C**), and E/A ratio (**D**) of 21-day-old male and female offspring from MUN and Control mothers. MUN, maternal undernutrition during gestation; C-on-C, control pups fostered by their mothers (6 males and 6 females); M-on-M, MUN rats fostered by their mothers (6 males and 6 females); C-on-M, control rats fostered by MUN mothers (6 males and 6 females); M-on-C, MUN rats fostered by a control mother (6 males and 6 females); LV, left ventricle. Data show the median and interquartile range [Q1; Q3], and the showed *p*-Value (P) was extracted from Dunnett's post-hoc pairwise comparison test when the Kruskal-Wallis test was P < 0.05. Dots show outliers.

Systolic function, assessed as ejection fraction, was not different between groups (Figure 5C). However, diastolic function, assessed as the ratio of E/A waves, was significantly reduced in M-on-M males compared to C-on-C counterparts. No significant differences were found in MUN males fostered by a C mother (M-on-C) compared to C-on-C (Figure 5D). Female M-on-M did not show significant differences in any of these parameters compared to C-on-C. However, the E/A ratio was significantly decreased in female M-on-C compared to female M-on-M (Figure 5D).

End-systolic and end-diastolic diameters were not significantly different between C-on-C and any of the other groups (Table 1).

3.5. Thoracic Aorta Structure

In M-on-M males' internal and external diameter, and wall cross-sectional area (CSA) were significantly larger compared to C-on-C. In M-on-C males the studied parameters were not different from C-on-C males and external diameter and CSA were significantly smaller compared to M-on-M rats (Figure 6).

Table 1. End left ventricle dimensions of heart cycles of the offspring at day 21.

Male	C-on-C	M-on-M	M-on-C	C-on-M	p-Value
End-diastolic diameter (mm)	2.92 [2.12; 3.07]	1.82 [1.75; 2.41]	2.55 [2.45; 2.55]	2.54 [2.49; 2.66]	0.358
End-systolic diameter (mm)	0.66 [0.49; 0.81]	0.36 [0.22; 0.36]	0.48 [0.44; 0.51]	0.84 [0.62; 1.06]	0.219
Female	**C-on-C**	**M-on-M**	**M-on-C**	**C-on-M**	**p-Value**
End-diastolic diameter (mm)	2.51 [2.15; 3.10]	2.55 [2.48; 2.70]	2.52 [2.30; 3.01]	2.74 [2.13; 2.90]	0.536
End-systolic diameter (mm)	0.62 [0.48; 0.71]	0.66 [0.66; 0.66]	0.70 [0.60; 0.89]	0.73 [0.62; 0.78]	0.528

Data show median and interquartile range [Q1; Q3]. The p-Value was extracted from the Kruskal-Wallis test. Control (C); Maternal undernutrition (MUN).

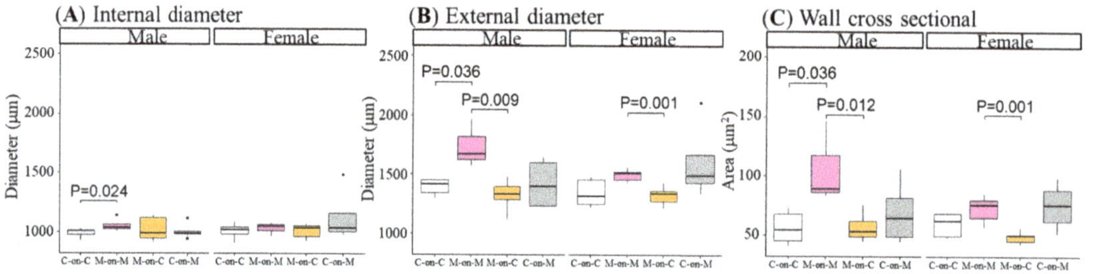

Figure 6. Internal (**A**) and external diameters (**B**), and wall cross-sectional (**C**) of the thoracic aorta in 21-day-old male and female offspring from MUN and Control mothers. MUN, maternal undernutrition during gestation; C-on-C, control pups fostered by their mothers (6 males and 6 females); M-on-M, MUN rats fostered by their mothers (6 males and 6 females); C-on-M, control rats fostered by MUN mothers (6 males and 6 females); M-on-C, MUN rats fostered by a control mother (6 males and 6 females). Data show the median and interquartile range [Q1; Q3], and the showed p-Value (p) was extracted from Dunnett's post-hoc pairwise comparison test when the Kruskal-Wallis test was $p < 0.05$. Dot show outliers.

M-on-M females tended to have larger parameters compared to C-on-C, without statistical significance. However, MUN females fostered by a control mother (M-on-C) had a smaller external diameter and CSA than M-on-M rats (Figure 6).

4. Discussion

In this study, we aimed to evaluate the impact of lactation environment on body, cardiovascular and adipose tissue growth in rats with LBW induced by undernutrition during fetal life (MUN offspring). The main findings of the study are summarized in Table 2. During lactation, MUN offspring accelerate body weight gain, while length growth is slower. In males, this growth pattern occurs during the exclusive lactation period, while in females takes place during the last week before weaning, when the rats suckle and eat chow. By weaning, MUN males but not females, exhibit diastolic dysfunction, and heart, aorta, and perivascular white adipocyte hypertrophy; a lower proportion of beige-type adipocytes was found in MUN offspring from both sexes. These sex-dependent alterations may set the basis for the observed hypertension development in MUN males and can predispose to obesity in males and females. We also demonstrated that nursing MUN rats with a controlling mother rescues the hypertrophy alterations, in parallel with a slower weight gain, while nursing control pups, one MUN mother did not have a hypertrophic effect. These data suggest that both MUN rats, in the fetal and perinatal periods, are required to develop phenotypic alterations. In addition, the alterations programmed during the fetal period may be counteracted during lactation by reducing the growth rate.

The period of exclusive lactation seems to be more relevant for the observed hypertrophy, even though no major modifications in breastmilk macronutrients were detected between MUN and Control dams, suggesting the role of breastmilk bioactive factors. In summary, the present data provide experimental evidence of the link between accelerated growth during early postnatal life in LBW individuals and higher-risk cardiometabolic diseases and support that modulation of growth during the lactation period can be an effective strategy to counteract alterations induced during fetal life.

Table 2. Summary of main findings.

		Male		Female	
		Alterations Induced by MUN	**Reversal by CF in Lactation**	**Alterations Induced by MUN**	**Reversal by CF in Lactation**
Body growth gain	Weight *from day 1 to 14*	Accelerated	Yes	ns	-
	Weight *from day 15 to 21*	ns	-	Accelerated	No
	Length *from day 1 to 14*	ns	-	ns	-
	Length *from day 15 to 21*	Decelerated	No	Decelerated	Yes
Adiposity	Size of adipocyte	Increased	Yes	ns	-
	% Mitotracker-positive	Decreased	Yes	Decreased	Yes
Heart structure and function	Posterior wall thickness	ns	-	ns	-
	Septum thickness	Increased	Yes	ns	-
	End-diastolic diameter	ns	-	ns	-
	End-systolic diameter	ns	-	ns	-
	Systolic function	ns	-	ns	-
	Diastolic function	Decreased	Yes	ns	-
Thoracic aorta structure	Internal diameter	Increased	Yes	ns	-
	External diameter	Increased	Yes	ns	-
	Wall cross-sectional	Increased	Yes	ns	-

Alterations induced by maternal undernutrition (MUN) are considered if M-on-M was different from C-on-C. Reversal by cross-fostering (CF) in lactation was considered if the alteration induced by MUN (M-on-C) is normalized, being similar to the C-on-C group. If no alterations were found, the effect of lactation was not considered (shown as -). No significant (ns).

Catch-up growth has been proposed to be deleterious for cardiometabolic health in individuals with LBW. We addressed the role of growth during lactation in a rat model of LBW induced by fetal undernutrition, which develops hypertension and cardiac alterations in adult life. In rats, suckling is the only source of nutrition until day 16 of lactation, while from day 17 to weaning there is an increasing content of chow in the stomach of the pups [28]. Therefore, we analyzed growth patterns during both periods of lactation. We found a sexual dimorphism in the growth pattern. In males, accelerated weight gain was observed during the period of exclusive lactation, while in female offspring catch-up growth was observed in the last week of lactation. We also observed that control rats fed on MUN mothers had a slower weight gain both in males and females. To explain these data, we considered differences in milk composition. However, we did not find significant variations in protein or fat content although a tendency toward lower values was observed in MUN dams. It is possible that the milk from MUN dams had lower macronutrient levels at birth, as we have evidenced in plasma [29], but they may normalize along lactation, since the rat returns to an *ad libitum* diet. In other rodent models of programming, milk protein content was found to be unaffected if the dam was fed with low protein [30,31] or with high energy diets [32] during gestation or lactation. We also observed a slower length gain during the last week of the lactation period, during which the rats suckle and also eat chow. We do not think this is related to the lower mineral content of the milk, since minerals seem to be relatively stable [13] and it is possible that there was a prior bone deficiency induced by undernutrition. We also discarded a lower milk yield due to compromised mammary gland development, observed in a rat model of fetal programming induced by uteroplacental insufficiency [33], since it would not explain the quicker weight gain in M-on-

M offspring. Growth acceleration in MUN males could be explained by hyperphagia which can be programmed [34], is associated with increased neonatal growth rates and visceral adiposity [35], and it has been described in offspring from protein-restricted dams [36]. We do not think dam behavioral alterations, which have been described in SHR and WKY cross-fostering [37], explain our data since all the MUN and Control dams accepted the pups from the other mother successfully. The main influence on growth rate was observed during the period of exclusive lactation; since no major alterations in macronutrients were observed, we suggest the role of milk bioactive components, such as hormones or growth factors. This hypothesis is supported by metabolomic analysis of milk from rats exposed to mild caloric restriction during lactation showing changes in several compounds related to metabolic pathways [38], and the evidence that leptin supplementation during the suckling period reverses some of the metabolic alterations induced by a moderate maternal calorie restriction during gestation [39]. It would be interesting to conduct an in-depth milk metabolomic study on the alterations induced by undernutrition in our rat model.

Rapid growth gain has been related to increased adiposity and obesity development, and therefore, we analyzed possible alterations in body fat. We found that, in parallel with accelerated weight gain, by weaning M-on-M males exhibited hypertrophied perivascular WAT adipocytes compared to controls, an alteration reversed by nursing MUN offspring on a control dam. Similar results were found in a model of obesity programming induced by monosodium glutamate, where cross-fostering also mitigates obesity development [22]. In addition to energy-storing WAT, adipose tissue also comprises thermogenic brown adipose tissue (BAT), and beige adipocytes, an inducible form of thermogenic adipocytes interspersed within WAT [40] recruited postnatally in a process called browning [41]. Browning or "beigeing" has recently gained attention, since beige adipose tissue has a larger energy expenditure capacity, and induced browning in newborn rats decreases adipogenesis in adult life, suggesting a possible way through which the neonatal period can influence obesity development [42]. Therefore, we analyzed the proportion of beige adipocytes within the perivascular WAT in our experimental model, using Mitotracker, a specific dye for mitochondria, since beige adipocytes possess abundant cristae-dense mitochondria [40,43]. MUN rats had a very small proportion of beige-type adipocytes, which may be related to mitochondrial alterations induced by undernutrition, since mitochondrion is a very sensitive organelle and programming has been demonstrated in response to several intrauterine stress factors [44]. A reduction in beige thermogenic tissue could disbalance energetics and facilitate lipid accumulation, leading to the observed increased WAT adipocyte size. It was interesting that beige-type adipocytes increased by fostering MUN rats on a control mother. It has been demonstrated that beige adipose tissue appears spontaneously in WAT during early postnatal development with a peak of expression observed at 21 days [45]. Therefore, the lactation period is an important window during which adipose tissue type may be modulated. It would be interesting to analyze in MUN rats the effect of the lactation environment on adipocyte progenitor markers, such as mesenchymal cell antigen 1 (MSCA1), which has been positively correlated to obesity, adipocyte hypertrophy, and inflammation in children [46]. It is interesting to note that, increased browning occurred in MUN rats exposed to a control lactation environment, while it decreased in control rats exposed to a MUN mother. Therefore, unlike cardiovascular hypertrophy, which seems to require the first hit being during intrauterine life, adipose tissue may be regulated by the lactation environment alone. This is evidenced by our previous study showing that accelerated growth during lactation induced by reducing litter size, without prior programming, also increases WAT deposits [20]. Prolonged retention of thermogenic beige adipocytes maintains high whole-body energy expenditure and protects mice from diet-induced obesity [47]. Therefore, it would be interesting to evaluate if cross-fostered MUN rats retain this characteristic in adult life. Our finding of beige adipocytes within the WAT perivascular tissue is interesting, since, in mice, beige adipocytes are enriched within subcutaneous fat depots, and are rarely detected in visceral depots [43]. Perivascular WAT also plays a role in vascular tone and maintenance of normal structure [43], which is dysregulated by excess fat, as we have reported in obese mice [48]. A reduced proportion of beige-adipocytes may play a role in vascular dysfunction, an alteration observed in several animal models of fetal programming [49]. These alterations may be due to the pro-oxidant and proinflammatory environment associated with excess fat, favoring the local release of vasoactive factors. Thus, alterations in the proportion of beige-to-WAT perivascular adipose tissue may contribute to the association between obesity and cardiovascular disease [50], an aspect that deserves further attention. The cardioprotective effect of beige adipose tissue is suggested by the fact that individuals with detectable thermogenic adipose tissues have lower odds of hypertension and coronary heart disease [51].

Accelerated body growth gain in MUN males was paralleled by a hypertrophic response in the cardiovascular system observed in the aorta and the heart. The characteristics of the structural alteration in the aorta are compatible with an outward hypertrophic remodeling, with an increase in wall mass due to expansion of the external diameter. Females show a better adaptation, although the data also evidence a tendency towards a larger growth, as we have previously described in resistance arteries [52]. Undernutrition induced marked heart alterations by weaning, in addition to ventricular hypertrophy, a reduced diastolic function was also observed, suggesting that cardiac tissue has a high susceptibility to programming, also observed in guinea pigs exposed to undernutrition, which show a permanent deficiency in cardiomyocyte number [53]. We did not explore molecular markers of hypertrophy, such as BNP; we have evidence that this factor is not elevated in MUN offspring by weaning, but it is increased in aging in MUN males along with hypertension development and further cardiac hypertrophy [18]. The role of sex in heart alterations induced by undernutrition may be related to the lower efficiency of MUN male placenta, associated with poor vascularization [29], which may be particularly detrimental for the heart due to the unique feature of cardiomyocytes as non-dividing cells [54]. Cardiovascular hypertrophy in MUN males was rescued by cross-fostering, reinforcing the role of lactation in reprogramming cardiovascular disease. The importance of this period for future cardiometabolic health is supported by data in the genetic model of essential hypertension, the SHR rat, which reduces blood pressure levels in adult life if cross-fostered to a control dam [6]. It is worth mentioning that this model of essential hypertension also exhibits sexual dimorphism and we have found several common features with the MUN model [55].

5. Conclusions

The lactation period acts as a second hit consolidating programming initiated by fetal undernutrition in a sex-dependent manner, inducing cardiovascular and adipocyte hypertrophy.

The alterations programmed during fetal life may be counteracted during the perinatal period avoiding accelerated growth and thus, lactation and can be a window of intervention to reverse fetal programming.

The exclusive lactation period seems to play a key role in organ hypertrophy, despite no differences in breastmilk macronutrients, suggesting the role of bioactive factors, which deserves in-depth analysis.

Author Contributions: Conceptualization, S.M.A., P.R.-R., I.M.-C. and D.R.-C.; methodology, P.R.-R., I.M.-C., S.P., B.Q.-V., A.G.-R. and S.C.; software, I.M.-C. and D.R.-C.; validation, P.R.-R., I.M.-C., D.R.-C., S.P., B.Q.-V., A.G.-R. and S.M.A.; formal analysis, D.R.-C.; investigation, P.R.-R., I.M.-C., S.P., B.Q.-V., A.G.-R. and S.C.; resources, S.M.A.; data curation, P.R.-R., I.M.-C., S.R. and S.P.; writing—original draft preparation, S.M.A. and D.R.-C.; writing—review and editing, P.R.-R., I.M.-C., S.P., B.Q.-V., S.R., A.G.-R., S.C., D.R.-C. and S.M.A.; visualization, P.R.-R., I.M.-C., D.R.-C. and S.M.A.; supervision, S.M.A.; funding acquisition, S.M.A. All authors have read and agreed to the published version of the manuscript.

Funding: This research was funded by the Ministerio de Ciencia, Innovación y Universidades (Spain; grant number RTI2018-097504-B-I00), co-financed with FEDER funds.

Institutional Review Board Statement: The animal study protocol was approved by the Institutional Review Board of Universidad Autónoma de Madrid and Comunidad Autónoma de Madrid (Spain, protocol code PROEX 004-19 and approved on 7 May 2019).

Informed Consent Statement: Not applicable.

Data Availability Statement: The data presented in this study are available on request from the corresponding author. The availability of the data is restricted to investigators based in academic institutions.

Conflicts of Interest: The funders had no role in the design of the study; in the collection, analysis, or interpretation of data; in the writing of the manuscript; or in the decision to publish the results.

References

1. Barker, D.; Osmond, C. Infant Mortality, Childhood Nutrition, and Ischaemic Heart Disease in England and Wales. *Lancet* **1986**, *1*, 1077–1081. [CrossRef]
2. Barker, D. Fetal Origins of Coronary Heart Disease. *BMJ* **1995**, *311*, 171–174. [CrossRef] [PubMed]
3. Hult, M.; Tornhammar, P.; Ueda, P.; Chima, C.; Bonamy, A.-K.E.; Ozumba, B.; Norman, M. Hypertension, Diabetes and Overweight: Looming Legacies of the Biafran Famine. *PLoS ONE* **2010**, *5*, e13582. [CrossRef] [PubMed]
4. Roseboom, T.J.; van der Meulen, J.H.; Osmond, C.; Barker, D.J.; Ravelli, A.C.; Schroeder-Tanka, J.M.; van Montfrans, G.A.; Michels, R.P.; Bleker, O.P. Coronary Heart Disease after Prenatal Exposure to the Dutch Famine, 1944–1945. *Heart* **2000**, *84*, 595–598. [CrossRef]
5. Wu, L.; Feng, X.; He, A.; Ding, Y.; Zhou, X.; Xu, Z. Prenatal Exposure to the Great Chinese Famine and Mid-Age Hypertension. *PLoS ONE* **2017**, *12*, e0176413. [CrossRef]
6. Ashton, N. Perinatal Development and Adult Blood Pressure. *Braz. J. Med. Biol. Res.* **2000**, *33*, 731–740. [CrossRef]
7. Morton, J.S.; Cooke, C.-L.; Davidge, S.T. In Utero Origins of Hypertension: Mechanisms and Targets for Therapy. *Physiol. Rev.* **2016**, *96*, 549–603. [CrossRef]
8. Singhal, A.; Cole, T.J.; Fewtrell, M.; Deanfield, J.; Lucas, A. Is Slower Early Growth Beneficial for Long-Term Cardiovascular Health? *Circulation* **2004**, *109*, 1108–1113. [CrossRef]
9. Eriksson, J.G. Early Growth and Coronary Heart Disease and Type 2 Diabetes: Findings from the Helsinki Birth Cohort Study (HBCS). *Am. J. Clin. Nutr.* **2011**, *94*, 1799S–1802S. [CrossRef]
10. Woo, J.G. Infant Growth and Long-Term Cardiometabolic Health: A Review of Recent Findings. *Curr. Nutr. Rep.* **2019**, *8*, 29–41. [CrossRef]
11. Kelishadi, R.; Haghdoost, A.A.; Jamshidi, F.; Aliramezany, M.; Moosazadeh, M. Low Birthweight or Rapid Catch-up Growth: Which Is More Associated with Cardiovascular Disease and Its Risk Factors in Later Life? A Systematic Review and Cryptanalysis. *Paediatr. Int. Child. Health* **2015**, *35*, 110–123. [CrossRef]
12. Kesavan, K.; Devaskar, S.U. Intrauterine Growth Restriction: Postnatal Monitoring and Outcomes. *Pediatr. Clin. N. Am.* **2019**, *66*, 403–423. [CrossRef] [PubMed]
13. Picó, C.; Reis, F.; Egas, C.; Mathias, P.; Matafome, P. Lactation as a Programming Window for Metabolic Syndrome. *Eur J. Clin. Investig.* **2021**, *51*, e13482. [CrossRef] [PubMed]
14. Gila-Diaz, A.; Arribas, S.M.; Algara, A.; Martín-Cabrejas, M.A.; López de Pablo, Á.L.; Sáenz de Pipaón, M.; Ramiro-Cortijo, D. A Review of Bioactive Factors in Human Breastmilk: A Focus on Prematurity. *Nutrients* **2019**, *11*, 1307. [CrossRef] [PubMed]
15. Wong, P.D.; Anderson, L.N.; Dai, D.D.W.; Parkin, P.C.; Maguire, J.L.; Birken, C.S.; TARGet Kids! Collaboration The Association of Breastfeeding Duration and Early Childhood Cardiometabolic Risk. *J. Pediatr.* **2018**, *192*, 80–85.e1. [CrossRef] [PubMed]
16. Pluymen, L.P.M.; Wijga, A.H.; Gehring, U.; Koppelman, G.H.; Smit, H.A.; van Rossem, L. Breastfeeding and Cardiometabolic Markers at Age 12: A Population-Based Birth Cohort Study. *Int. J. Obes.* **2019**, *43*, 1568–1577. [CrossRef]
17. Quigley, M.; Embleton, N.D.; McGuire, W. Formula versus Donor Breast Milk for Feeding Preterm or Low Birth Weight Infants. *Cochrane Database Syst. Rev.* **2018**, *6*, CD002971. [CrossRef]
18. Rodríguez-Rodríguez, P.; López De Pablo, A.L.; García-Prieto, C.F.; Somoza, B.; Quintana-Villamandos, B.; Gómez De Diego, J.J.; Gutierrez-Arzapalo, P.Y.; Ramiro-Cortijo, D.; González, M.C.; Arribas, S.M. Long Term Effects of Fetal Undernutrition on Rat Heart. Role of Hypertension and Oxidative Stress. *PLoS ONE* **2017**, *12*, e0171544. [CrossRef]
19. Gutiérrez-Arzapalo, P.Y.; Rodríguez-Rodríguez, P.; Ramiro-Cortijo, D.; López de Pablo, A.L.; López-Giménez, M.R.; Condezo-Hoyos, L.; Greenwald, S.E.; González, M.d.C.; Arribas, S.M. Role of Fetal Nutrient Restriction and Postnatal Catch-up Growth on Structural and Mechanical Alterations of Rat Aorta. *J. Physiol.* **2018**, *596*, 5791–5806. [CrossRef] [PubMed]
20. Munoz-Valverde, D.; Rodríguez-Rodríguez, P.; Gutierrez-Arzapalo, P.Y.; López de Pablo, A.L.; Carmen González, M.; López-Giménez, R.; Somoza, B.; Arribas, S.M. Effect of Fetal Undernutrition and Postnatal Overfeeding on Rat Adipose Tissue and Organ Growth at Early Stages of Postnatal Development. *Physiol. Res.* **2015**, *64*, 547–559. [CrossRef]
21. Rodríguez-Rodríguez, P.; López de Pablo, A.L.; Condezo-Hoyos, L.; Martín-Cabrejas, M.A.; Aguilera, Y.; Ruiz-Hurtado, G.; Gutierrez-Arzapalo, P.Y.; Ramiro-Cortijo, D.; Fernández-Alfonso, M.S.; González, M.D.C.; et al. Fetal Undernutrition Is Associated with Perinatal Sex-Dependent Alterations in Oxidative Status. *J. Nutr. Biochem.* **2015**, *26*, 1650–1659. [CrossRef]

22. Miranda, R.A.; da Silva Franco, C.C.; de Oliveira, J.C.; Barella, L.F.; Tófolo, L.P.; Ribeiro, T.A.; Pavanello, A.; da Conceição, E.P.S.; Torrezan, R.; Armitage, J.; et al. Cross-Fostering Reduces Obesity Induced by Early Exposure to Monosodium Glutamate in Male Rats. *Endocrine* **2017**, *55*, 101–112. [CrossRef]
23. Herreid, E.O.; Harmon, C. A Study of Methods of Obtaining Milk Samples for Estimating Milk Fat by the Mojonnier Method. *J. Dairy Sci.* **1944**, *27*, 33–38. [CrossRef]
24. Bradford, M.M. A Rapid and Sensitive Method for the Quantitation of Microgram Quantities of Protein Utilizing the Principle of Protein-Dye Binding. *Anal. Biochem.* **1976**, *72*, 248–254. [CrossRef]
25. Quintana-Villamandos, B.; Gomez de Diego, J.J.; Delgado-Martos, M.J.; Muñoz-Valverde, D.; Soto-Montenegro, M.L.; Desco, M.; Delgado-Baeza, E. Dronedarone Produces Early Regression of Myocardial Remodelling in Structural Heart Disease. *PLoS ONE* **2017**, *12*, e0188442. [CrossRef] [PubMed]
26. Schindelin, J.; Arganda-Carreras, I.; Frise, E.; Kaynig, V.; Longair, M.; Pietzsch, T.; Preibisch, S.; Rueden, C.; Saalfeld, S.; Schmid, B.; et al. Fiji: An Open-Source Platform for Biological-Image Analysis. *Nat. Methods* **2012**, *9*, 676–682. [CrossRef]
27. Chazotte, B. Labeling Mitochondria with MitoTracker Dyes. *Cold Spring Harb. Protoc.* **2011**, *2011*, 990–992. [CrossRef] [PubMed]
28. Henning, S.J.; Chang, S.S.; Gisel, E.G. Ontogeny of Feeding Controls in Suckling and Weanling Rats. *Am. J. Physiol* **1979**, *237*, R187–R191. [CrossRef]
29. Phuthong, S.; Reyes-Hernández, C.G.; Rodríguez-Rodríguez, P.; Ramiro-Cortijo, D.; Gil-Ortega, M.; González-Blázquez, R.; González, M.C.; López de Pablo, A.L.; Arribas, S.M. Sex Differences in Placental Protein Expression and Efficiency in a Rat Model of Fetal Programming Induced by Maternal Undernutrition. *Int. J. Mol. Sci* **2020**, *22*, 237. [CrossRef] [PubMed]
30. Prentice, A.M.; Goldberg, G.R.; Prentice, A. Body Mass Index and Lactation Performance. *Eur. J. Clin. Nutr.* **1994**, *48* (Suppl. 3), S78–S86; discussion S86–S89.
31. Kugananthan, S.; Gridneva, Z.; Lai, C.T.; Hepworth, A.R.; Mark, P.J.; Kakulas, F.; Geddes, D.T. Associations between Maternal Body Composition and Appetite Hormones and Macronutrients in Human Milk. *Nutrients* **2017**, *9*, 252. [CrossRef] [PubMed]
32. Palou, M.; Torrens, J.M.; Castillo, P.; Sánchez, J.; Palou, A.; Picó, C. Metabolomic Approach in Milk from Calorie-Restricted Rats during Lactation: A Potential Link to the Programming of a Healthy Phenotype in Offspring. *Eur. J. Nutr* **2020**, *59*, 1191–1204. [CrossRef] [PubMed]
33. Briffa, J.F.; O'Dowd, R.; Moritz, K.M.; Romano, T.; Jedwab, L.R.; McAinch, A.J.; Hryciw, D.H.; Wlodek, M.E. Uteroplacental Insufficiency Reduces Rat Plasma Leptin Concentrations and Alters Placental Leptin Transporters: Ameliorated with Enhanced Milk Intake and Nutrition. *J. Physiol.* **2017**, *595*, 3389–3407. [CrossRef]
34. Langley-Evans, S.C.; Bellinger, L.; McMullen, S. Animal Models of Programming: Early Life Influences on Appetite and Feeding Behaviour. *Matern. Child. Nutr.* **2005**, *1*, 142–148. [CrossRef]
35. de Blasio, M.J.; Gatford, K.L.; Robinson, J.S.; Owens, J.A. Placental Restriction of Fetal Growth Reduces Size at Birth and Alters Postnatal Growth, Feeding Activity, and Adiposity in the Young Lamb. *Am. J. Physiol. Regul. Integr. Comp. Physiol.* **2007**, *292*, R875–R886. [CrossRef]
36. Bellinger, L.; Sculley, D.V.; Langley-Evans, S.C. Exposure to Undernutrition in Fetal Life Determines Fat Distribution, Locomotor Activity and Food Intake in Ageing Rats. *Int J. Obes.* **2006**, *30*, 729–738. [CrossRef]
37. Gouldsborough, I.; Black, V.; Johnson, I.T.; Ashton, N. Maternal Nursing Behaviour and the Delivery of Milk to the Neonatal Spontaneously Hypertensive Rat. *Acta Physiol. Scand.* **1998**, *162*, 107–114. [CrossRef] [PubMed]
38. Konieczna, J.; García, A.P.; Sánchez, J.; Palou, M.; Palou, A.; Picó, C. Oral Leptin Treatment in Suckling Rats Ameliorates Detrimental Effects in Hypothalamic Structure and Function Caused by Maternal Caloric Restriction during Gestation. *PLoS ONE* **2013**, *8*, e81906. [CrossRef]
39. Schuster, S.; Hechler, C.; Gebauer, C.; Kiess, W.; Kratzsch, J. Leptin in Maternal Serum and Breast Milk: Association with Infants' Body Weight Gain in a Longitudinal Study over 6 Months of Lactation. *Pediatr. Res.* **2011**, *70*, 633–637. [CrossRef] [PubMed]
40. Ikeda, K.; Maretich, P.; Kajimura, S. The Common and Distinct Features of Brown and Beige Adipocytes. *Trends Endocrinol. Metab.* **2018**, *29*, 191–200. [CrossRef]
41. Kuryłowicz, A.; Puzianowska-Kuźnicka, M. Induction of Adipose Tissue Browning as a Strategy to Combat Obesity. *Int. J. Mol. Sci* **2020**, *21*, 6241. [CrossRef]
42. Ying, T.; Golden, T.; Cheng, L.; Ishibashi, J.; Seale, P.; Simmons, R.A. Neonatal IL-4 Exposure Decreases Adipogenesis of Male Rats into Adulthood. *Am. J. Physiol. Endocrinol. Metab.* **2021**, *320*, E1148–E1157. [CrossRef]
43. Koenen, M.; Hill, M.A.; Cohen, P.; Sowers, J.R. Obesity, Adipose Tissue and Vascular Dysfunction. *Circ. Res.* **2021**, *128*, 951–968. [CrossRef]
44. Rodríguez-Rodríguez, P.; Ramiro-Cortijo, D.; Reyes-Hernández, C.G.; López de Pablo, A.L.; González, M.C.; Arribas, S.M. Implication of Oxidative Stress in Fetal Programming of Cardiovascular Disease. *Front. Physiol.* **2018**, *9*, 602. [CrossRef]
45. Chabowska-Kita, A.; Kozak, L.P. The Critical Period for Brown Adipocyte Development: Genetic and Environmental Influences. *Obesity* **2016**, *24*, 283–290. [CrossRef]
46. Hanschkow, M.; Boulet, N.; Kempf, E.; Bouloumié, A.; Kiess, W.; Stein, R.; Körner, A.; Landgraf, K. Expression of the Adipocyte Progenitor Markers *MSCA1* and *CD36* Is Associated With Adipose Tissue Function in Children. *J. Clin. Endocrinol. Metab.* **2022**, *107*, e836–e851. [CrossRef]
47. Loncar, D. Convertible Adipose Tissue in Mice. *Cell Tissue Res.* **1991**, *266*, 149–161. [CrossRef]

48. Gil-Ortega, M.; Martín-Ramos, M.; Arribas, S.M.; González, M.C.; Aránguez, I.; Ruiz-Gayo, M.; Somoza, B.; Fernández-Alfonso, M.S. Arterial Stiffness Is Associated with Adipokine Dysregulation in Non-Hypertensive Obese Mice. *Vascul Pharmacol.* **2016**, *77*, 38–47. [CrossRef]
49. Feng, X.; Yu, T.; Zhang, Y.; Li, L.; Qu, M.; Wang, J.; Dong, F.; Zhang, L.; Wang, F.; Zhang, F.; et al. Prenatal High-Sucrose Diet Induced Vascular Dysfunction in Thoracic Artery of Fetal Offspring. *Mol. Nutr. Food Res.* **2021**, *65*, e2100072. [CrossRef]
50. Villarroya, F.; Cereijo, R.; Gavaldà-Navarro, A.; Villarroya, J.; Giralt, M. Inflammation of Brown/Beige Adipose Tissues in Obesity and Metabolic Disease. *J. Intern. Med.* **2018**, *284*, 492–504. [CrossRef]
51. Becher, T.; Palanisamy, S.; Kramer, D.J.; Eljalby, M.; Marx, S.J.; Wibmer, A.G.; Butler, S.D.; Jiang, C.S.; Vaughan, R.; Schöder, H.; et al. Brown Adipose Tissue Is Associated with Cardiometabolic Health. *Nat. Med.* **2021**, *27*, 58–65. [CrossRef]
52. Gutiérrez-Arzapalo, P.Y.; Rodríguez-Rodríguez, P.; Ramiro-Cortijo, D.; Gil-Ortega, M.; Somoza, B.; de Pablo, Á.L.L.; González, M.C.; Arribas, S.M. Fetal Undernutrition Induces Resistance Artery Remodeling and Stiffness in Male and Female Rats Independent of Hypertension. *Biomedicines* **2020**, *8*, 424. [CrossRef]
53. Masoumy, E.P.; Sawyer, A.A.; Sharma, S.; Patel, J.A.; Gordon, P.M.K.; Regnault, T.R.H.; Matushewski, B.; Weintraub, N.L.; Richardson, B.; Thompson, J.A.; et al. The Lifelong Impact of Fetal Growth Restriction on Cardiac Development. *Pediatr. Res.* **2018**, *84*, 537–544. [CrossRef]
54. Arima, Y.; Fukuoka, H. Developmental Origins of Health and Disease Theory in Cardiology. *J. Cardiol.* **2020**, *76*, 14–17. [CrossRef]
55. Vieira-Rocha, M.S.; Sousa, J.B.; Rodríguez-Rodríguez, P.; Arribas, S.M.; Diniz, C. Elevated Vascular Sympathetic Neurotransmission and Remodelling Is a Common Feature in a Rat Model of Foetal Programming of Hypertension and SHR. *Biomedicines* **2022**, *10*, 1902. [CrossRef]

biomedicines

MDPI

Article

Activation of RAAS Signaling Contributes to Hypertension in Aged *Hyp* Mice

Nejla Latic, Ana Zupcic, Danny Frauenstein and Reinhold G. Erben *

Department of Biomedical Sciences, University of Veterinary Medicine, Veterinaerplatz 1, 1210 Vienna, Austria; nejla.latic@vetmeduni.ac.at (N.L.); ana.zupcic@vetmeduni.ac.at (A.Z.); danny.frauenstein@gmail.com (D.F.)
* Correspondence: reinhold.erben@vetmeduni.ac.at; Tel.: +43-1-250-77-4550

Abstract: High circulating levels of fibroblast growth factor-23 (FGF23) are associated with left ventricular hypertrophy as well as increased morbidity and mortality in patients suffering from chronic kidney disease. However, the mechanisms underlying this association are controversial. Here, we aimed to further characterize the cardiovascular sequelae of long term endogenous FGF23 hypersecretion using 14-month-old male *Hyp* mice as a model of FGF23 excess. *Hyp* mice were characterized by a ~10-fold increase in circulating intact FGF23, hypophosphatemia, increased serum aldosterone, but normal kidney function, relative to wildtype (WT) controls. Cardiovascular phenotyping did not reveal any evidence of left ventricular hypertrophy or functional impairment in 14-month-old *Hyp* mice. Fractional shortening, ejection fraction, molecular markers of hypertrophy (*Anp*, *Bnp*), and intracardiac markers of contractility and diastolic function were all unchanged in these animals. However, intraarterial catheterization revealed an increase in systolic, diastolic, and mean arterial pressure of ~12 mm Hg in aged *Hyp* mice relative to WT controls. Hypertension in *Hyp* mice was associated with increased peripheral vascular resistance. To test the hypothesis that a stimulation of the renin–angiotensin–aldosterone system (RAAS) contributes to hypertension in aged *Hyp* mice, we administered the angiotensin receptor blocker losartan (30 mg/kg twice daily) or the mineralocorticoid receptor antagonist canrenone (30 mg/kg once daily) to aged *Hyp* and WT mice over 5 days. Both drugs had minor effects on blood pressure in WT mice, but reduced blood pressure and peripheral vascular resistance in *Hyp* mice, suggesting that a stimulation of the RAAS contributes to hypertension in aged *Hyp* mice.

Keywords: fibroblast growth factor-23; left ventricular hypertrophy; hypertension; renin–angiotensin–aldosterone system; losartan; canrenone; *Hyp* mice; X-linked hypophosphatemia

Citation: Latic, N.; Zupcic, A.; Frauenstein, D.; Erben, R.G. Activation of RAAS Signaling Contributes to Hypertension in Aged *Hyp* Mice. *Biomedicines* **2022**, *10*, 1691. https://doi.org/10.3390/biomedicines10071691

Academic Editor: Ramón C. Hermida

Received: 7 June 2022
Accepted: 9 July 2022
Published: 13 July 2022

1. Introduction

The most frequent inherited renal phosphate-wasting disease in humans is X-linked hypophosphatemia (XLH). The murine homolog is *Hyp* (hypophosphatemia). XLH patients and *Hyp* mice lose phosphate via the urine and show impaired bone mineralization as a consequence of hypophosphatemia and alterations in the organic bone matrix. The disease is caused by loss-of-function mutations in *PHEX* (Phosphate-regulating endopeptidase homolog, X-linked) in mice and men. By mechanisms that are poorly understood, *PHEX* mutations lead to excessive secretion of the bone-derived, phosphotropic hormone fibroblast growth factor-23 (FGF23) [1–3].

FGF23 is a 32 kDa glycoprotein circulating in the bloodstream. Only the intact molecule is biologically active. FGF23 signals through FGF receptors (FGFRs), with FGFR1c being the most important receptor under physiological conditions. High affinity binding of FGF23 to FGFRs in the cell membrane requires the presence of the co-receptor protein αKlotho [4,5]. The major site of action of FGF23 is the kidney, where it inhibits tubular reabsorption of phosphate and suppresses vitamin D hormone ($1,25(OH)_2D$) production in proximal renal tubules [6]. Excessive concentrations of intact FGF23 lead to renal phosphate wasting, hypophosphatemia, and impaired bone mineralization in patients with normal kidney function [7].

Although FGF23 mainly acts on epithelial cells in the kidney under physiological conditions, other organs such as the cardiovascular system may become targets of FGF23 signaling in pathological settings. The cardiovascular sequelae of chronically elevated FGF23 are still controversial. Studies in chronic kidney disease (CKD) patients have shown that elevated FGF23 levels are independently associated with CKD progression, left ventricular (LV) hypertrophy, cardiovascular risk, and all-cause mortality [8–10]. A key question in this context is whether FGF23 is only a biomarker of impaired phosphate metabolism or chronic inflammation or whether there is a causal link between FGF23 and increased cardiovascular risk. The myocardium does not express αKlotho [10,11], so the heart is likely not a target of FGF23 action under normal conditions. However, at high circulating concentrations such as those found in CKD patients, FGF23 may promote LV hypertrophy by a direct, αKlotho-independent, FGFR4-mediated action on cardiomyocytes [10–12]. Alternatively, FGF23 excess may cause LV hypertrophy by contributing to volume overload through its sodium-conserving effect [13].

Whether LV hypertrophy is typically associated with other diseases (than CKD) characterized by FGF23 excess, such as XLH, is unclear. Some studies find a high incidence of LV hypertrophy in XLH patients [14,15], but others report no such association [16]. In contrast, the data from experimental studies are more clear-cut: several independent studies failed to find LV hypertrophy in *Hyp* mice [17,18] or in mice with a loss-of-function point mutation in the *Phex* gene [19]. Although LV hypertrophy appears to be absent at least in younger *Hyp* mice up to 9 months of age, we and others found mild hypertension in these mice [6,13]. The pathophysiological mechanisms underlying hypertension in *Hyp* mice are not entirely clear, but they may be related to an FGF23-mediated upregulation in renal sodium-chloride cotransporter (NCC) abundance [13]. Injection of recombinant FGF23 for five days into wild-type mice led to an increase in systolic, diastolic, and mean arterial pressure [13]. *Hyp* mice are characterized by chronic elevation of FGF23 and may, therefore, serve as a model to investigate the pathophysiology of hypertension in diseases associated with excessive FGF23 secretion.

An improved understanding of the mechanistic link between chronically elevated FGF23 and LV hypertrophy or other untoward cardiovascular side effects such as hypertension is of major importance, not only for XLH patients but for all diseases characterized by chronic elevations of circulating intact FGF23, such as CKD. Here, we sought to further elucidate the long term cardiovascular sequelae of elevated circulating levels of FGF23 in 14-month-old *Hyp* mice as a model of FGF23 excess. We found that aged *Hyp* mice lacked LV hypertrophy but were characterized by increased serum aldosterone and hypertension that could be rescued by inhibitors of the renin–angiotensin–aldosterone system (RAAS).

2. Materials and Methods

2.1. Animals

The study was undertaken in accordance with prevailing EU and national guidelines for animal care and welfare and in compliance with ARRIVE (Animal Research: Reporting of In Vivo Experiments) guidelines. All animal procedures were approved by the Ethics and Animal Welfare Committee of the University of Veterinary Medicine Vienna, Austria, and by the Austrian Federal Ministry of Education, Science and Research (permit number BMWFW-68.205/0054-II/3b/2013 and 2021-0.331.140).

Male wild-type (WT) controls and *Hyp* mice were bred by mating WT females with *Hyp* males on C57BL/6 background in our animal facility. Tail length at the time of weaning was used for genotyping. Animals were kept at 24 °C with a 12/12-h light/dark cycle, and were housed in stable groups of 2–5 mice from the time of weaning. They were fed a normal mouse diet (V1124-000, Sniff, Soest, Germany) containing 1.0% calcium, 0.7% phosphorus, and 1000 IU vitamin D/kg, and had access to food and tap water ad libitum. At necropsy, the mice were exsanguinated from the abdominal vena cava under general anesthesia with ketamine/medetomidine (100/0.25 mg/kg i.p.) for serum collection. No mice were excluded from the study.

2.2. Losartan and Canrenone Treatment

Some *Hyp* and WT mice were randomized to treatment with losartan (Sandoz, Austria, 50 mg) or vehicle (0.9% NaCl) via oral gavage twice a day or with subcutaneous injections of canrenone (Aldactone, Reimser, Germany, 200 mg/10 mL) or vehicle (0.9% NaCl) daily. Both treatments were given for 5 days.

2.3. Biochemical Analysis

Serum sodium, phosphate, and calcium were measured using a Cobas c111 analyzer (Roche, Mannheim, Germany). Serum intact FGF23 (Kainos, Tokyo, Japan), aldosterone (NovaTec Immundiagnostica, Dietzenbach, Germany), and intact PTH (Immutopics Inc., San Clemente, CA, USA) were detected using commercially available ELISA kits according to the manufacturer's instructions. Serum was extracted with diethylether and re-suspended in steroid-free serum (DRG Diagnostics, Marburg, Germany) for the aldosterone ELISA.

2.4. Echocardiography

Echocardiography was performed one day before the necropsies using a 14 MHz linear transducer (Siemens Accuson s2000, Munich, Germany) for evaluation of cardiac function. Mice were under 1% isoflurane anesthesia with a stable body temperature of 37 °C. M-mode in short-axis at the level of the papillary muscles was used to evaluate LV thickness, fractional shortening, and internal diameters in systole and diastole. At least four cardiac cycles were analyzed for each parameter.

2.5. Central Arterial and Cardiac Pressure Measurements

Central arterial pressure was measured by inserting a micro-tip catheter (1.4 Millar Instruments, Houston, TX, USA) into the ascending aorta via the right carotid artery under 1.5% isoflurane anesthesia. The catheter was then further advanced into the left ventricle to obtain cardiac pressure parameters. Traces were recorded for at least three minutes and analyzed via LabChart7 software (ADInstruments, Dunedin, New Zealand).

2.6. Augmentation Index

The aortic augmentation index was identified from the late systolic portion of the arterial pressure wave as described previously [20]. The augmentation index was defined as the height from the augmentation point to the systolic peak of the pressure wave divided by the pulse pressure and was expressed as a percentage.

2.7. Western Blot

Fresh frozen kidneys were homogenized in RIPA lysis buffer supplemented with phosphatase inhibitor cocktail (Roche) and protease inhibitor cocktail (Roche). Kidney tissue homogenates were mixed with Laemmli sample buffer, fractionated on SDS-PAGE (20 µg/well), and transferred to a nitrocellulose membrane (Thermo Fisher Scientific, Waltham, Massachusetts, USA). Immunoblots were incubated overnight at 4 °C with primary antibodies, including monoclonal rat anti-human Klotho (1:500, KO603, Trans Genic Inc., Tokyo, Japan), polyclonal rabbit anti-phospho-NCC (pThr53, 1:1000, Novus Biologicals, Littleton, CO, USA), and monoclonal mouse anti-β-actin (1:5000, Sigma, St. Louis, MS, USA) in 2% (*w*/*v*) bovine serum albumin (BSA, Sigma) and washed in TBS-T buffer (150 mM NaCl, 10 mM Tris/HCl pH 7.4, 0.2% *v*/*v* Tween-20). After washing, membranes were incubated with horseradish peroxidase-conjugated secondary antibodies (Amersham Biosciences, Buckinghamshire, UK). Specific signal was visualized using ECL kit (Amersham Life Sciences, Arlington Heights, IL, USA). The protein bands were quantified by Image Studio Lite 5.2 software (LI-COR, Bad Homburg, Germany).

2.8. Histological Evaluation

Hearts were fixed in 4% paraformaldehyde, paraffin-embedded, and cut into 5 µm sections. Fibrotic tissue was visualized by staining with picrosirius red according to a standard protocol. Total collagen was quantified using ImageJ software and was expressed as the ratio of collagen-stained area to total muscle area of the left ventricle and septum. For the analysis of cardiomyocyte size, cardiac sections were stained with FITC-labeled wheat germ agglutinin. At least ten random areas of the heart were measured, and only cardiomyocytes with well-defined borders and visible nuclei were used. Images were obtained by the Zeiss LSM 880 Airyscan confocal microscope and analyzed using semi-automated Image J software. All histological images were analyzed by two independent investigators in a blinded manner.

2.9. Immunofluorescence

Kidneys fixed in paraformaldehyde (PFA)-fixed were cut in 5-µm-thick paraffin sections. Sections were dewaxed, demasked for 20 min with proteinase K, and, after washing, pretreated with 10% normal goat serum in PBST for 60 min at room temperature (RT). Without rinsing, sections were incubated with polyclonal rabbit anti-pNCC (Novus Biologicals, 1:200) antibody at 4 °C overnight. After washing, sections were incubated for 1 h with goat anti-rabbit Alexa 594 (1:500, Invitrogen, Waltham, MA, USA). Nuclear staining was performed with DAPI (1:1000) for 5 min. Controls were performed by omitting primary antibodies. The slides were analyzed on a Zeiss LSM 880 Airyscan confocal microscope equipped with a 63× oil immersion lens (NA 1.3).

2.10. RNA Isolation and Quantitative RT-PCR

Total RNA was isolated from snap-frozen hearts after homogenization in a 24 Fast Prep machine using TRI Reagent. The nucleic acid concentration and integrity were determined by electrophoresis (Agilent Tapestation). Only samples that had a RIN value above seven were used. Two micrograms of RNA were transcribed into cDNA using the High Capacity cDNA Reverse Transcription Kit (Applied Biosystems, Waltham, MA, USA). Quantitative RT-PCR was performed on a qTOWER3/G qPCR device (Analytic Jena, Jena, Germany). The qPCR performed in a volume of 15 µL was composed of 1× PCR buffer B2 (Tris-HCl, $(NH_4)_2SO_4$, and Tween-20; Solis Biodyne, Tartu, Estonia), 3.5 mM $MgCl_2$, 200 µM dNTP mix (Solis Biodyne), 250 nM of each primer, and either 200 nM hydrolysis probe or 0.4 × EvaGreen (Biotium, Fremont, CA, USA), 1 U HOT FIREPol® DNA polymerase (Solis Biodyne), and 2 µL template DNA. For mouse primer sequences, see Supplementary Table S1. Cycling conditions consisted of an initial 15-min incubation step at 95 °C for polymerase activation and template denaturation, followed by 45 cycles of 95 °C denaturation for 15 s and 60 °C annealing and elongation for 60 s. All samples were measured in triplicate and normalized to two housekeeping genes (*Dpm1* and *Txnl4a*). The qPCR results were gained and primarily evaluated with the software qPCRsoft 4.1 (V4.1.3.0 Analytic Jena) and then analyzed using the standard delta delta Cq method.

2.11. Statistical Analysis

Statistical analysis was performed using Graph Pad Prism 9 or SPSS. The data were analyzed by two-sided t-test for two groups, one-way or two-way analysis of variance (ANOVA) to assess the influence of genotype and treatment, as well as their two-way interactions followed by the Student–Newman–Keuls multiple comparison test when comparing more than two groups; *p*-values of less than 0.05 were considered significant. Data are presented as scatter dot plots with bars depicting means ± SEM.

3. Results

Aged *Hyp* mice show a stimulation of the RAAS and a downregulation of renal αKlotho protein abundance, but they have normal kidney function.

As expected, 14-month-old male *Hyp* mice were characterized by an approximately 10-fold increase in serum intact FGF23 when compared to age- and sex-matched WT control animals (Figure 1A). The elevated circulating intact FGF23 levels in aged *Hyp* mice were associated with hypophosphatemia (Figure 1B). Serum calcium tended to be lower, whereas serum intact PTH tended to be higher in *Hyp* mice, but both effects did not reach statistical significance (Figure 1C,D). We reported earlier that 3-month-old *Hyp* mice were characterized by lower serum aldosterone levels relative to WT mice [13]. However, in aged *Hyp* mice, serum aldosterone was not down-, but rather up-regulated (Figure 1E). Aldosterone and FGF23 are known regulators of the sodium chloride cotransporter NCC activity in distal convoluted renal tubules [13]. Therefore, we quantified renal pNCC protein abundance by Western blotting and by immunofluorescence analysis. Similar to younger *Hyp* mice [13], renal pNCC levels were up-regulated and pNCC immunofluorescence tended to be higher in 14-month-old *Hyp* mice, relative to WT controls (Figure 1F,G). The abundance of αKlotho protein was distinctly down-regulated by about 50% in the kidneys of *Hyp* mice (Figure 1H), which may be a counter-regulatory mechanism to protect against the chronically elevated circulating FGF23 levels. However, renal function as evidenced by serum creatinine concentration and by renal creatinine clearance remained unchanged in aged *Hyp* mice (Figure 1I,J).

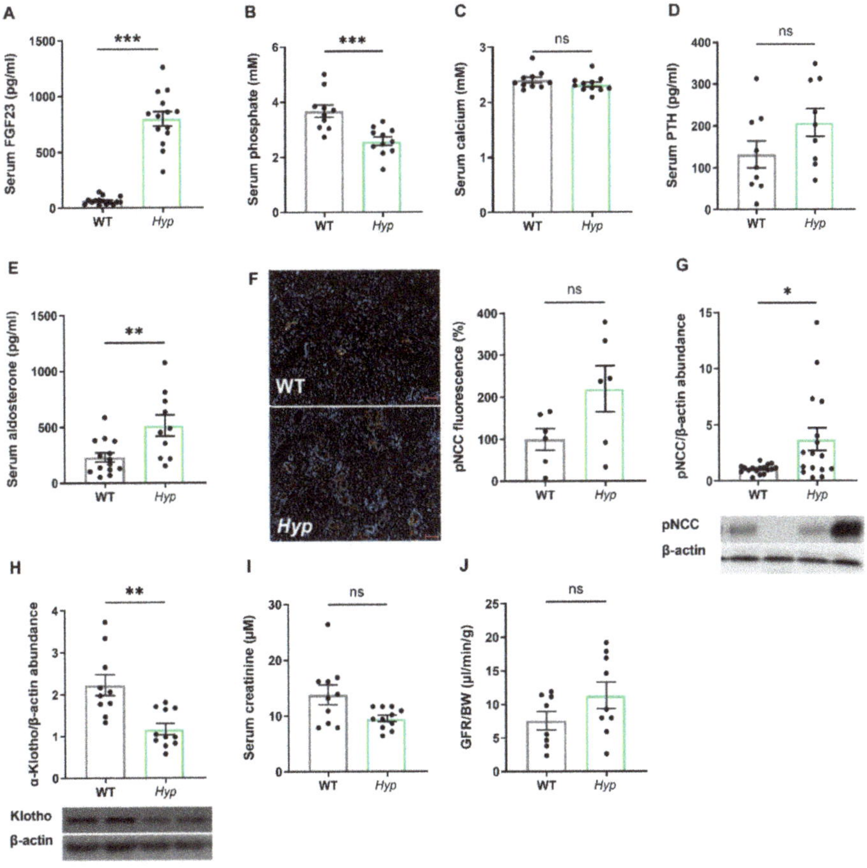

Figure 1. Aged *Hyp* mice show a stimulation of the RAAS and a down-regulation of renal αKlotho protein abundance, but have normal kidney function. (A) Serum intact FGF23 levels

(n = 14–15), (**B**,**C**) serum phosphate and calcium concentrations (n = 10–11), and (**D**,**E**) serum intact PTH and aldosterone levels (n = 9–14) in 14-month-old male wildtype (WT) and *Hyp* mice. (**F**) Left: representative images of immunofluorescent staining of kidney paraffin sections with an anti-pNCC antibody (original magnification 400×). Right: quantification of anti-pNCC immunofluorescence in the kidney in 14-month-old male WT and *Hyp* mice (n = 6). Western blot quantification of (**G**) pNCC and (**H**) α-Klotho from kidney cortex homogenates (n = 6–11), as well as (**I**) serum creatinine levels (n = 10–11) and (**J**) glomerular filtration rate per gram body weight as measured by renal creatinine clearance (n = 8–9) in 14-month-old male *Hyp* mice and WT controls. Bars in (**A–J**) represent mean ± SEM for WT and *Hyp* mice. * $p < 0.05$, ** $p < 0.01$, *** $p < 0.001$ by Student's *t*-test. ns, not significant.

3.1. Elevated Levels of Circulating FGF23 Are Associated with Mild Hypertension in Aged Hyp Mice but Do Not Cause LVH

Several earlier reports have failed to find LV hypertrophy in *Hyp* mice of up to nine months of age [17,18]. However, data about potential changes in cardiovascular function in aged *Hyp* mice as a model of long term FGF23 excess are scarce. Therefore, we examined the cardiovascular phenotype of aged *Hyp* mice by intraarterial and left ventricular catheterization as well as by echocardiography. The HW/BW ratio was distinctly increased in *Hyp* mice relative to WT controls (Figure 2A). However, this increase was mainly driven by lower body weight in *Hyp* mice, and not by a higher heart mass (Figure 2B,C), questioning the relevance of the HW/BW ratio as a read-out for heart hypertrophy in this animal model. Notably, intraarterial catheterization revealed a higher mean arterial pressure (MAP) of about 12 mmHg in aged *Hyp* mice compared to WT controls (Figure 2D). Despite the presence of hypertension, LV function as evidenced by fractional shortening and ejection fraction measured by echocardiography was actually improved in *Hyp* compared with WT mice (Figure 2E,F). In agreement with the echocardiography data, LV contractility as measured by Max dP/dt during LV catheterization was unchanged in 14-month-old *Hyp* mice (Figure 2G). Moreover, there was no difference in mean cardiomyocyte size as measured by wheat germ agglutinin (WGA) staining or in LV collagen content as measured by picrosirius red staining between WT and *Hyp* mice (Figure 2H,I). The absence of LV hypertrophy in *Hyp* mice was further confirmed by qRT-PCR analysis of typical markers of hypertrophy, such as *Anp* and *Bnp*, which remained unaltered between the genotypes (Figure 2J,K). Taken together, these data confirm the presence of mild hypertension in aged *Hyp* mice but strongly argue against LV hypertrophy and functional impairment in these mice.

To further address the question of what drives hypertension in aged *Hyp* mice, we measured the augmentation index in the carotid artery by pulse wave analysis. The latter analysis revealed an increase in augmentation index in aged *Hyp* mice relative to WT controls (Figure 2L). This finding may point to increased peripheral vascular resistance in aged *Hyp* mice. In combination with the increase in aldosterone levels found in *Hyp* mice (Figure 1E), we hypothesized that hypertension in aged *Hyp* mice may be due to crosstalk between FGF23 and RAAS signaling in the kidney and blood vessels, leading to a combination of volume overload through increased aldosterone and FGF23 secretion and increased peripheral vascular resistance by elevated angiotensin II levels.

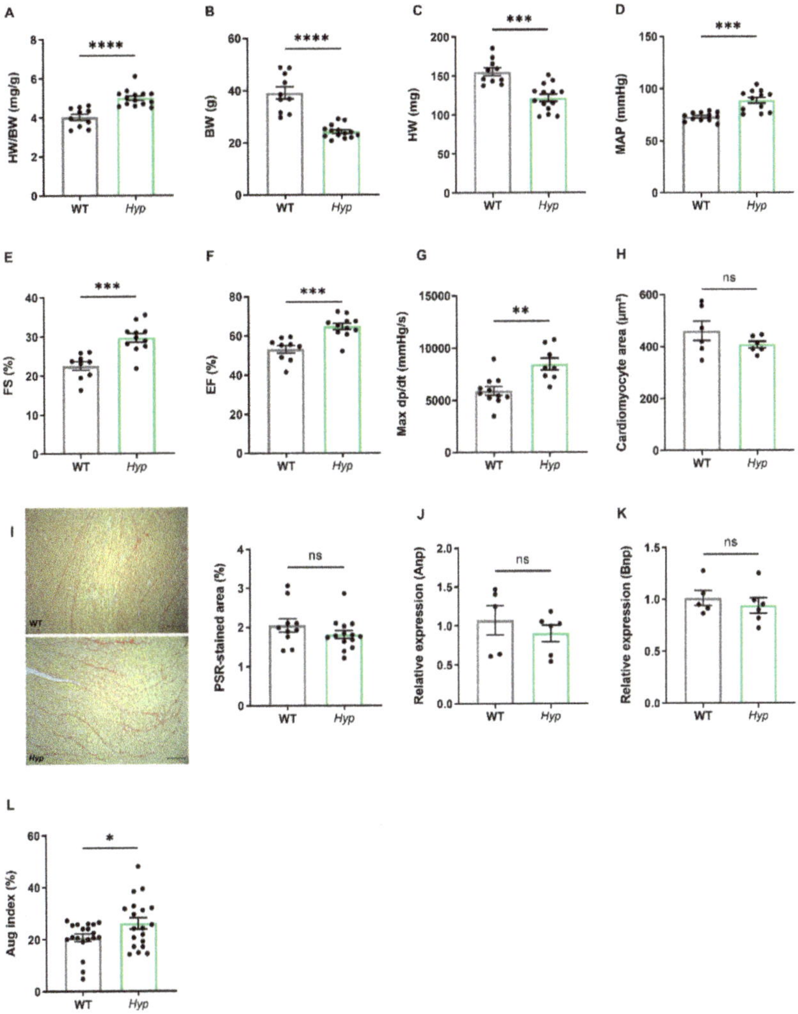

Figure 2. *Hyp* mice are hypertensive but do not develop LV hypertrophy. (**A**) Heart/body weight ratio (HW/BW), (**B**) body weight (BW), and (**C**) heart weight (HW) (n = 4–14) in 14-month-old male WT and *Hyp* mice. (**D**) Mean arterial pressure measured by arterial catheterization (n = 12), (**E**,**F**) fractional shortening (FS) and ejection fraction (EF) measured by echocardiography (n = 9–11), and (**G**) LV contractility (Max dP/dt) measured by intracardiac catheterization (n = 12) in 14-month-old male WT and *Hyp* mice. (**H**) Quantification of mean cardiomyocyte size after FITC-WGA staining (n = 6), (**I**) left, representative images of collagen staining using picrosirius red (PSR) in cardiac paraffin sections (bar = 100 µm), right, quantification of fibrosis as measured by PSR-stained area (n = 9–12), (**J**,**K**) relative mRNA expression of markers of hypertrophy, (**J**) atrial natriuretic peptide (Anp) and (**K**) brain natriuretic peptide (Bnp) (n = 5), and (**L**) augmentation index (AI) measured by pulse wave analysis (n = 19) in 14-month-old male WT and *Hyp* mice. Bars in (**A**–**L**) represent mean ± SEM. * $p < 0.05$, ** $p < 0.01$, *** $p < 0.001$, **** $p < 0.0001$ by Student's *t*-test. ns, not significant.

3.2. Inhibition of RAAS Signaling Normalizes Blood Pressure in Hyp Mice

To test this hypothesis, we administered the aldosterone antagonist canrenone and the angiotensin receptor blocker losartan over five days to 12- to 14-month-old male *Hyp* and

WT mice. Intraarterial catheterization confirmed hypertension in vehicle-treated *Hyp* mice compared to WT controls. Interestingly, daily subcutaneous injections of 30 mg/kg canrenone led to a distinct decrease in systolic, diastolic, and mean arterial pressure in *Hyp* mice relative to vehicle-treated *Hyp* mice (Figure 3A–C). In contrast, WT mice injected with canrenone showed only minor reductions in mean, systolic, and diastolic blood pressure (Figure 3A–C). Losartan was administered via oral gavage at the dose of 30 mg/kg twice daily, and the effects of losartan were examined one hour after the last administration, due to a shorter half-life of this drug compared with canrenone. Similar to canrenone, losartan caused a pronounced decrease in systolic, diastolic, and mean arterial pressure in *Hyp* mice, relative to vehicle-treated *Hyp* mice, but had only negligible effects in WT animals (Figure 3D–F).

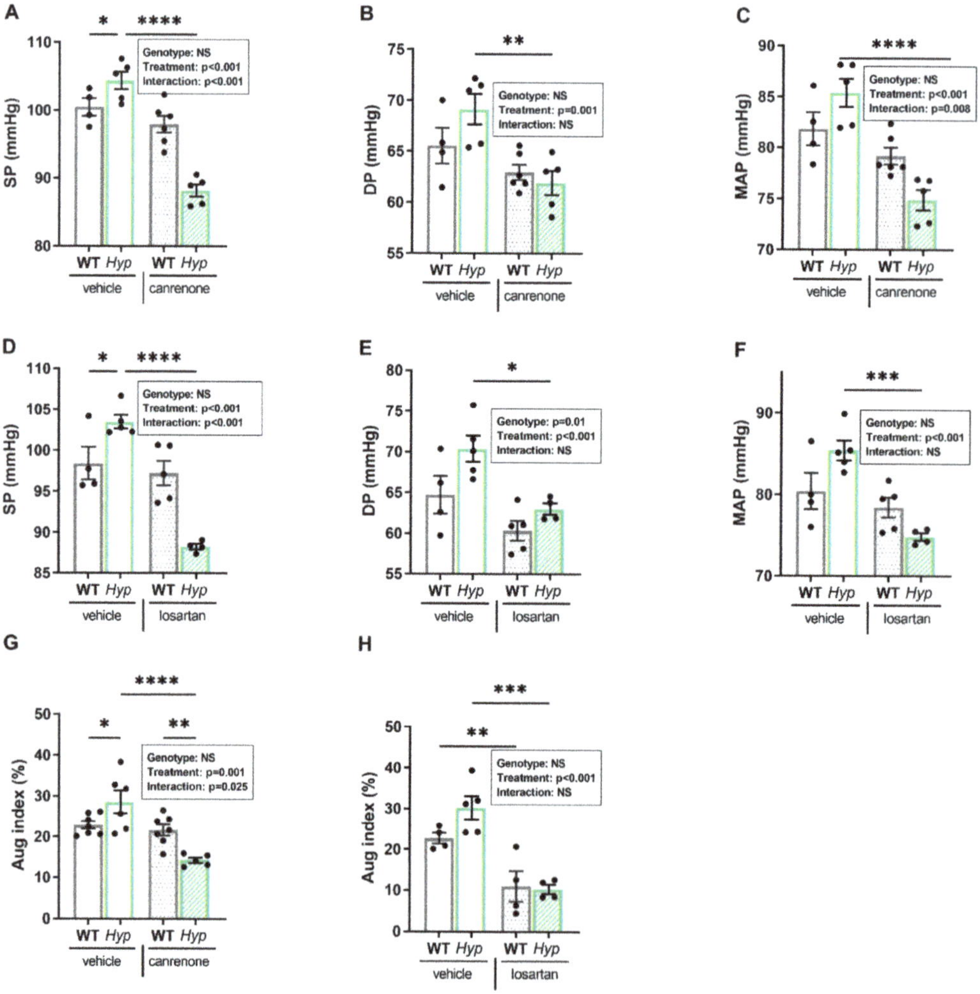

Figure 3. Inhibition of RAAS signaling normalizes blood pressure in *Hyp* mice. (**A**,**D**) Systolic (SP), (**B**,**E**) diastolic (DP), and (**C**,**F**) mean arterial blood pressure (MAP), as well as (**G**,**H**) augmentation index (AI) in 12–14-month-old male WT and *Hyp* mice treated over 5 days with 30 mg/kg canrenone (s.c. once daily) or 30 mg/kg losartan (gavage twice daily) (n = 4–8). Bars represent mean ± SEM. * $p < 0.05$, ** $p < 0.01$, *** $p < 0.001$, **** $p < 0.0001$ vs. vehicle by one-way ANOVA followed by Student–Newman–Keuls post-hoc test. Insets show results of two-way ANOVA.

To determine whether the beneficial effect of RAAS inhibition on hypertension was associated with a decrease in peripheral vascular resistance, we analyzed the augmentation index in WT and *Hyp* mice post-treatment. Interestingly, both treatments reduced the elevated augmentation index in *Hyp* mice. Canrenone did not alter the augmentation index in WT mice, but significantly lowered the augmentation index in *Hyp* mice, whereas losartan had a similar effect in both genotypes (Figure 3G,H). Two-way ANOVA revealed a significant two-way interaction between genotype and treatment for systolic pressure ($p < 0.001$), mean arterial pressure ($p < 0.01$), and augmentation index ($p < 0.05$) in animals treated with canrenone, whereas this interaction was only found significant for the reduction in systolic pressure ($p < 0.01$) in *Hyp* mice treated with losartan (Figure 3A–H). Collectively, our data suggest that the blood pressure-lowering response of *Hyp* mice to both canrenone and losartan is exaggerated compared with WT mice, supporting our hypothesis that activation of RAAS signaling contributes to hypertension in aged *Hyp* mice. Because both canrenone and losartan reduced blood pressure and augmentation index in aged *Hyp* mice, it is likely that the contributing effect of the RAAS to the development of hypertension is mediated through a combination of increased peripheral resistance together with increased blood volume.

4. Discussion

The central aim of this study was to characterize the pathophysiological role of long term endogenous FGF23 hypersecretion in the cardiovascular system of 14-month-old male *Hyp* mice. We found that aged *Hyp* mice were mildly hypertensive but did not develop LV hypertrophy. Hypertension in aged *Hyp* mice was associated with increases in serum aldosterone levels, in vascular peripheral resistance, and in renal pNCC abundance. Administration of the angiotensin II receptor blocker losartan and the mineralocorticoid receptor blocker canrenone rescued the cardiovascular phenotype by lowering blood pressure and vascular resistance in *Hyp* mice.

Elevated circulating FGF23 levels are associated with accelerated disease progression, morbidity, and/or mortality in several clinical disorders, including CKD but also cardiac failure [8–10,21]. It was proposed that excessive FGF23 causes LV hypertrophy by Klotho-independent binding to FGFR4 and subsequent activation of the calcineurin/NFAT pathway in cardiomyocytes [11,12]. However, it is still unclear whether this disease mechanism is relevant in the setting of normal kidney function, because evidence from clinical studies in XLH patients and mouse models of XLH suggests that chronic FGF23 excess does not invariably cause LV hypertrophy [15–18]. In our study, an increase in circulating intact FGF23 levels of ~10-fold was not sufficient to induce an increase in LV size or deterioration of LV function in 14-month-old *Hyp* mice when compared to WT littermates. Fractional shortening, ejection fraction, and molecular markers of hypertrophy (*Anp*, *Bnp*) all remained unchanged in these animals. This is in line with previous studies in animal models of XLH that did not find any association between increased levels of FGF23 and LV hypertrophy [17,19].

The current study has shown that aged *Hyp* mice are characterized by a small increase in systolic, diastolic, and mean arterial blood pressure, corroborating earlier studies in younger *Hyp* mice [13,22]. Although hypertension is not a universal complication in XLH patients [14], early-onset hypertension is frequently found in adult XLH patients [23]. Based on the sodium-conserving function of FGF23, elevated circulating FGF23 may predispose subjects to the development of hypertension through volume expansion [13]. However, an increase in serum intact FGF23 levels of about 6-fold did not increase blood pressure in mice with a loss-of-function point mutation in the *Phex* gene that are characterized by normal kidney function [19]. Rather, the latter study reported that systolic pressure was actually slightly reduced in aged $Phex^{C733R}$ male mice relative to WT controls. A possible explanation for these discrepant findings may be the greater increase in FGF23 serum levels in *Hyp* mice. However, elevated circulating FGF23 levels per se may not be sufficient to cause hypertensive changes in the absence of other contributing factors.

Our data have revealed for the first time that serum aldosterone as well as augmentation index were increased in aged *Hyp* mice. Although we did not measure renin activity and angiotensin II levels, these findings suggest a general stimulation of the RAAS in aged *Hyp* mice. In contrast, we found lower aldosterone levels in 3-month-old *Hyp* mice [13]. We do not have a conclusive explanation for the discrepancy between young and aged *Hyp* mice regarding aldosterone secretion. There may be age-related changes in aldosterone secretion in *Hyp* mice or differences in the interaction between FGF23 signaling and the RAAS in young, growing vs. aged, non-growing *Hyp* mice. Nevertheless, the findings in the current study led us to hypothesize that hypertension in aged *Hyp* mice may be caused by increased RAAS signaling alone or in combination with FGF23-mediated increased sodium absorption, leading to a combination of volume overload and increased peripheral vascular resistance. Aldosterone, similar to FGF23, enhances sodium and water reabsorption in the distal nephron, indirectly increasing blood pressure [24–26]. Furthermore, it was shown previously that Ang II administration to *Hyp* mice resulted in additive effects on blood pressure [21]. We used short term (5-day) treatment with the mineralocorticoid receptor blocker canrenone and the angiotensin II receptor type 1 antagonist losartan as tools to dissect the renal and cardiovascular effects of aldosterone and angiotensin II. However, both drugs lowered blood pressure and decreased peripheral resistance in aged *Hyp* mice. Therefore, it is likely that a combination of volume effects and increased peripheral vascular resistance is involved in the hypertension-promoting effects of RAAS stimulation in aged *Hyp* mice. Interestingly, these drugs had little effect in WT mice, further corroborating that augmented RAAS signaling contributes to the development of hypertension in *Hyp* mice.

Although our study has provided novel insights into the pathogenesis of hypertension in *Hyp* mice, a key question in this context remains unanswered: what is driving the stimulation of RAAS in aged *Hyp* mice? It has been suggested that FGF23 hypersecretion may stimulate the RAAS through the suppression of vitamin D hormone production in CKD [27,28]. However, whether RAAS is regulated by vitamin D remains a controversial issue [29], and it is currently not known if this proposed mechanism may also be relevant for *Hyp* mice. It is clear that further experimentation is required to define the molecular basis of the crosstalk between FGF23 and the RAAS. This is not only relevant for XLH patients, but also has much broader implications for diseases characterized by a combination of chronic FGF23 hypersecretion and increased RAAS signaling, such as CKD [30].

Supplementary Materials: The following supporting information can be downloaded at: https://www.mdpi.com/article/10.3390/biomedicines10071691/s1, Table S1. Mouse primer sequences for quantitative real-time PCR analysis

Author Contributions: N.L. and R.G.E. conceived and designed the studies; N.L., A.Z. and D.F. performed experiments and analyzed the data; N.L. and R.G.E. wrote the manuscript; all authors discussed and reviewed the manuscript. All authors have read and agreed to the published version of the manuscript.

Funding: Open Access Funding by the University of Veterinary Medicine Vienna.

Institutional Review Board Statement: All animal procedures were approved by the Ethics and Animal Welfare Committee of the University of Veterinary Medicine Vienna, Austria, and by the Austrian Federal Ministry of Education, Science and Research (permit number BMWFW-68.205/0054-II/3b/2013 and 2021-0.331.140).

Informed Consent Statement: Not applicable.

Data Availability Statement: All data generated or analyzed in this study are included in this article.

Acknowledgments: We thank C. Schüler, C. Bergow, A. Petric, and N. Ginner for excellent technical assistance, and Ute Zeitz for help with animal breeding.

Conflicts of Interest: The authors have declared that no conflict of interest exists.

References

1. Jonsson, K.B.; Zahradnik, R.; Larsson, T.; White, K.E.; Sugimoto, T.; Imanishi, Y.; Yamamoto, T.; Hampson, G.; Koshiyama, H.; Ljunggren, O.; et al. Fibroblast growth factor 23 in oncogenic osteomalacia and X-linked hypophosphatemia. *N. Engl. J. Med.* **2003**, *348*, 1656–1663. [CrossRef] [PubMed]
2. Tenenhouse, H.S. X-linked hypophosphataemia: A homologous disorder in humans and mice. *Nephrol. Dial. Transpl.* **1999**, *14*, 333–341. [CrossRef] [PubMed]
3. Aono, Y.; Hasegawa, H.; Yamazaki, Y.; Shimada, T.; Fujita, T.; Yamashita, T.; Fukumoto, S. Anti-FGF-23 neutralizing antibodies ameliorate muscle weakness and decreased spontaneous movement of Hyp mice. *J. Bone Miner. Res.* **2011**, *26*, 803–810. [CrossRef] [PubMed]
4. Urakawa, I.; Yamazaki, Y.; Shimada, T.; Iijima, K.; Hasegawa, H.; Okawa, K.; Fujita, T.; Fukumoto, S.; Yamashita, T. Klotho converts canonical FGF receptor into a specific receptor for FGF23. *Nature* **2006**, *444*, 770–774. [CrossRef]
5. Chen, G.; Liu, Y.; Goetz, R.; Fu, L.; Jayaraman, S.; Hu, M.-C.; Moe, O.W.; Liang, G.; Li, X.; Mohammadi, M. α-Klotho is a non-enzymatic molecular scaffold for FGF23 hormone signalling. *Nature* **2018**, *553*, 461–466. [CrossRef] [PubMed]
6. Martin, A.; David, V.; Quarles, L.D. Regulation and Function of the FGF23/Klotho Endocrine Pathways. *Physiol. Rev.* **2012**, *92*, 131–155. [CrossRef]
7. Huang, X.; Jiang, Y.; Xia, W. FGF23 and Phosphate Wasting Disorders. *Bone Res.* **2013**, *1*, 120–132. [CrossRef]
8. Isakova, T.; Xie, H.; Yang, W.; Xie, D.; Anderson, A.H.; Scialla, J.; Wahl, P.; Gutiérrez, O.M.; Steigerwalt, S.; He, J.; et al. Fibroblast growth factor 23 and risks of mortality and end-stage renal disease in patients with chronic kidney disease. *JAMA* **2011**, *305*, 2432–2439. [CrossRef]
9. Chonchol, M.; Greene, T.; Zhang, Y.; Hoofnagle, A.N.; Cheung, A.K. Low Vitamin D and High Fibroblast Growth Factor 23 Serum Levels Associate with Infectious and Cardiac Deaths in the HEMO Study. *J. Am. Soc. Nephrol.* **2016**, *27*, 227–237. [CrossRef]
10. Faul, C.; Amaral, A.P.; Oskouei, B.; Hu, M.C.; Sloan, A.; Isakova, T.; Gutiérrez, O.M.; Aguillon-Prada, R.; Lincoln, J.; Hare, J.M.; et al. FGF23 induces left ventricular hypertrophy. *J. Clin. Investig.* **2011**, *121*, 4393–4408. [CrossRef] [PubMed]
11. Grabner, A.; Amaral, A.P.; Schramm, K.; Singh, S.; Sloan, A.; Yanucil, C.; Li, J.; Shehadeh, L.A.; Hare, J.M.; David, V.; et al. Activation of Cardiac Fibroblast Growth Factor Receptor 4 Causes Left Ventricular Hypertrophy. *Cell Metab.* **2015**, *22*, 1020–1032. [CrossRef] [PubMed]
12. Han, X.; Cai, C.; Xiao, Z.; Quarles, L.D. FGF23 induced left ventricular hypertrophy mediated by FGFR4 signaling in the myocardium is attenuated by soluble Klotho in mice. *J. Mol. Cell. Cardiol.* **2020**, *138*, 66–74. [CrossRef] [PubMed]
13. Andrukhova, O.; Slavic, S.; Smorodchenko, A.; Zeitz, U.; Shalhoub, V.; Lanske, B.; Pohl, E.E.; Erben, R.G. FGF23 regulates renal sodium handling and blood pressure. *EMBO Mol. Med.* **2014**, *6*, 744–759. [CrossRef] [PubMed]
14. Nehgme, R.; Fahey, J.T.; Smith, C.; Carpenter, T.O. Cardiovascular abnormalities in patients with X-linked hypophosphatemia. *J. Clin. Endocrinol. Metab.* **1997**, *82*, 2450–2454. [CrossRef] [PubMed]
15. Hernández-Frías, O.; Gil-Peña, H.; Pérez-Roldán, J.M.; González-Sanchez, S.; Ariceta, G.; Chocrón, S.; Loza, R.; de la Cerda Ojeda, F.; Madariaga, L.; Vergara, I.; et al. Risk of cardiovascular involvement in pediatric patients with X-linked hypophosphatemia. *Pediatr. Nephrol.* **2019**, *34*, 1077–1086. [CrossRef]
16. Takashi, Y.; Kinoshita, Y.; Hori, M.; Ito, N.; Taguchi, M.; Fukumoto, S. Patients with FGF23-related hypophosphatemic rickets/osteomalacia do not present with left ventricular hypertrophy. *Endocr. Res.* **2017**, *42*, 132–137. [CrossRef]
17. Liu, E.S.; Thoonen, R.; Petit, E.; Yu, B.; Buys, E.S.; Scherrer-Crosbie, M.; Demay, M.B. Increased Circulating FGF23 Does Not Lead to Cardiac Hypertrophy in the Male Hyp Mouse Model of XLH. *Endocrinology* **2018**, *159*, 2165–2172. [CrossRef]
18. Leifheit-Nestler, M.; Richter, B.; Basaran, M.; Nespor, J.; Vogt, I.; Alesutan, I.; Voelkl, J.; Lang, F.; Heineke, J.; Krick, S.; et al. Impact of Altered Mineral Metabolism on Pathological Cardiac Remodeling in Elevated Fibroblast Growth Factor 23. *Front. Endocrinol.* **2018**, *9*, 333. [CrossRef]
19. Pastor-Arroyo, E.-M.; Gehring, N.; Krudewig, C.; Costantino, S.; Bettoni, C.; Knöpfel, T.; Sabrautzki, S.; Lorenz-Depiereux, B.; Pastor, J.; Strom, T.M.; et al. The elevation of circulating fibroblast growth factor 23 without kidney disease does not increase cardiovascular disease risk. *Kidney Int.* **2018**, *94*, 49–59. [CrossRef]
20. Andrukhova, O.; Slavic, S.; Zeitz, U.; Riesen, S.C.; Heppelmann, M.S.; Ambrisko, T.D.; Markovic, M.; Kuebler, W.M.; Erben, R.G. Vitamin D Is a Regulator of Endothelial Nitric Oxide Synthase and Arterial Stiffness in Mice. *Mol. Endocrinol.* **2014**, *28*, 53–64. [CrossRef]
21. Pi, M.; Ye, R.; Han, X.; Armstrong, B.; Liu, X.; Chen, Y.; Sun, Y.; Quarles, L.D. Cardiovascular Interactions between Fibroblast Growth Factor-23 and Angiotensin II. *Sci. Rep.* **2018**, *8*, 12398. [CrossRef] [PubMed]
22. Han, X.; Ross, J.; Kolumam, G.; Pi, M.; Sonoda, J.; King, G.; Quarles, L.D. Cardiovascular Effects of Renal Distal Tubule Deletion of the FGF Receptor 1 Gene. *J. Am. Soc. Nephrol.* **2018**, *29*, 69–80. [CrossRef] [PubMed]
23. Nakamura, Y.; Takagi, M.; Takeda, R.; Miyai, K.; Hasegawa, Y. Hypertension is a characteristic complication of X-linked hypophosphatemia. *Endocr. J.* **2017**, *64*, 283–289. [CrossRef] [PubMed]
24. Goodfriend, T.L. Aldosterone—A Hormone of Cardiovascular Adaptation and Maladaptation. *J. Clin. Hypertens.* **2006**, *8*, 133–139. [CrossRef] [PubMed]
25. Zhao, D.; Seth, D.M.; Navar, L.G. Enhanced Distal Nephron Sodium Reabsorption in Chronic Angiotensin II–Infused Mice. *Hypertension* **2009**, *54*, 120–126. [CrossRef]

26. Chen, S.Y.; Bhargava, A.; Mastroberardino, L.; Meijer, O.C.; Wang, J.; Buse, P.; Firestone, G.L.; Verrey, F.; Pearce, D. Epithelial sodium channel regulated by aldosterone-induced protein sgk. *Proc. Natl. Acad. Sci. USA* **1999**, *96*, 2514–2519. [CrossRef]
27. Freundlich, M.; Cuervo, C.; Abitbol, C.L. Fibroblast growth factor 23 and tubular sodium handling in young patients with incipient chronic kidney disease. *Clin. Kidney J.* **2020**, *13*, 389–396. [CrossRef]
28. Freundlich, M.; Gamba, G.; Rodriguez-Iturbe, B. Fibroblast growth factor 23—Klotho and hypertension: Experimental and clinical mechanisms. *Pediatric Nephrol.* **2021**, *36*, 3007–3022. [CrossRef]
29. Latic, N.; Erben, R.G. Vitamin D and Cardiovascular Disease, with Emphasis on Hypertension, Atherosclerosis, and Heart Failure. *Int. J. Mol. Sci.* **2020**, *21*, 6483. [CrossRef]
30. Epstein, M.; Freundlich, M. The intersection of Mineralocorticoid Receptor (MR) activation and the FGF23—Klotho cascade. A Duopoly that promotes renal and cardiovascular injury. *Nephrol. Dial. Transpl.* **2022**, *37*, 211–221. [CrossRef]

biomedicines

MDPI

Article

Long-Term Excessive Dietary Phosphate Intake Increases Arterial Blood Pressure, Activates the Renin–Angiotensin–Aldosterone System, and Stimulates Sympathetic Tone in Mice

Nejla Latic [1], Mirko Peitzsch [2], Ana Zupcic [1], Jens Pietzsch [3,4] and Reinhold G. Erben [1,*]

1 Department of Biomedical Sciences, University of Veterinary Medicine, 1210 Vienna, Austria
2 Institute of Clinical Chemistry and Laboratory Medicine, University Hospital Carl Gustav Carus, 01307 Dresden, Germany
3 Helmholtz-Zentrum Dresden-Rossendorf, Institute of Radiopharmaceutical Cancer Research, Department of Radiopharmaceutical and Chemical Biology, 01328 Dresden, Germany
4 School of Science, Faculty of Chemistry and Food Chemistry, Technische Universität Dresden, 01062 Dresden, Germany
* Correspondence: reinhold.erben@vetmeduni.ac.at; Tel.: +43-1-250-774-550

Abstract: Increased dietary phosphate intake has been associated with severity of coronary artery disease, increased carotid intima–media thickness, left ventricular hypertrophy (LVH), and increased cardiovascular mortality and morbidity in individuals with normal renal function as well as in patients suffering from chronic kidney disease. However, the underlying mechanisms are still unclear. To further elucidate the cardiovascular sequelae of long-term elevated phosphate intake, we maintained male C57BL/6 mice on a calcium, phosphate, and lactose-enriched diet (CPD, 2% Ca, 1.25% P, 20% lactose) after weaning them for 14 months and compared them with age-matched male mice fed a normal mouse diet (ND, 1.0% Ca, 0.7% P). Notably, the CPD has a balanced calcium/phosphate ratio, allowing the effects of elevated dietary phosphate intake largely independent of changes in parathyroid hormone (PTH) to be investigated. In agreement with the rationale of this experiment, mice maintained on CPD for 14 months were characterized by unchanged serum PTH but showed elevated concentrations of circulating intact fibroblast growth factor-23 (FGF23) compared with mice on ND. Cardiovascular phenotyping did not provide evidence for LVH, as evidenced by unchanged LV chamber size, normal cardiomyocyte area, lack of fibrosis, and unchanged molecular markers of hypertrophy (*Bnp*) between the two groups. However, intra-arterial catheterization revealed increases in systolic pressure, mean arterial pressure, and pulse pressure in mice fed the CPD. Interestingly, chronically elevated dietary phosphate intake stimulated the renin–angiotensin–aldosterone system (RAAS) as evidenced by increased urinary aldosterone in animals fed the CPD, relative to the ND controls. Furthermore, the catecholamines epinephrine, norepinephrine, and dopamine as well as the catecholamine metabolites metanephrine. normetanephrine and methoxytyramine as measured by mass spectrometry were elevated in the urine of mice on CPD, relative to mice on the ND. These changes were partially reversed by switching 14-month-old mice on CPD back to ND for 2 weeks. In conclusion, our data suggest that excess dietary phosphate induces a rise in blood pressure independent of secondary hyperparathyroidism, and that this effect may be mediated through activation of the RAAS and stimulation of the sympathetic tone.

Keywords: hypertension; left ventricular hypertrophy; renin–angiotensin–aldosterone system; cardiovascular disease; catecholamines; liquid chromatography–tandem mass spectrometry

Citation: Latic, N.; Peitzsch, M.; Zupcic, A.; Pietzsch, J.; Erben, R.G. Long-Term Excessive Dietary Phosphate Intake Increases Arterial Blood Pressure, Activates the Renin–Angiotensin–Aldosterone System, and Stimulates Sympathetic Tone in Mice. *Biomedicines* **2022**, *10*, 2510. https://doi.org/10.3390/biomedicines10102510

Academic Editors: Josef Zicha and Ivana Vaněčková

Received: 22 July 2022
Accepted: 5 October 2022
Published: 7 October 2022

1. Introduction

Phosphate is part of the fundamental chemical necessary for cellular structure, signaling and energy production, making it essential for various biological processes. Phosphate metabolism is mainly regulated by the vitamin D hormone ($1{,}25(OH)_2D_3$) and by two major phosphaturic hormones, fibroblast growth factor 23 (FGF23) and parathyroid hormone (PTH). FGF23 and PTH act on proximal tubules in the kidney to increase renal phosphate excretion by inhibiting reabsorption of filtered phosphate. Therefore, FGF23 and PTH lower phosphate levels in the blood. $1{,}25(OH)_2D_3$ on the other hand stimulates phosphate absorption from the intestine, thereby increasing serum phosphate levels [1].

Despite its importance for various processes in the body, the accumulation of phosphate in the blood, hyperphosphatemia, can have deleterious effects. Epidemiological and observational studies have reported an association between increased serum phosphate levels and dietary phosphate load with left ventricular hypertrophy (LVH), cardiac calcification, as well as morbidity and mortality in patients suffering from chronic kidney disease (CKD) [2–5]. Cardiovascular events are the most frequent cause of death in CKD patients [6]. As the glomerular filtration rate decreases in CKD, the kidney is unable to adequately excrete phosphate, leading to hyperphosphatemia. Hyperphosphatemia has been identified as an independent risk factor for accelerated cardiovascular disease (CVD) development in these patients, but the exact mechanisms have not yet been elucidated [7]. Interestingly, high serum phosphate levels are also associated with an increased risk for CVD in adults with no history of CVD or CKD [8–10].

High dietary phosphate may also be linked to hypertension. Hypertension is a major health burden affecting about 1.3 billion people worldwide [11]. The existing epidemiological data on the effect of dietary phosphate on hypertension are inconsistent. While some studies have found a clear association between phosphate intake and hemodynamic parameters, others have not observed changes in blood pressure that were attributable to a high dietary phosphate load [10,12,13]. A recent intervention study shed additional light on this issue. Mohammad and coworkers examined the effects of a high phosphate diet in healthy subjects with normal renal function, and found that a 6-week dietary phosphate load induced small, but significant increases in systolic and diastolic blood pressure that were accompanied by activation of the sympathetic tone, relative to subjects receiving the low phosphate diet [10].

It still remains an open question whether the untoward effects of excess dietary phosphate on the cardiovascular system are direct or indirect. Experimental studies have attempted to clarify the mechanisms involved. Studies supporting a direct role of phosphate postulate that hyperphosphatemia leads to arterial calcification by directly stimulating the differentiation of vascular smooth muscle cells into osteoblast-like cells [14–16]. Furthermore, a study in rat aortas demonstrated that an increase in extracellular phosphate is followed by an increase in oxidative stress, a decrease in nitric oxide production, and inhibitory phosphorylation of endothelial nitric oxide synthase (eNOS) [17]. On the other hand, phosphate may act indirectly by inducing changes in the endocrine regulators of phosphate homeostasis, FGF23, PTH and $1{,}25(OH)_2D_3$. An increase in serum phosphate stimulates FGF23 and PTH secretion, while at the same time inhibiting production of $1{,}25(OH)_2D_3$. In this context, it is established that FGF23 causes LVH via FGF receptor 4-dependent activation of the calcineurin-NFAT signaling pathway [17]. PTH may cause hypertension by its direct effects on arteries and myocytes to promote arterial stiffness and LVH [18]. Indeed, Bozic and coworkers found an increase in blood pressure and LVH in normotensive and spontaneously hypertensive rats with normal kidney function that was driven by an increase in PTH [19].

It is clear that elucidation of the mechanisms underlying the potential hypertensive and hypertrophy-promoting effects of a high phosphate intake may have major implications for the prevention of CVD and its sequelae in humans. Previous experimental studies in rodents mostly used a phosphate-enriched diet to investigate the effects of an increased dietary phosphate intake on the cardiovascular system or the kidney [19–22]. However, an isolated increase in dietary phosphate leads to secondary hyperparathyroidism, making it difficult to dissect the effects of phosphate from that of increased PTH on cardiovascular endpoints. In this study, we sought to further explore the cardiovascular sequelae of long-term elevated phosphate intake in aged mice, largely independent of changes in PTH. To this end, we maintained male C57BL/6 mice on a calcium, phosphate, and lactose-enriched diet (CPD) with a balanced calcium/phosphate ratio for 14 months and compared them with age-matched mice fed a normal mouse diet. We found that 14-month-old mice on a CPD were characterized by hypertension and increased arterial stiffness, and that these changes were associated with increased urinary aldosterone excretion and augmented sympathetic tone.

2. Material and Methods

2.1. Animals

The study was undertaken in accordance with prevailing EU and national guidelines for animal care and welfare and in compliance with ARRIVE guidelines. All animal procedures were approved by the Ethics and Animal Welfare Committee of the University of Veterinary Medicine Vienna, Austria, and by the Austrian Federal Ministry of Education, Science and Research (permit number BMWFW-68.205/0188-WF/V/3b/2017).

All experiments were performed using 12–14-month-old male C57BL/6 mice. Animals were kept on 24 °C with a 12/12-h light/dark cycle and were housed in stable groups of 2–5 mice from the time of weaning. Starting from weaning, the mice were fed either a normal mouse diet (V1124-000, Sniff, Soest, Germany) containing 1.0% calcium, 0.7% phosphorus, and 1000 IU vitamin D/kg, or a CPD containing 2% Ca, 1.25% P, and 20% lactose. The Ca/P ratio of both diets was similar (1.43 for ND and 1.60 for CPD), and both diets had the same energy content (14 MJ/kg) and the same nutrient distribution (carbohydrates, protein, fat). All animals had access to food and tap water ad libitum. Before necropsy, urine was collected overnight in metabolic cages. At necropsy, the mice were exsanguinated from the abdominal vena cava under general anesthesia with ketamine/medetomidine (100/0.25 mg/kg i.p.) for serum collection. Necropsies were performed between 9 a.m. and 3 p.m. for all mice. Some mice maintained on a CPD until 12–14 months of age were switched to ND for two weeks to assess the effect of the diet switch on urinary excretion of catecholamines. Investigators were not blinded to the group allocation of the animals.

2.2. Biochemical Analysis

Serum and urinary phosphate, calcium, sodium, and creatinine were measured using a Cobas c111 analyzer (Roche, Mannheim, Germany). Serum intact FGF23 (Kainos, Tokyo, Japan), serum and urinary aldosterone (NovaTec Immundiagnostica, Dietzenbach, Germany), serum intact PTH (Immutopics Inc, San Clemente, CA, USA), and serum renin (Abbexa Ltd., Cambridge, UK) were detected using commercially available ELISA kits according to the manufacturer's instructions. Serum was extracted with diethylether and re-suspended in steroid-free serum (DRG Diagnostics, Marburg, Germany) for the aldosterone ELISA.

2.3. RNA Isolation and Quantitative RT-PCR

Total RNA isolation and quantitative RT-PCR analysis were performed as described previously [23]. All samples were measured in triplicate and normalized to two housekeeping genes (*Dpm1 and Txnl4a*). The qPCR results were obtained and evaluated with the software qPCRsoft 4.1 (V4.1.3.0, Analytik Jena, Jena, Germany), and then analyzed using the standard delta delta Cq method.

2.4. Echocardiography

Echocardiography was performed one day before the necropsies using a 14MHz linear transducer (Siemens Accuson s2000, Munich, Germany) for evaluation of cardiac function. Mice were under 1% isoflurane anesthesia with a stable body temperature of 37 °C. M-mode in short-axis at the level of the papillary muscles was used to evaluate LV thickness, fractional shortening, and internal diameters in systole and diastole. At least four cardiac cycles were analyzed for each parameter.

2.5. Central Arterial and Cardiac Pressure Measurements and Augmentation Index

Central arterial pressure was measured by inserting a micro-tip catheter (1.4 Millar Instruments, Houston, TX, USA) into the ascending aorta via the right carotid artery under 1.5% isoflurane anesthesia. The catheter was then further advanced into the left ventricle to obtain cardiac pressure parameters. Traces were recorded for at least three minutes and analyzed via LabChart7 software (ADInstruments, Dunedin, New Zealand). The aortic augmentation index was identified from the late systolic portion of the arterial pressure wave as described previously [24]. The augmentation index was defined as the height from the augmentation point to the systolic peak of the pressure wave divided by the pulse pressure, and was expressed as a percentage.

2.6. Histological Evaluation

Hearts were fixed in 4% paraformaldehyde, and then paraffin-embedded and cut into 5 µm sections. Cardiac sections were stained with FITC-labeled wheat germ agglutinin for the analysis of cardiomyocyte size. At least 10 random areas of the heart were measured and only cardiomyocytes with well-defined borders and visible nuclei were used. Images were obtained by the Zeiss LSM 880 Airyscan confocal microscope (Zeiss, Oberkochen, Germany) and analyzed using Image J software. Fibrotic tissue was visualized by staining with picrosirius red according to a standard protocol. Total collagen was quantified using ImageJ software and was expressed as the ratio of collagen-stained area to total muscle area of the left ventricle and septum. All histological images were analyzed by two independent investigators in a blinded manner.

2.7. Urinary Catecholamine Measurement

Mice were placed in individual cages for spontaneous urine collection. At least 150 ul was collected and used for further analysis. The urinary concentration of the catecholamines and catecholamine metabolites epinephrine, norepinephrine, dopamine, normetanephrine, metanephrine, and methoxytyramine were measured using liquid chromatography–tandem mass spectrometry (LC–MS/MS) as described previously [25].

2.8. Statistical Analysis

Statistical planning of the experiment was based on a minimum group size of 10 mice per diet, and was performed based on the variance of the main target variable, mean arterial blood pressure. Since the mice were bred in our own animal facility, and the experiments were built up using cohorts of mice, the number of mice per diet group ranged between 12–16. A different set of mice (n = 6 per group) was used for measurement of urinary catecholamines, and for the diet switch experiment. One animal was excluded from the analysis of urinary catecholamines because this animal, for unknown reasons, showed about 4-fold lower values compared with the rest of the group for all metabolites measured. Some values for specific measurements were excluded from the analysis if the data acquired were out of the defined range for a specific measurement.

Statistical analysis was performed using Graph Pad Prism 9 (GraphPad Software, San Diego, CA, USA). The data were analyzed by two-sided Mann–Whitney U-test for comparison of mice maintained on ND or CPD, or by Pearson's correlation analysis, followed by linear regression analysis. Urinary levels of catecholamines and their metabolites before and after switching the diet were analyzed using the Wilcoxon matched-pairs test. p values of less than 0.05 were considered significant. Data are presented as scatter dot plots with bars depicting means ± SEM.

3. Results

3.1. Mice Maintained on CPD for 14 Months have Increased Levels of FGF23, but Normal Kidney Function

To establish a model of long-term excessive dietary phosphate intake not associated with secondary hyperparathyroidism, we fed male WT mice a calcium, phosphate, and lactose enriched diet (CPD) after weaning for 14 months and compared them with age-matched controls fed a normal mouse diet (ND). Both diets had a similar Ca/P ratio (1.43 for ND and 1.60 for CPD), the same energy content, and the same distribution of major nutrients (carbohydrates, protein, fat). The high lactose content of the CPD is used to facilitate paracellular, vitamin D independent, absorption of calcium and phosphate in the gut [26]. Mice fed the CPD had comparable body weight (BW), relative to animals fed the ND (Figure 1A). Furthermore, mice on the CPD were normocalcemic, normophosphatemic, normonatremic, and showed unchanged serum levels of creatinine, suggesting normal kidney function (Figure 1B–E). Urinary phosphate excretion was significantly increased in the CPD mice compared to ND controls (Figure 1F). As expected, mice kept on CPD did not present with elevated levels of intact PTH (Figure 1G), but showed increased concentrations of circulating intact FGF23 (Figure 1H). In accordance with the notion that the rise in circulating intact FGF23 occurred as a compensatory mechanism to enhance urinary phosphate excretion in response to an increase in dietary phosphate intake, we found a positive, albeit statistically non-significant, correlation between FGF23 and urinary phosphate excretion (Figure 1I). The correlation between PTH and urinary phosphate excretion was much weaker and actually negative (Figure 1I). These findings support the idea that in our model of dietary phosphate excess separates between the two phosphaturic hormones, resulting in a selective upregulation of FGF23, because a rise in PTH is largely prevented by the normal Ca/P ratio of the CPD.

Hence, we successfully established a mouse model of long-term excessive dietary phosphate intake, characterized by normophosphatemia, normocalcemia, elevated intact FGF23, but largely normal PTH.

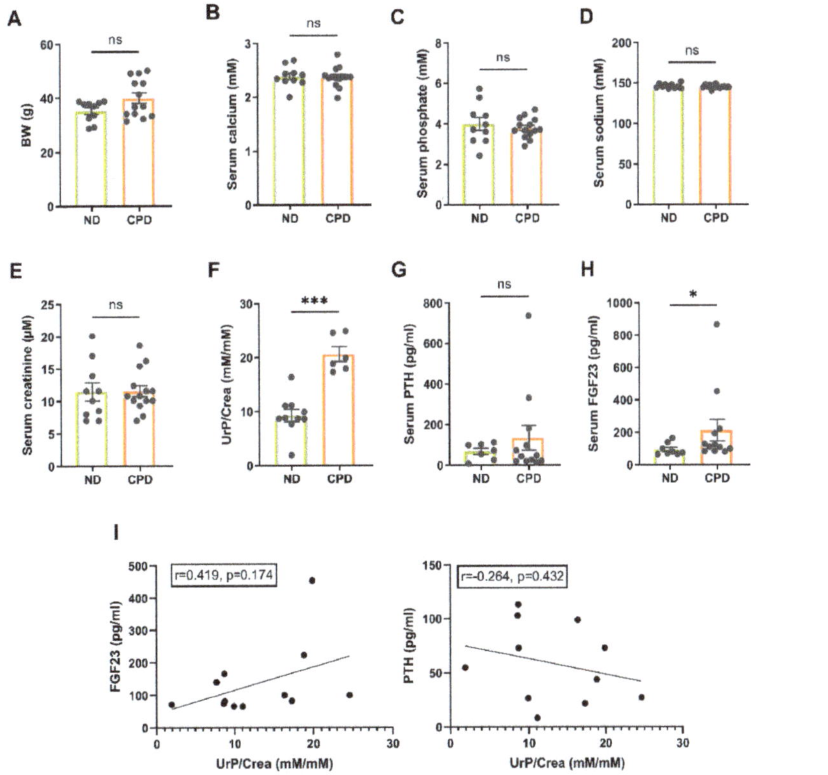

Figure 1. Mice maintained on CPD for 14 months have increased levels of FGF23, but normal kidney function. (**A**) Mice fed the CPD have comparable body weight (BW) to animals fed the ND (n = 10–13).

(**B–E**) CPD mice are normocalcemic, normophosphatemic, and normonatremic (n = 10–14). (**E**) Serum creatinine levels are unchanged between CPD and ND mice (n = 10–14). (**F**) Urinary phosphate excretion relative to creatinine (UrP/Crea) is significantly increased in CPD mice compared to ND controls (n = 6–10). (**G**) Serum PTH levels are not changed between ND and CPD mice (n = 7–12). (**H**) CPD mice present with increased concentrations of circulating intact FGF23 (n = 8–12). (**I**) Correlation analysis of serum FGF23 and PTH with urinary phosphate excretion (n = 11–12). Bars in (**A–H**) represent mean ± SEM for ND and CPD mice. * $p < 0.05$, *** $p < 0.001$ by Mann–Whitney U-test. Insets show Pearson correlation coefficients. ns, not significant.

3.2. High Phosphate Intake in Aged Mice Leads to Hypertension, but Not Left Ventricular Hypertrophy

Several earlier studies have reported LV hypertrophy in mice fed a high phosphate diet [27,28]. To assess cardiovascular function in our model of long-term dietary phosphate excess, we phenotyped the mice via echocardiography, intraarterial and cardiac catheterization, heart histology, and expression analysis of cardiac hypertrophy markers. Heart weight-to-body weight ratio was not increased in mice on CPD, relative to ND controls (Figure 2A). Moreover, left ventricular function was unaltered between the groups, as evidenced by unchanged ejection fraction (EF), MaxdP/dt, end-diastolic pressure (EDP), and LV internal diameter (LVID) (Figure 2B–E). Furthermore, mice fed the CPD diet did not present with LV hypertrophy as evidenced by unchanged cardiomyocyte area measured by wheat germ agglutinin (WGA) staining, unchanged LV collagen content measured by picrosirius red staining, and similar cardiac mRNA expression of the molecular marker of hypertrophy, *Bnp* (Figure 2F–H). Notably, cardiovascular phenotyping revealed a significant increase in systolic and mean arterial pressure as well as pulse pressure in mice fed CPD, relative to ND control mice (Figure 2I–K). To further elucidate the factors that may cause hypertension in mice on CPD, we measured augmentation index. The augmentation index is an indirect measure of arterial stiffness. Arterial stiffness increases with age [29], and is associated with elevations in systolic and diastolic blood pressure [30,31]. Pressure wave analysis revealed increased arterial stiffness as shown by an increase in the augmentation index in the CPD group, suggesting increased peripheral resistance in these mice (Figure 2L).

Together, these data demonstrate increased arterial blood pressure and increased arterial stiffness in aged mice on CPD, but strongly argue against LV hypertrophy and LV functional impairment in these mice.

3.3. Chronically Elevated Dietary Phosphate Intake Stimulates the Renin–Angiotensin–Aldosterone System and Increases Sympathetic Activity

Next, we examined two major regulators of arterial stiffness and vascular tone, the renin–angiotensin–aldosterone system (RAAS) and the sympathetic nervous system in our model of dietary phosphate excess. We reported previously that CKD mice fed the CPD have higher aldosterone levels compared with CKD mice on ND [32]. Besides the fact that the RAAS is a known regulator of blood pressure and vascular tone, it has been suggested that the RAAS is also an important determinant of arterial stiffness [33]. Notably, we found that urinary aldosterone secretion was increased in mice on CPD (Figure 3A). Serum aldosterone tended to be higher in mice on CPD, but this effect did not reach statistical significance (Figure 3B). Serum renin concentration did not show differences between mice on CPD and ND (Figure 3C).

The sympathetic nervous system is another known modulator of arterial stiffness, and has been shown to interact with the RAAS in individuals with hypertension [34–36]. To test whether mice on CPD have increased sympathetic tone, we measured urinary catecholamines and their metabolites, using mass spectrometry in 12–14 month-old mice [25]. Interestingly, norepinephrine (NEPI), dopamine (DA), metanephrine (MN), normetanephrine (NMN), and methoxytyramine (MTY) were all significantly elevated in mice fed the CPD, relative to the control mice on ND (Figure 3D–I). Epinephrine (EPI) tended to be higher in CPD mice, but this effect did not reach statistical significance (Figure 3E).

To determine if the observed effect was indeed caused by CPD and if it was reversible, we switched aged mice maintained on CPD since weaning to ND for two weeks, and subsequently repeated the measurements. In most mice, the urinary concentration of catecholamines and catecholamine metabolites decreased after switching them from CPD to ND (Figure 3J–O). However, this effect reached statistical significance only for dopamine (DA), normetanephrine (NMN), and methoxytyramine (MTY) (Figure 3L,M,O).

In conclusion, our data show that excess dietary phosphate induces increased blood pressure and increased arterial stiffness, and that these effects are accompanied by an activation of the RAAS and a stimulation of the sympathetic tone.

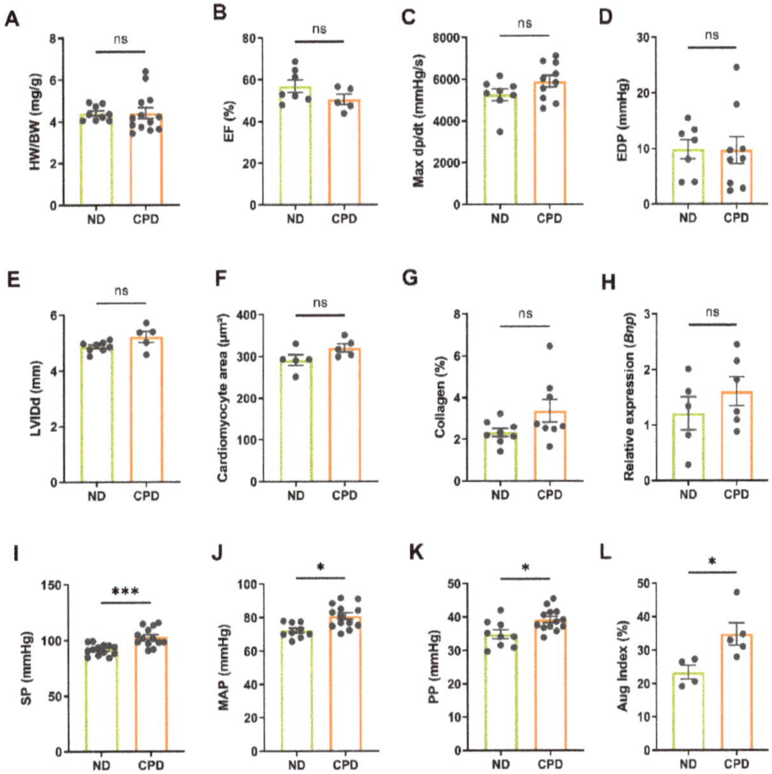

Figure 2. High phosphate intake in aged mice leads to hypertension, but not to left ventricular hypertrophy. (**A**) Heart weight-to-body weight ratio (HW/BW) is not increased in mice on CPD, relative to ND controls (n = 10–13). (**B–E**) Ejection fraction (EF), MaxdP/dt, end-diastolic pressure (EDP), and LV internal diameter (LVID) are unchanged between CPD and ND mice (n = 5–10). (**F**) Cardiomyocyte size is unchanged in CPD and ND mice (n = 5). Collagen content (**G**) and cardiac mRNA expression of *Bnp* (**H**) is similar between ND and CPD mice (n = 5–9). (**I–K**) Systolic (SP), mean arterial (MAP), and pulse pressure (PP) are significantly increased in mice fed CPD, relative to ND (n = 11–15). (**L**) CPD mice have increased arterial stiffness as shown by an increase in the augmentation index (Aug Index) (n = 4–5). Bars in (**A–L**) represent mean ± SEM for ND and CPD mice. * $p < 0.05$, *** $p < 0.001$ by Mann–Whitney U-test. ns, not significant.

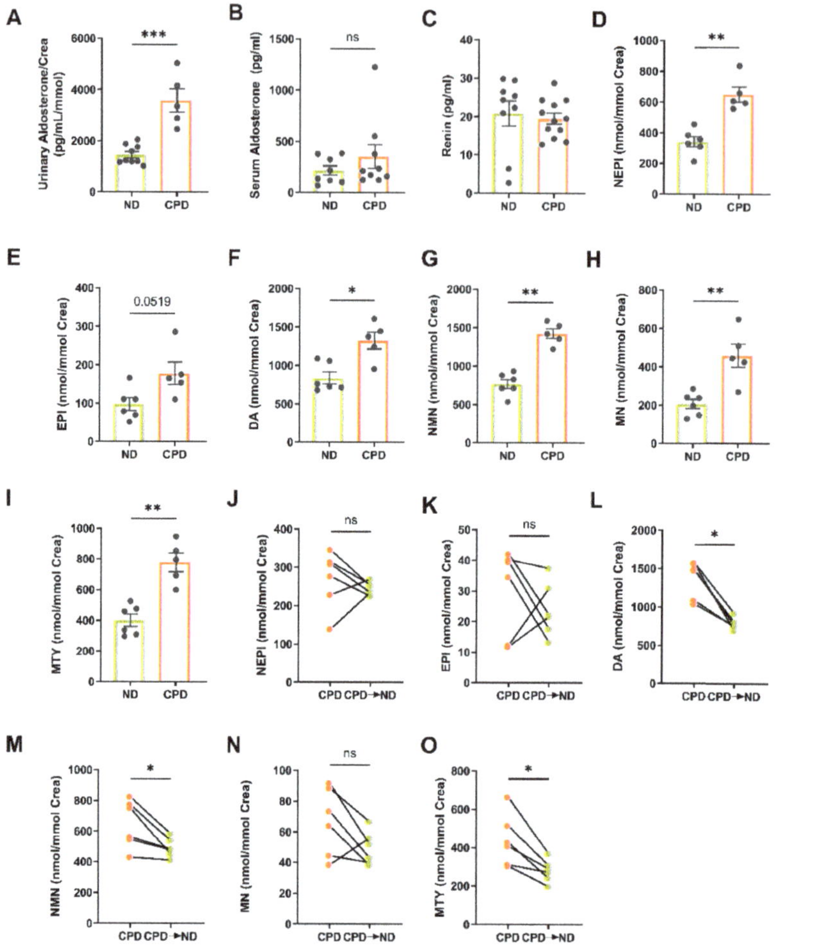

Figure 3. Chronically elevated dietary phosphate intake stimulates the renin–angiotensin–aldosterone system (RAAS) and increases sympathetic activity. (**A**) Urinary aldosterone secretion is increased in CPD mice (n = 5–9). (**B**) Serum aldosterone and (**C**) serum renin concentrations are unchanged between mice fed CPD and ND (n = 8–12). (**D–I**) Urinary excretion of the catecholamines norepinephrine (NEPI), dopamine (DA) and the catecholamine metabolites normetanephrine (NMN), metanephrine (MN) and methoxytyramine (MTY) are significantly elevated in mice fed the CPD compared to mice on ND, while epinephrine (EPI) did not reach statistical significance (n = 5–6). Levels of (**J**) norepinephrine, (**K**) epinephrine and (**N**) metanephrine are not altered after switching the mice from CPD to ND for two weeks. The diet switch lowered levels of (**L**) dopamine, (**M**) normetanephrine and (**O**) metoxytyramine (n = 6). Bars in (**A–I**) represent mean ± SEM for ND and CPD mice. * $p < 0.05$, ** $p < 0.01$, *** $p < 0.001$ by Mann–Whitney U-test or Wilcoxon matched-pairs test. ns, not significant.

4. Discussion

The main goal of this study was to evaluate for the first time the effects of long-term excess dietary phosphate intake on the cardiovascular system in aged, healthy mice, independent of secondary hyperparathyroidism. Here, we show that chronic exposure to high dietary phosphate in 14-month-old mice caused increased blood pressure and increased arterial stiffness despite normophosphatemia, and that these changes were associated with an activation of the RAAS and a stimulation of the sympathetic tone.

Male C57BL/6 mice maintained for 14 months on CPD in our study demonstrated an increase in systolic, mean arterial, and pulse pressure, independent of hyperphosphatemia or increased intact PTH. Furthermore, the mice on CPD were characterized by enhanced arterial stiffness as evidenced by an increase in augmentation index. Increased arterial stiffness is a precursor for hypertension [37,38]. In line with our findings, the recent intervention study by Mohammad and coworkers [10] reported a small increase in arterial blood pressure in healthy individuals in response to a high phosphate diet. However, the latter investigators failed to detect changes in arterial elasticity induced by dietary phosphate loading [10]. Previous studies examining the acute effects of a phosphate load in non-CKD patients also failed to observe changes in augmentation index or pulse wave velocity [39]. The reason for these discrepancies is currently unclear. It is conceivable that arterial stiffening induced by excessive dietary phosphate in healthy subjects is only seen after long-term exposure to increased dietary phosphate.

To further decipher the pathophysiological process(es) by which phosphate causes an increase in blood pressure, we measured aldosterone levels in serum and urine as well as serum renin concentration, and assessed urinary secretion of catecholamines. We found urinary aldosterone/creatinine concentrations to be significantly elevated in mice fed the CPD. This finding suggests that activation of RAAS may at least partially be responsible for the increased arterial blood pressure in the CPD group. Interestingly, serum renin concentration remained unchanged in mice on CPD, suggesting that the CPD-induced activation of RAAS occurs downstream of renin secretion. In addition, we found an increase in urinary concentration of catecholamines and their metabolites in mice fed the high phosphate diet, suggesting sympathoadrenergic activation. Similarly, healthy subjects placed on a high phosphate diet showed an increase in urinary metanephrine and normetanephrine excretion [10]. Catecholamines are endogenous neurotransmitters and hormones derived from tyrosine metabolism, and are essential for maintenance of metabolic and cardiovascular homeostasis and for adaptation to stressors [40,41]. The most abundant catecholamines in circulation are norepinephrine, epinephrine, and dopamine [42]. Catecholamines have many cardiovascular and metabolic actions, including increasing the heart rate, blood pressure, myocardial contractility, and cardiac conduction velocity. The *O*-methylated metabolites of norepinephrine, epinephrine and dopamine are normetanephrine (NMN), metanephrine (MN), and methoxytyramine (MTY), respectively [41]. Metanephrine and normetanephrine are referred to as metanephrines. It has been proposed that metanephrines can be used as markers of sympathetic activity [43]. Indeed, two independent studies have found a strong association between urinary metanephrine levels and systolic blood pressure in humans [41,44]. Therefore, the partially reversible increase in urinary excretion of catecholamines induced by the CPD in our study may contribute to increased arterial blood pressure and arterial stiffness, in addition to RAAS activation.

It is clear that the key question is the mechanism for how dietary phosphate may influence cardiovascular function. The effect of phosphate on the cardiovascular system may be direct, or mediated indirectly through one of the hormones involved in its homeostasis. Our study suggests that the untoward cardiovascular effects of a high dietary phosphate intake are independent of PTH. It is of course tempting to speculate that the increase in circulating FGF23 induced by excessive phosphate intake in CPD mice upregulated RAAS activity and sympathetic tone, thereby causing elevated blood pressure and increased arterial stiffness. As a matter of fact, there is accumulating evidence suggesting that FGF23 and the RAAS may interact [45]. In this context, it has been shown in mice that FGF23 suppresses the renal angiotensin converting enzyme-2 (ACE2), shifting the balance between the vasodilatory angiotensin (1–7) and the ACE1-produced angiotensin-2 towards the vasoconstrictive and prohypertensive angiotensin-2 [46,47]. Hence, FGF23 may directly activate the classical, prohypertensive RAAS by increasing angiotensin-2 production, which may subsequently induce a rise in peripheral resistance and aldosterone secretion. In line with this notion, we found unchanged renin concentrations in CPD mice, relative to mice kept on ND. However, it is unclear whether this is also true for humans because a recent study in hemodialysis patients with LVH failed to find an association between FGF23 levels and components of the RAAS [48].

Based on our data and the study by Mohammad and coworkers in healthy humans, high dietary phosphate intake stimulates sympathoadrenergic activity. However, the mechanistic link between FGF23 signaling and sympathoadrenergic activity has yet to be defined. It is interesting to note in this context that FGF23 has been shown to stimulate rostral ventrolateral medulla presympathetic neuron activity in the brainstem [49]. Conversely, it is also possible that sympathetic activation might be responsible for the changes in FGF23 serum levels in mice fed the CPD. In the latter context, Kawai and coworkers reported that acute dosing with the beta-adrenergic agonist isoproterenol increased *Fgf23* transcription in bone, and that this effect was reverted after administration of the beta blocker propranolol [50]. On the other hand, normalization of PTH and FGF23 in healthy adults on a high phosphate diet did not normalize blood pressure in these individuals, they rather remained hypertensinogenic, suggesting that the effects of phosphate are not mediated by FGF23 or PTH. Thus, the question of how dietary phosphate influences cardiovascular function remains controversial.

In conclusion, here we present a novel mouse model of long-term dietary phosphate excess characterized by increased blood pressure and arterial stiffness, normophosphatemia, normal PTH, but elevated FGF23. Our model may be useful to disentangle the complex relationship between phosphate, calcium-regulating hormones, FGF23, RAAS, sympathoadrenergic activity, and the cardiovascular system in health and disease.

Author Contributions: N.L., J.P. and R.G.E. conceived and designed the studies; N.L., M.P., A.Z. and J.P. performed experiments and analyzed the data; N.L. and R.G.E. wrote the manuscript. All authors have read and agreed to the published version of the manuscript.

Funding: This work was supported in part by the Collaborative Research Center Transregio 205 "The Adrenal: Central Relay in Health and Disease" (CRC/TRR 205/1,2; M.P. and J.P.), and by Open Access Funding by the University of Veterinary Medicine Vienna.

Institutional Review Board Statement: The study was undertaken in accordance with prevailing EU and national guidelines for animal care and welfare and in compliance with ARRIVE guidelines. All animal procedures were approved by the Ethics and Animal Welfare Committee of the University of Veterinary Medicine Vienna, Austria, and by the Austrian Federal Ministry of Education, Science and Research (permit number BMWFW-68.205/0188-WF/V/3b/2017).

Informed Consent Statement: Not applicable.

Data Availability Statement: All data generated or analyzed in this study are included in this article.

Acknowledgments: We thank Christiane Schüler, Claudia Bergow, Alexandra Petric, and Nikole Ginner for excellent technical assistance, and Ute Zeitz for help with animal breeding.

Conflicts of Interest: The authors have declared that no conflict of interest exists.

References

1. Peacock, M. Phosphate Metabolism in Health and Disease. *Calcif. Tissue Res.* **2020**, *108*, 3–15. [CrossRef]
2. Foley, R.N.; Parfrey, P.S.; Sarnak, M.J. Clinical epidemiology of cardiovascular disease in chronic renal disease. *Am. J. Kidney Dis.* **1998**, *32*, S112–S119. [CrossRef] [PubMed]
3. Go, A.S.; Chertow, G.M.; Fan, D.; McCulloch, C.E.; Hsu, C.Y. Chronic Kidney Disease and the Risks of Death, Cardiovascular Events, and Hospitalization. *N. Engl. J. Med.* **2004**, *351*, 1296–1305. [CrossRef]
4. Chang, A.R.; Grams, M.E. Serum Phosphorus and Mortality in the Third National Health and Nutrition Examination Survey (NHANES III): Effect Modification by Fasting. *Am. J. Kidney Dis.* **2014**, *64*, 567–573. [CrossRef] [PubMed]
5. Wang, Q.; Cui, Y.; Yogendranath, P.; Wang, N. Blood pressure and heart rate variability are linked with hyperphosphatemia in chronic kidney disease patients. *Chrono-Int.* **2018**, *35*, 1329–1334. [CrossRef] [PubMed]
6. Jankowski, J.; Floege, J.; Fliser, D.; Böhm, M.; Marx, N. Cardiovascular Disease in Chronic Kidney Disease. *Circulation* **2021**, *143*, 1157–1172. [CrossRef] [PubMed]

7. Shang, D.; Xie, Q.; Ge, X.; Yan, H.; Tian, J.; Kuang, D.; Hao, C.H.; Zhu, T. Hyperphosphatemia as an independent risk factor for coronary artery calcification progression in peritoneal dialysis patients. *BMC Nephrol.* **2015**, *16*, 107. [CrossRef] [PubMed]
8. Dhingra, R.; Sullivan, L.; Fox, C.S.; Wang, T.J.; D'Agostino, R.B.; Gaziano, J.M.; Vasan, R.S. Relations of Serum Phosphorus and Calcium Levels to the Incidence of Cardiovascular Disease in the Community. *Arch. Intern. Med.* **2007**, *167*, 879–885. [CrossRef] [PubMed]
9. Foley, R.N.; Collins, A.J.; Herzog, C.A.; Ishani, A.; Kalra, P.A. Serum Phosphate and Left Ventricular Hypertrophy in Young Adults: The Coronary Artery Risk Development in Young Adults Study. *Kidney Blood Press. Res.* **2009**, *32*, 37–44. [CrossRef] [PubMed]
10. Mohammad, J.; Scanni, R.; Bestmann, L.; Hulter, H.N.; Krapf, R. A Controlled Increase in Dietary Phosphate Elevates BP in Healthy Human Subjects. *J. Am. Soc. Nephrol.* **2018**, *29*, 2089–2098. [CrossRef] [PubMed]
11. Mills, K.T.; Stefanescu, A.; He, J. The global epidemiology of hypertension. *Nat. Rev. Nephrol.* **2020**, *16*, 223–237. [CrossRef] [PubMed]
12. Patel, R.K.; Jeemon, P.; Stevens, K.K.; Mccallum, L.; Hastie, C.E.; Schneider, A.; Jardine, A.G.; Mark, P.B.; Padmanabhan, S. Association between serum phosphate and calcium, long-term blood pressure, and mortality in treated hypertensive adults. *J. Hypertens.* **2015**, *33*, 2046–2053. [CrossRef]
13. E Olivo, R.; Hale, S.L.; Diamantidis, C.J.; A Bhavsar, N.; Tyson, C.C.; Tucker, K.L.; Carithers, T.; Kestenbaum, B.; Muntner, P.; Tanner, R.M.; et al. Dietary Phosphorus and Ambulatory Blood Pressure in African Americans: The Jackson Heart Study. *Am. J. Hypertens.* **2018**, *32*, 94–103. [CrossRef] [PubMed]
14. Moe, S.M.; Chen, N.X. Pathophysiology of Vascular Calcification in Chronic Kidney Disease. *Circ. Res.* **2004**, *95*, 560–567. [CrossRef]
15. Jono, S.; McKee, M.D.; Murry, C.E.; Shioi, A.; Nishizawa, Y.; Mori, K.; Morii, H.; Giachelli, C.M. Phosphate Regulation of Vascular Smooth Muscle Cell Calcification. *Circ. Res.* **2000**, *87*, E10–E17. [CrossRef]
16. Giachelli, C.M. Vascular Calcification: In Vitro Evidence for the Role of Inorganic Phosphate. *J. Am. Soc. Nephrol.* **2003**, *14*, S300–S304. [CrossRef]
17. Van, T.V.; Watari, E.; Taketani, Y.; Kitamura, T.; Shiota, A.; Tanaka, T.; Tanimura, A.; Harada, N.; Nakaya, Y.; Yamamoto, H.; et al. Dietary phosphate restriction ameliorates endothelial dysfunction in adenine-induced kidney disease rats. *J. Clin. Biochem. Nutr.* **2012**, *51*, 27–32. [CrossRef]
18. Sancho, J.J.; Rouco, J.; Riera-Vidal, R.; Sitges-Serra, A. Long-term effects of parathyroidectomy for primary hyperparathyroidism on arterial hypertension. *World J. Surg.* **1992**, *16*, 732–735. [CrossRef] [PubMed]
19. Bozic, M.; Panizo, S.; Sevilla, M.A.; Riera, M.; Soler, M.J.; Pascual, J.; Lopez, I.; Freixenet, M.; Fernandez, E.; Valdivielso, J.M. High phosphate diet increases arterial blood pressure via a parathyroid hormone mediated increase of renin. *J. Hypertens.* **2014**, *32*, 1822–1832. [CrossRef]
20. Jowsey, J.; Reiss, E.; Canterbury, J.M. Long-Term Effects of High Phosphate Intake on Parathyroid Hormone Levels and Bone Metabolism. *Acta Orthop. Scand.* **1974**, *45*, 801–808. [CrossRef]
21. Demeter, J.G.; A De Jong, S.; Oslapas, R.; Ernst, K.; Hessel, P.; Jarosz, H.; Smith, M.; Nayyar, R.; Lawrence, A.M.; Paloyan, E. High phosphate diet-induced primary hyperparathyroidism: An animal model. *Surgery* **1991**, 110.
22. Richter, B.; Kapanadze, T.; Weingärtner, N.; Walter, S.; Vogt, I.; Grund, A.; Schmitz, J.; Bräsen, J.H.; Limbourg, F.P.; Haffner, D.; et al. High phosphate-induced progressive proximal tubular injury is associated with the activation of Stat3/Kim-1 signaling pathway and macrophage recruitment. *Faseb J.* **2022**, *36*, e22407. [CrossRef] [PubMed]
23. Latic, N.; Zupcic, A.; Frauenstein, D.; Erben, R.G. Activation of RAAS Signaling Contributes to Hypertension in Aged Hyp Mice. *Biomedicines* **2022**, *10*, 1691. [CrossRef]
24. Andrukhova, O.; Slavic, S.; Zeitz, U.; Riesen, S.C.; Heppelmann, M.S.; Ambrisko, T.D.; Markovic, M.; Kuebler, W.M.; Erben, R.G. Vitamin D Is a Regulator of Endothelial Nitric Oxide Synthase and Arterial Stiffness in Mice. *Mol. Endocrinol.* **2014**, *28*, 53–64. [CrossRef]
25. Peitzsch, M.; Pelzel, D.; Glöckner, S.; Prejbisz, A.; Fassnacht, M.; Beuschlein, F.; Januszewicz, A.; Siegert, G.; Eisenhofer, G. Simultaneous liquid chromatography tandem mass spectrometric determination of urinary free metanephrines and catecholamines, with comparisons of free and deconjugated metabolites. *Clin. Chim. Acta* **2013**, *418*, 50–58. [CrossRef]
26. Kollenkirchen, U.; Fox, J.; Walters, M.R. Normocalcemia without hyperparathyroidism in vitamin D-deficient rats. *J. Bone Miner. Res.* **1991**, *6*, 273–278. [CrossRef] [PubMed]
27. Hu, M.C.; Shi, M.; Cho, H.J.; Adams-Huet, B.; Paek, J.; Hill, K.; Shelton, J.; Amaral, A.P.; Faul, C.; Taniguchi, M.; et al. Klotho and Phosphate Are Modulators of Pathologic Uremic Cardiac Remodeling. *J. Am. Soc. Nephrol.* **2015**, *26*, 1290–1302. [CrossRef]
28. Grabner, A.; Schramm, K.; Silswal, N.; Hendrix, M.; Yanucil, C.; Czaya, B.; Singh, S.; Wolf, M.; Hermann, S.; Stypmann, J.; et al. FGF23/FGFR4-mediated left ventricular hypertrophy is reversible. *Sci. Rep.* **2017**, *7*, 1993. [CrossRef] [PubMed]
29. Fantin, F.; Mattocks, A.; Bulpitt, C.J.; Banya, W.; Rajkumar, C. Is augmentation index a good measure of vascular stiffness in the elderly? *Age Ageing* **2006**, *36*, 43–48. [CrossRef]
30. Blacher, J.; Safar, M.E. Large-artery stiffness, hypertension and cardiovascular risk in older patients. *Nat. Clin. Pract. Cardiovasc. Med.* **2005**, *2*, 450–455. [CrossRef]
31. Safar, M.E. Arterial stiffness as a risk factor for clinical hypertension. *Nat. Rev. Cardiol.* **2017**, *15*, 97–105. [CrossRef] [PubMed]

32. Radloff, J.; Latic, N.; Pfeiffenberger, U.; Schüler, C.; Tangermann, S.; Kenner, L.; Erben, R.G. A phosphate and calcium-enriched diet promotes progression of 5/6-nephrectomy-induced chronic kidney disease in C57BL/6 mice. *Sci. Rep.* **2021**, *11*, 14868. [CrossRef] [PubMed]
33. Mahmud, A.; Feely, J. Review: Arterial stiffness and the renin-angiotensin-aldosterone system. *J. Renin-Angiotensin-Aldosterone Syst.* **2004**, *5*, 102–108. [CrossRef] [PubMed]
34. Bühler, F.R.; Laragh, J.H.; Baer, L.; Vaughan, E.D.; Brunner, H.R. Propranolol Inhibition of Renin Secretion. *N. Engl. J. Med.* **1972**, *287*, 1209–1214. [CrossRef]
35. Miller, A.J.; Arnold, A.C. The renin–angiotensin system in cardiovascular autonomic control: Recent developments and clinical implications. *Clin. Auton. Res.* **2018**, *29*, 231–243. [CrossRef]
36. Iliescu, R.; Lohmeier, T.E.; Tudorancea, I.; Laffin, L.; Bakris, G.L. Renal denervation for the treatment of resistant hypertension: Review and clinical perspective. *Am. J. Physiol. Physiol.* **2015**, *309*, F583–F594. [CrossRef]
37. Tanaka, H.; Safar, M.E. Influence of lifestyle modification on arterial stiffness and wave reflections*. *Am. J. Hypertens.* **2005**, *18*, 137–144. [CrossRef]
38. Kaess, B.M.; Rong, J.; Larson, M.G.; Hamburg, N.M.; Vita, J.A.; Levy, D.; Benjamin, E.J.; Vasan, R.S.; Mitchell, G.F. Aortic Stiffness, Blood Pressure Progression, and Incident Hypertension. *JAMA* **2012**, *308*, 875–881. [CrossRef]
39. Martínez-Moreno, J.M.; Herencia, C.; de Oca Addy, M.; Díaz-Tocados, J.M.; Vergara, N.; Gómez-Luna, M.J.; López-Argüello, S.D.; Camargo, A.; Peralbo-Santaella, E.; Rodríguez-Ortiz, M.E.; et al. High phosphate induces a pro-inflammatory response by vascular smooth muscle cells and modulation by vitamin D derivatives. *Clin. Sci.* **2017**, *131*, 1449–1463. [CrossRef]
40. Andreis, D.T.; Singer, M. Catecholamines for inflammatory shock: A Jekyll-and-Hyde conundrum. Intensiv. *Care Med.* **2016**, *42*, 1387–1397. [CrossRef]
41. Parasiliti-Caprino, M.; Obert, C.; Lopez, C.; Bollati, M.; Bioletto, F.; Bima, C.; Egalini, F.; Berton, A.; Prencipe, N.; Settanni, F.; et al. Association of Urine Metanephrine Levels with CardiometaBolic Risk: An Observational Retrospective Study. *J. Clin. Med.* **2021**, *10*, 1967. [CrossRef]
42. Grouzmann, E.; Lamine, F. Determination of catecholamines in plasma and urine. *Best Pract. Res. Clin. Endocrinol. Metab.* **2013**, *27*, 713–723. [CrossRef] [PubMed]
43. Eisenhofer, G.; Friberg, P.; Pacak, K.; Goldstein, D.S.; Murphy, D.L.; Tsigos, C.; Quyyumi, A.A.; Brunner, H.G.; Lenders, J.W. Plasma metadrenalines: Do they provide useful information about sympatho-adrenal function and catecholamine metabolism? *Clin. Sci.* **1995**, *88*, 533–542. [CrossRef] [PubMed]
44. Coulson, J.M. The relationship between blood pressure variability and catecholamine metabolites: A pilot study. *J. Hum. Hypertens.* **2014**, *29*, 50–52. [CrossRef]
45. Freundlich, M.; Gamba, G.; Rodriguez-Iturbe, B. Fibroblast growth factor 23—Klotho and hypertension: Experimental and clinical mechanisms. *Pediatr. Nephrol.* **2020**, *36*, 3007–3022. [CrossRef]
46. Pi, M.; Ye, R.; Han, X.; Armstrong, B.; Liu, X.; Chen, Y.; Sun, Y.; Quarles, L.D. Cardiovascular Interactions between Fibroblast Growth Factor-23 and Angiotensin II. *Sci. Rep.* **2018**, *8*, 12398. [CrossRef]
47. Dai, B.; David, V.; Martin, A.; Huang, J.; Li, H.; Jiao, Y.; Gu, W.; Quarles, L.D. A Comparative Transcriptome Analysis Identifying FGF23 Regulated Genes in the Kidney of a Mouse CKD Model. *PLoS ONE* **2012**, *7*, e44161. [CrossRef] [PubMed]
48. Dörr, K.; Kammer, M.; Reindl-Schwaighofer, R.; Lorenz, M.; Marculescu, R.; Poglitsch, M.; Beitzke, D.; Oberbauer, R. The Effect of FGF23 on Cardiac Hypertrophy Is Not Mediated by Systemic Renin-Angiotensin- Aldosterone System in Hemodialysis. *Front. Med.* **2022**, 9. [CrossRef]
49. Oshima, N.; Onimaru, H.; Yamagata, A.; Ito, S.; Imakiire, T.; Kumagai, H. Rostral ventrolateral medulla neuron activity is suppressed by Klotho and stimulated by FGF23 in newborn Wistar rats. *Auton. Neurosci.* **2020**, *224*, 102640. [CrossRef]
50. Kawai, M.; Kinoshita, S.; Shimba, S.; Ozono, K.; Michigami, T. Sympathetic Activation Induces Skeletal Fgf23 Expression in a Circadian Rhythm-dependent Manner. *J. Biol. Chem.* **2014**, *289*, 1457–1466. [CrossRef]

biomedicines

MDPI

Article

Sacubitril/Valsartan and Ivabradine Attenuate Left Ventricular Remodelling and Dysfunction in Spontaneously Hypertensive Rats: Different Interactions with the Renin–Angiotensin–Aldosterone System

Fedor Simko [1,2,3,*,†], Tomas Baka [1,†], Peter Stanko [1], Kristina Repova [1], Kristina Krajcirovicova [1], Silvia Aziriova [1], Oliver Domenig [4], Stefan Zorad [3], Michaela Adamcova [5,‡] and Ludovit Paulis [1,6,‡]

1 Institute of Pathophysiology, Faculty of Medicine, Comenius University, 81108 Bratislava, Slovakia; dr.tomas.baka@gmail.com (T.B.); pete.stanko@gmail.com (P.S.); kristina.repova@fmed.uniba.sk (K.R.); krikratina@gmail.com (K.K.); silvia.aziriova@gmail.com (S.A.); ludovit.paulis@gmail.com (L.P.)
2 3rd Department of Internal Medicine, Faculty of Medicine, Comenius University, 83305 Bratislava, Slovakia
3 Institute of Experimental Endocrinology, Biomedical Research Center, Slovak Academy of Sciences, 84505 Bratislava, Slovakia; stefan.zorad@savba.sk
4 Attoquant Diagnostics, 1110 Vienna, Austria; oliver.domenig@attoquant.com
5 Department of Physiology, Faculty of Medicine in Hradec Kralove, Charles University, 50003 Hradec Kralove, Czech Republic; adamcova@lfhk.cuni.cz
6 Institute of Normal and Pathological Physiology, Centre of Experimental Medicine, Slovak Academy of Sciences, 81371 Bratislava, Slovakia
* Correspondence: fedor.simko@fmed.uniba.sk
† These authors contributed equally to this work.
‡ These authors contributed equally to this work.

Citation: Simko, F.; Baka, T.; Stanko, P.; Repova, K.; Krajcirovicova, K.; Aziriova, S.; Domenig, O.; Zorad, S.; Adamcova, M.; Paulis, L. Sacubitril/Valsartan and Ivabradine Attenuate Left Ventricular Remodelling and Dysfunction in Spontaneously Hypertensive Rats: Different Interactions with the Renin–Angiotensin–Aldosterone System. *Biomedicines* **2022**, *10*, 1844. https://doi.org/10.3390/biomedicines10081844

Academic Editor: Elena Kaschina

Received: 13 June 2022
Accepted: 27 July 2022
Published: 31 July 2022

Abstract: This study investigated whether sacubitril/valsartan and ivabradine are able to prevent left ventricular (LV) fibrotic remodelling and dysfunction in a rat experimental model of spontaneous hypertension (spontaneously hypertensive rats, SHRs) and whether this potential protection is associated with RAAS alterations. Five groups of three-month-old male Wistar rats and SHRs were treated for six weeks as follows: untreated Wistar controls, Wistar plus sacubitril/valsartan, SHR, SHR plus sacubitril/valsartan, and SHR plus ivabradine. The SHRs developed a systolic blood pressure (SBP) increase, LV hypertrophy and fibrosis, and LV systolic and diastolic dysfunction. However, no changes in serum RAAS were observed in SHRs compared with the controls. Elevated SBP in SHRs was decreased by sacubitril/valsartan but not by ivabradine, and only sacubitril/valsartan attenuated LV hypertrophy. Both sacubitril/valsartan and ivabradine reduced LV collagen content and attenuated LV systolic and diastolic dysfunction. Sacubitril/valsartan increased the serum levels of angiotensin (Ang) II, Ang III, Ang IV, Ang 1-5, Ang 1-7, and aldosterone, while ivabradine did not affect the RAAS. We conclude that the SHR is a normal-to-low serum RAAS model of experimental hypertension. While the protection of the hypertensive heart in SHRs by sacubitril/valsartan may be related to an Ang II blockade and the protective Ang 1-7, the benefits of ivabradine were not associated with RAAS modulation.

Keywords: SHR; sacubitril/valsartan; ARNI; ivabradine; remodelling; cardiac dysfunction; fibrosis; renin–angiotensin–aldosterone system; angiotensin II; angiotensin 1-7

1. Introduction

Left ventricular (LV) hypertrophy in hypertension is considered to be a compensatory reaction to a chronically increased haemodynamic burden. LV mass enlargement supports the heart's performance without increasing wall tension. However, a hypertensive heart is associated with fibrotic rebuilding of the LV, resulting in a deterioration of cardiac function and a worsening prognosis. It is generally believed that curbing pathological cardiac remodelling reduces the transition from a hypertensive heart to heart failure (HF). Thus, the search for novel therapeutic strategies against the consequences of haemodynamic overload-induced cardiac remodelling in various models of experimental hypertension and in clinical conditions is unremitting [1]. Hypertensive heart disease involves the structural remodelling of the musculature and collagenous and non-collagenous matrix. Myocardial hypertrophy is determined by pressure or volume overload, which induces the compensatory growth of cardiomyocytes. The structural homogeneity may be disturbed by the imbalance of two groups of substances: by increased levels of angiotensin II, aldosterone, endothelin, and catecholamines which represent stimulators of pathologic growth with fibrocyte proliferation and an overabundance of collagen; or by a reduced production of nitric oxide (NO), natriuretic peptides, bradykinin, and prostaglandins with the opposite effect on growth and proliferation. The absolute or relative overproduction of angiotensin (Ang) II and aldosterone governs the development of pathologic fibrosis associated with deteriorated heart function and rhythm disturbances [2,3]. Thus, blocking the renin–angiotensin–aldosterone system (RAAS) by angiotensin-converting enzyme (ACE) inhibitors, angiotensin II type 1 receptor (AT1R) blockers, or aldosterone receptor blockers enables the attenuation of the vasoconstrictor, pro-inflammatory, and pro-proliferative actions.

During the past decade, two novel approaches to HF management with different mechanisms of action have been introduced. Neprilysin is an enzyme expressed in the cell membrane of various tissues that splits atrial and brain natriuretic peptides (ANP and BNP, respectively). The inhibition of neprilysin by sacubitril enhances circulating ANP and BNP levels with vasodilative, diuretic, and antiproliferative actions. Since neprilysin's substrates include both natriuretic peptides (NP) and Ang II, its inhibition increases not only the level of beneficial NP but also the concentration of adverse Ang II, potentially counterbalancing the desirable vasodilative effects of NP. To avoid this, sacubitril, an inhibitor of neprilysin, was combined with the AT1R blocker valsartan to attenuate Ang II effects [4,5]. The PARADIGM-HF study involving heart failure patients with systolic dysfunction showed that the combination of neprilysin inhibition by sacubitril and the AT1R blocker valsartan, i.e., sacubitril/valsartan (ARNI), reduced morbidity and mortality more effectively than the ACE inhibitor enalapril [6]. Thus, ARNI is becoming the cornerstone of HF therapy. Moreover, in the PARALLAX trial comprising HF patients with a preserved LV ejection fraction, sacubitril/valsartan resulted in a significantly greater decrease in plasma N-terminal pro-brain natriuretic peptide levels compared with a standard treatment affecting the renin–angiotensin system [7]. Thus, the combination of neprilysin with a renin–angiotensin system blockade may be of potential benefit in hearts with not only systolic but also diastolic LV dysfunction.

Ivabradine is a selective inhibitor of the I_f current in the sinoatrial node, which is responsible for pacemaking. Ivabradine reduces the heart rate (HR) without the negative inotropic effect inherent to beta-blockers. In the SHIFT study, ivabradine decreased the composite end-point of mortality and hospitalisations for HF, and it is recommended for patients with systolic HF and a HR above 70 bpm despite treatment with or in case of intolerance of beta-blockers [8].

It is generally accepted that cardiovascular protection is achieved by interfering with the excessive neurohumoral activation seen in chronic HF. Indeed, modulation of the RAAS, whose chronic activation induces a pathologic remodelling of the target organs, is pivotal in HF management. Moreover, neprilysin activity is linked to RAAS modulation: while neprilysin participates in Ang I degradation, ANP and BNP inhibit the release of renin [9].

However, data regarding the complex interference of ARNI or ivabradine with the RAAS are sparse. Thus, the aim of this study was to show in a rat experimental model of spontaneous hypertension (spontaneously hypertensive rats, SHRs) whether ARNI or ivabradine are able to protect a hypertensive heart and whether this potential protection is due to their interaction with the deleterious classical ACE/Ang II/AT1R pathway and the protective alternative ACE2/Ang 1-7/Mas receptor (MasR) pathway of the renin–angiotensin system.

2. Materials and Methods

2.1. Animals and Treatment

Twelve-week-old male Wistar rats and age- and weight-matched male SHRs (Department of Toxicology and Laboratory Animals Breeding, Slovak Academy of Sciences, Dobra Voda, Slovak Republic) were randomly divided into five groups (15 per group) and treated for six weeks as follows: Wistar rats with no treatment (C); Wistar rats treated with ARNI (68 mg/kg/day; Novartis, Basel, Switzerland) (ARNI); SHRs with no treatment (SHR); SHRs treated with ARNI (68 mg/kg/day) (SHR + ARNI); and SHRs treated with ivabradine (10 mg/kg/day; Servier, Suresnes, France) (SHR + IVA). The therapeutics were dissolved in drinking water and their concentration was adjusted to daily water consumption. The natural water consumption was 12–13 mL per 100 g body weight. To ensure that all of the water-therapeutics solutions were drunk by a particular rat, only 10 mL per 100 g body weight of solution was offered. The solutions were prepared by dissolving the appropriate amount of therapeutics in water, while no additional substance was added. The rats were housed in individual cages, fed a regular pellet diet ad libitum and maintained under standard laboratory conditions (12:12-h light–dark cycle, 22 ± 2 °C temperature, and 55 ± 10% humidity). The study was conducted in conformity with the Guide for the Care and Use of Laboratory Animals published by the US National Institutes of Health (NIH publication no. 85-23, revised 1996). The protocol was approved by the Ethics Committee of the Institute of Pathophysiology, Faculty of Medicine, Comenius University, Bratislava, Slovak Republic (approval number: 809/19-221/3; approval date: 23 April 2019).

Systolic blood pressure (SBP) and HR were measured twice before treatment and once a week during treatment by non-invasive tail-cuff plethysmography (Hugo-Sachs Elektronik, Freiburg, Germany). After six weeks of treatment, the rats were euthanised by isoflurane inhalation. Body weight (BW), heart weight, and left ventricular weight (LVW) were measured, and the LVW/BW ratio was subsequently calculated. LV samples were frozen at −80 °C and, later, hydroxyproline concentrations were measured. Blood samples were collected from the abdominal aorta during euthanasia. Serum obtained by centrifuging the blood samples at 2000× *g* for 15 min was stored at −80 °C for subsequent angiotensin and aldosterone analysis.

2.2. Determination of Hydroxyproline in the Left Ventricle

Collagenous proteins in the LV were isolated by treating LV samples stepwise with different buffers, as described previously [10]. Briefly, CH3COOH-pepsin buffer (pH 1.4, 24 h at 4 °C) was used to extract soluble collagenous proteins, and 1.1 mol/L NaOH (45 min at 105 °C) was used to extract the remaining insoluble collagenous proteins. The hydrolysed samples were oxidised by chloramine T added to an acetate–citrate buffer at pH 6.0. After incubation for 20 min at room temperature, the reaction was stopped by adding 20 volumes of Ehrlich's reagent to the mixture. The samples were then incubated at 65 °C for 15 min, and the hydroxyproline concentration (a marker of fibrosis) in the LV was measured in both collagenous fractions using spectrophotometry at 550 nm. The hydroxyproline content in the LV was subsequently calculated and expressed as mg per total weight of the LV.

2.3. Determination of Serum Angiotensins and Aldosterone Concentration and the Markers of Renin and ACE Activities

Serum samples from six animals per group that were not subject to prior echocardiography were used for angiotensin and aldosterone analyses. Equilibrium Ang peptide and aldosterone levels were determined by mass spectrometry, as described previously [11]. Briefly, the equilibrium peptide levels were stabilised by equilibration of the conditioned serum at 37 °C for 60 min. Thereafter, the stabilised samples were spiked with internal standards for each angiotensin metabolite (isotopes labelled Ang I, Ang II, Ang 1-7, Ang 1-5,

Ang 2-8, and Ang 3-8) at concentrations of 200 pg/mL, and for aldosterone (deuterated aldosterone) at a concentration of 500 pg/mL. After a C18-based solid-phase extraction, the samples were analysed by LC–MS/MS using a reversed-phase analytical column (Acquity UPLC® C18, Waters Corp., Milford, MA, USA) operating in line with a XEVO TQ-S triple quadrupole mass spectrometer (Waters Corp.) in MRM mode. The peptide recovery of the sample preparation (for each Ang metabolite in each sample) was corrected using internal standards. The corresponding response factors determined with appropriate calibration curves in the original sample matrix, which integrated signals exceeding a signal-to-noise ratio of 10, were used to assess Ang peptide concentrations. The Ang 1-5/Ang 1-7 ratio, a marker of Ang 1-7 cleavage to Ang 1-5, was subsequently calculated.

The marker of renin activity (RA-S) was subsequently calculated as the sum of Ang I and Ang II. Indeed, in previous studies, the sum of Ang I and Ang II obtained from the above equilibrium analysis was shown to be closely correlated with the measured renin activity, independent of species or treatment [12].

The marker of ACE activity (ACE-S) was subsequently calculated as the Ang II/Ang I ratio. It provides information about the expected ACE activity [13].

The aldosterone/Ang II ratio (AA2 ratio) was calculated to assess adrenal responsiveness following Ang II signalling resulting in the release of aldosterone [14].

2.4. Echocardiography

After six weeks of treatment, transthoracic echocardiography was performed on seven animals per group using a 14-MHz matrix probe (M12L) coupled with a GE Medical Vivid 7 Dimension System (GE Medical Systems CZ Ltd., Prague, Czech Republic), as described previously [15]. Briefly, the animals were anesthetised throughout the protocol by applying isoflurane (2.5% inspiratory concentration at a flow rate of 2 L/min) during spontaneous breathing. After placing the rat in the supine position on a warming pad (38 °C), the thoracic wall was shaved. The HR and body temperature were monitored throughout the protocol. To assess the LV systolic function, the LV end-systolic and end-diastolic internal diameters were measured from the anatomical M-mode images in a long-axis view using the leading-edge method. Subsequently, the left ventricular fractional shortening (LVFS) and ejection fraction (LVEF, using the Teichholz formula) were determined. To assess the LV diastolic function, the diastolic transmitral peak early (E) and late (A) filling velocities were measured from the two-dimensionally guided Doppler spectra of mitral inflow in the apical four-chamber view, and the E/A ratio was then calculated. The maximal velocities of the early (Em) and late (Am) diastolic wall movement waves at the level of the septal mitral annulus were determined by tissue Doppler imaging from the apical four-chamber view; the E/Em ratio was subsequently calculated. Echocardiography was performed by an experienced echocardiographer blinded to the group identity. All measurements were averaged over three consecutive cardiac cycles.

2.5. Statistical Analysis

The results are presented as means ± SEM. Data distribution was assessed by a Shapiro–Wilk normality test. A two-way, repeated-measures analysis of variance (ANOVA) followed by multiple comparisons with a Bonferroni post-hoc test was used for the statistical analysis of SBP and HR data. A one-way, two-tailed ANOVA followed by multiple comparisons with a Bonferroni post-hoc test was used for the statistical analysis of the remaining data, including the heart weights, LV hydroxyproline concentrations and contents, serum Ang and aldosterone levels, and echocardiography. The differences were considered significant if $p < 0.05$. The statistical analysis was conducted using GraphPad Prism 9 for Windows (GraphPad Software, La Jolla, CA, USA).

3. Results

3.1. Haemodynamics and Heart Weights

The SBP was 131.71 ± 3.71 mmHg in the control group, and ARNI decreased ($p < 0.05$) it by 13% after six weeks of treatment. In the SHR group, SBP was higher than in controls by 39% (182.89 ± 4.22 mmHg, $p < 0.05$ vs. C), and ARNI decreased ($p < 0.05$) it by 23%. Ivabradine did not affect SBP in SHRs (Figure 1A).

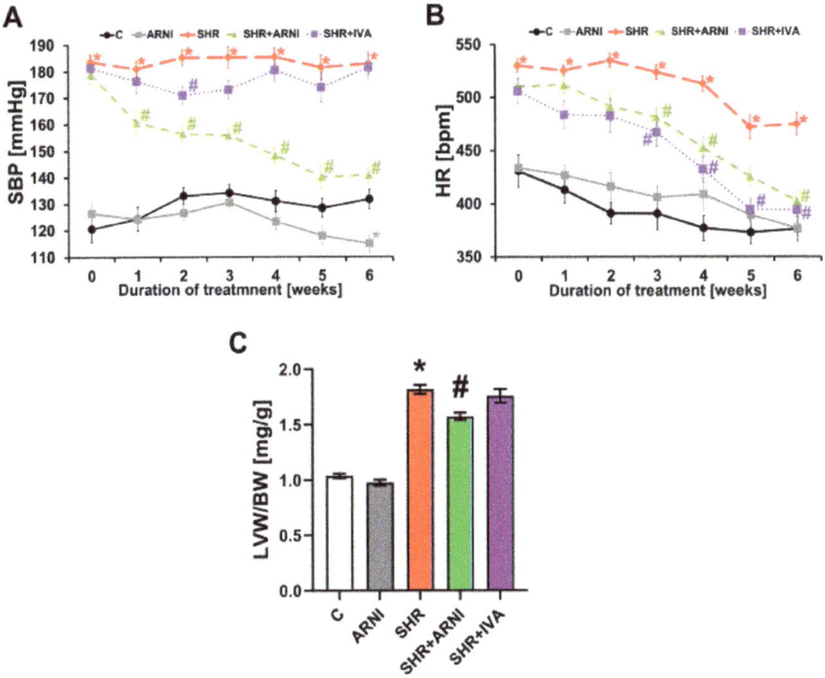

Figure 1. Effect of ARNI and ivabradine on systolic blood pressure (SBP) (**A**) and heart rate (HR) (**B**) throughout the experiment, and the relative weight of the left ventricle (left ventricular weight/body weight; LVW/BW) (**C**) in SHRs after six weeks of treatment. ARNI, sacubitril/valsartan; C, Wistar controls; IVA, ivabradine; SHRs, spontaneously hypertensive rats. Results are presented as means ± SEM. $n = 15$ per group. Repeated measures ANOVA (**A**,**B**) or one-way, two-tailed ANOVA (**C**) followed by multiple comparisons with a Bonferroni post-hoc test; * $p < 0.05$ vs. C; # $p < 0.05$ vs. SHR.

The HR was 375.93 ± 11.61 bpm in the control group, and ARNI did not affect it after six weeks of treatment. In the SHR group, the HR was higher than in controls by 26% (474.95 ± 10.53 bpm, $p < 0.05$ vs. C), and ARNI and ivabradine decreased it ($p < 0.05$) by 15% and 17%, respectively (Figure 1B).

The LVW/BW ratio was 1.04 ± 0.02 mg/g in the control group, and ARNI did not affect it after six weeks of treatment. In the SHR group, the LVW/BW ratio was higher than in controls by 75% (1.82 ± 0.04 mg/g, $p < 0.05$ vs. C), and ARNI decreased it ($p < 0.05$) by 13%. Ivabradine did not affect the LVW/BW ratio in SHRs (Figure 1C).

3.2. Hydroxyproline Concentration and Content in Soluble and Insoluble Collagen and Total Hydroxyproline in the Left Ventricle

The hydroxyproline concentrations in the soluble collagenous protein were 0.174 ± 0.008 mg/g and 0.199 ± 0.013 mg/g in the control and SHR groups, respectively (ns). After six weeks of treatment, none of the therapeutics affected the hydroxyproline concentration in the soluble collagenous protein (Figure 2A).

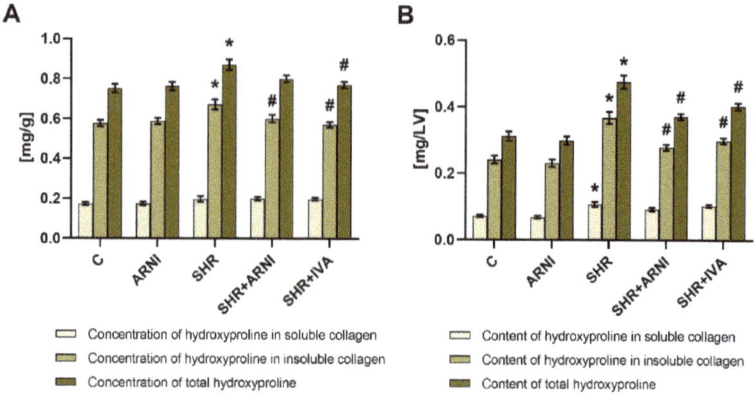

Figure 2. Effect of ARNI and ivabradine on hydroxyproline concentration in soluble and insoluble collagenous proteins and on the total hydroxyproline concentration (**A**), and on hydroxyproline content in the soluble and insoluble collagenous proteins, and on the total hydroxyproline content (**B**) in the left ventricle in SHRs after six weeks of treatment. ARNI, sacubitril/valsartan; C, Wistar controls; IVA, ivabradine; SHRs, spontaneously hypertensive rats. Results are presented as means ± SEM. n = 15 per group. One-way, two-tailed ANOVA followed by multiple comparisons with a Bonferroni post-hoc test; * $p < 0.05$ vs. C; # $p < 0.05$ vs. SHR.

The hydroxyproline concentration in the insoluble collagenous protein was 0.578 ± 0.017 mg/g in the control group, and ARNI had no effect after six weeks of treatment. In the SHR group, the hydroxyproline concentration in the insoluble collagenous protein was higher than in controls by 16% (0.673 ± 0.025 mg/g, $p < 0.05$ vs. C), and ARNI and ivabradine decreased it ($p < 0.05$) by 11% and 15%, respectively (Figure 2A).

The total hydroxyproline concentration was 0.752 ± 0.022 mg/g in the control group, and ARNI had no effect after six weeks of treatment. In the SHR group, the total hydroxyproline concentration was higher than in controls by 16% (0.871 ± 0.026 mg/g, $p < 0.05$ vs. C), and ivabradine decreased it ($p < 0.05$) by 11%; ARNI had no significant effect (Figure 2A).

The hydroxyproline content in the soluble collagenous protein was 0.072 ± 0.003 mg/LV in the control group, and ARNI had no effect after six weeks of treatment. In the SHR group, the hydroxyproline content in the soluble collagenous protein was higher than in controls by 50% (0.108 ± 0.007 mg/LV, $p < 0.05$ vs. C), and none of the therapeutics had a significant effect (Figure 2B).

The hydroxyproline content in the insoluble collagenous protein was 0.241 ± 0.012 mg/LV in the control group, and ARNI had no effect after six weeks of treatment. In the SHR group, the hydroxyproline content in the insoluble collagenous protein was higher than in controls by 53% (0.368 ± 0.018 mg/LV, $p < 0.05$ vs. C), and ARNI and ivabradine decreased it ($p < 0.05$) by 24% and 19%, respectively (Figure 2B).

The total hydroxyproline content was 0.313 ± 0.014 mg/LV in the control group, and ARNI had no effect after six weeks of treatment. In the SHR group, the total hydroxyproline content was higher than in controls by 52% (0.476 ± 0.019 mg/LV, $p < 0.05$ vs. C), and ARNI and ivabradine decreased it ($p < 0.05$) by 22% and 15%, respectively (Figure 2B).

3.3. Serum Concentration of Angiotensins and Aldosterone, and the Markers of Renin and ACE Activities

The mean serum angiotensin and aldosterone concentrations in the study groups after six weeks of treatment are schematically depicted in Figure 3.

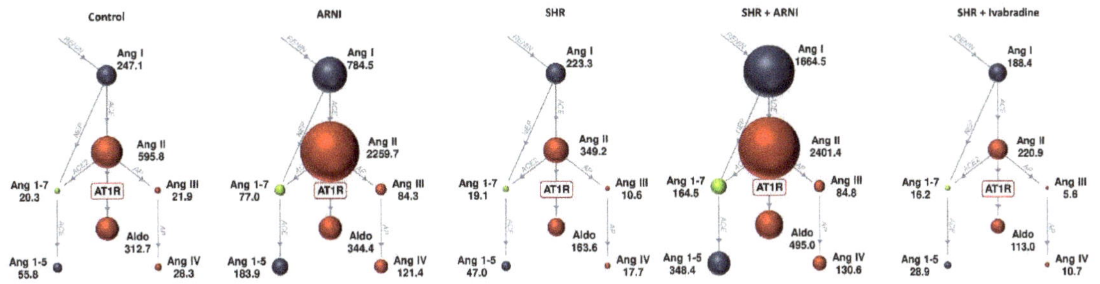

Figure 3. Schematic depicting the serum angiotensin and aldosterone concentrations in the study groups after six weeks of treatment. Results are presented as means. ACE, angiotensin-converting enzyme; Aldo, aldosterone; Ang, angiotensin; AP, aminopeptidase; ARNI, sacubitril/valsartan; AT1R, angiotensin II type 1 receptor; NEP, neprilysin (neutral endopeptidase); SHRs, spontaneously hypertensive rats. $n = 6$ per group.

The serum equilibrium level of Ang 1-10 (Ang I) was 247.15 ± 40.13 pmol/L and 223.32 ± 47.4 pmol/L in the control and SHR groups, respectively (ns); ARNI increased it by 217% (ns) and 645% ($p < 0.05$) in the control and SHR groups, respectively (Figure 4A).

The level of Ang 1-8 (Ang II) was 595.77 ± 87.89 pmol/L and 349.22 ± 79.1 pmol/L in the control and SHR groups, respectively (ns); ARNI increased it by 279% ($p < 0.05$) and 588% ($p < 0.05$) in the control and SHR groups, respectively, after six weeks of treatment (Figure 4B).

The level of Ang 2-8 (Ang III) was 21.88 ± 3.22 pmol/L and 10.63 ± 3.88 pmol/L in the control and SHR groups, respectively (ns); ARNI increased it by 285% (ns) and 697% ($p < 0.05$) in the control and SHR groups, respectively (Figure 4C).

The level of Ang 3-8 (Ang IV) was 28.27 ± 4.49 pmol/L and 17.72 ± 4.0 pmol/L in the control and SHR groups, respectively (ns); ARNI increased it by 329% (ns) and 637% ($p < 0.05$) in the control and SHR groups, respectively (Figure 4D).

The level of Ang 1-7 was 20.33 ± 2.66 pmol/L and 19.07 ± 4.69 pmol/L in the control and SHR groups, respectively (ns); ARNI increased it by 279% (ns) and 763% ($p < 0.05$) in the control and SHR groups, respectively (Figure 4E).

The level of Ang 1-5 was 55.75 ± 6.95 pmol/L and 46.97 ± 11.54 pmol/L in the control and SHR groups, respectively (ns); ARNI increased it by 230% (ns) and 642% ($p < 0.05$) in the control and SHR groups, respectively (Figure 4F).

None of the Ang levels in SHR were significantly affected by ivabradine after six weeks of treatment (Figure 4A–F).

The Ang 1-5/Ang 1-7 ratio was 2.78 ± 0.23 and 2.73 ± 0.32 in the control and SHR groups, respectively (ns); ivabradine decreased it by 32% ($p < 0.05$) in the SHR group. ARNI had no significant effect on the Ang 1-5/Ang 1-7 ratio in either controls or SHRs (Figure 4G).

The marker of renin activity (RA-S; Ang I + Ang II) was 842.88 ± 125.38 pmol/L and 572.5 ± 125.96 pmol/L in the control and SHR groups, respectively (ns); ARNI increased it by 261% (ns) and 610% ($p < 0.05$) in the control and SHR groups, respectively. Ivabradine had no significant effect on the marker of renin activity in SHRs after six weeks of treatment (Figure 4H).

The marker of ACE activity (ACE-S; Ang II/Ang I ratio) was 2.47 ± 0.20 and 1.53 ± 0.08 in the control and SHR groups, respectively ($p < 0.05$). ARNI and ivabradine had no significant effect on ACE-S in SHRs (Figure 4I).

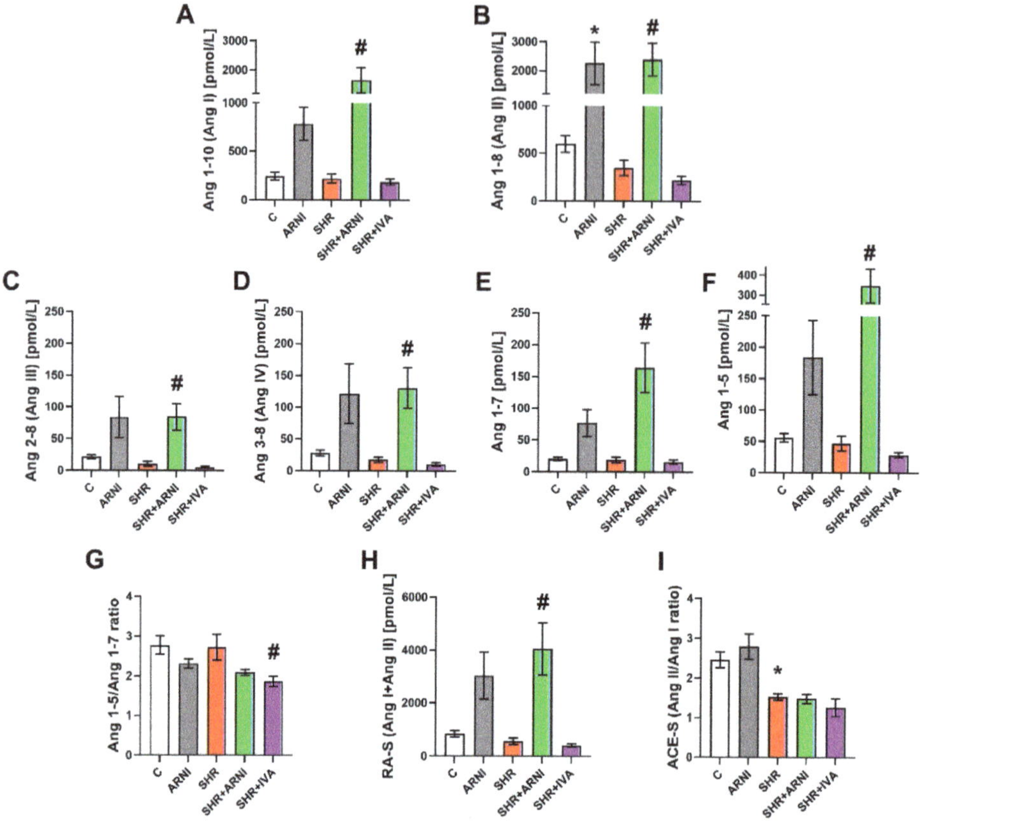

Figure 4. Effect of ARNI and ivabradine on the serum level of angiotensin 1-10 (Ang I) (**A**), angiotensin 1-8 (Ang II) (**B**), angiotensin 2-8 (Ang III) (**C**), angiotensin 3-8 (Ang IV) (**D**), angiotensin 1-7 (**E**), angiotensin 1-5 (**F**), angiotensin 1-5/angiotensin 1-7 ratio (**G**), marker of renin activity (RA-S; Ang I + Ang II) (**H**), and marker of angiotensin-converting enzyme activity (ACE-S; Ang II/Ang I) (**I**) in SHRs after six weeks of treatment. Ang, angiotensin; ARNI, sacubitril/valsartan; C, Wistar controls; IVA, ivabradine; SHRs, spontaneously hypertensive rats. Results are presented as means ± SEM. $n = 6$ per group. One-way, two-tailed ANOVA followed by multiple comparisons with a Bonferroni post-hoc test; * $p < 0.05$ vs. C; # $p < 0.05$ vs. SHR.

The serum concentration of aldosterone was 312.73 ± 66.99 fmol/mL and 163.63 ± 29.02 fmol/mL in the control and SHR groups, respectively (ns); ARNI increased it by 10% (ns) and 203% ($p < 0.05$) in the control and SHR groups, respectively (Figure 5A). The aldosterone/Ang II ratio (AA2 ratio) was 0.55 ± 0.14 and 0.52 ± 0.08 in the control and SHR groups, respectively (ns); ARNI decreased it by 61% (ns) and 52% (ns) in the control and SHR groups, respectively (Figure 5B). Ivabradine had no effect on the serum concentration of aldosterone and the AA2 ratio in SHRs after six weeks of treatment (Figure 5A,B).

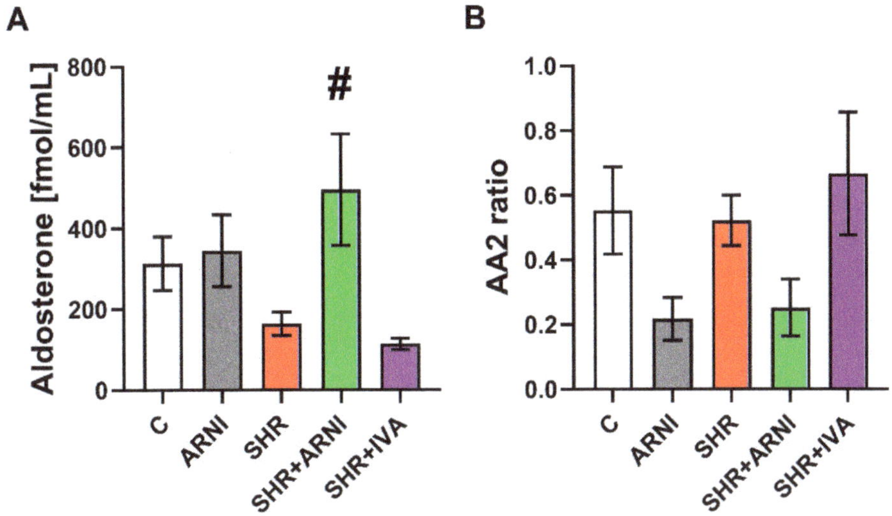

Figure 5. Effect of ARNI and ivabradine on the serum level of aldosterone (**A**) and the aldosterone/angiotensin II ratio (AA2 ratio) (**B**) in SHRs after six weeks of treatment. ARNI, sacubitril/valsartan; C, Wistar controls; IVA, ivabradine; SHRs, spontaneously hypertensive rats. Results are presented as means ± SEM. $n = 6$ per group. One-way, two-tailed ANOVA followed by multiple comparisons with a Bonferroni post-hoc test; # $p < 0.05$ vs. SHR.

3.4. *Echocardiography*

The LVEF was 72.52 ± 1.55% in the control group, and ARNI had no effect after six weeks of treatment. In the SHR group, the LVEF was lower than controls by 13% (63.0 ± 2.12%, $p < 0.05$ vs. C), and ARNI and ivabradine increased it ($p < 0.05$) by 12% and 10%, respectively (Figure 6A).

The LVFS was 37.38 ± 1.29% in the control group, and ARNI had no effect after six weeks of treatment. In the SHR group, the LVFS was lower than controls by 20% (30.0 ± 1.39%, $p < 0.05$ vs. C), and ARNI and ivabradine increased it ($p < 0.05$) by 17% and 14%, respectively (Figure 6B).

The E/A ratio was 1.42 ± 0.1 in the control group, and ARNI had no effect after six weeks of treatment. In the SHR group, the E/A ratio was higher than controls by 53% (2.17 ± 0.12%, $p < 0.05$ vs. C), and ivabradine decreased it ($p < 0.05$) by 24%. ARNI had no significant effect on the E/A ratio in SHRs after six weeks of treatment (Figure 6C).

The E/Em ratio was 10.58 ± 0.76 in the control group, and ARNI had no effect after six weeks of treatment. In the SHR group, the E/Em ratio was higher than controls by 132% (24.58 ± 2.01, $p < 0.05$ vs. C), and ARNI and ivabradine decreased it ($p < 0.05$) by 28% and 42%, respectively (Figure 6D).

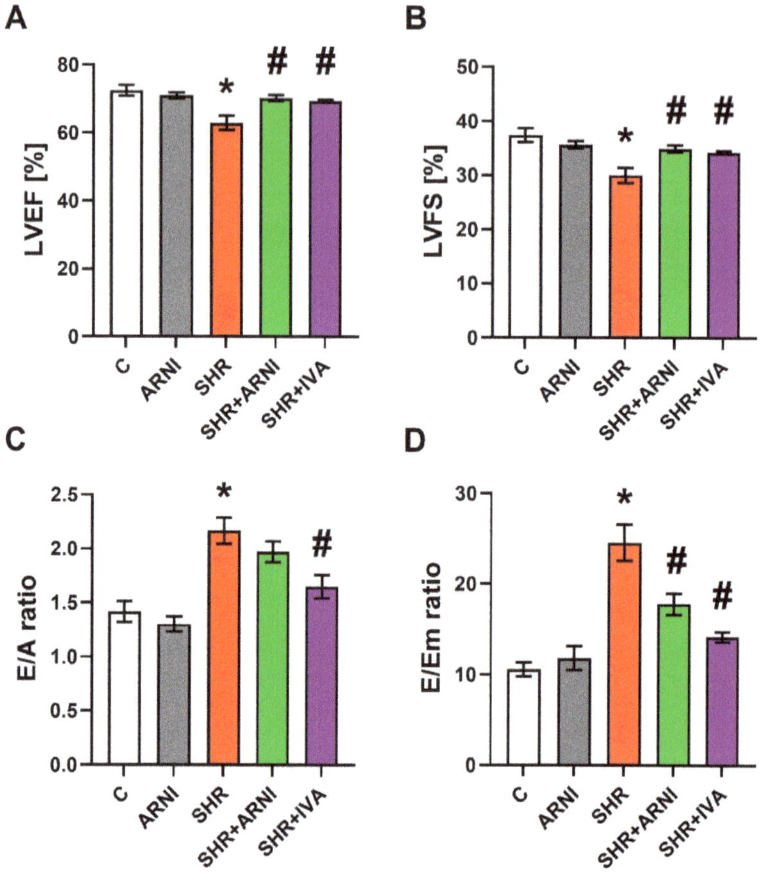

Figure 6. Effect of ARNI and ivabradine on left ventricular ejection fraction (LVEF) (**A**), left ventricular fractional shortening (LVFS) (**B**), the ratio of the diastolic transmitral peak early and late filling velocities (E/A ratio) (**C**), and the ratio of the diastolic transmitral peak early filling velocity and the maximal velocity of early diastolic wall movement wave at the level of mitral annulus (E/Em ratio) (**D**) in SHRs after six weeks of treatment. ARNI, sacubitril/valsartan; C, Wistar controls; IVA, ivabradine; SHRs, spontaneously hypertensive rats. Results are presented as means ± SEM. $n = 7$ per group. One-way, two-tailed ANOVA followed by multiple comparisons with a Bonferroni post-hoc test; * $p < 0.05$ vs. C; # $p < 0.05$ vs. SHR.

4. Discussion

The effects of the neprilysin inhibitor/AT1R blocker sacubitril/valsartan (ARNI) and ivabradine on SBP, HR, myocardial remodelling, LV systolic and diastolic function, and the RAAS were investigated in SHRs.

The SHR, a commonly employed rat experimental model of spontaneous hypertension, mimics primary hypertension with target organ damage in humans. The mechanisms underlying the development of primary hypertension are complex and comprise several potential players. Endothelial dysfunction in conduit and resistance arteries is frequently considered to contribute to the BP increase in the SHR. However, disturbed endothelial function has been described mainly in aged but not young SHRs, suggesting that endothelial dysfunction is more a consequence than a cause of elevated BP [16]. Data regarding the participation of the RAAS in hypertension pathophysiology in the SHR are driven mainly by the impact of targeted inhibition of the presumably deleterious classical ACE/Ang II/AT1R pathway [17] or by activation of the supposedly beneficial alternative ACE2/Ang 1-7/MasR pathway [18,19]. Data characterising the RAAS in the untreated SHR varies considerably in different laboratories. The serum level of renin, Ang II [20], and aldosterone [20,21], AT1R expression in mesenteric and coronary arteries [19], and heart expression of Mas-related G protein-coupled receptor D (MrgD) were higher in SHRs compared to Wistar rats, while the MasR expression in arteries was lower in SHRs [19]. Moreover, SHRs showed an altered circadian gene expression affecting the transcriptional regulation of clock-controlled genes for aldosterone and corticosterone [22]. In our experiment with three-month-old male SHRs, the levels of Ang I (Ang 1-10), Ang II (Ang 1-8), Ang III (Ang 2-8), Ang IV (Ang 3-8), Ang 1-7, and Ang 1-5 did not show significant changes compared to Wistar controls, corresponding with the unchanged marker of renin activity (RA-S). The trend towards reduced Ang II levels, albeit non-significant, corresponded with the decreased marker of ACE activity (ACE-S), calculated as the Ang II/Ang I ratio, and with the trend toward serum aldosterone concentration reduction. These data suggest that SHR is a normal-to-low renin and normal-to-low angiotensin/aldosterone model of hypertension. Thus, the underlying mechanism of hypertension and target organ damage in the SHR remains elusive. Importantly, increased renal sympathetic activity has been previously reported in the SHR [23]. Indeed, in our experiment, HR was significantly elevated in SHRs by approximately 100 bpm during the entire course of the six-week experiment, indicating activation of the sympathetic nervous system. In previous experiments, renal sympathetic denervation reduced intrarenal norepinephrine, the renal tissue protein of Ang II, aldosterone, and AT1R [24], and ameliorated renal fibrosis and dysfunction along with the delayed onset of hypertension in the SHR [25]. These results correspond with the findings of the higher activity of tyrosine hydroxylase, the rate-limiting enzyme in the synthesis of catecholamines, observed in the heart and kidney of SHRs compared with Wistar rats [26]. The above data suggest that sympathetic system activation plays a crucial role in BP elevation in the SHR. Along with sympathetic activation, the local renin–angiotensin system may also be activated in the kidney [25], brain [21], or other organs. In addition, other neurohumoral alterations, such as an elevated endothelin 1 level [20] or reduced endothelial nitric oxide synthase (eNOS) activity and nitric oxide (NO) bioavailability [16,25], may also contribute to BP elevation in the SHR. Furthermore, hypertensive encephalopathy due to a higher mineralocorticoid receptor expression and their activation by endogenous corticosterone may participate in hippocampal neuroinflammation, which may potentially contribute to BP dysregulation and hypertension [27].

In our study, the dual inhibitor of the endopeptidase neprilysin and the AT1R, sacubitril/valsartan (ARNI), significantly reduced systolic BP as well as LV mass in SHRs after six weeks of treatment. Moreover, ARNI significantly reduced the LV concentration of insoluble collagen, and numerically also the total collagen, and significantly decreased their LV contents. ARNI completely prevented the deterioration of LV systolic function and attenuated the deterioration of diastolic function in SHRs. This anti-remodelling nature of ARNI is in agreement with data from other laboratories and large clinical studies. Sacubitril/valsartan prevented myocardial fibrosis and remodelling and improved cardiac function after myocardial infarction in mice [28] and rats [29,30], and in streptozotocin-induced diabetic hearts in mice [31]; it reduced cardiomyocyte size in Ang II-induced cardiac hypertrophy in mice [32], attenuated LV fibrosis and dysfunction in high-salt diet-induced diastolic dysfunction in rats [33], and reduced BP and prevented stroke in stroke-prone hypertensive rats [34]. A meta-analysis of clinical studies from 2010 to 2019 revealed that ARNI exerted reverse remodelling in terms of reduced LV size and hypertrophy compared with ACE inhibitors or AT1R blockers in patients with HF with a reduced LV ejection fraction [35].

While the classical ACE/Ang II/AT1R pathway is considered to be deleterious when chronically activated by stimulating vasoconstriction, proliferation, and inflammation, the alternative ACE2/Ang 1-7/Ang 1-5/MasR seems to be a counterbalancing pathway by reducing oxidative stress and inflammation, inducing vasodilatation, and inhibiting myocyte growth and fibrotic proliferation [36–39]. In our experiment, ARNI, via AT1R blockade by valsartan, enhanced the levels of Ang II and Ang I and increased Ang 1-7 and Ang 1-5, along with Ang III and Ang IV. The marker of ACE activity (ACE-S), calculated as the Ang II/Ang I ratio, provides information about the expected ACE activity. ACE-S was lower in SHR and remained unaffected by ARNI. On the other hand, Ang 1-7 was remarkably increased by ARNI in SHRs and also stabilised, as shown by a trend to a reduced Ang 1-5/Ang 1-7 ratio. Ang 1-7 is considered to be a decisive player in damaged heart protection [40]. The increase in Ang 1-7 levels has to be considered with regard to the simultaneous AT1R blockade by valsartan, rendering Ang II ineffective and Ang 1-7 as the dominant effector in the RAAS. The results on serum aldosterone levels are puzzling. Although the AT1R blocker moiety of ARNI effectively blocked AT1R signalling in the adrenal glands, as indicated by a decreased aldosterone/Ang II ratio (AA2 ratio), the actual aldosterone level was increased by ARNI in SHRs, suggesting an important role of stimuli (such as potassium levels, NO availability, ACTH release or sympathetic activity) [41,42] for aldosterone production different from Ang II. Thus, in SHRs, the presumably protective effect of increased Ang 1-7 may be partly counterbalanced by elevated aldosterone levels. The increased serum levels of aldosterone with ARNI treatment might have determined the only mild-to-moderate antifibrotic effect of ARNI in this study. One could hypothesise that the potential combination of ARNI with a mineralocorticoid receptor antagonist would be a beneficial strategy in the treatment of hypertensive hearts. Additionally, the inhibition of neprilysin by sacubitril, with the subsequent enhancement of ANP and BNP exerting vasodilative and antiproliferative effects [43], might have also contributed to the protection by ARNI in SHRs.

Elevated HR is a risk factor in healthy individuals and various cardiovascular pathologies [44]. Ivabradine selectively inhibits the I_f current of the pacemaker cells in the sinoatrial node, thus reducing HR without negative inotropy [45]. Although HR reduction coming with an improvement in myocardial energy balance seems to be the principal factor underlying the protection offered by ivabradine in HF patients, a number of pleiotropic effects, including antioxidative, anti-inflammatory, and neurohumoral action modulation, may contribute to heart protection [45]. Indeed, in aortic constriction-induced LV hypertrophy in mice, ivabradine attenuated LV hypertrophy, fibrosis, and dysfunction [46]. In isoproterenol-induced heart damage, ivabradine reduced LV fibrotic remodelling and improved survival [47]. In our previous experiment with L-NAME-induced hypertension, ivabradine improved the systolic and diastolic function of the remodelled LV [48]. In line with the above data, in this experiment, ivabradine reduced fibrosis of the LV and improved systolic and diastolic function in SHRs. However, ivabradine did not affect the serum concentrations of Ang II and Ang 1-7 or other downstream products of the classical and alternative pathways of the renin–angiotensin system. Interestingly, even the serum levels of aldosterone, which were reduced by ivabradine in L-NAME-induced hypertension [48] and supposedly contributed to the hypertensive heart protection by ivabradine, were unchanged in SHRs. There are two potential factors underlying the apparent protection by ivabradine in SHRs to consider. First, HR reduction associated with an improvement of myocardial energy balance in a haemodynamically overloaded hypertensive heart could result in improved LV contractility and relaxation. Second, ivabradine may provide a benefit via its presumable pleiotropic sympatholytic effect. Indeed, the pre-treatment of rats exposed to handling stress with ivabradine was associated with a reduced release of adrenaline and noradrenaline into the blood stream [49], and changes to heart rate variability (HRV) in a rat model of doxorubicin-induced HF indicated an improved autonomic imbalance by ivabradine [50]. The HRV analysis in a SHIFT Holter sub-study showed an ivabradine-mediated shift toward a more prominent parasympathetic tone [51].

Limitations: For the pathophysiological implications, the local tissue concentration of RAAS peptides remains important. It seems, however, that the local RAAS remains largely dependent on the circulating RAAS peptides, as previously discussed for the brain

RAAS [52]. In fact, the concentration of renin and angiotensinogen in the heart tissue is very low compared to plasma [53]. Additionally, we have recently shown [48] that apart from Ang 1-10 and Ang 1-8, the tissue concentrations of other RAAS components remain low and the LV concentration of Ang II correlates with the circulating Ang II levels.

5. Conclusions

We conclude that the SHR is a normal-to-low RAAS model of experimental hypertension. Recent drugs from the portfolio of HF management—ARNI and ivabradine—exerted the attenuation of LV remodelling and dysfunction in the SHR. Considering the changes to the RAAS, the cardiovascular protection by ARNI may be related to the Ang II blockade and the protective nature of Ang 1-7, while the cardiovascular protection by ivabradine was not associated with the modification of RAAS in the SHR.

Author Contributions: Conceptualization, F.S.; methodology, F.S., T.B., P.S., O.D. and M.A.; validation and formal analysis, F.S., T.B., S.Z., M.A. and L.P.; investigation, resources, data curation, T.B., P.S., K.R., K.K., S.A., O.D. and M.A.; writing—original draft preparation, F.S.; writing—review and editing, all authors; visualization, T.B., P.S., K.R. and O.D.; supervision, F.S., K.K., M.A. and L.P.; project administration, F.S. and K.K.; funding acquisition, F.S. and S.Z. All authors have read and agreed to the published version of the manuscript.

Funding: This work was supported by research grants VEGA 1/0035/19 and APVV-20-0421.

Institutional Review Board Statement: The animal study protocol was approved by the Ethics Committee of the Institute of Pathophysiology, Faculty of Medicine, Comenius University, Bratislava, Slovak Republic (approval number: 809/19-221/3; approval date: 23 April 2019).

Informed Consent Statement: Not applicable.

Data Availability Statement: Data supporting the reported results are available from the corresponding author per request.

Conflicts of Interest: The authors declare no conflict of interest. The funders had no role in the design of the study; in the collection, analyses, or interpretation of data; in the writing of the manuscript; or in the decision to publish the results.

References

1. Simko, F.; Pechanova, O. Remodelling of the Heart and Vessels in Experimental Hypertension: Advances in Protection. *J. Hypertens.* **2010**, *28* (Suppl. 1), S1–S6. [CrossRef] [PubMed]
2. Weber, K.T. Fibrosis and Hypertensive Heart Disease. *Curr. Opin. Cardiol.* **2000**, *15*, 264–272. [CrossRef] [PubMed]
3. Simko, F.; Simko, J. The Potential Role of Nitric Oxide in the Hypertrophic Growth of the Left Ventricle. *Physiol. Res.* **2000**, *49*, 37–46. [PubMed]
4. McMurray, J.J.V. Neprilysin Inhibition to Treat Heart Failure: A Tale of Science, Serendipity, and Second Chances. *Eur. J. Heart Fail.* **2015**, *17*, 242–247. [CrossRef] [PubMed]
5. Simko, F.; Dukat, A. Inhibition of Neprilysin—New Horizon in Heart Failure Therapy. *Cardiol. Lett.* **2016**, *25*, 273–276.
6. Packer, M.; McMurray, J.J.V.; Desai, A.S.; Gong, J.; Lefkowitz, M.P.; Rizkala, A.R.; Rouleau, J.L.; Shi, V.C.; Solomon, S.D.; Swedberg, K.; et al. Angiotensin Receptor Neprilysin Inhibition Compared with Enalapril on the Risk of Clinical Progression in Surviving Patients with Heart Failure. *Circulation* **2015**, *131*, 54–61. [CrossRef] [PubMed]
7. Pieske, B.; Wachter, R.; Shah, S.J.; Baldridge, A.; Szeczoedy, P.; Ibram, G.; Shi, V.; Zhao, Z.; Cowie, M.R.; PARALLAX Investigators and Committee members. Effect of Sacubitril/Valsartan vs Standard Medical Therapies on Plasma NT-ProBNP Concentration and Submaximal Exercise Capacity in Patients with Heart Failure and Preserved Ejection Fraction: The PARALLAX Randomized Clinical Trial. *JAMA* **2021**, *326*, 1919–1929. [CrossRef]
8. Swedberg, K.; Komajda, M.; Böhm, M.; Borer, J.S.; Ford, I.; Dubost-Brama, A.; Lerebours, G.; Tavazzi, L.; SHIFT Investigators. Ivabradine and Outcomes in Chronic Heart Failure (SHIFT): A Randomised Placebo-Controlled Study. *Lancet* **2010**, *376*, 875–885. [CrossRef]
9. Burnett, J.C.; Granger, J.P.; Opgenorth, T.J. Effects of Synthetic Atrial Natriuretic Factor on Renal Function and Renin Release. *Am. J. Physiol.* **1984**, *247*, F863–F866. [CrossRef] [PubMed]
10. Reddy, G.K.; Enwemeka, C.S. A Simplified Method for the Analysis of Hydroxyproline in Biological Tissues. *Clin. Biochem.* **1996**, *29*, 225–229. [CrossRef]
11. Basu, R.; Poglitsch, M.; Yogasundaram, H.; Thomas, J.; Rowe, B.H.; Oudit, G.Y. Roles of Angiotensin Peptides and Recombinant Human ACE2 in Heart Failure. *J. Am. Coll. Cardiol.* **2017**, *69*, 805–819. [CrossRef] [PubMed]

12. Pavo, N.; Goliasch, G.; Wurm, R.; Novak, J.; Strunk, G.; Gyöngyösi, M.; Poglitsch, M.; Säemann, M.D.; Hülsmann, M. Low- and High-Renin Heart Failure Phenotypes with Clinical Implications. *Clin. Chem.* **2018**, *64*, 597–608. [CrossRef] [PubMed]
13. Guo, Z.; Poglitsch, M.; Cowley, D.; Domenig, O.; McWhinney, B.C.; Ungerer, J.P.J.; Wolley, M.; Stowasser, M. Effects of Ramipril on the Aldosterone/Renin Ratio and the Aldosterone/Angiotensin II Ratio in Patients with Primary Aldosteronism. *Hypertension* **2020**, *76*, 488–496. [CrossRef] [PubMed]
14. Burrello, J.; Buffolo, F.; Domenig, O.; Tetti, M.; Pecori, A.; Monticone, S.; Poglitsch, M.; Mulatero, P. Renin-Angiotensin-Aldosterone System Triple-A Analysis for the Screening of Primary Aldosteronism. *Hypertension* **2020**, *75*, 163–172. [CrossRef] [PubMed]
15. Liu, J.; Rigel, D.F. Echocardiographic Examination in Rats and Mice. In *Cardiovascular Genomics: Methods and Protocols*; Methods in Molecular Biology Series; Humana: Totowa, NJ, USA, 2009; Volume 573, pp. 139–155. [CrossRef]
16. Bernatova, I.; Conde, M.V.; Kopincova, J.; González, M.C.; Puzserova, A.; Arribas, S.M. Endothelial Dysfunction in Spontaneously Hypertensive Rats: Focus on Methodological Aspects. *J. Hypertens. Suppl.* **2009**, *27*, S27–S31. [CrossRef] [PubMed]
17. Simko, F.; Pechanova, O.; Pelouch, V.; Krajcirovicova, K.; Mullerova, M.; Bednarova, K.; Adamcova, M.; Paulis, L. Effect of Melatonin, Captopril, Spironolactone and Simvastatin on Blood Pressure and Left Ventricular Remodelling in Spontaneously Hypertensive Rats. *J. Hypertens. Suppl.* **2009**, *27*, S5–S10. [CrossRef] [PubMed]
18. Stoyell-Conti, F.F.; Chabbra, A.; Puthentharayil, J.; Rigatto, K.; Speth, R.C. Chronic Administration of Pharmacological Doses of Angiotensin 1-7 and Iodoangiotensin 1-7 Has Minimal Effects on Blood Pressure, Heart Rate, and Cognitive Function of Spontaneously Hypertensive Rats. *Physiol. Rep.* **2021**, *9*, e14812. [CrossRef] [PubMed]
19. Zhang, F.; Tang, H.; Sun, S.; Luo, Y.; Ren, X.; Chen, A.; Xu, Y.; Li, P.; Han, Y. Angiotensin-(1-7) Induced Vascular Relaxation in Spontaneously Hypertensive Rats. *Nitric Oxide Biol. Chem.* **2019**, *88*, 1–9. [CrossRef]
20. Jing, Y.; Hu, J.; Zhao, J.; Yang, J.; Huang, N.; Song, P.; Xu, J.; Zhang, M.; Li, P.; Yin, Y. Experimental Study of Blood Pressure and Its Impact on Spontaneous Hypertension in Rats with Xin Mai Jia. *Biomed. Pharmacother.* **2019**, *112*, 108689. [CrossRef]
21. Wang, X.; Zhu, Y.; Wang, S.; Wang, Z.; Sun, H.; He, Y.; Yao, W. Effects of Eplerenone on Cerebral Aldosterone Levels and Brain Lesions in Spontaneously Hypertensive Rats. *Clin. Exp. Hypertens.* **2020**, *42*, 531–538. [CrossRef]
22. Tanaka, S.; Ueno, T.; Tsunemi, A.; Nagura, C.; Tahira, K.; Fukuda, N.; Soma, M.; Abe, M. The Adrenal Gland Circadian Clock Exhibits a Distinct Phase Advance in Spontaneously Hypertensive Rats. *Hypertens. Res.* **2019**, *42*, 165–173. [CrossRef] [PubMed]
23. Sun, Z.; Zhang, Z. Historic Perspectives and Recent Advances in Major Animal Models of Hypertension. *Acta Pharmacol. Sin.* **2005**, *26*, 295–301. [CrossRef]
24. Qin, F.; Li, J.; Dai, Y.-F.; Zhong, X.-G.; Pan, Y.-J. Renal Denervation Inhibits the Renin-Angiotensin-Aldosterone System in Spontaneously Hypertensive Rats. *Clin. Exp. Hypertens.* **2022**, *44*, 83–92. [CrossRef]
25. Wang, M.; Han, W.; Zhang, M.; Fang, W.; Zhai, X.; Guan, S.; Qu, X. Long-Term Renal Sympathetic Denervation Ameliorates Renal Fibrosis and Delays the Onset of Hypertension in Spontaneously Hypertensive Rats. *Am. J. Transl. Res.* **2018**, *10*, 4042–4053. [PubMed]
26. Lee, Y.H.; Kim, Y.G.; Moon, J.-Y.; Kim, J.S.; Jeong, K.-H.; Lee, T.W.; Ihm, C.-G.; Lee, S.H. Genetic Variations of Tyrosine Hydroxylase in the Pathogenesis of Hypertension. *Electrolyte Blood Press. EBP* **2016**, *14*, 21–26. [CrossRef]
27. Brocca, M.E.; Pietranera, L.; de Kloet, E.R.; De Nicola, A.F. Mineralocorticoid Receptors, Neuroinflammation and Hypertensive Encephalopathy. *Cell. Mol. Neurobiol.* **2019**, *39*, 483–492. [CrossRef] [PubMed]
28. Liu, J.; Zheng, X.; Zhang, C.; Zhang, C.; Bu, P. Lcz696 Alleviates Myocardial Fibrosis after Myocardial Infarction through the SFRP-1/Wnt/β-Catenin Signaling Pathway. *Front. Pharmacol.* **2021**, *12*, 724147. [CrossRef]
29. Kompa, A.R.; Lu, J.; Weller, T.J.; Kelly, D.J.; Krum, H.; von Lueder, T.G.; Wang, B.H. Angiotensin Receptor Neprilysin Inhibition Provides Superior Cardioprotection Compared to Angiotensin Converting Enzyme Inhibition after Experimental Myocardial Infarction. *Int. J. Cardiol.* **2018**, *258*, 192–198. [CrossRef] [PubMed]
30. von Lueder, T.G.; Wang, B.H.; Kompa, A.R.; Huang, L.; Webb, R.; Jordaan, P.; Atar, D.; Krum, H. Angiotensin Receptor Neprilysin Inhibitor LCZ696 Attenuates Cardiac Remodeling and Dysfunction after Myocardial Infarction by Reducing Cardiac Fibrosis and Hypertrophy. *Circ. Heart Fail.* **2015**, *8*, 71–78. [CrossRef] [PubMed]
31. Suematsu, Y.; Miura, S.-I.; Goto, M.; Matsuo, Y.; Arimura, T.; Kuwano, T.; Imaizumi, S.; Iwata, A.; Yahiro, E.; Saku, K. LCZ696, an Angiotensin Receptor-Neprilysin Inhibitor, Improves Cardiac Function with the Attenuation of Fibrosis in Heart Failure with Reduced Ejection Fraction in Streptozotocin-Induced Diabetic Mice. *Eur. J. Heart Fail.* **2016**, *18*, 386–393. [CrossRef] [PubMed]
32. Tashiro, K.; Kuwano, T.; Ideishi, A.; Morita, H.; Idemoto, Y.; Goto, M.; Suematsu, Y.; Miura, S.-I. Sacubitril/Valsartan Inhibits Cardiomyocyte Hypertrophy in Angiotensin II-Induced Hypertensive Mice Independent of a Blood Pressure-Lowering Effect. *Cardiol. Res.* **2020**, *11*, 376–385. [CrossRef] [PubMed]
33. Zhang, W.; Liu, J.; Fu, Y.; Ji, H.; Fang, Z.; Zhou, W.; Fan, H.; Zhang, Y.; Liao, Y.; Yang, T.; et al. Sacubitril/Valsartan Reduces Fibrosis and Alleviates High-Salt Diet-Induced HFpEF in Rats. *Front. Pharmacol.* **2020**, *11*, 600953. [CrossRef] [PubMed]
34. Rubattu, S.; Cotugno, M.; Forte, M.; Stanzione, R.; Bianchi, F.; Madonna, M.; Marchitti, S.; Volpe, M. Effects of Dual Angiotensin Type 1 Receptor/Neprilysin Inhibition vs. Angiotensin Type 1 Receptor Inhibition on Target Organ Injury in the Stroke-Prone Spontaneously Hypertensive Rat. *J. Hypertens.* **2018**, *36*, 1902–1914. [CrossRef] [PubMed]
35. Wang, Y.; Zhou, R.; Lu, C.; Chen, Q.; Xu, T.; Li, D. Effects of the Angiotensin-Receptor Neprilysin Inhibitor on Cardiac Reverse Remodeling: Meta-Analysis. *J. Am. Heart Assoc.* **2019**, *8*, e012272. [CrossRef] [PubMed]
36. Steckelings, U.M.; Sumners, C. Correcting the Imbalanced Protective RAS in COVID-19 with Angiotensin AT2-Receptor Agonists. *Clin. Sci.* **2020**, *134*, 2987–3006. [CrossRef] [PubMed]

37. Hrenak, J.; Simko, F. Renin-Angiotensin System: An Important Player in the Pathogenesis of Acute Respiratory Distress Syndrome. *Int. J. Mol. Sci.* **2020**, *21*, 8038. [CrossRef]
38. Simko, F.; Baka, T. Angiotensin-Converting Enzyme Inhibitors and Angiotensin II Receptor Blockers: Potential Allies in the COVID-19 Pandemic Instead of a Threat? *Clin. Sci.* **2021**, *135*, 1009–1014. [CrossRef] [PubMed]
39. Simko, F.; Hrenak, J.; Adamcova, M.; Paulis, L. Renin-Angiotensin-Aldosterone System: Friend or Foe-The Matter of Balance. Insight on History, Therapeutic Implications and COVID-19 Interactions. *Int. J. Mol. Sci.* **2021**, *22*, 3217. [CrossRef]
40. Wang, K.; Basu, R.; Poglitsch, M.; Bakal, J.A.; Oudit, G.Y. Elevated Angiotensin 1-7/Angiotensin II Ratio Predicts Favorable Outcomes in Patients with Heart Failure. *Circ. Heart Fail.* **2020**, *13*, e006939. [CrossRef] [PubMed]
41. Mulrow, P.J. Angiotensin II and Aldosterone Regulation. *Regul. Pept.* **1999**, *80*, 27–32. [CrossRef]
42. Rebuffat, P.; Malendowicz, L.K.; Nussdorfer, G.G.; Mazzocchi, G. Stimulation of Endogenous Nitric Oxide Production Is Involved in the Inhibitory Effect of Adrenomedullin on Aldosterone Secretion in the Rat. *Peptides* **2001**, *22*, 923–926. [CrossRef]
43. Repova, K.; Aziriova, S.; Krajcirovicova, K.; Simko, F. Cardiovascular Therapeutics: A New Potential for Anxiety Treatment? *Med. Res. Rev.* **2022**, *42*, 1202–1245. [CrossRef] [PubMed]
44. Simko, F.; Baka, T.; Paulis, L.; Reiter, R.J. Elevated Heart Rate and Nondipping Heart Rate as Potential Targets for Melatonin: A Review. *J. Pineal Res.* **2016**, *61*, 127–137. [CrossRef]
45. Simko, F.; Baka, T. Ivabradine and Blood Pressure Reduction: Underlying Pleiotropic Mechanisms and Clinical Implications. *Front. Cardiovasc. Med.* **2021**, *8*, 607998. [CrossRef]
46. Yu, Y.; Hu, Z.; Li, B.; Wang, Z.; Chen, S. Ivabradine Improved Left Ventricular Function and Pressure Overload-Induced Cardiomyocyte Apoptosis in a Transverse Aortic Constriction Mouse Model. *Mol. Cell. Biochem.* **2019**, *450*, 25–34. [CrossRef] [PubMed]
47. Simko, F.; Baka, T.; Repova, K.; Aziriova, S.; Krajcirovicova, K.; Paulis, L.; Adamcova, M. Ivabradine Improves Survival and Attenuates Cardiac Remodeling in Isoproterenol-Induced Myocardial Injury. *Fundam. Clin. Pharmacol.* **2021**, *35*, 744–748. [CrossRef] [PubMed]
48. Simko, F.; Baka, T.; Poglitsch, M.; Repova, K.; Aziriova, S.; Krajcirovicova, K.; Zorad, S.; Adamcova, M.; Paulis, L. Effect of Ivabradine on a Hypertensive Heart and the Renin-Angiotensin-Aldosterone System in L-NAME-Induced Hypertension. *Int. J. Mol. Sci.* **2018**, *19*, 3017. [CrossRef] [PubMed]
49. Ondicova, K.; Hegedusova, N.; Tibensky, M.; Mravec, B. Ivabradine Reduces Baseline and Stress-Induced Increase of Heart Rate and Blood Pressure and Modulates Neuroendocrine Stress Response in Rats Depending on Stressor Intensity. *Gen. Physiol. Biophys.* **2019**, *38*, 165–173. [CrossRef]
50. El-Naggar, A.E.; El-Gowilly, S.M.; Sharabi, F.M. Possible Ameliorative Effect of Ivabradine on the Autonomic and Left Ventricular Dysfunction Induced by Doxorubicin in Male Rats. *J. Cardiovasc. Pharmacol.* **2018**, *72*, 22–31. [CrossRef]
51. Böhm, M.; Borer, J.S.; Camm, J.; Ford, I.; Lloyd, S.M.; Komajda, M.; Tavazzi, L.; Talajic, M.; Lainscak, M.; Reil, J.-C.; et al. Twenty-Four-Hour Heart Rate Lowering with Ivabradine in Chronic Heart Failure: Insights from the SHIFT Holter Substudy. *Eur. J. Heart Fail.* **2015**, *17*, 518–526. [CrossRef]
52. van Thiel, B.S.; Góes Martini, A.; Te Riet, L.; Severs, D.; Uijl, E.; Garrelds, I.M.; Leijten, F.P.J.; van der Pluijm, I.; Essers, J.; Qadri, F.; et al. Brain Renin-Angiotensin System: Does It Exist? *Hypertension* **2017**, *69*, 1136–1144. [CrossRef] [PubMed]
53. Raizada, V.; Skipper, B.; Luo, W.; Griffith, J. Intracardiac and Intrarenal Renin-Angiotensin Systems: Mechanisms of Cardiovascular and Renal Effects. *J. Investig. Med.* **2007**, *55*, 341–359. [CrossRef] [PubMed]

biomedicines

Article

How Effective Is a Late-Onset Antihypertensive Treatment? Studies with Captopril as Monotherapy and in Combination with Nifedipine in Old Spontaneously Hypertensive Rats

Christina Hawlitschek [1], Julia Brendel [1], Philipp Gabriel [1], Katrin Schierle [2], Aida Salameh [3], Heinz-Gerd Zimmer [1] and Beate Rassler [1,*]

1 Carl-Ludwig-Institute of Physiology, University of Leipzig, 04103 Leipzig, Germany
2 Institute of Pathology, University of Leipzig, 04103 Leipzig, Germany
3 Department of Pediatric Cardiology, Heart Centre, University of Leipzig, 04289 Leipzig, Germany
* Correspondence: beate.rassler@medizin.uni-leipzig.de; Tel.: +49-341-9715565

Abstract: Background: A major problem in the treatment of human hypertension is the late diagnosis of hypertension and, hence, the delayed start of treatment. Very often, hypertension has existed for a long time and cardiac damage has already developed. Therefore, we tested whether late-onset antihypertensive treatment is effective in lowering blood pressure (BP) and in reducing or even preventing left ventricular hypertrophy and fibrosis. Methods: Twenty-one male 60-week-old spontaneously hypertensive rats (SHR) were included. Fourteen rats received oral treatment with captopril (CAP) either as monotherapy or combined with nifedipine (CAP + NIF) over 22 weeks. Seven untreated SHR served as controls. We examined the therapeutic effects on BP, heart weight and histological and biochemical markers of left ventricular remodeling and fibrosis. Results: At 82 weeks of age, BP was reduced in the CAP and CAP + NIF groups by 44 and 51 mmHg, respectively ($p < 0.001$), but not in untreated controls. Despite the late therapy start, cardiac hypertrophy and fibrosis were attenuated compared to controls. Both treatments reduced heart weight by 1.2 mg/g (25%, $p = 0.001$) and collagens I and III by 66% and 60%, respectively ($p < 0.001$), thus proving nearly equivalent cardioprotective efficacy. Conclusion: These data clearly emphasize the benefit of antihypertensive treatment in reducing BP and mitigating the development of cardiac damage even when treatment is started late in life.

Keywords: old SHR; antihypertensive therapy; blood pressure monitoring; LV hypertrophy; ECM markers; cardiac fibrosis

Citation: Hawlitschek, C.; Brendel, J.; Gabriel, P.; Schierle, K.; Salameh, A.; Zimmer, H.-G.; Rassler, B. How Effective Is a Late-Onset Antihypertensive Treatment? Studies with Captopril as Monotherapy and in Combination with Nifedipine in Old Spontaneously Hypertensive Rats. *Biomedicines* **2022**, *10*, 1964. https://doi.org/10.3390/biomedicines10081964

Academic Editors: Josef Zicha and Ivana Vaněčková

Received: 29 June 2022
Accepted: 9 August 2022
Published: 12 August 2022

1. Introduction

Arterial hypertension is the main risk factor for morbidity and mortality of cerebrovascular and cardiovascular diseases. The number of hypertensive adults aged 30–79 years has almost doubled in the last 30 years to more than 1.2 billion people worldwide, but only 59% of women and 49% of men with hypertension are diagnosed, and another 11–12% are diagnosed but not treated [1]. The problem of inadequate antihypertensive therapy is exacerbated by the patients' inconsistent drug intake and inconsequent blood pressure control by the patients and their therapists. In less than 40% of diagnosed patients, hypertension has been controlled to below 140/90 mmHg [1]. Untreated or insufficiently treated arterial hypertension leads to cardiac hypertrophy and may progress to cardiac remodeling, fibrosis, dilatation, and in the final stages, to heart failure. A long history of undetected and untreated hypertension is a risk factor as high as advanced age, particularly if hypertension has persisted for some time and the first cardiac sequelae may have already occurred. The earlier in the course of pathophysiological development an efficient antihypertensive therapy is started, the more successful it can be in preventing complications [2,3].

Spontaneously hypertensive rats (SHR) are an established animal model for studying the causes and development of hypertension. This is due to their prehypertensive stage and their clinical complication pattern, which is similar to that of humans including LV hypertrophy and dilatation and often resulting in heart failure [4]. Upregulation of renin-angiotensin-aldosterone-system (RAAS)-related key genes and increased intracellular calcium concentration belong to the main factors responsible for the increased blood pressure (BP) in adult SHR [5,6]. Correspondingly, in numerous studies in SHR, profound antihypertensive and cardioprotective effects were achieved by antagonizing the RAAS and reducing intracellular calcium concentration [2,7–9]. The antagonization of angiotensin II (AT II) and the blockade of calcium channels have also proved to be effective antihypertensive therapies in humans and are recommended by the guidelines of the European Societies of Cardiology and Hypertension as preferred treatments of human hypertension [10]. In particular, the combination of two antihypertensive drugs is highly recommended, as better antihypertensive effects can be achieved with lower drug doses, and thus with fewer side effects and better tolerability [10,11].

One of the major problems in human hypertension and its therapy is the fact that hypertension is often diagnosed late in life when the BP has been elevated for many years. Many of those patients present an advanced degree of hypertension, and the first cardiovascular complications have already developed. The therapy has to start immediately with strict control of patients' compliance and BP development. However, even if antihypertensive therapy is well-adjusted and controlled, a late-onset treatment may be less effective than a treatment started in the very early stages of hypertension. Studies in SHR showed that antihypertensive treatment starting in a very early phase of life (between four and 10 weeks of age) reduced blood pressure (BP) even to normotension [3,12] and prevented left ventricular hypertrophy and fibrosis [7,13,14]. In contrast, the same treatment starting at a later stage of life (24 to 30 weeks of age) had a much lower antihypertensive effect [3,15]. For treatment of human hypertension, it is important to clarify whether and to what extent a late-onset antihypertensive therapy is able to significantly reduce BP and, even more importantly, to attenuate cardiac hypertrophy and fibrosis and to delay transition into cardiac failure.

We investigated these questions in an animal model of old SHR. Rats aged 24–30 weeks can be considered to be adult. At this age, normotensive rats are still in the first half of their regular life span, which is estimated to be 2.5–3.5 years [16]. Our interest was focused on an antihypertensive therapy starting in the period of senescence. The present study was designed to investigate the efficacy of such a late-onset antihypertensive therapy on BP development and on cardiac remodeling in 60-week-old SHR. In this context, two therapeutic regimens, a monotherapy and a combination therapy, were compared. In a preceding study on seven-week-old SHR, we systematically tested the antihypertensive and cardioprotective effects of several classes of antihypertensive drugs as single-drugs and as combination treatments. Specifically, captopril (CAP) as an antagonist of the RAAS and nifedipine (NIF) as a calcium channel blocker were applied. CAP as monotherapy and a combination of CAP and NIF proved to be the most effective treatments with respect to BP lowering and to preventing cardiac hypertrophy and fibrosis [14]. These two therapeutic regimens were chosen for the present study to treat 60-week-old SHR over a period of 22 weeks. We investigated BP, heart weight (HW), biochemical markers of cardiac hypertrophy and cardiac fibrosis as well as cardiac histology.

We hypothesized that these treatments would decrease BP as well as biochemical and histological markers of cardiac hypertrophy and fibrosis in old SHR as well. The results of this study might have importance for antihypertensive treatment in humans. In particular, one of the main reasons for patients' poor adherence to antihypertensive therapy is that they do not understand the serious consequences of chronically elevated BP. The results of studies like this may help to improve the education of patients and the strict control of antihypertensive therapy.

2. Materials and Methods

2.1. Animal Model

All experiments were performed on 21 male SHR supplied by Charles River, Sulzfeld, Germany. The animals were fed a standard pellet diet (Altromin C100, Altromin GmbH, Lage, Germany) and had free access to tap water. All animal protocols were approved by the state agency (Landesdirektion Sachsen, number and date of approval: TVV 36/08; 13 May 2009) in accordance with the Guide for the Care and Use of Laboratory Animals published by the National Institutes of Health and with the "European Convention for the Protection of Vertebrate Animals used for Experimental and other Scientific Purposes" [17].

2.2. Experimental Protocol

The 22-week study phase was preceded by a two-week adaptation period to accustomize the animals to drug-free tablets and to the procedure of BP measurement. The animals were 60.5 $\pm$ 0.25 weeks of age (BW 404 $\pm$ 31 g) at the beginning and 81.9 $\pm$ 0.44 weeks at the end of the 22-week study phase. They were subdivided into three groups (n = seven per group). The group size was calculated for a medium effect size of systolic blood pressure (SBP) reduction (with Cohen's f = 0.25). Two groups received antihypertensive treatment with CAP (60 mg kg^{-1} d^{-1}, Axxora, Lörrach, Germany) or CAP + NIF (CAP 60 + NIF 10 mg kg^{-1} d^{-1}, Sigma-Aldrich Chemie, Steinheim, Germany), while the third group received drug-free tablets and served as untreated control (CTRL). The drugs were added to commercially available rodent sweets (Vitakraft-Werke, Wührmann & Sohn GmbH & Co., KG, Bremen, Germany) and formed to tablets. The tablets were given into the cages for oral uptake along with chow once daily between 9:00 a.m. and 10:00 a.m. After the two-week adaptation period, non-invasive BP measurements were carried out every two to three weeks. At the end of the experimental period, animals were sacrificed, and their hearts were removed for further analyses.

2.3. Non-Invasive Blood Pressure Measurement

SBP was measured non-invasively using the tail-cuff-method (TSE Blood Pressure Monitor, Series 209002, TSE Systems GmbH, Bad Homburg, Germany). Measurements were performed in awake animals about 1 to 2 h after the administration of the drugs or the drug-free tablets. The animals were placed on a heated plate (36 °C) and were allowed to move relatively freely, and were only held by the experimenter's hand on their back or tail. We avoided the use of a conventional restraint box to minimize stress to the animals. After six to eight preliminary tests, the animals were familiar with the environment and the experimenters. The procedure of SBP measurement was well tolerated by the animals. To ensure reproducible results, for each SBP measurement a mean was calculated from two to three tests with each test containing 10 single readings.

2.4. Further Analyses on Heart Tissue

After extraction, HW was determined as a measure of cardiac hypertrophy. As body weight (BW) developed differently in the SHR groups, the ratio of HW/BW was calculated by relating HW to baseline BW. The apex was separated and fixated in formalin for histological examination. Pieces of the LV were frozen and stored at −80 °C for biochemical analyses.

2.5. Ribonuclease Protection Assay

A ribonuclease protection assay was performed as previously described [14] to determine mRNA expression of atrial natriuretic peptide (ANP), transforming growth factor-β_1 (TGF-β_1), TGF-β_2 and TGF-β_3, matrix metalloproteinase 2 (MMP-2), tissue inhibitor of metalloproteinases 2 (TIMP-2) and collagen types I (Coll I) and III (Coll III) in the LV.

In brief, total RNA was isolated according to the method of Chomczynski and Sacchi [18] using TRIZOL (Invitrogen GmbH, Karlsruhe, Germany) as described in the manufacturer's protocol. For investigation of TGF-β, we isolated 2.5 µg of total TGF-β

RNA, and for the extracellular matrix (ECM) markers ANP, MMP-2, TIMP-2, Coll I, and Coll III, 5 μg of total RNA was isolated. The isolated RNA was hybridized overnight with the template sets rTGF-β and ECM-3, respectively, and labelled with an RiboQuant® In vitro Transcription Kit (Pharmingen, Hamburg, Germany) and [α-32P]-UTP as described by the manufacturer. After hybridization and digestion of the unprotected probes, the protected radioactive RNA was displayed on a denaturing polyacrylamide gel. For densitometric evaluation we used the Molecular Imager (BioRad, München, Germany). mRNA expression was semi-quantitatively determined in percent of glyceraldehyde-3-phosphate dehydrogenase (GAP-DH) mRNA.

2.6. Zymography

LV myocardial MMP activity was measured as described elsewhere [19]. Briefly, extracellular proteins of frozen tissue were extracted by the addition of extraction buffer and then directly used for electrophoresis. The gels were run at 200 V at 4 °C. Following electrophoresis, the gels were stained in Coomassie Blue G-250 and scanned using the EagleEye II imaging system (Stratagene, Heidelberg, Germany) for relative lytic activity after normalization to the amount of protein extract loaded onto the gel. The protein activity was related to the median of the CTRL animals (CTRL = 1.0). For more details see [20].

2.7. ELISA

To determine the protein content of TGF-β_1 and TGF-β_2 in the LV, the immunoassay was performed according to the Quantikine® ELISA Mouse/Rat/Porcine/Canine TGF-β_1 Immunoassay Protocol (R&D Systems, Inc., Minneapolis, MN, USA) and the Quantikine® Human TGF-β_2 Immunoassay Protocol (R&D Systems, Inc., Minneapolis, MN, USA). Polyclonal secondary rabbit-antibody (Dianova, Hamburg, Germany) conjugated to horseradish peroxidase specific for TGF-β_1 was used. Absorbance was measured at a wavelength of 450 nm. The TGF-β_1 and TGF-β_2 concentrations were calculated by the I-smart program (Packard Instruments Company, Inc., Downers Grove, IL, USA). The protein concentrations were related to the median of the CTRL animals (Ctrl = 1.0).

2.8. Histological Investigation of the Heart

The apex of the heart was cut off, fixated in formalin and embedded in paraffin. Sections of 8–9 μm thickness were stained with hematoxylin—eosin and Masson's trichrome. For the evaluation of cardiac fibrosis, sections were examined with an Axioscope microscope (Carl Zeiss, Oberkochen, Germany), digitized with AxioCam MRc 5 (Carl Zeiss, Oberkochen, Germany) and evaluated using the Photoshop CS6 image processing program. A score ranging from 0 to 3 was applied to assess the histological degree of fibrosis: 0 = no signs of fibrosis; 0.5 = marginal perivascular fibrosis and/or marginal interstitial fibrosis; 1 = perivascular fibrosis + mild interstitial fibrosis; 2 = perivascular fibrosis + moderate interstitial fibrosis; 3 = fibrosis of the entire heart [14].

2.9. Statistical Analysis

Statistical analyses were carried out with the software package SigmaPlot Version 14.0 (Systat Software GmbH, Erkrath, Germany) for Windows. The groups were statistically compared using Analysis of Variance (ANOVA) procedures. Firstly, a Shapiro-Wilk test of normality was performed. If data were normally distributed, we used a One Way ANOVA with a post-hoc test according to Fisher's LSD method. This applied to SBP, HW and quantitative evaluations from histology. These data are presented as means ± SEM. If the data were not normally distributed, a Kruskal-Wallis ANOVA on ranks with a post-hoc test according to Dunn's method was applied. These data (the results of the biochemical analyses) are given as medians [25th/75th percentile]. As age, BW and SBP at baseline might have an effect on the results, we additionally performed an analysis of covariance (ANCOVA) to assess the influence of these covariates and their interaction with treatment effects on the results. p values < 0.05 were considered significant.

3. Results

3.1. Systolic Blood Pressure

Baseline SBP as measured at 60 weeks of age was similar in all three groups (CTRL 231 ± 9 mmHg, CAP 228 ± 13 mmHg, CAP + NIF 245 ± 8 mmHg). During the observation period, SBP increased slightly in CTRL rats (239 ± 5 mmHg; $p > 0.05$). CAP and CAP + NIF induced a significant SBP reduction by about 20% after 22 weeks of therapy ($p < 0.001$; Figure 1). SBP measured at 82 weeks of age was not dependent on age, BW and SBP at baseline.

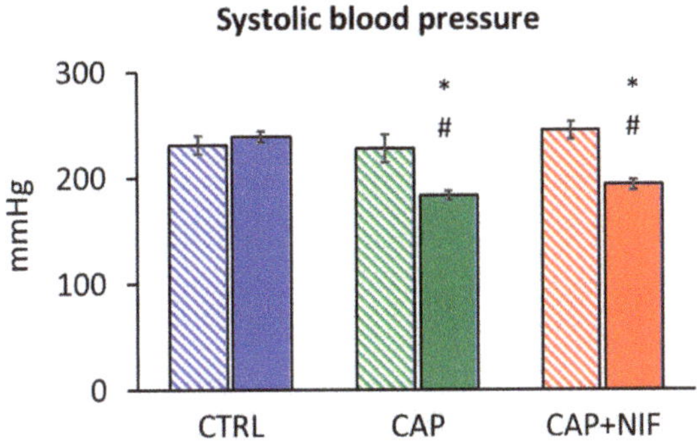

Figure 1. Systolic blood pressure of old SHR expressed as mean ± SEM at baseline (60 weeks of age; hatched columns) and at the final measurement after 22 weeks of experiment (82 weeks of age; filled columns). Significance marks: # significant vs. baseline measurement ($p < 0.001$); * significant vs. CTRL ($p < 0.001$).

3.2. Cardiac Hypertrophy

3.2.1. Heart Weight

At the end of the experimental period, the HW of 82-week-old CTRL rats was 1860 ± 93 mg. Both forms of therapy counteracted the development of cardiac hypertrophy and reduced HW significantly by 26% (CAP) and 23% (CAP + NIF). ANCOVA revealed that the treatment effects on HW did not depend on age, BW and SBP at baseline. HW/BW in untreated SHR was 4.7 ± 0.25 mg/g and was 34% higher than in the treated animals (3.5 ± 0.2 mg/g in both groups, $p = 0.001$; Figure 2a). Notably, the final BW of CTRL rats at 82 weeks of age was 378 ± 37 g and was about 15% lower than that of treated SHR ($p = 0.08$; data not shown).

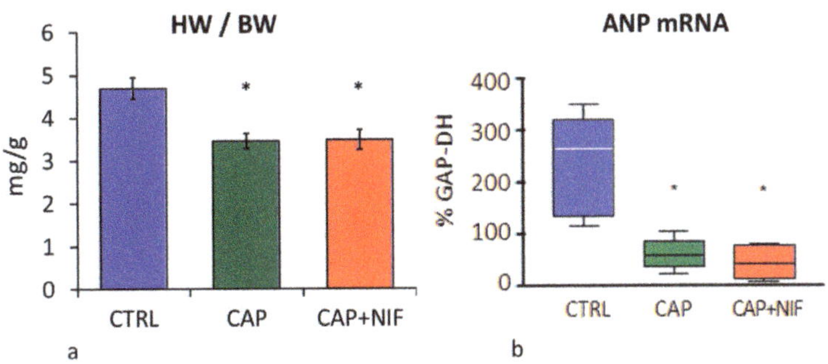

Figure 2. (**a**): Heart weight-to-body weight ratio (HW/BW; in mg/g) expressed as mean ± SEM; (**b**): mRNA expression of atrial natriuretic peptide (ANP, in % of GAP-DH) expressed as median (line in the box) with 25th/75th percentiles (boxes) and 10th/90th percentiles (whiskers). All values were obtained from 82-week-old SHR. Significance marks: * significant vs. CTRL ($p \leq 0.001$).

3.2.2. ANP mRNA Expression

ANP mRNA expression in the LV of untreated 82-week-old SHR was 264 [154/294] percent of GAP-DH mRNA expression. Both treatments resulted in a significant reduction of ANP mRNA expression to 60 [48/68] (CAP) and 43 [15/78] (CAP + NIF) percent of GAP-DH ($p < 0.001$; Figure 2b). ANP mRNA expression was not dependent on the covariates age, BW and SBP at baseline.

3.3. Cardiac Fibrosis

3.3.1. Biochemical Markers of Cardiac Remodeling

TGF-β: In the LV of untreated SHR, mRNA expression of TGF-β_1 was 2.68 [2.10/2.74] percent of GAP-DH. Treatments with CAP and CAP + NIF decreased TGF-β_1 to 1.77 [1.64/2.11] and 1.19 [1.14/1.98], respectively, but this effect was significant only with CAP + NIF ($p = 0.04$). TGF-β_2 and TGF-β_3 were not significantly reduced with either treatment (Figure 3a–c). The changes in TGF-β_1 and TGF-β_3 were dependent on SBP at baseline.

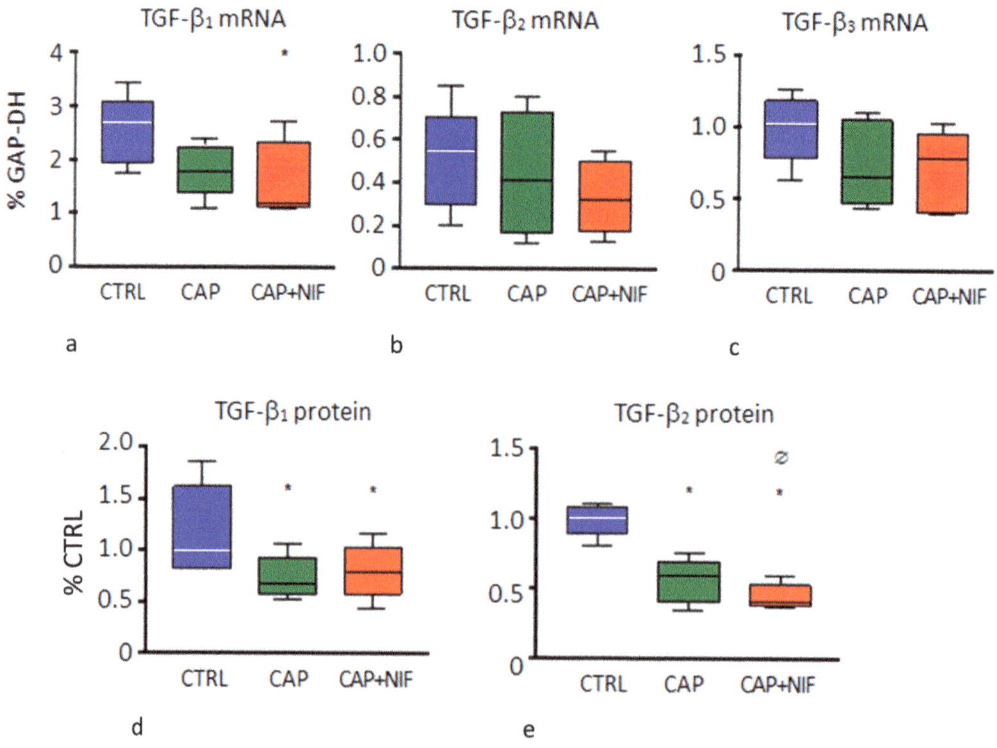

Figure 3. (**a**–**c**): mRNA expression of TGF-β isoforms (in % of GAP-DH) in the LV. (**d**,**e**): Protein concentrations of TGF-β_1 and TGF-β_2 (in % of CTRL) in the LV. Data are given as median (line in the box) with 25th/75th percentiles (boxes) and 10th/90th percentiles (whiskers). All values were obtained from 82-week-old SHR. Significance marks: * significant vs. CTRL; Ø significant vs. CAP.

However, both therapies significantly reduced the protein concentrations of TGF-β_1 and TGF-β_2 in the LV. TGF-β_1 concentration decreased by 33% (CAP) and 21% (CAP + NIF) compared to CTRL ($p = 0.03$). For TGF-β_2, the combination therapy proved to be more effective than monotherapy ($p < 0.001$), and reduced the protein concentration by 41% (CAP) and by 60% (CAP + NIF) compared to CTRL ($p < 0.001$; Figure 3d,e).

MMP-2 and TIMP-2: The expression of MMP-2 and TIMP-2 mRNA in the LV of CTRL rats was 5.58% [4.43/8.98] and 24.9% [13.7/28.8], respectively. Both therapies resulted in a significant reduction of mRNA expressions of MMP-2 (CAP by 53%, CAP + NIF by 46% of CTRL; $p = 0.001$) and TIMP-2 (CAP by 57%, CAP by 59% of CTRL; $p < 0.001$). Both treatments also decreased the activity of MMP-2 by 41% (CAP) and 51% (CAP + NIF) of CTRL ($p = 0.03$; Figure 4). ANCOVA results showed that these changes did not depend on age, BW and SBP at baseline.

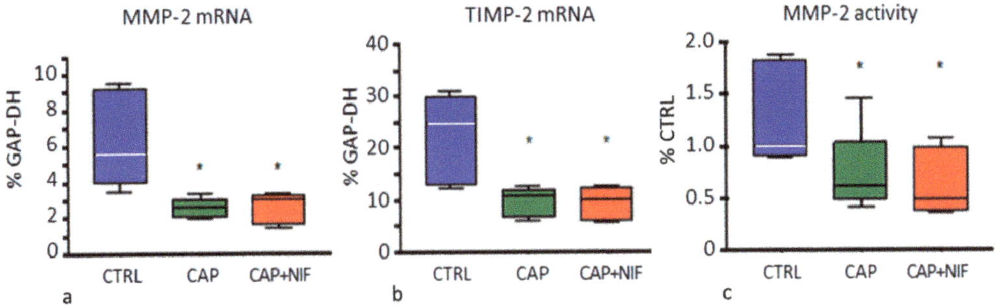

Figure 4. (**a**,**b**): MMP-2- and TIMP2 mRNA expression (in % of GAP-DH) in the LV. (**c**): MMP-2 activity (in % of CTRL) in the LV. Data are given as median (line in the box) with 25th/75th percentiles (boxes) and 10th/90th percentiles (whiskers). All values were obtained from 82-week-old SHR. Significance marks: * significant vs. CTRL.

Collagens: In the LV of CTRL rats, mRNA expression of Coll I and Coll III was 13.8% [9.7/16.7] and 24.8% [22.4/26.8], respectively. Both treatments significantly reduced Coll I to 4.5–4.8% (i.e., about 34% of CTRL, $p < 0.001$) and Coll III to 9.5–10.5% (i.e., about 40% of CTRL, $p < 0.001$; Figure 5). ANCOVA revealed that the effect of the treatments was not dependent on the values of the covariates age, BW and SBP at baseline.

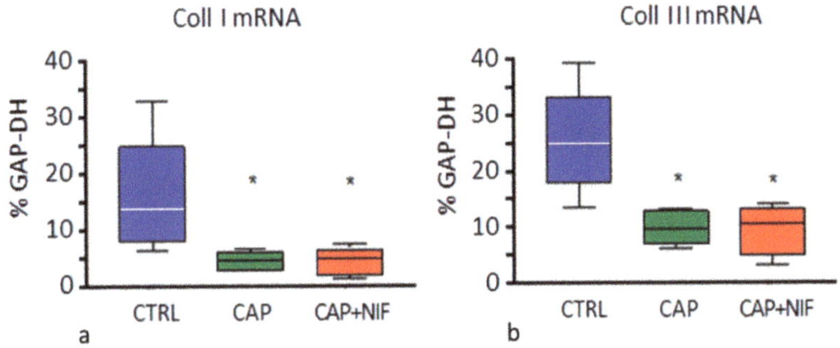

Figure 5. Coll I (**a**) and Coll III (**b**) mRNA expression (in % of GAP-DH) in the LV. Data are given as median (line in the box) with 25th/75th percentiles (boxes) and 10th/90th percentiles (whiskers). All values were obtained from 82-week-old SHR. Significance marks: * significant vs. CTRL ($p < 0.001$).

3.3.2. Histological Manifestation of Cardiac Fibrosis

The hearts of the untreated CTRL rats showed a marked interstitial and perivascular fibrosis (fibrosis degree 2.3 ± 0.11; Figure 6A). Both treatments significantly reduced the degree of perivascular and interstitial fibrosis. CAP therapy (fibrosis degree 1.8 ± 0.05; $p = 0.001$ vs. CTRL) proved to be slightly superior to CAP + NIF treatment (fibrosis degree 2.1 ± 0.07; $p < 0.001$ vs. CAP; Figure 6B,C). The histological degree of fibrosis did not depend on baseline age (about 60 weeks), but there was an interaction with BW and SBP at baseline.

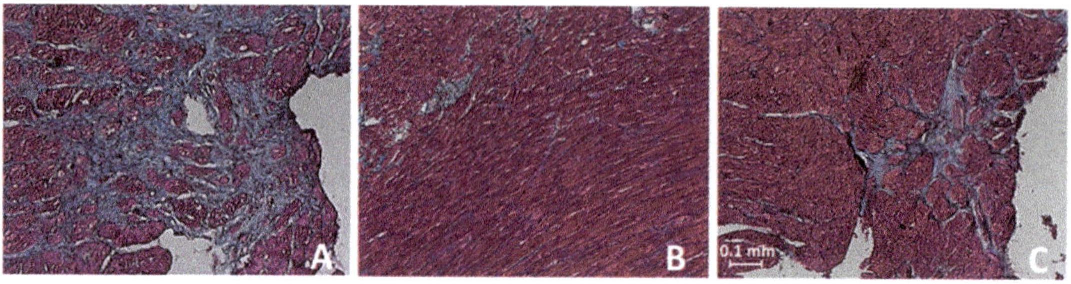

Figure 6. Hearts of 82-week-old CTRL and treated SHR in trichrome staining (10-fold magnification; the scale bar in part C represents 0.1 mm). (**A**): CTRL, perivascular and marked interstitial fibrosis, fibrosis degree of the group 2.3 ± 0.11; (**B**): CAP, perivascular and slight interstitial fibrosis, fibrosis degree of the group 1.8 ± 0.05; (**C**): CAP + NIF, perivascular and interstitial fibrosis, fibrosis degree of the group 2.1 ± 0.07 (data are given as means ± SEM).

4. Discussion

In the present study, we have shown that even a late start of antihypertensive therapy can achieve a significant reduction of SBP in old SHR with the long persistence of hypertension. At that age, cardiac hypertrophy, cardiac remodeling and fibrosis, as consequences of hypertension, have already developed; they could not be prevented but their progression was delayed and attenuated by both therapies. This corresponds to the functional effects of late-onset CAP and CAP + NIF treatments in old SHR as detected by echocardiography. Compared to untreated old SHR, both types of treatment significantly delayed the increase in LV wall thickness and preserved the systolic and diastolic pumping function for a significantly longer period of time [21].

4.1. Development of Hypertension in SHR in Early and Later Stages of Life

In SHR, hypertension develops in a very early period of life. Seven-week-old SHR showed 50% higher SBP values than normotensive Wistar-Kyoto rats (WKY) of the same age, and during the following three weeks, their SBP increased significantly to almost 170% of WKY [14]. During this early stage, SHR developed functional and morphological vascular changes [22,23] which are associated with vascular hypertrophy and increasing vascular resistance, thus leading to cardiac pressure overload [24]. The development of cardiac hypertrophy in SHR starts between four [25] and twelve weeks of age [26]. Our previous study on young SHR showed that mild cardiac hypertrophy and fibrosis were already present at 10 weeks of age [14]. As hypertension progresses, SHR develops increasing structural changes in the heart between the sixth and 24th months of life, which are associated with a further increase in cardiac hypertrophy [27]. In the context of hypertrophy and remodeling, numerous fetal genes are re-expressed, such as ANP [28,29].

The transition to heart failure in SHR is associated with the development of fibrosis resulting from the increased expression of genes encoding extracellular matrix (ECM) proteins and from the increased turnover of matrix proteins through increased MMP activity [20]. TGF-β as an important regulator of MMPs and TIMPs plays a crucial role in ECM remodeling [30,31]. The upregulation of TGF-β in the LV, particularly of TGF-β_1, is associated with increased activity of MMP-2 and enhanced synthesis of collagens and other extracellular matrix proteins [4,8]. Of note, activation of the RAAS upregulates TGF-β_1 synthesis and results in myocardial fibrosis. Correspondingly, RAAS antagonization attenuates these processes [8], which was also confirmed by the present results.

Significant collagen deposition in the heart of SHR has been found as early as at 8 weeks of age [32], and its amount further increased during adulthood [33]. The remodeling processes led to reduced contractile myocardial function, ventricular fibrosis and heart failure at an advanced age (>20 months) in untreated SHR [4,34]. This was also confirmed in an echocardiography study on old SHR. Both echocardiographic examination and heart catheterization showed marked LV dysfunction in untreated animals [21]. We assume that the loss in BW observed in the SHR CTRL group of the present study might be a consequence of this severe cardiac dysfunction. A comparison of the present results with those of the previous study in young SHR [14] reveals significant increases both in SBP and in biochemical and histological markers of ECM remodeling in the old SHR CTRL group (Table 1). In particular, it confirms a significant increase in markers of cardiac hypertrophy (HW/BW, ANP mRNA expression). Note that these markers did not depend on age, BW and SBP at the beginning of the experiment. The expression of TGF-β isoforms and TIMP-2 was also significantly higher in old than in young SHR, while MMP-2 expression was only slightly elevated. Coll I and III mRNA were equal in old and young SHR. However, considering the significantly higher histological degree of fibrosis in old SHR and the marked effects of treatment both in young and old SHR, this finding may suggest that collagen mRNA expression in SHR is elevated from a young age and does not adequately reflect the degree of collagen deposition in the heart. This corresponds with our observation that collagen mRNA in 82-week-old SHR is not dependent on age, BW, and SBP at baseline.

Table 1. Comparison of systolic blood pressure (SBP) and parameters of cardiac hypertrophy and ECM remodeling in young (10-week-old) and old (82-week-old) SHR.

Parameter	Young SHR (10 Weeks)	Old SHR (82 Weeks)	p Value
final SBP	202 ± 6	239 ± 5	0.002
HW/BW	3.50 ± 0.07	4.70 ± 0.25	0.003
ANP mRNA	11.0 [7.5/20.3]	264 [154/293]	0.008
TGF-β1 mRNA	1.37 [1.34/1.54]	2.68 [2.10/2.74]	0.004
TGF-β2 mRNA	0.16 [0.14/0.18]	0.56 [0.55/0.72]	0.004
TGF-β3 mRNA	0.64 [0.54/0.67]	1.02 [0.93/1.12]	0.02
MMP-2 mRNA	3.92 [3.57/4.69]	5.58 [4.43/8.98]	0.222
TIMP-2 mRNA	8.01 [6.95/10.2]	24.9 [13.7/28.8]	0.008
Coll I mRNA	16.3 [9.6/16.6]	13.8 [9.7/16.7]	1.0
Coll III mRNA	20.9 [20.1/22.4]	24.8 [22.4/26.8]	0.548
Degree of fibrosis	1.47 ± 0.12	2.30 ± 0.11	<0.001

Data from young SHR are from [14]. Normally distributed parameters are given as mean ± SEM, and statistics were performed using a t-test. Parameters that were not normally distributed are given as median [25th/75th percentile], statistics were calculated using Mann-Whitney's U test. HW/BW heart weight-to-body weight ratio; ANP atrial natriuretic peptide; TGF-β transforming growth factor-β; MMP-2 matrix metalloproteinase 2; TIMP-2 tissue inhibitor of metalloproteinases 2; Coll I, Coll III collagen types I and III.

4.2. *Effects of Antihypertensive Therapies*

The degree of BP reduction is associated with the age at which treatment is implemented. The early start of antihypertensive therapy does not only reduce BP but has also cardioprotective effects, which are well-documented [2,7,13]. Early antihypertensive treatment can lead to normotension, whereas late therapy-onset results only in incomplete blood pressure regulation [35]. Moreover, early antihypertensive therapy can prevent or, at least, minimize secondary damage such as cardiac hypertrophy and remodeling, which again maintain and increase hypertension [2,3]. This was fully confirmed in our previous study on young SHR [14]. Harrap and co-workers [36] suspected a critical phase with an increased sensitivity to pharmacological influences between the 6th and 10th week of life in the development of hypertension in SHR. Intervention in the blood pressure regulation during this time may change the long-term course of disease development [36]. In young rats, RAAS antagonists are particularly effective, even if the therapy is stopped after a defined period of treatment [3,35,36]. In contrast, therapy onset at a higher age cannot completely reverse the development of hypertension. A study of Kost et al. [35] showed that SHR, which received captopril at the age of 24 weeks, had higher BP values than SHR treated with captopril at the age of four weeks. These results are in line with the findings of the present study: In old SHR, therapy with CAP or CAP + NIF over 22 weeks significantly reduced SBP by about 20%. However, these values were with 183 and 194 mmHg, respectively, about 25% higher than in 10-week-old SHR after only three weeks with CAP or CAP + NIF treatment [14]. Of note, in that study on young SHR, the relative SBP reduction (in % of baseline) was with 18% and 3%, respectively, lower than in the present old SHR study with the same doses of CAP and CAP + NIF. This clearly shows that the doses of antihypertensive drugs are also effective in old SHR indicating that their final SBP remains elevated mainly due to their higher baseline SBP. An interesting question is whether increased doses of antihypertensive drugs might induce a stronger SBP reduction.

There are only a few therapy studies in old SHR with late onset of antihypertensive treatment. Ito et al. [37], Susic et al. [38] and Mukherjee and Sen [39] examined old SHR at the age between 12 to 22 months during treatment with RAAS antagonists over different time intervals (between 4 and 12 weeks) and showed reduced signs for hypertrophy. Only Brooks et al. [40] treated SHR aged 12, 18 and 21 months with captopril 2 g/L in drinking water over a very long period of four to 12 months. They showed attenuated signs of heart failure and decreased ventricular hypertrophy even when heart failure had already started to develop.

In the present study, antihypertensive therapy with CAP or CAP + NIF over 22 weeks significantly reduced HW and HW/BW. Correspondingly, mRNA expression of the hypertrophy marker ANP in the LV decreased by 77 and 84%, respectively, compared to untreated CTRL (Figure 2). However, HW/BW was still about 12–14% above the values of 10-week-old SHR treated with CAP and CAP + NIF over three weeks. ANP mRNA of treated old SHR was even about tenfold of the values in corresponding young animals [14].

In addition, both types of treatment diminished ECM remodeling in the heart of old SHR. AT II plays an important role in the development of hypertension-induced cardiac fibrosis [8,41] by stimulating the production of IL-6 and TGF-β_1 in cardiac fibroblasts [42]. A significant reduction of TGF-β_1 expression by CAP was also shown by Brooks and co-workers [40] in 12 to 24-month-old SHR. As a result of the reduced TGF-β production, which was predominantly due to the effect of CAP, we observed a significant reduction of mRNA expressions of MMP-2, TIMP-2, Coll I and Coll III, as well as of MMP-2 activity by about 50–60% in the LV of old SHR (Figures 3–5). Consequently, a significant decrease of histological signs of fibrosis was found in CAP and CAP + NIF animals (Figure 6). However, this effect was much lower than in young SHR: in our previous study on young SHR, fibrosis was attenuated by 60% with CAP + NIF treatment and completely abrogated by CAP treatment [14]. In contrast, in old SHR the fibrosis degree only decreased by 9% and 22%, respectively, despite a much longer period of treatment. As indicated by ANCOVA results, the high SBP at the beginning of treatment might have contributed to this reduced therapy effect. Consequently, functional deterioration due to fibrotic impediment could be delayed but not completely prevented.

An echocardiographic study in old SHR demonstrated that with antihypertensive treatment, systolic and diastolic pump function in treated SHR was maintained for a longer period of time. In these animals, a slight deterioration of cardiac function began around the 76th week of life, while untreated rats showed a progressive decline of systolic and diastolic function throughout the experimental period [21]. Once structural changes in LV have occurred, therapy with inhibitors of AT converting enzyme or calcium antagonists can no longer reverse them but delay and attenuate the deterioration of cardiac function [43]. Nevertheless, even in long-standing hypertension, which may be diagnosed late, immediate therapy can significantly lower SBP, delay the development of hypertrophy and significantly reduce fibrosis, as demonstrated in the present study. Thus, the onset of heart failure can be prevented or at least delayed.

4.3. Comparison of Monotherapy CAP with Combination Therapy CAP + NIF

Previous studies in rats have shown that RAAS antagonists have superior antihypertensive effects compared to other therapies, e.g., calcium antagonists [44]. RAAS antagonists play an important role in lowering HW, preventing cardiac hypertrophy and mitigating structural remodeling of the cardiovascular system and terminal organ damage in young SHR [2,7,13]. In our previous study on young SHR, a monotherapy with the calcium antagonist NIF proved to be effective, but was less effective than CAP [14].

In the present study on old SHR, monotherapy with CAP and the combination of CAP + NIF proved to be equivalent. Previous results of preliminary examinations in young SHR showed a particularly high antihypertensive and cardioprotective effect of therapies with RAAS antagonists [14]. Two therapies from that study were applied in the present study to old SHR (CAP and CAP + NIF). These therapeutic regimens induced a significant reduction in SBP and an attenuated and delayed development of cardiac hypertrophy. An echocardiographic examination in old SHR demonstrated that rats treated with CAP or CAP + NIF exhibited a stable LV function, while untreated SHR showed clear signs of LV insufficiency [21].

4.4. Limitations of the Study

The main limitation of this study is the low number of experimental animals. The group size was calculated for a medium effect size of SBP reduction (with Cohen's $f = 0.25$). However, clinical symptoms of decompensated heart failure and also the BW reduction of CTRL rats showed high interindividual differences. These characteristics correspond to observations in human pathology. The number of animals to be examined should be kept as low as possible, since only as little SHR as necessary should be exposed to the suffering caused by the consequences of hypertension. The calculated sample size of 21 animals was considered representative.

A further limitation is that we did not measure tibia length. Treated SHR slightly increased their body weight over the experimental period, while untreated CTRL animals showed a weight loss of 11% relative to baseline. Thus, the ratio of heart weight to tibia length would be more adequate than HW/BW in reflecting cardiac hypertrophy.

With respect to the similar efficacy of CAP monotherapy and CAP + NIF combination therapy, further studies should include a direct comparison of monotherapies with CAP and with NIF in old SHR. Further research might also examine the question of whether a late-onset therapy with higher doses of antihypertensive drugs may result in the further reduction of SBP and better cardioprotective effects.

5. Conclusions

Two therapy regimes, a monotherapy with CAP and a combination therapy with CAP + NIF, proved to be equally effective at reducing SBP and delaying the development of cardiac secondary damage in old SHR. Long-term-treatment significantly reduced SBP as well as cardiac hypertrophy and fibrosis. However, the levels of young SHR with early-starting treatment were not achieved. These findings are of importance with regard to the late diagnosis of arterial hypertension in humans, which is often made in the elderly. Consequent antihypertensive therapy, even if it starts at an advanced age, can reduce cardiac stress and thus cardiac secondary diseases.

Author Contributions: B.R., H.-G.Z., C.H. and J.B. contributed conception and design of the study; C.H., J.B. and B.R. performed the animal experiments including blood pressure measurements. C.H. prepared the histological slices; C.H., K.S. and P.G. evaluated and took photographs from the histological slices. C.H. performed the molecular biological analyses; C.H. and A.S. evaluated the molecular biological data. C.H. and B.R. performed the statistics. C.H., J.B. and P.G. created figures and tables. C.H. and B.R. drafted the original manuscript; A.S. and H.-G.Z. reviewed and edited the manuscript. All authors have read and agreed to the published version of the manuscript.

Funding: This research received no external funding.

Institutional Review Board Statement: This manuscript does not contain clinical studies or patient data. The study was conducted according to the guidelines of the Declaration of Helsinki. All animal protocols were approved by the state agency (Landesdirektion Sachsen, number and date of approval: TVV 36/08; 13 May 2009) in accordance with the Guide for the Care and Use of Laboratory Animals published by the National Institutes of Health and with the "European Convention for the Protection of Vertebrate Animals used for Experimental and other Scientific Purposes" [17].

Informed Consent Statement: Not applicable.

Data Availability Statement: Not applicable.

Acknowledgments: We gratefully appreciate the valuable technical assistance of Brigitte Mix, Ursula Vogt and Grit Marx. The authors acknowledge support from the German Research Foundation (DFG) and Universität Leipzig within the program of Open Access Publishing.

Conflicts of Interest: The authors declare that they have no conflict of interest.

Abbreviations

ANP	atrial natriuretic peptide
AT	angiotensin
BP	blood pressure
BW	body weight
CAP	captopril (60 mg kg^{-1} d^{-1})
CAP + NIF	captopril plus nifedipine (60 + 10 mg kg^{-1} d^{-1})
Coll I	collagen type I
Coll III	collagen type III
CTRL	untreated hypertensive control rats
ECM	extracellular matrix
GAP-DH	glyceraldehyde-3-phosphate dehydrogenase
HW	heart weight
LV	left ventricle, left ventricular
MMP-2	matrix metalloproteinase 2
NIF	nifedipine (10 mg kg^{-1} d^{-1})
RAAS	renin-angiotensin-aldosterone-system
SBP	systolic blood pressure
SHR	spontaneously hypertensive rats
TGF-β	Transforming Growth Factor-β
TIMP-2	tissue inhibitor of metalloproteinases 2

References

1. NCD Risk Factor Collaboration (NCD-RisC). Worldwide trends in hypertension prevalence and progress in treatment and control from 1990 to 2019: A pooled analysis of 1201 population-representative studies with 104 million participants. *Lancet* **2021**, *398*, 957–980. [CrossRef]
2. Demirci, B.; McKeown, P.P.; Bayraktutan, U. Blockade of angiotensin II provides additional benefits in hypertension- and ageing-related cardiac and vascular dysfunctions beyond its blood pressure-lowering effects. *J. Hypertens.* **2005**, *23*, 2219–2227. [CrossRef]
3. Zicha, J.; Dobesová, Z.; Kunes, J. Late blood pressure reduction in SHR subjected to transient captopril treatment in youth: Possible mechanisms. *Physiol. Res.* **2008**, *57*, 495–498. [CrossRef] [PubMed]
4. Boluyt, M.O.; Bing, O.H. Matrix gene expression and decompensated heart failure: The aged SHR model. *Cardiovasc. Res.* **2000**, *46*, 239–249. [CrossRef]
5. Frank, K.F.; Bölck, B.; Brixius, K.; Kranias, E.G.; Schwinger, R.H.G. Modulation of SERCA: Implications for the Failing Human Heart. *Basic Res. Cardiol.* **2002**, *97*, I72–I78. [CrossRef] [PubMed]
6. Williamson, C.R.; Khurana, S.; Nguyen, P.; Byrne, C.J.; Tai, T.C. Comparative Analysis of Renin-Angiotensin System (RAS)-Related Gene Expression between Hypertensive and Normotensive Rats. *Med. Sci. Monit. Basic Res.* **2017**, *23*, 20–24. [CrossRef]
7. Rocha, W.A.; Lunz, W.; Baldo, M.P.; Pimentel, E.B.; Dantas, E.M.; Rodrigues, S.L.; Mill, J.G. Kinetics of cardiac and vascular remodeling by spontaneously hypertensive rats after discontinuation of long-term captopril treatment. *Braz. J. Med. Biol. Res.* **2010**, *43*, 390–396. [CrossRef]

8. Zhang, Y.; Shao, L.; Ma, A.; Guan, G.; Wang, J.; Wang, Y.; Tian, G. Telmisartan delays myocardial fibrosis in rats with hypertensive left ventricular hypertrophy by TGF- b1/Smad signal pathway. *Hypertens. Res.* **2014**, *37*, 43–49. [CrossRef] [PubMed]
9. Aritomi, S.; Harada, E.; Sugino, K.; Nishimura, M.; Nakamura, T.; Takahara, A. Comparison of the cardioprotective and renoprotective effects of the L/N-type calcium channel blocker cilnidipine in adriamycin-treated spontaneously-hypertensive rats. *Clin. Exp. Pharmacol. Physiol.* **2015**, *42*, 344–352. [CrossRef]
10. European Society of Cardiology/European Society of Hypertension (ESC/ESH). 2018 ESC/ESH Guidelines for the management of arterial hypertension. *Eur. Heart J.* **2018**, *39*, 3021–3104. [CrossRef] [PubMed]
11. Ruzicka, M.; Leenen, F.H. Combination therapy as first-line treatment of arterial hypertension. *Can. J. Cardiol.* **2002**, *18*, 1317–1327.
12. Paulis, L.; Lísková, S.; Pintérová, M.; Dobesová, Z.; Kunes, J.; Zicha, J. Nifedipine-sensitive noradrenergic vasoconstriction is enhanced in spontaneously hypertensive rats: The influence of chronic captopril treatment. *Acta Physiol.* **2007**, *191*, 255–266. [CrossRef]
13. Hale, T.M.; Robertson, S.J.; Burns, K.D.; deBlois, D. Short-term ACE inhibition confers long-term protection against target organ damage. *Hypertens. Res.* **2012**, *35*, 604–610. [CrossRef]
14. Hawlitschek, C.; Brendel, J.; Gabriel, P.; Schierle, K.; Salameh, A.; Zimmer, H.G.; Rassler, B. Antihypertensive and cardioprotective effects of different monotherapies and combination therapies in young spontaneously hypertensive rats—A pilot study. *Saudi J. Biol. Sci.* **2022**, *29*, 339–345. [CrossRef] [PubMed]
15. Hojná, S.; Kadlecová, M.; Dobesová, Z.; Valousková, V.; Zicha, J.; Kunes, J. The participation of brain NO synthase in blood pressure control of adult spontaneously hypertensive rats. *Mol. Cell. Biochem.* **2007**, *297*, 21–29. [CrossRef] [PubMed]
16. Sengupta, P. The Laboratory Rat: Relating Its Age with Human's. *Int. J. Prev. Med.* **2013**, *4*, 624–630.
17. Council of Europe. European Convention for the Protection of Vertebrate Animals Used for Experimental and Other Scientific Purposes (ETS No 123) Strasbourg 18.III. (1986). Text Amended according to the Provisions of the Protocol (ETS No. 170) as of Its Entry into Force on 2 December 2005. Available online: https://rm.coe.int/168007a67b (accessed on 27 June 2022).
18. Chomczynski, P.; Sacchi, N. Single-step method of RNA isolation by acid guanidinium thiocyanate phenol-chloroform extraction. *Anal. Biochem.* **1987**, *162*, 156–159. [CrossRef]
19. Tyagi, S.C.; Matsubara, L.; Weber, K.T. Direct extraction and estimation of collagenase(s) activity by zymography in microquantities of rat myocardium and uterus. *Clin. Biochem.* **1993**, *26*, 191–198. [CrossRef]
20. Briest, W.; Hölzl, A.; Raßler, B.; Deten, A.; Leicht, M.; Baba, H.A.; Zimmer, H.G. Cardiac remodeling after long-term norepinephrine treatment in rats. *Cardiovasc. Res.* **2001**, *52*, 265–273. [CrossRef]
21. Zimmer, J.; Hawlitschek, C.; Rabald, S.; Hagendorff, A.; Zimmer, H.G.; Rassler, B. Effects of late-onset and long-term captopril and nifedipine treatment in aged spontaneously hypertensive rats: Echocardiographic studies. *Hypertens. Res.* **2015**, *38*, 716–722. [CrossRef] [PubMed]
22. Anishchenko, A.M.; Aliev, O.I.; Sidekhmenova, A.V.; Shamanaev, A.Y.; Plotnikov, M.B. Dynamics of Blood Pressure Elevation and Endothelial Dysfunction in SHR Rats During the Development of Arterial Hypertension. *Bull. Exp. Biol. Med.* **2015**, *159*, 591–593. [CrossRef]
23. Bencze, M.; Behuliak, M.; Vavrínová, A.; Zicha, J. Altered contractile responses of arteries from spontaneously hypertensive rat: The role of endogenous mediators and membrane depolarization. *Life Sci.* **2016**, *166*, 46–53. [CrossRef]
24. Waleska, C.D.; Marcelo, E.S. Animal models for the study of arterial hypertension. *J. Biosci.* **2011**, *36*, 731–737.
25. Kokubo, M.; Uemura, A.; Matsubara, T.; Murohara, T. Noninvasive evaluation of the time course of change in cardiac function in spontaneously hypertensive rats by echocardiography. *Hypertens. Res.* **2005**, *28*, 601–609. [CrossRef]
26. Van Empel, V.P.; De Windt, L.J. Myocyte hypertrophy and apoptosis: A balancing act. *Cardiovasc. Res.* **2004**, *63*, 487–499. [CrossRef] [PubMed]
27. Engelmann, G.L.; Vitullo, J.C.; Gerrity, R.G. Morphometric analysis of cardiac hypertrophy during development maturation and senescence in spontaneously hypertensive rats. *Circ. Res.* **1987**, *60*, 487–494. [CrossRef]
28. Day, M.L.; Schwartz, D.; Wiegand, R.C.; Stockman, P.T.; Brunnert, S.R.; Tolunay, H.E.; Currie, M.G.; Standaert, D.G.; Needleman, P. Ventricular Atriopeptin. Unmasking of Messenger RNA and Peptide Synthesis by Hypertrophy or Dexamethasone. *Hypertension* **1987**, *9*, 485–491. [CrossRef]
29. Du, X.J. Divergence of hypertrophic growth and fetal gene profile: The influence of β-blockers. *Br. J. Pharmacol.* **2007**, *152*, 169–171. [CrossRef]
30. Hall, M.C.; Young, D.A.; Waters, J.G.; Rowan, A.D.; Chantry, A.; Edwards, D.R.; Clark, I.M. The Comparative Role of Activator Protein 1 and Smad Factors in the Regulation of Timp-1 and MMP-1 Gene Expression by Transforming Growth Factor-β1. *J. Biol. Chem.* **2003**, *278*, 10304–10313. [CrossRef]
31. Wu, J.; Jackson-Weaver, O.; Xu, J. The TGFβ superfamily in cardiac dysfunction. *Acta Biochim. Biophys. Sin.* **2018**, *50*, 323–335. [CrossRef]
32. Perrucci, G.L.; Barbagallo, V.A.; Corlianò, M.; Tosi, D.; Santoro, R.; Nigro, P.; Poggio, P.; Bulfamante, G.; Lombardi, F.; Pompilio, G. Integrin αvβ5 in vitro inhibition limits pro-fibrotic response in cardiac fibroblasts of spontaneously hypertensive rats. *J. Transl. Med.* **2018**, *16*, 352. [CrossRef] [PubMed]
33. Huang, A.; Li, H.; Zeng, C.; Chen, W.; Wei, L.; Liu, Y.; Qi, X. Endogenous CCN5 Participates in Angiotensin II/TGF-β1 Networking of Cardiac Fibrosis in High Angiotensin II-Induced Hypertensive Heart Failure. *Front. Pharmacol.* **2020**, *11*, 1235. [CrossRef] [PubMed]

34. Conrad, C.H.; Brooks, W.W.; Hayes, J.A.; Sen, S.; Robinson, K.G.; Bing, O.H. Myocardial fibrosis and stiffness with hypertrophy and heart failure in the spontaneously hypertensive rat. *Circulation* **1995**, *91*, 161–170. [CrossRef]
35. Kost, C.K., Jr.; Li, P.; Jackson, E.K. Blood Pressure after Captopril Withdrawal from Spontaneously Hypertensive Rats. *Hypertension* **1995**, *25*, 82–87. [CrossRef] [PubMed]
36. Harrap, S.B.; Van der Merwe, W.M.; Griffin, S.A.; Macpherson, F.; Lever, A.F. Brief angiotensin converting enzyme inhibitor treatment in young spontaneously hypertensive rats reduces blood pressure long-term. *Hypertension* **1990**, *16*, 603–614. [CrossRef] [PubMed]
37. Ito, N.; Ohishi, M.; Yamamoto, K.; Tatara, Y.; Shiota, A.; Hayashi, N.; Komai, N.; Yanagitani, Y.; Rakugi, H.; Ogihara, T. Renin-angiotensin inhibition reverses advanced cardiac remodeling in aging spontaneously hypertensive rats. *Am. J. Hypertens.* **2007**, *20*, 792–799. [CrossRef]
38. Susic, D.; Varagic, J.; Frohlich, E.D. Pharmacologic agents on cardiovascular mass coronary dynamics and collagen in aged spontaneously hypertensive rats. *J. Hypertens.* **1999**, *17*, 1209–1215. [CrossRef]
39. Mukherjee, D.; Sen, S. Collagen phenotypes during development and regression of myocardial hypertrophy in spontaneously hypertensive rats Collagen phenotypes during development and regression of myocardial hypertrophy in spontaneously hypertensive rats. *Circ. Res.* **1990**, *67*, 1474–1480. [CrossRef]
40. Brooks, W.W.; Bing, A.H.L.; Robinson, K.G.; Slawsky, M.T.; Chaletsky, D.M.; Conrad, C.H. Effect of angiotensin-converting enzyme inhibition on myocardial fibrosis and function in hypertrophied and failing myocardium from the spontaneously hypertensive rat. *Circulation* **1997**, *96*, 4002–4010. [CrossRef]
41. Qi, G.; Jia, L.; Li, Y.; Bian, Y.; Cheng, J.; Li, H.; Xiao, C.; Du, J. Angiotensin II infusion- induced inflammation monocytic fibroblast precursor infiltration and cardiac fibrosis are pressure dependent. *Cardiovasc. Toxicol.* **2011**, *11*, 157–167. [CrossRef]
42. Ma, F.; Li, Y.; Jia, L.; Han, Y.; Cheng, J.; Li, H.; Qi, Y.; Du, J. Macrophage-Stimulated Cardiac Fibroblast Production of IL-6 Is Essential for TGF β/Smad Activation and Cardiac Fibrosis Induced by Angiotensin II. *PLoS ONE* **2012**, *7*, e35144. [CrossRef] [PubMed]
43. Araujo, I.G.; Trindade, D.C.; Mecawi, A.S.; Sonoda-Cortes, R.; Werneck-de-Castro, J.P.S.; Costa-e-Sousa, R.H.; Reis, L.C.; Olivares, E.L. Inhibition of brain renin- Angiotensin system improves diastolic cardiac function following myocardial infarction in rats. *Clin. Exp. Pharmacol. Physiol.* **2009**, *36*, 803–809. [CrossRef] [PubMed]
44. Ziegelhöffer-Mihalovicova, B.; Arnold, N.; Marx, G.; Tannapfel, A.; Zimmer, H.G.; Rassler, B. Effects of salt loading and various therapies on cardiac hypertrophy and fibrosis in young spontaneously hypertensive rats. *Life Sci.* **2006**, *79*, 838–846. [CrossRef] [PubMed]

biomedicines

MDPI

Article

How Do Young and Old Spontaneously Hypertensive Rats Respond to Antihypertensive Therapy? Comparative Studies on the Effects of Combined Captopril and Nifedipine Treatment

Beate Rassler *, Christina Hawlitschek, Julia Brendel and Heinz-Gerd Zimmer

Carl-Ludwig-Institute of Physiology, University of Leipzig, 04103 Leipzig, Germany
* Correspondence: beate.rassler@medizin.uni-leipzig.de; Tel.: +49-341-9715565

Abstract: Numerous studies on the effects of antihypertensive treatment in young spontaneously hypertensive rats (SHRs) have shown that early-onset therapy may effectively reduce their blood pressure (BP) even to normotensive values. In contrast, only a few studies investigated the effects of treatment started at an advanced age. These studies revealed that antihypertensive effects are lower in adult or even in senescent SHRs compared with young SHRs. Even more, prevention of cardiac sequelae of hypertension such as hypertrophy and fibrosis is less effective when treatment starts late in life. Because, in patients, combination therapies with calcium antagonists are favored, we studied the efficacy of a combination therapy with captopril and nifedipine in young and old SHRs. We directly compared the treatment effects on BP as well as on cardiac hypertrophy and remodeling between these two animal cohorts. With antihypertensive treatment, significantly lower BP values were achieved in young SHRs despite a shorter treatment period compared with old SHRs. Although treatment effects on cardiac hypertrophy were greater in old than in young SHRs, cardiac fibrosis was significantly attenuated only in young but not in old SHRs. The results emphasize the value of antihypertensive therapy and particularly accentuate the importance of an early-onset therapy. With respect to problems such as late diagnosis and poor therapy adherence, these results may have great importance for the treatment of human hypertension.

Keywords: young SHR; old SHR; antihypertensive therapy; combination therapy; treatment effect; systolic blood pressure; LV hypertrophy; cardiac fibrosis

Citation: Rassler, B.; Hawlitschek, C.; Brendel, J.; Zimmer, H.-G. How Do Young and Old Spontaneously Hypertensive Rats Respond to Antihypertensive Therapy? Comparative Studies on the Effects of Combined Captopril and Nifedipine Treatment. *Biomedicines* **2022**, *10*, 3059. https://doi.org/10.3390/biomedicines10123059

Academic Editors: Josef Zicha and Ivana Vaněčková

Received: 20 October 2022
Accepted: 21 November 2022
Published: 28 November 2022

Publisher's Note: MDPI stays neutral with regard to jurisdictional claims in published maps and institutional affiliations.

1. Introduction

With a global prevalence of about 40% in the adult population [1], arterial hypertension is the main risk factor for morbidity and mortality of cardiovascular diseases and is considered to be the leading cause of death in industrialized countries [2–4]. In the overwhelming majority of patients, hypertension manifests in adulthood. The number of hypertensive adults aged 30–79 years in the world is more than 1.2 billion people at present [5]. The prevalence of hypertension increases with advancing age to 60–70% in people aged over 60–65 years [3,6]. From all hypertensive patients, only about 50% are diagnosed, and in about 40–50% of diagnosed patients, treatment is lacking or insufficient to stabilize blood pressure (BP) at values below 140/90 mmHg [5,7–9]. Untreated or insufficiently treated arterial hypertension leads to constant pressure load of the left ventricle (LV), which first results in cardiac hypertrophy. With long maintenance of hypertension, it may progress to cardiac remodeling, fibrosis, dilatation, and finally heart failure. One of the main causes of lacking or insufficient treatment of diagnosed hypertension is poor adherence or even non-adherence to treatment [3]. The term adherence denotes the extent to which the patient complies with the therapy recommendation. Non-adherence to antihypertensive therapy affects 10–80% of all hypertensive patients, preferably elderly patients [9,10]. Poor adherence to antihypertensive treatment is associated with elevated BP values, an increased cardiovascular risk, and a poor prognosis in these patients [3,9].

Although the prevalence of hypertension is significantly higher in persons above 60 years, about 13% of younger adults between 20 and 40 years of age worldwide are hypertensive with an average BP above 140/90 mmHg [11]. In a study on young adults, only 11% of the hypertensive persons received antihypertensive medication at the time of enrolment [12]. Hypertension is less frequently diagnosed in young than in old adults [11]. Younger adults may be in their prehypertensive state, which may remain undetected or underestimated for a longer period of time. However, four out of five prehypertensive people aged 40–49 years would become hypertensive in the next 10 years [13]. Prehypertensive and even hypertensive young people are often not aware of their BP [14]. However, several studies have demonstrated that elevated BP at youth correlates with higher rate of LV hypertrophy and cardiovascular mortality at an older age [15–17]. Therefore, the ESC/ESH recommends that all adults ≥18 years should have their office BP measured and recorded, and be aware of their BP [3].

Several years ago, the SPRINT study was the first randomized trial that clearly demonstrated the benefits of a systolic BP (SBP) reduction below previous target values to 120 mmHg [18,19]. A change in BP of 20/10 mmHg is associated with a 50% difference in cardiovascular risk [9]. In patients who were at increased risk of cardiovascular disease, antihypertensive therapy targeting SBP below 120 mmHg resulted in lower rates of major adverse cardiovascular events and mortality compared with target SBP of less than 140 mmHg [20]. Consequently, the BP targets have been revised in the recent European, U.S., and WHO guidelines [3,21,22]. Early initiation of therapy and faster achievement of the recommended BP targets (within 6 months, ideally even within 3 months) are associated with a significant reduction in heart failure and other major cardiovascular events [23]. For initiation of antihypertensive therapy, a combination treatment of a blocker of the renin-angiotensin system (RAS; either an angiotensin-converting enzyme (ACE) inhibitor or an angiotensin receptor blocker) with a calcium channel blocker (CCB) or diuretic is the preferred recommendation for most hypertensive patients. Combination therapy, preferably as single-pill combination, has been proven to be more effective in fast achievement of the recommended BP targets and to be associated with a better adherence of patients to treatment [3,23].

Purpose of the Present Study

To prevent the development of cardiovascular complications, adequate therapy should start as early as possible. In animal models, efficient antihypertensive therapy started in their prehypertensive state proved to be more efficient than the same therapy started in adulthood when hypertension was manifest [24,25]. In humans, there is still much controversy about the treatment of younger adults with low-risk grade 1 hypertension and about the efficacy of antihypertensive treatment in young hypertensives compared with old patients. Conventional clinical outcome studies are difficult to perform in people before age 40, hence there is a lack of randomized clinical trials for young hypertensive patients [3,11]. Consequently, it is not clear whether the effects of antihypertensive treatments in old and young people are comparable. Even in animal models, there are only very few studies comparing treatment effects in young and old animals. Previously, we had performed studies on antihypertensive treatments in young (aged 7–10 weeks) [26] and old (aged 60–82 weeks) spontaneously hypertensive rats (SHRs) [27]. The present study was designed to directly compare the treatment effects achieved in young and old SHRs. The relative reduction of SBP, parameters of LV hypertrophy and LV remodeling and fibrosis induced by a combination of the ACE inhibitor captopril and the CCB nifedipine was compared using a two-way ANOVA with the factors treatment and age. The purpose of this analysis is to show how effectively the reduction in SBP and attenuation of cardiac hypertrophy and remodeling can be achieved with early-onset compared with late-onset therapy.

2. Materials and Methods

We re-evaluated data from two previously published studies on male SHRs [26,27]. The raw data were obtained from these studies and were re-analyzed using a new statistical approach allowing a direct comparison between the data from both animal studies. All animal protocols were approved by the state agency (Landesdirektion Sachsen, number and date of approval: TVV 36/08; 13 May 2009) in accordance with the Guide for the Care and Use of Laboratory Animals published by the National Institutes of Health and with the "European Convention for the Protection of Vertebrate Animals used for Experimental and other Scientific Purposes".

In brief, we analyzed data from 12 young SHRs aged 7–10 weeks (ySHRs) and fourteen old SHRs aged 60–82 weeks (oSHRs). The study on ySHRs [26] was designed as a pilot study to identify the most appropriate treatments to be applied to oSHRs. The ySHRs were intended to be in their adolescence throughout the experiment. For this reason, the experimental period was defined from 7 to 10 weeks of age. In contrast, the old SHRs should be treated over a long period of time, which was set to 22 weeks [27].

All animals were cultured at Charles River, Sulzfeld, Germany. Both groups were subdivided at random into two subgroups each, one remaining untreated and serving as control (yCtrl, n = 6; oCtrl, n = 7) and the other one being treated with a combination of captopril (60 mg kg^{-1} d^{-1}, Axxora, Lörrach, Germany) and nifedipine (10 mg kg^{-1} d^{-1}, Sigma-Aldrich Chemie, Steinheim, Germany). The treated groups were labeled as yC+N (n = 6) and oC+N (n = 7). ySHRs were 7 weeks of age at the beginning of the treatment period, which lasted 3 weeks. In oSHRs, the treatment period started at 60 weeks of age and lasted 22 weeks. The medication (or placebo in Ctrl animals) was administered as tablets for oral uptake given into the cages along with chow once daily between 9:00 a.m. and 10:00 a.m.

Before and during the experimental period, we regularly measured systolic blood pressure (SBP) using the tail-cuff-method (TSE Blood Pressure Monitor, Series 209002, TSE Systems GmbH, Bad Homburg, Germany; for more details, see [26,27]). Values from the final SBP measurement performed in the last experimental week were taken for the present re-evaluation.

At the end of the experiments, the animals were sacrificed and their hearts were extracted. We determined heart weight (HW) as a measure of cardiac hypertrophy. We analyzed the absolute HW as well as HW normalized to final body weight (HW/BW). Pieces of the left ventricle (LV) were obtained to perform ribonuclease protection assays to determine several markers of LV hypertrophy and remodeling. mRNA expressions of atrial natriuretic peptide (ANP) and collagen types I (Coll 1) and III (Coll 3) were included into the present analysis. Histological preparations were made from the cardiac apex. A score ranging from 0 (= no fibrosis) to 3 (= fibrosis of the entire heart) was applied to assess the histological degree of fibrosis (for more details, see [26,27]). The degree of fibrosis was also included in the present analysis.

Statistical analyses were carried out with the software package SigmaPlot Version 14.0 (Systat Software GmbH, Erkrath, Germany) for Windows. First, we directly compared final SBP, HW, HW/BW, ANP mRNA, Coll 1 mRNA, Coll 3 mRNA, and degree of fibrosis among all four groups (yCtrl, yC+N, oCtrl, oC+N) using a two-way ANOVA with the factors A = treatment (Ctrl or C+N) and B = age (y or o). As a post-hoc test, an all pairwise multiple comparison procedure according to the Holm–Sidak method was applied.

In addition, we assessed the treatment effect for each parameter and compared the treatment effects in young versus old SHRs. The treatment effect was calculated for each treated animal as the difference between the average value of the respective Ctrl group (y or o) and the individual value of the treated animal. The difference was normalized to the average yCtrl or oCtrl value (given in %). A positive value of the treatment effect marks an improvement (i.e., reduction) in the parameter. For the comparison of young versus old, we first performed a Shapiro–Wilk test for normality. In the case of normal distribution, we performed a t-test, otherwise a Mann–Whitney rank sum test was applied. p-values < 0.05 were considered significant.

3. Results

3.1. Treatment Effects on SBP

SBP of young SHRs at baseline (age 7 weeks) was 173.8 ± 3.1 and 157.5 ± 3.7 mmHg for Ctrl and C+N animals, respectively (Table 1). This difference was not significant. The original experiment [26] included a total of 42 ySHRs with an average SBP at baseline of 171.2 ± 3.5 mmHg. In untreated ySHRs, SBP increased continuously over the 3-week period of observation to 201.7 ± 6.5 mmHg at 10 weeks of age. This was significantly higher than SBP of age-matched SHRs after 3 weeks of C+N treatment. In the first two weeks of treatment, SBP decreased to 115.8 ± 2.4 mmHg. SBP of age-matched Wistar–Kyoto rats (WKY) was with 123.1 ± 0.9 mmHg in a similar range [26]. However, in the final treatment week, SBP re-increased to 152.5 ± 2.8 mmHg. The difference to untreated ySHRs was with 49.2 ± 2.8 mmHg significant ($p < 0.001$) corresponding to a treatment effect of 24.4 ± 1.4% of untreated Ctrls.

Table 1. SBP, HR, and BW of young and old SHRs.

Animal Group	SBP [mmHg]		HR [min^{-1}]		BW [g]	
	Baseline	Final	Baseline	Final	Baseline	Final
yCtrl	173.8 ± 3.1	201.7 ± 6.5 °	384.2 ± 10.3	416.5 ± 13.5	185.7 ± 4.2	309.7 ± 7.3 °°
yC+N	157.5 ± 3.7	152.5 ± 2.8	396.9 ± 9.1	397.7 ± 9.2	175.8 ± 7.3	308.0 ± 7.3 °°
oCtrl	231.0 ± 8.7	239.0 ± 5.1	413.5 ± 14.3	400.2 ± 20.6	396.7 ± 9.7	377.6 ± 37.2
oC+N	244.5 ± 8.1	193.6 ± 4.8 °	401.0 ± 14.6	409.2 ± 6.7	414.7 ± 8.6	418.6 ± 5.4 *

Data are given as means ± SEM. Baseline values were obtained at 7 weeks of age in young SHRs and at 60 weeks of age in old SHRs. Final values were obtained at 10 weeks of age in young SHRs and at 82 weeks of age in old SHRs. Treatment groups: y/oCtrl: untreated animals, y/oC+N: animals treated with captopril + nifedipine. * significant vs. corresponding Ctrl group ($p < 0.05$); ° significant vs. baseline value ($p < 0.01$); °° significant vs. baseline value ($p < 0.001$).

Age had a significant effect on SBP ($p < 0.001$). In untreated oSHRs aged 60 weeks, SBP was 231.0 ± 8.7 mmHg and hardly rose during the following 22 weeks to 239.0 ± 5.1 mmHg at week 82 of age ($p < 0.001$ compared with untreated ySHRs). Treatment with C+N over 22 weeks significantly reduced SBP by 45.4 ± 4.8 mmHg to 193.6 ± 4.8 mmHg ($p < 0.001$; Table 1); this is, however significantly higher than in yC+N rats ($p < 0.001$; Figure 1A). The treatment effect is 19.0 ± 2.0% of the untreated oCtrls. However, despite a much longer period of therapy in oSHRs, this treatment effect is lower than in ySHRs ($p = 0.059$; Figure 1B).

3.2. Treatment Effects on Cardiac Hypertrophy

3.2.1. Heart Weight

The relative heart weight (HW/BW) of ySHRs was 3.50 ± 0.07 mg/g without therapy and 3.06 ± 0.06 mg/g ($p = 0.331$) after 3 weeks of C+N therapy, which is a treatment effect of 12.6 ± 1.6%. In untreated oSHRs, HW/BW was significantly higher (5.51 ± 0.51 mg/g, $p < 0.001$), but 22 weeks of C+N treatment significantly reduced HW/BW to 3.44 ± 0.20 mg/g ($p < 0.001$), which is in a similar range as HW/BW of yCtrl rats (Figure 2A). The treatment effect was with 37.4 ± 3.7%, significantly greater than in ySHRs ($p < 0.005$; Figure 2B).

However, the body weight (BW) of the animals changed in a different way; while BW in 10-week-old SHRs was similar in the untreated and treated groups (309.7 ± 7.3 and 308.0 ± 7.3 g, respectively), 82-week-old untreated SHRs had a considerably lower BW (377.6 ± 37.2 g) than age-matched C+N-treated rats (418.6 ± 5.4 g; $p = 0.042$; Table 1). Moreover, SHRs aged 7–10 weeks are still juvenile, and their HW/BW ratio is substantially different from that of old SHRs aged 60–82 weeks. Therefore, analysis of the absolute HW may provide more realistic information about the treatment effects (Figure 2C,D). In ySHRs, HW was 1095 ± 40 mg in untreated SHRs and 934 ± 28 mg in C+N-treated animals ($p = 0.136$), which means a treatment effect of 14.7 ± 2.6%. In 82-week-old untreated SHRs, HW was significantly higher than in young animals ($p < 0.001$). Twenty-two weeks of C+N treatment significantly decreased HW from 1860 ± 93 mg to 1440 ± 81 mg ($p < 0.001$) corresponding to a treatment effect of 26.8 ± 1.3%. Although this treatment effect was significantly greater than in ySHRs ($p = 0.035$), HW of treated oSHRs remained significantly higher than in ySHRs after only 3 weeks of treatment ($p < 0.001$).

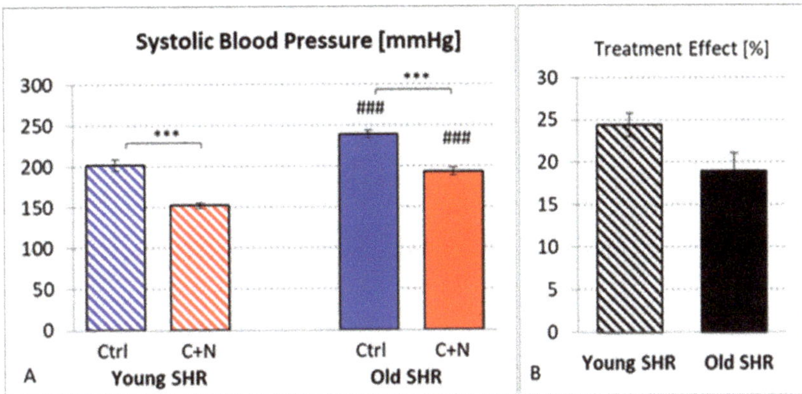

Figure 1. (**A**) Systolic blood pressure (in mmHg) at 10 (young SHR) or 82 (old SHR) weeks of age in untreated (Ctrl) and treated (captopril + nifedipine, C+N) SHRs. (**B**) Treatment effects in young and old SHRs (in % of corresponding average Ctrl value). Significance marks: ⌈*⌉ significant difference between marked groups, *** $p < 0.001$; ### significant difference to corresponding young SHR group, $p < 0.001$.

To assess to what extent the treatment effect on HW is mediated by SBP reduction, we further calculated HW reduction per unit SBP reduction. While in yC+N rats, HW decreased by 3.5 ± 0.7 mg/mmHg, this ratio was significantly greater in oC+N animals (9.7 ± 2.1 mg/mmHg, $p = 0.035$; data not shown).

3.2.2. ANP mRNA

ANP as a marker of ventricular hypertrophy in 10-week-old untreated SHRs was 13.3 ± 4.0. After 3 weeks of treatment, age-matched SHRs presented a value of 4.3 ± 0.8 ($p = 0.775$), which is a treatment effect of 67.8 ± 6.0%. There was a significant effect of age on ANP expression ($p < 0.001$): In 82-week-old untreated SHRs, it was 235.4 ± 43.8, and 22 weeks of treatment decreased it to 45.7 ± 10.9; $p < 0.001$; Figure 3A). The treatment effect was 80.6 ± 4.6%, which is in the same range as in ySHRs ($p = 0.118$; Figure 3B) despite a much longer period of treatment.

3.3. *Treatment Effects on Cardiac Remodeling and Fibrosis*

3.3.1. mRNA Expression of Collagens I and III

With transition to fibrosis, expression of collagen mRNA can be observed. In untreated ySHRs, the mRNA expression of Coll 1 and Coll 3 was 13.9 ± 2.1 and 22.5 ± 1.9, respectively. Treatment had no significant effect; the mRNA expression of the two collagens was even slightly increased to 15.4 ± 3.4 and 23.4 ± 3.7, respectively.

In 82-week-old untreated SHRs, mRNA expression of Coll 1 and Coll 3 was in a similar range, with 15.9 ± 4.6 and 25.4 ± 4.2, respectively. However, 22 weeks of treatment significantly reduced the collagen mRNA expressions to 4.1 ± 0.8 (Coll 1; $p = 0.007$) and 9.4 ± 1.5 (Coll 3; $p < 0.001$). The treatment effects were significantly higher than in young animals (Coll 1: 74.5 ± 5.2%, $p = 0.003$; Coll 3: 63.1 ± 6.0%, $p = 0.001$; Figure 4).

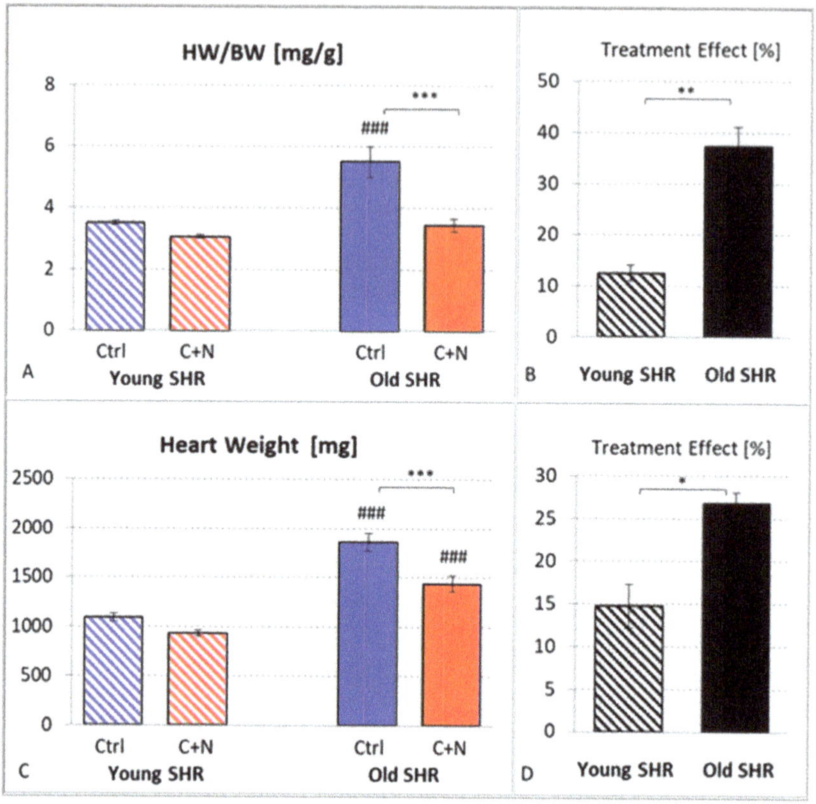

Figure 2. (**A**) Heart weight related to body weight in untreated (Ctrl) and treated (C+N) young and old SHRs. (**B**) Treatment effects on HW/BW in young and old SHRs (in % of corresponding average Ctrl value). (**C**) Absolute heart weight in untreated (Ctrl) and treated (C+N) young and old SHRs. (**D**): Treatment effects on heart weight in young and old SHRs (in % of corresponding average Ctrl value). Significance marks: ⌈*⌉ significant difference between marked groups, * $p < 0.05$, ** $p < 0.01$, *** $p < 0.001$; ### significant difference to corresponding young SHR group, $p < 0.001$.

3.3.2. Fibrosis

Mild to moderate histological signs of fibrosis were already observed in 10-week-old untreated SHRs (fibrosis degree: 1.47 ± 0.12). Three weeks of treatment reduced the degree of fibrosis to 0.59 ± 0.17; $p < 0.001$. The treatment effect was $58.0 \pm 11.9\%$.

Age had a significant effect on the degree of fibrosis. Untreated SHRs aged 82 weeks presented moderate to severe signs of fibrosis (fibrosis degree: 2.30 ± 0.11), which was significantly higher than in ySHRs ($p < 0.001$). After 22 weeks of treatment, there were still signs of moderate fibrosis (fibrosis degree: 2.13 ± 0.07) corresponding to a treatment effect of $7.5 \pm 3.1\%$, which was significantly smaller than that in ySHRs ($p < 0.001$; Figure 5).

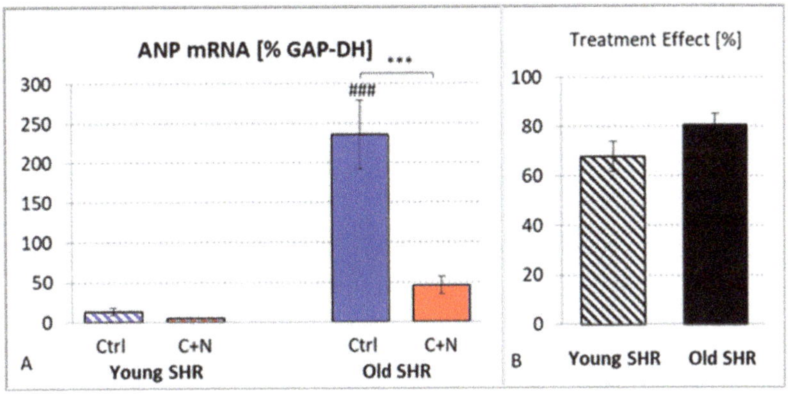

Figure 3. (**A**) mRNA expression of atrial natriuretic peptide in untreated (Ctrl) and treated (C+N) young and old SHRs. (**B**) Treatment effects in young and old SHRs (in % of corresponding average Ctrl value). Significance marks: ⌈*⌉ significant difference between marked groups, *** $p < 0.001$; ### significant difference to corresponding young SHR group, $p < 0.001$.

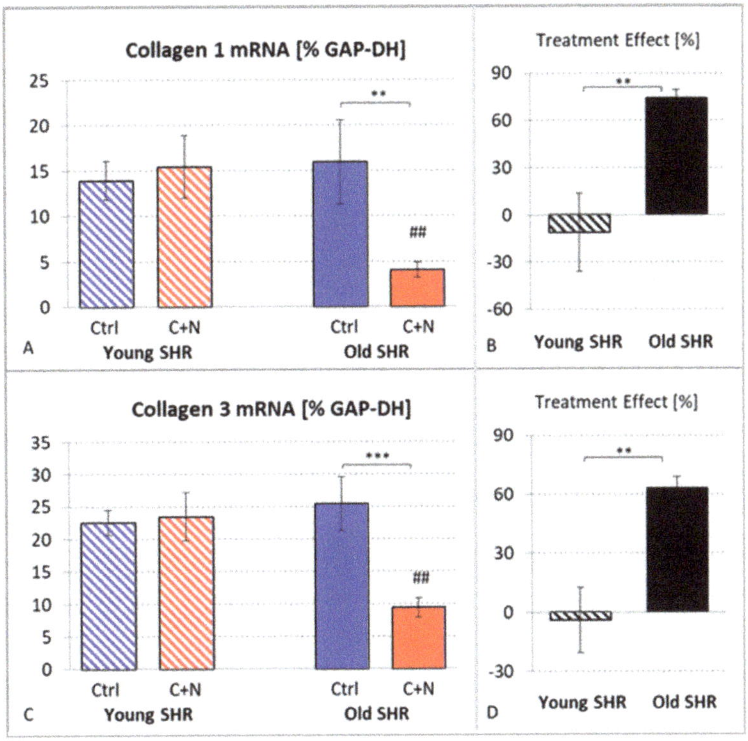

Figure 4. (**A**) mRNA expression of collagen type I in untreated (Ctrl) and treated (C+N) young and old SHRs. (**B**) Treatment effects on Coll 1 in young and old SHRs (in % of corresponding average Ctrl value). (**C**) mRNA expression of collagen type III in untreated (Ctrl) and treated (C+N) young and old SHRs. (**D**) Treatment effects on Coll 3 in young and old SHRs (in % of corresponding average Ctrl value). Significance marks: ⌈*⌉ significant difference between marked groups, ** $p < 0.01$, *** $p < 0.001$; ## significant difference to corresponding young SHR group, $p < 0.01$.

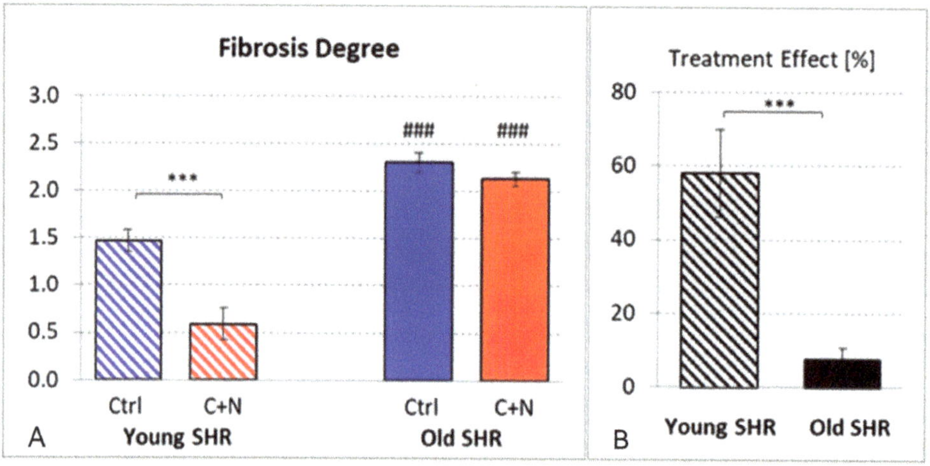

Figure 5. (**A**) Degree of fibrosis in untreated (Ctrl) and treated (C+N) young and old SHRs. (**B**) Treatment effects in young and old SHRs (in % of corresponding average Ctrl value). Significance marks: ⌈*⌉ significant difference between marked groups, *** $p < 0.001$; ### significant difference to corresponding young SHR group, $p < 0.001$.

4. Discussion

It is widely accepted that antihypertensive treatment should start as early in life as possible [3]. Epidemiological studies in humans have shown that young people with high BP were at higher risk of LV hypertrophy and cardiovascular disease events compared with normotensive persons [15–17]. Animal studies have shown that early-onset treatment exerts better effects with respect to SBP and cardioprotection than a therapy started at an advanced age [24,25]. However, only few studies have directly compared the effects of antihypertensive treatment in young and senescent SHRs.

4.1. *Effects on SBP*

The present results showed that treatment effects on SBP were greater in ySHRs than in oSHRs, thus confirming the experience from earlier studies [24,25]. Early treatment of SHR with RAS antagonists is able to reduce SBP even to normotensive values [24,28]. However, in their studies, the treatment started in the prehypertensive stage of life, that is, at or before the age of 4 weeks, and lasted 6–8 weeks. At 7 weeks of age, SBP of untreated SHRs ranged between 154 and 187 mmHg [26] and increased to 210–220 mmHg by week 12–15 [29]. For comparison, at that age, SBP of normotensive WKY ranges between 115 and 137 mmHg [24,26]. With further aging up to senescence, SBP increases only slightly; in untreated SHRs aged 60–82 weeks, it was about 240 mmHg (see Table 1). In contrast, SBP of old WKY remained in the range of young WKY [24]. The present results show that the treatment effect significantly depends on age; despite a much shorter interval of therapy (3 weeks only) treatment with C+N between the 7th and 10th week of life reduced SBP even more than a 22-week therapy with the same medication between the 60th and 82nd week of life.

Antihypertensive treatment started early in life has not only stronger but also more persistent effects on SBP. In young SHRs aged 4–6 weeks, transient antihypertensive therapy over 4–6 weeks resulted in a significant SBP reduction, which remained until age 24–30 weeks. In contrast, if such a treatment was initiated during adulthood (between week 20 and 24), the effect on SBP was significantly lower or even completely abolished [25,30–32]. Antagonists of the RAS not only exert vasodilatory effects, thus directly decreasing BP. In addition, they prevent or attenuate structural changes of resistance vessels. This effect, however, is only present in adolescent and young adult SHRs up to about 20 weeks of age, and is most pronounced at

4–10 weeks of age [33]. Of note, vascular antihypertrophic effects of RAS antagonists were also demonstrated in both young and old WKY [24]. Nifedipine is also able to reduce vascular hypertrophy in SHRs [34]. Prevention of vascular hypertrophy may explain the more potent antihypertensive effect of early-onset treatment and the long-term maintenance of BP reduction even after withdrawal of the therapy.

The effect of age found in our study cannot be differentiated from an effect of SBP at treatment onset; while SBP of all ySHRs presented here was lower than 200 mmHg at baseline, all of the oSHRs had a baseline SBP above 200 mmHg [26,27]. There was a significant positive correlation between age and SBP at baseline ($r = 0.83$; $p < 0.001$). These results strongly suggest that greater treatment effects and even normotension might be achieved the earlier in life the treatment is started.

4.2. *Cardioprotective Effects*

The therapy effects were not confined to antihypertensive effects but also included cardioprotective effects. With C+N treatment, cardiac hypertrophy and remodeling were differentially attenuated in young and old SHRs.

4.2.1. Effects on Cardiac Hypertrophy and Remodeling in Young SHRs

In genetic forms of hypertension such as in SHRs or in humans with essential hypertension, vascular hypertrophy, and consequently elevated vascular resistance, cause pressure load to the LV, which in turn induces hypertrophy and remodeling of the LV myocardium. In a previous study on 6-month-old SHRs, both total peripheral resistance (TPR) and HW/BW were significantly higher than in age-matched WKY rats [35]. A similar observation has been made in the ySHR study: TPR of yCtrl rats was also significantly higher than in age matched WKY rats (0.20 ± 0.01 and 0.12 ± 0.02 mmHg·min·kg·mL^{-1}, respectively; $p = 0.01$; unpublished data).

In SHR, development of cardiac hypertrophy starts after the prehypertensive stage between the 4th and 12th week of life [36,37]. At this stage, HW/BW of SHRs is about 13% higher than that of age-matched WKY [24]. Several studies showed that antihypertensive therapy with RAS antagonists started between the 4th and 14th week of life reduced HW/BW or LVW/BW by 19–27% [24,29,30,38,39]. Our study on young SHRs provided similar results: three weeks of C+N treatment reduced HW/BW by about 14%, but not to the level of normotensive WKY rats [26]. Without treatment, cardiac hypertrophy progresses over time, but even when treatment with RAS antagonists is started in adulthood (between week 24 and 34 of life), significant attenuation of cardiac hypertrophy can be achieved [24,40].

However, with progress of cardiac hypertrophy, remodeling processes including profibrotic processes will develop. Perrucci and co-workers demonstrated cardiac fibrosis and significantly increased collagen deposition in the hearts of 8-week-old untreated SHRs [41]. The progredient increase in blood pressure and the development of LV hypertrophy and fibrosis are accompanied by a deterioration of LV function. Echocardiographic examination demonstrated significant systolic and diastolic dysfunction in untreated SHRs even at 2–3 months of age [37]. With enalapril treatment over 14 weeks, deterioration of ejection fraction and fractional shortening as well as development of cardiac fibrosis were prevented or at least attenuated [42]. This is in line with our observations on ySHRs; at 10 weeks of age, they presented mild cardiac fibrosis, which was significantly reduced by more than 50% with three weeks of C+N treatment [26].

4.2.2. Effects on Cardiac Hypertrophy and Remodeling in Old SHRs

With progress of age, the HW/BW differences between SHRs and WKY increased up to 29% at week 83 [24]. In the advanced stage of hypertension at week 82, cardiac hypertrophy of our untreated oSHRs had achieved a markedly higher degree than in young animals, as reflected by the significantly higher values of ANP mRNA expression (20 times higher than in yCtrls) and HW/BW (more than 30% higher than in yCtrls). Angiotensin II exerts many prohypertrophic effects on the heart. Consequently, prevention or attenuation of cardiac hypertrophy and transition into heart failure is a major goal of ACE inhibitor therapy [43]. Even in normotensive WKY rats, RAS antagonists can significantly reduce HW/BW [24]. In senescent SHRs at more than 80 weeks of age, short-term treatment with RAS antagonists has only weak effects: after 8 weeks of losartan therapy, HW/BW was only reduced by 7% [24]. In contrast, chronic captopril treatment administered from 14 to 24 months of age-reduced LVW/BW by about 30% [44], which is similar to the results of the present study. Direct antihypertrophic effects on cardiomyocytes have also been demonstrated for nifedipine [45]. Thus, cardioprotective effects of C+N treatment are not only based on antihypertensive effects, but in addition on direct antihypertrophic effects. This is also reflected in the decrease in HW related to the decrease in SBP, which was even higher in oSHRs than in ySHRs.

As cardiac hypertrophy progresses, remodeling and accumulation of extracellular matrix (ECM) advance too. Increased mRNA levels of ECM molecules such as transforming growth factor-beta and tissue inhibitor of metalloproteinases 2 have been found in 82-week-old untreated SHRs [27]. Of note, the levels of collagen mRNA and protein develop with aging in a nonsynchronous way: in young rats, collagen mRNA in the heart is relatively high, but does not result in collagen accumulation. In contrast, collagen mRNA in hearts of old rats is hardly elevated, but induces a two-fold increase in collagen protein content [46]. This may explain the similar levels of collagen mRNA in old and young untreated SHRs in the present study, which were associated with different histological degrees of fibrosis. While C+N treatment in ySHRs did not decrease collagen mRNA expression, but significantly reduced fibrosis, the opposite effect was achieved in oSHRs. Analysis of mRNA expression can only reflect a moment within a process developing over a longer period of time. In particular, accumulation of collagen is not linearly related to elevated levels of collagen mRNA. Hence, we consider the histological degree of fibrosis to be the more meaningful parameter with respect to cardiac remodeling. Even though the treatment effect on collagen mRNA was much greater in oSHRs compared with ySHRs, attenuation of fibrosis was only weak as the intervention was initiated in an advanced stage of cardiac remodeling. An early start of antihypertensive treatment may attenuate and delay the process of remodeling more effectively.

The transition into fibrosis is associated with further functional deterioration and finally leads to cardiac failure. At 80 weeks of age, diastolic function and compliance of the LV in untreated SHRs were impaired compared with age-matched WKY [47]. Treatment significantly improved ejection fraction index, but not to the level of normotensive WKY [44]. An echocardiographic study on old SHRs demonstrated a deterioration of both systolic and diastolic LV function between week 60 and 82 without treatment. Moreover, LV catheterization under anesthesia revealed signs of cardiac failure and LV decompensation. All of these changes were attenuated and delayed, but not completely prevented by antihypertensive treatment [48]. These results are in line with the weak antifibrotic effect of a late-onset treatment.

4.2.3. Early-Onset versus Late-Onset Treatment

Similar to its effects on SBP, early-onset antihypertensive treatment also has long-lasting effects on cardiac hypertrophy and remodeling. Antagonization of prohypertrophic Angiotensin II effects in a sensitive period of life (up to age 20 weeks) makes even a transient treatment with ACE inhibitors effective with regard to prevention of LV hypertrophy [29,39].

The duration of treatment is an important factor with regard to the treatment effect. This is well illustrated by a comparison of our results with those of Demirci and co-workers: They applied losartan at the same dose over the same period of time to young and old SHRs and observed a lower treatment effect in the old animals [24]. In contrast, we administered the same medication (C+N) to young and old SHRs, but a more than seven-fold treatment interval was necessary to achieve a similar antihypertrophic treatment effect in oSHRs [26,27]. Lifelong antihypertensive treatment started at one month of age doubled the lifespan of stroke-prone SHRs to 30 months, which is identical to that of normotensive WKY. In contrast, 80% of untreated SHRs had died after 15 months [49]. These findings emphasize the importance of a long-lasting antihypertensive therapy, and this is the more important the later in life treatment is initiated.

4.2.4. Applications to Antihypertensive Treatment in Humans

Our results on animals clearly show the advantages of an early start of an effective antihypertensive therapy, in particular with respect to preventing cardiac remodeling and thus transition into cardiac failure. This emphasizes the importance of starting BP control during early adulthood. Important aspects of antihypertensive treatment in humans are fast achievement of target BP values and adequate persistence and adherence to therapy. A combination of two or three drugs from different drug classes facilitates the adjustment of treatment to the individual patient as it allows dose reduction of the single drugs and minimization of adverse drug effects, thus ensuring the patient's adherence to therapy. Consequently, a combination of two or more, ideally as fixed-dose single-pill administration, as initial therapy of hypertension is currently recommended in most guidelines [4,22].

4.3. Limitations of the Study

There are several limitations to this study. An important limitation is the lack of functional measurements and investigation of vascular changes. The animal studies were designed to analyze cardiac but not vascular sequelae of hypertension. Unfortunately, only systolic BP values were measured. We are aware of the fact that diastolic BP would provide relevant data characterizing the vascular treatment effects. This limitation is because of the method of non-invasive BP measurement with a tail cuff device that did not allow measurement of diastolic BP. Data on TPR, which may reflect vascular changes, were only obtained from the ySHRs. In addition, we have compared changes in cardiac tissue with echocardiographic data from oSHRs reported previously [48].

Another limitation is in the analysis of collagen mRNA expression to characterize the development of fibrosis. Previous studies on rats stimulated with norepinephrine showed a good correlation between LV collagen mRNA expression and collagen amount in the histological specimen [50,51]. Fibrotic processes include enhanced collagen turnover and an increased collagen accumulation. We assume that mRNA expression does not adequately reflect the accumulation of collagen. Moreover, collagen mRNA and protein content develop with aging in a nonsynchronous way [46]. Collagen protein concentration or hydroxyproline content would more adequately reflect the intensity of fibrotic processes. We consider the histological degree of fibrosis to be a reliable and meaningful marker of fibrosis.

5. Conclusions

As elevated BP is the leading cause of premature death and is the major risk factor for a variety of cardiovascular disease events [3], early and effective antihypertensive therapy is of paramount importance for the patients. The present data of an animal study show that even a late-onset treatment of hypertension has both antihypertensive and cardioprotective effects. However, in those animals, SBP remained at significantly higher levels than in ySHRs, which received the same medication administered over a much shorter period of time at an early stage of life. Consequently, cardiac remodeling and transition into fibrosis were delayed but not prevented in the old animals. In contrast, fibrosis was significantly attenuated in ySHRs after only three weeks of therapy. This result gives strong reason to believe that a life-long antihypertensive therapy started at an early stage in life may prevent severe cardiac remodeling and transition into cardiac failure at advanced age.

The results emphasize that antihypertensive therapy should start as early as possible and be maintained as long as BP is elevated, usually life-long. Continuous BP control and strict adherence to treatment are inevitable. Early reduction in SBP, preferably to normotensive or near-normotensive values, contributes to prevent cardiac complications of hypertension.

Author Contributions: B.R., H.-G.Z., C.H. and J.B. contributed to the conception and design of the study. C.H., J.B. and B.R. performed the animal experiments including blood pressure measurements and the analyses on heart tissue. B.R. performed the statistics and created figures and tables. H.-G.Z. and B.R. drafted the original manuscript. All authors contributed to the final version of the manuscript. All authors have read and agreed to the published version of the manuscript.

Funding: This research received no external funding.

Institutional Review Board Statement: This manuscript does not contain clinical studies or patient data. The study was conducted according to the guidelines of the Declaration of Helsinki. All animal protocols were approved by the state agency (Landesdirektion Sachsen, number and date of approval: TVV 36/08; 13 May 2009) in accordance with the Guide for the Care and Use of Laboratory Animals published by the National Institutes of Health and with the "European Convention for the Protection of Vertebrate Animals used for Experimental and other Scientific Purposes".

Data Availability Statement: Not applicable.

Acknowledgments: The authors acknowledge support from the German Research Foundation (DFG) and Universität Leipzig within the program of Open Access Publishing.

Conflicts of Interest: The authors declare no conflict of interest.

Abbreviations

ACE	angiotensin-converting enzyme
ANP	atrial natriuretic peptide
BP	blood pressure
BW	body weight
C+N	captopril + nifedipine
CCB	calcium channel blocker
Coll 1	collagen type I
Coll 3	collagen type III
Ctrl	control
GAP-DH	glyceraldehyde-3-phosphate dehydrogenase
HW	heart weight
LV	left ventricle/left ventricular
LVW	left ventricular weight
o	old rats (aged 60–82 weeks)
RAS	renin-angiotensin system
SBP	systolic blood pressure
SHR	spontaneously hypertensive rats
TPR	total peripheral resistance
WKY	Wistar–Kyoto rats
y	young rats (aged 7–10 weeks)

References

1. Mazur, I.; Belenichev, I.; Kucherenko, L.; Bukhtiyarova, N.; Puzyrenko, A.; Khromylova, O.; Bidnenko, O.; Gorchakova, N. Antihypertensive and cardioprotective effects of new compound 1-(b-phenylethyl)-4-amino-1,2,4-triazolium bromide (Hypertril). *Eur. J. Pharmacol.* **2019**, *853*, 336–344. [CrossRef] [PubMed]
2. He, F.J.; MacGregor, G.A. Blood pressure is the most important cause of death and disability in the world. *Eur. Heart J. Suppl.* **2007**, *9*, B23–B28. [CrossRef]
3. Williams, B.; Mancia, G.; Spiering, W.; Agabiti Rosei, E.; Azizi, M.; Burnier, M.; Clement, D.L.; Coca, A.; de Simone, G.; Dominiczak, A.; et al. ESC Scientific Document Group. 2018 ESC/ESH Guidelines for the management of arterial hypertension. *Eur. Heart J.* **2018**, *39*, 3021–3104. [CrossRef] [PubMed]
4. Zhou, B.; Perel, P.; Mensah, G.A.; Ezzati, M. Global epidemiology, health burden and effective interventions for elevated blood pressure and hypertension. *Nat. Rev. Cardiol.* **2021**, *18*, 785–802. [CrossRef]

5. NCD Risk Factor Collaboration (NCD-RisC). Worldwide trends in hypertension prevalence and progress in treatment and control from 1990 to 2019: A pooled analysis of 1201 population-representative studies with 104 million participants. *Lancet* **2021**, *398*, 957–980. [CrossRef]
6. Briasoulis, A.; Agarwal, V.; Tousoulis, D.; Stefanadis, C. Effects of antihypertensive treatment in patients over 65 years of age: A meta-analysis of randomised controlled studies. *Heart* **2014**, *100*, 317–323. [CrossRef]
7. Tiffe, T.; Wagner, M.; Rücker, V.; Morbach, C.; Gelbrich, G.; Störk, S.; Heuschmann, P.U. Control of cardiovascular risk factors and its determinants in the general population-findings from the STAAB cohort study. *BMC Cardiovasc. Disord.* **2017**, *17*, 276. [CrossRef]
8. Volpe, M.; Gallo, G.; Tocci, G. Is early and fast blood pressure control important in hypertension management? *Int. J. Cardiol.* **2018**, *254*, 328–332. [CrossRef]
9. Unger, T.; Borghi, C.; Charchar, F.; Khan, N.A.; Poulter, N.R.; Prabhakaran, D.; Ramirez, A.; Schlaich, M.; Stergiou, G.S.; Tomaszewski, M.; et al. 2020 International Society of Hypertension Global Hypertension Practice Guidelines. *Hypertension* **2020**, *75*, 1334–1357. [CrossRef]
10. Raja, W.; Ayub, T.; Jeelani, A.; Khan, S.M.S. Adherence to antihypertensive therapy and its determinants among patients attending primary care hospitals of Kashmir, India. *J. Family Med. Prim. Care* **2021**, *10*, 4153–4159. [CrossRef]
11. Hinton, T.C.; Adams, Z.H.; Baker, R.P.; Hope, K.A.; Paton, J.F.R.; Hart, E.C.; Nightingale, A.K. Investigation and Treatment of High Blood Pressure in Young People: Too Much Medicine or Appropriate Risk Reduction? *Hypertension* **2020**, *75*, 16–22. [CrossRef]
12. Yano, Y.; Reis, J.P.; Colangelo, L.A.; Shimbo, D.; Viera, A.J.; Allen, N.B.; Gidding, S.S.; Bress, A.P.; Greenland, P.; Muntner, P.; et al. Association of Blood Pressure Classification in Young Adults Using the 2017 American College of Cardiology/American Heart Association Blood Pressure Guideline With Cardiovascular Events Later in Life. *JAMA* **2018**, *320*, 1774–1782. [CrossRef]
13. Moreira, L.B.; Fuchs, S.C.; Wiehe, M.; Gus, M.; Moraes, R.S.; Fuchs, F.D. Incidence of hypertension in Porto Alegre, Brazil: A population-based study. *J. Hum. Hypertens.* **2008**, *22*, 48–50. [CrossRef]
14. Gooding, H.C.; McGinty, S.; Richmond, T.K.; Gillman, M.W.; Field, A.E. Hypertension awareness and control among young adults in the national longitudinal study of adolescent health. *J. Gen. Intern. Med.* **2014**, *29*, 1098–1104. [CrossRef]
15. Drukteinis, J.S.; Roman, M.J.; Fabsitz, R.R.; Lee, E.T.; Best, L.G.; Russell, M.; Devereux, R.B. Cardiac and systemic hemodynamic characteristics of hypertension and prehypertension in adolescents and young adults: The Strong Heart Study. *Circulation* **2007**, *115*, 221–227. [CrossRef]
16. Sundström, J.; Neovius, M.; Tynelius, P.; Rasmussen, F. Association of blood pressure in late adolescence with subsequent mortality: Cohort study of Swedish male conscripts. *BMJ* **2011**, *342*, d643. [CrossRef]
17. Yano, Y.; Stamler, J.; Garside, D.B.; Daviglus, M.L.; Franklin, S.S.; Carnethon, M.R.; Liu, K.; Greenland, P.; Lloyd-Jones, D.M. Isolated systolic hypertension in young and middle-aged adults and 31-year risk for cardiovascular mortality: The Chicago Heart Association Detection Project in Industry study. *J. Am. Coll. Cardiol.* **2015**, *65*, 327–335. [CrossRef]
18. SPRINT Research Group; Wright, J.T., Jr.; Williamson, J.D.; Whelton, P.K.; Snyder, J.K.; Sink, K.M.; Rocco, M.V.; Reboussin, D.M.; Rahman, M.; Oparil, S.; et al. A Randomized Trial of Intensive versus Standard Blood-Pressure Control. *N. Engl. J. Med.* **2015**, *373*, 2103–2116. [CrossRef]
19. Volpe, M.; Citoni, B.; Coluccia, R.; Battistoni, A.; Tocci, G. Hypertension Across the Atlantic: A Sprint or a Marathon? *High Blood Press. Cardiovasc. Prev.* **2017**, *24*, 99–102. [CrossRef]
20. SPRINT Research Group; Lewis, C.E.; Fine, L.J.; Beddhu, S.; Cheung, A.K.; Cushman, W.C.; Cutler, J.A.; Evans, G.W.; Johnson, K.C.; Kitzman, D.W.; et al. Final Report of a Trial of Intensive versus Standard Blood-Pressure Control. *N. Engl. J. Med.* **2021**, *384*, 1921–1930. [CrossRef]
21. Whelton, P.K.; Carey, R.M.; Aronow, W.S.; Casey, D.E., Jr.; Collins, K.J.; Dennison Himmelfarb, C.; DePalma, S.M.; Gidding, S.; Jamerson, K.A.; Jones, D.W.; et al. 2017 ACC/AHA/AAPA/ABC/ACPM/AGS/APhA/ASH/ASPC/NMA/PCNA Guideline for the Prevention, Detection, Evaluation, and Management of High Blood Pressure in Adults: Executive Summary: A Report of the American College of Cardiology/American Heart Association Task Force on Clinical Practice Guidelines. *Hypertension* **2018**, *71*, 1269–1324. [CrossRef] [PubMed]
22. Al-Makki, A.; DiPette, D.; Whelton, P.K.; Murad, M.H.; Mustafa, R.A.; Acharya, S.; Beheiry, H.M.; Champagne, B.; Connell, K.; Cooney, M.T.; et al. Hypertension Pharmacological Treatment in Adults: A World Health Organization Guideline Executive Summary. *Hypertension* **2022**, *79*, 293–301. [CrossRef] [PubMed]
23. Volpe, M.; Gallo, G. To whom recommend intensive treatment for hypertension? *Eur. Heart J. Suppl.* **2020**, *22*, E167–E172. [CrossRef]
24. Demirci, B.; McKeown, P.P.; Bayraktutan, U. Blockade of angiotensin II provides additional benefits in hypertension- and ageing-related cardiac and vascular dysfunctions beyond its blood pressure-lowering effects. *J. Hypertens.* **2005**, *23*, 2219–2227. [CrossRef] [PubMed]
25. Zicha, J.; Dobesova, Z.; Kunes, J. Late blood pressure reduction in SHR subjected to transient captopril treatment in youth: Possible mechanisms. *Physiol. Res.* **2008**, *57*, 495–498. [CrossRef] [PubMed]
26. Hawlitschek, C.; Brendel, J.; Gabriel, P.; Schierle, K.; Salameh, A.; Zimmer, H.G.; Rassler, B. Antihypertensive and cardioprotective effects of different monotherapies and combination therapies in young spontaneously hypertensive rats: A pilot study. *Saudi J. Biol. Sci.* **2022**, *29*, 339–345. [CrossRef]

27. Hawlitschek, C.; Brendel, J.; Gabriel, P.; Schierle, K.; Salameh, A.; Zimmer, H.G.; Rassler, B. How Effective Is a Late-Onset Antihypertensive Treatment? Studies with Captopril as Monotherapy and in Combination with Nifedipine in Old Spontaneously Hypertensive Rats. *Biomedicines* **2022**, *10*, 1964. [CrossRef]
28. Paulis, L.; Líšková, S.; Pintérová, M.; Dobešová, Z.; Kuneš, J.; Zicha, J. Nifedipine-sensitive noradrenergic vasoconstriction is enhanced in spontaneously hypertensive rats: The influence of chronic captopril treatment. *Acta Physiol.* **2007**, *191*, 255–266. [CrossRef]
29. Hale, T.M.; Robertson, S.J.; Burns, K.D.; deBlois, D. Short-term ACE inhibition confers long-term protection against target organ damage. *Hypertens. Res.* **2012**, *35*, 604–610. [CrossRef]
30. Harrap, S.B.; Van der Merwe, W.M.; Griffin, S.A.; Macpherson, F.; Lever, A.F. Brief angiotensin converting enzyme inhibitor treatment in young spontaneously hypertensive rats reduces blood pressure long-term. *Hypertension* **1990**, *16*, 603–614. [CrossRef]
31. Kost, C.K.; Li, P.; Jackson, E.K. Blood Pressure After Captopril Withdrawal From Spontaneously Hypertensive Rats. *Hypertension* **1995**, *25*, 82–87. [CrossRef]
32. Zicha, J.; Dobešová, Z.; Behuliak, M.; Pintérová, M.; Kuneš, J.; Vaněčková, I. Nifedipine-sensitive blood pressure component in hypertensive models characterized by high activity of either sympathetic nervous system or renin-angiotensin system. *Physiol. Res.* **2014**, *63*, 13–26. [CrossRef]
33. Zicha, J.; Kunes, J. Ontogenetic aspects of hypertension development: Analysis in the rat. *Physiol. Rev.* **1999**, *79*, 1227–1282. [CrossRef]
34. Lemay, J.; Tea, B.S.; Hamet, P.; deBlois, D. Regression of neointimal lesions in the carotid artery of nifedipine-treated SHR and WKY rats: Possible role of apoptosis. *J. Vasc. Res.* **2001**, *38*, 462–470. [CrossRef]
35. Ziegelhöffer-Mihalovicova, B.; Arnold, N.; Marx, G.; Tannapfel, A.; Zimmer, H.G.; Rassler, B. Effects of salt loading and various therapies on cardiac hypertrophy and fibrosis in young spontaneously hypertensive rats. *Life Sci.* **2006**, *79*, 838–846. [CrossRef]
36. van Empel, V.P.; De Windt, L.J. Myocyte hypertrophy and apoptosis: A balancing act. *Cardiovasc. Res.* **2004**, *63*, 487–499. [CrossRef]
37. Kokubo, M.; Uemura, A.; Matsubara, T.; Murohara, T. Noninvasive evaluation of the time course of change in cardiac function in spontaneously hypertensive rats by echocardiography. *Hypertens. Res.* **2005**, *28*, 601–609. [CrossRef]
38. Asai, T.; Kushiro, T.; Fujita, H.; Kanmatsuse, K. Different effects on inhibition of cardiac hypertrophy in spontaneously hypertensive rats by monotherapy and combination therapy of adrenergic receptor antagonists and/or the angiotensin II type 1 receptor blocker under comparable blood pressure reduction. *Hypertens. Res.* **2005**, *28*, 79–87. [CrossRef]
39. Rocha, W.A.; Lunz, W.; Baldo, M.P.; Pimentel, E.B.; Dantas, E.M.; Rodrigues, S.L.; Mill, J.G. Kinetics of cardiac and vascular remodeling by spontaneously hypertensive rats after discontinuation of long-term captopril treatment. *Braz.J. Med. Biol. Res.* **2010**, *43*, 390–396. [CrossRef]
40. Hojná, S.; Kadlecová, M.; Dobešová, Z.; Valoušková, V.; Zicha, J.; Kuneš, J. The participation of brain NO synthase in blood pressure control of adult spontaneously hypertensive rats. *Mol. Cell Biochem.* **2007**, *297*, 21–29. [CrossRef]
41. Perrucci, G.L.; Barbagallo, V.A.; Corlianò, M.; Tosi, D.; Santoro, R.; Nigro, P.; Poggio, P.; Bulfamante, G.; Lombardi, F.; Pompilio, G. Integrin amb5 in vitro inhibition limits pro-fibrotic response in cardiac fibroblasts of spontaneously hypertensive rats. *J. Transl. Med.* **2018**, *16*, 352. [CrossRef] [PubMed]
42. Huang, A.; Li, H.; Zeng, C.; Chen, W.; Wei, L.; Liu, Y.; Qi, X. Endogenous CCN5 Participates in Angiotensin II/TGF-β1 Networking of Cardiac Fibrosis in High Angiotensin II-Induced Hypertensive Heart Failure. *Front. Pharmacol.* **2020**, *11*, 1235. [CrossRef] [PubMed]
43. Schlüter, K.D.; Wenzel, S. Angiotensin II: A hormone involved in and contributing to pro-hypertrophic cardiac networks and target of anti-hypertrophic cross-talks. *Pharmacol. Ther.* **2008**, *119*, 311–325. [CrossRef] [PubMed]
44. Pfeffer, J.M.; Pfeffer, M.A.; Mirsky, I.; Braunwald, E. Regression of left ventricular hypertrophy and prevention of left ventricular dysfunction by captopril in the spontaneously hypertensive rat. *Proc. Natl. Acad. Sci. USA* **1982**, *79*, 3310–3314. [CrossRef] [PubMed]
45. Ago, T.; Yang, Y.; Zhai, P.; Sadoshima, J. Nifedipine inhibits cardiac hypertrophy and left ventricular dysfunction in response to pressure overload. *J. Cardiovasc. Transl. Res.* **2010**, *3*, 304–313. [CrossRef]
46. Besse, S.; Robert, V.; Assayag, P.; Delcayre, C.; Swynghedauw, B. Nonsynchronous changes in myocardial collagen mRNA and protein during aging: Effect of DOCA-salt hypertension. *Am. J. Physiol.* **1994**, *267*, H2237–H2244. [CrossRef]
47. Slama, M.; Ahn, J.; Varagic, J.; Susic, D.; Frohlich, E.D. Long-term left ventricular echocardiographic follow-up of SHR and WKY rats: Effects of hypertension and age. *Am. J. Physiol. Heart Circ. Physiol.* **2004**, *286*, H181–H185. [CrossRef]
48. Zimmer, J.; Hawlitschek, C.; Rabald, S.; Hagendorff, A.; Zimmer, H.G.; Rassler, B. Effects of late-onset and long-term captopril and nifedipine treatment in aged spontaneously hypertensive rats: Echocardiographic studies. *Hypertens. Res.* **2015**, *38*, 716–722. [CrossRef]
49. Linz, W.; Jessen, T.; Becker, R.H.; Schölkens, B.A.; Wiemer, G. Long-term ACE inhibition doubles lifespan of hypertensive rats. *Circulation* **1997**, *96*, 3164–3172. [CrossRef]
50. Briest, W.; Hölzl, A.; Rassler, B.; Deten, A.; Leicht, M.; Baba, H.A.; Zimmer, H.-G. Cardiac remodeling after long term norepinephrine treatment in rats. *Cardiovasc. Res.* **2001**, *52*, 265–273. [CrossRef]
51. Briest, W.; Hölzl, A.; Rassler, B.; Deten, A.; Baba, H.A.; Zimmer, H.-G. Significance of matrix metalloproteinases in norepinephrine-induced remodelling of rat hearts. *Cardiovasc. Res.* **2003**, *57*, 379–387. [CrossRef]

biomedicines MDPI

Article

Cardiac Cx43 Signaling Is Enhanced and TGF-β1/SMAD2/3 Suppressed in Response to Cold Acclimation and Modulated by Thyroid Status in Hairless SHR^{M}

Katarina Andelova [1], Barbara Szeiffova Bacova [1], Matus Sykora [1], Stanislav Pavelka [2,3], Hana Rauchova [3] and Narcis Tribulova [1,*]

[1] Centre of Experimental Medicine, v.v.i., Slovak Academy of Sciences, 84104 Bratislava, Slovakia Republic; katarina.andelova@savba.sk (K.A.); barbara.bacova@savba.sk (B.S.B.); matus.sykora@savba.sk (M.S.)
[2] Institute of Molecular Genetics, v.v.i., Academy of Sciences of the Czech Republic, 14220 Prague, Czech Republic; stanislav.pavelka@img.cas.cz
[3] Institute of Physiology, v.v.i., Academy of Sciences of the Czech Republic, 14220 Prague, Czech Republic; hana.rauchova@fgu.cas.cz
* Correspondence: narcisa.tribulova@savba.sk

Abstract: The hearts of spontaneously hypertensive rats (SHR) are prone to malignant arrhythmias, mainly due to disorders of electrical coupling protein Cx43 and the extracellular matrix. Cold acclimation may induce cardio-protection, but the underlying mechanisms remain to be elucidated. We aimed to explore whether the adaptation of 9-month-old hairless SHR^{M} to cold impacts the fundamental cardiac pro-arrhythmia factors, as well as the response to the thyroid status. There were no significant differences in the registered biometric, redox and blood lipids parameters between hairless (SHR^{M}) and wild type SHR. Prominent findings revealed that myocardial Cx43 and its variant phosphorylated at serine 368 were increased, while an abnormal cardiomyocyte Cx43 distribution was attenuated in hairless SHR^{M} vs. wild type SHR males and females. Moreover, the level of β-catenin, ensuring mechanoelectrical coupling, was increased as well, while extracellular matrix collagen-1 and hydroxyproline were lower and the TGF-β1 and SMAD2/3 pathway was suppressed in hairless SHR^{M} males compared to the wild type strain. Of interest, the extracellular matrix remodeling was less pronounced in females of both hypertensive strains. There were no apparent differences in response to the hypothyroid or hyperthyroid status between SHR strains concerning the examined markers. Our findings imply that hairless SHR^{M} benefit from cold acclimation due to the attenuation of the hypertension-induced adverse downregulation of Cx43 and upregulation of extracellular matrix proteins.

Keywords: hairless SHR^{M}; cold acclimation; cardiac Cx43; extracellular matrix; thyroid hormones

Citation: Andelova, K.; Szeiffova Bacova, B.; Sykora, M.; Pavelka, S.; Rauchova, H.; Tribulova, N. Cardiac Cx43 Signaling Is Enhanced and TGF-β1/SMAD2/3 Suppressed in Response to Cold Acclimation and Modulated by Thyroid Status in Hairless SHR^{M}. *Biomedicines* **2022**, *10*, 1707. https://doi.org/10.3390/biomedicines10071707

Academic Editor: Julie Chan

Received: 6 June 2022
Accepted: 7 July 2022
Published: 14 July 2022

1. Introduction

The regulation of body temperature is fundamental to the maintenance of homeostasis, and adaptive thermogenesis, tightly regulated by the central nervous system, is essential for survival [1,2]. Catecholamines and thyroid hormones (TH) are the most important factors impacting thermogenesis. TH activate thermogenesis in brown adipose tissue, and white adipose tissue can undergo browning via adrenergic stimulation [3]. The adaptation to cold results in the deiodination of thyroxine (T_4) and causes an increase of the triiodothyronine (T_3) levels in blood. Higher T_3 leads to an increase in expression of uncoupling protein 2, following an increase in heat production [4,5].

Cardiovascular system responses to thermoregulatory challenges and cold acclimation may induce cardio-protection [6,7]. However, there is still a gap in the knowledge on the underlying mechanisms by which a temperature challenge or adaptation to cold induces cardiac responses. Moreover, molecular mechanisms linked with cardio-protection to maintain heart function in response to cold adaptation are poorly elucidated as well.

The heart is an electromechanical pump, and gap junction channels formed by connexin-43 (Cx43) ensure coupling between cardiac myocytes to enable the transmission of electrical and molecular signals, resulting in coordinated contractions [8,9], while the downregulation of Cx43 and/or channel dysfunction, as well as abnormal Cx43 topology, jeopardize synchronous cardiac electromechanical functions and render the heart prone to developing arrhythmias, including potentially lethal ventricular fibrillation (VF) [10,11]. Moreover, reduced inter-myocyte coupling and communication activate pro-fibrotic signaling, followed by excessive extracellular matrix (ECM) deposition [12], that profoundly contribute to arrhythmogenesis and heart mechanical failure [13].

Of interest, an increased myocardial expression of Cx43 was demonstrated in hibernators during cold acclimatization and hibernation [14]. The upregulation of Cx43 appears to be a mechanism by which the hibernators avoid lethal arrhythmia during hibernation and arousal [15], despite marked changes in body temperature, which, in non-hibernating subjects and humans, induce VF. The questions arise as to whether the upregulation of cardiac Cx43 in response to cold adaptation is a general phenomenon occurring in non-hibernating animals and whether there are sex-related differences.

Genetically, hairless strains of rodents represent rewarding models to study adaptive thermogenesis. It has been reported that hairless coisogenic spontaneously hypertensive rats (SHR^M), harboring the mutant desmoglein-4 gene, exhibit increased thermogenesis due to a metabolic adaptation to cold [16]. The standard ambient temperature of 22 °C lies well below thermoneutrality for such rats due to the diminished insulating capacity. However, the response of the heart to cold acclimation was not investigated in this strain. It would be of a great interest, since hypertension increases the arrhythmia risk, and SHR has been shown as prone to VF due to the downregulation of myocardial Cx43, its abnormal topology and ECM alterations (fibrosis) [13]. Moreover, SHR differ in Cx43 and ECM responses to altered the thyroid status when compared to normotensive rats [17].

Taken together, we aimed to investigate whether there are specific arrhythmia-related differences between the hearts of hairless vs. wild type SHR and in the response to the induced hyperthyroid or hypothyroid status. In this context, our main objective was to explore the expression of myocardial Cx43 and its topology, as well as the key markers of fibrogenesis in hairless SHR^M males and females. Additionally, the expression levels of the selected proteins that may modulate Cx43 channel's function were examined in the heart of both SHR strains.

2. Materials and Methods

2.1. Animals and Experimental Design

The experiments were performed on 9-month-old male and female hairless coisogenic spontaneously hypertensive rats (SHR) harboring the mutant desmoglein 4 (Dsg4) gene (SHR^M), as well as age and sex-matched wild type SHR. Besides, male and female Wistar Kyoto rats (WKY) were used as a reference normotensive strain. Animals were obtained from the Institute of Physiology, v.v.i., Academy of Sciences of the Czech Republic, Prague. The Dsg4 mutation was originally found in the SHR.BN-chr.1 congenic strain and was transferred on the SHR genetic background by backcross breeding [16]. The animals were housed at 22 °C with 12-h light/dark cycles and ad libitum access to tap water and standard laboratory chow. Experiments were performed in agreement with the Animal Protection Laws of the Czech Republic. The maintenance and handling of the animals were performed in accordance with the "Guide for the Care and Use of Laboratory Animals" published by the U.S. National Institutes of Health (NIH Publication, 8th ed., revised 2011) and approved by the Ethics Committee of the Institute of Physiology v.v.i, Academy of Sciences of the Czech Republic, Prague.

Male and female SHR were randomly divided into 6 experimental groups: male SHR (n = 6), female SHR (n = 6), male hyperthyroid SHR TH (n = 5), female hyperthyroid SHR

TH (n = 5), male hypothyroid SHR HY (n = 5) and female hypothyroid SHR HY (n = 5). Male and female SHR^M were randomly divided into 6 experimental groups: male SHR^M (n = 6), female SHR^M (n = 6), male hyperthyroid SHR^M TH (n = 5), female hyperthyroid SHR^M TH (n = 5), male hypothyroid SHR^M HY (n = 5) and female hypothyroid SHR^M HY (n = 5). The hyperthyroid status was established by the intraperitoneal injection of 3,3′,5-triiodo-L-thyronine (T_3; Sigma Aldrich, St. Louis, MO, USA) at 0.15 mg/kg b.w. three times weekly, and the hypothyroid status was induced by a 0.05% solution of methimazole (Sigma Aldrich, St. Louis, MO, USA) in the drinking water. Wistar Kyoto (male WKY, n = 5; female WKY, n = 5) nontreated normotensive rats were used as the control reference strain. At the end of the experiment, the animals were euthanized with 100 mg/kg b.w. of ketamine (Narketan; Vetoquinol UK Ltd., Towcester, UK), followed by 10 mg/kg b.w. of myorelaxant xylazine (Xylapan; Vetoquinol UK Ltd., Towcester, UK), and the body weight was registered. Then, the chest was opened, and the heart was quickly excised into ice-cold saline, followed by weight registration and heart tissue sampling. Blood samples were taken from the thoracic aorta. All samples were stored in a freezer at −80 °C.

2.2. *Biometric, Blood Samples and Cardiac Left Ventricular Tissue Parameters Monitoring*

Body weight, heart and left ventricular weight of rats were registered at the end of the experiment. The systemic blood pressure (BP) was monitored via the cannulated carotid artery, as previously described [18], in WKY, SHR and SHR^M males and females. Postprandial levels of triglycerides (TG), total cholesterol (TC) and high-density lipoproteins (HDL) were assessed using available kits (Pliva-Lachema Diagnostika, Brno, Czech Republic). Low-density lipoprotein-cholesterol (LDL) was estimated indirectly using the concentration relations: LDL = TC − (TG/5) − HDL. Concentrations of the total L-thyroxine (tT_4) and 3,3′,5-triiodo-L-thyronine (tT_3) in blood sera of the rats were determined by radioimmunoassay (RIA) using commercial RIA kits with $[^{125}I]$-T_4 and $[^{125}I]$-T_3 tracers (Immunotech/Beckman Coulter Co., Prague, Czech Republic), as described previously [19]. The left ventricular thiobarbituric acid reactive substances (TBARS) were determined as described previously [17], and the content of reduced glutathione (GSH) was assessed in the left ventricular tissue, as described previously [20].

2.3. *Myocardial Histology and Enzyme Histochemistry*

Conventional hematoxylin–eosin and Van Gieson staining of heart left ventricular tissue sections were used for a light microscopic examination of the myocardial structure (Zeiss Apotome 2 microscope: Carl Zeiss, Jena, Germany). Catalytic enzyme histochemistry was performed according to [21] using 10-μm-thick heart cryostat sections to demonstrate the activities of capillary endothelium-related alkaline phosphatase (AP, E.C.3.1.3.1) with naphthol AS-MX phosphate as a substrate and dipeptidyl peptidase-4 (DPP4, E.C.3.4.15.4) with glycyl-L-proline-4-methoxy-beta naphthylamide as a substrate to detect functional arteriolar and venular capillary network. For quantification of intensity of AP and DPP4 histochemical reactions (corresponding to the enzyme activity), randomly selected areas (15 per heart) of positive signal were analyzed and defined as the number of pixels with a code lower than 128 on the "0–255 RGB color scale". The total number of positive pixels was expressed as a total integral optical density per area (IOD) (Image-Pro Plus).

2.4. *Immunofluorescence Labelling of Myocardial Cx43, Cadherin, β-Catenin and Quantitative Image Analysis*

Cryostat sections from the left ventricle were used for the in situ immunodetection of Cx43, cadherin and β-catenin. The tissue sections were fixed in ice-cold methanol, permeabilized in 0.3% Triton X-100, washed in phosphate-buffered saline (PBS) and blocked with 1% bovine serum. The sections were incubated overnight with primary anti-Cx43 antibody (1:500, MAB3068, CHEMICON International, Inc., Temecula, CA, USA), anti-cadherin (1:300, sc-7939, Santa Cruz, Dallas, TX, USA) and with anti-β-catenin (1:250, sc-7963, Santa Cruz, Dallas, TX, USA) at 4 °C, washed in PBS and subsequently incubated for one and a half hours with secondary antibodies conjugated with anti-mouse FITC-fluorescein isothiocyanate (1:500, Jackson Immuno Research Labs, West Grove, PA, USA) or with anti-rabbit Alexa Fluor 594 (1:500, Jackson Immuno Research Laboratory Labs, West Grove, PA, USA).

The immunostaining of Cx43, cadherin and β-catenin was examined using a Zeiss Apotome 2 microscope (Carl Zeiss, Jena, Germany), and digital images were used for the quantitative analysis (Image-Pro Plus). Ten randomly selected areas were examined per heart. The quantification of Cx43 located on lateral sides of the cardiomyocytes (as a marker of the pathophysiological topology of Cx43) was performed as previously described [22]. After manual delineation of terminal intercalated disc-related Cx43 immunolabeling, the difference between the total IOD and IOD of terminal Cx43 corresponded to lateral Cx43. It was expressed as a percentage calculated from the ratio of lateral topology divided by the total IOD.

2.5. Determination of Collagen Content by Hydroxyproline Measurement

The hydroxyproline content in the left ventricle tissue was estimated by the spectrophotometric method, as described previously [23]. Samples were hydrolyzed in 6 M HCl for 3 h at 130 °C. Dried samples were treated at room temperature with chloramine T in the acetate–citrate buffer (pH 6.0), and the reaction was stopped after 20 min by adding Ehrlich's reagent solution. It was followed by incubation at 65 °C for 15 min. The concentration of hydroxyproline was measured spectrophotometrically at 550 nm and expressed in mg per total weight of the left ventricle [22].

2.6. Determination of Myocardial Protein Levels by Western Blotting

Frozen left ventricular tissue was powdered in liquid nitrogen and homogenized in SB20 lysis buffer (20% SDS, 10 mmol/L EDTA and 100 mmol/L Tris, pH 6.8). To analyze the active form of collagen-1, the samples were prepared in Laemmli buffer without 2-mercaptoethanol and loaded onto gels without denaturation. For other proteins, the tissue lysate was diluted in the Laemmli sample buffer and boiled for 5 min, and an equal amount of protein was loaded in each well, followed by separation on SDS-PAGE 10% bis-acrylamide gels at a constant voltage of 120 V (Mini-Protean TetraCell, Bio-Rad, Hercules, CA, USA), as previously described [17]. Subsequently, the proteins were transferred to a nitrocellulose membrane (0.2-μm pore size, Advantec, Tokyo, Japan) and blocked for 4 h with 5% fat-free milk in Tris-buffered saline containing 0.1% Tween 20 (TBST). The membrane was incubated overnight with the primary antibodies listed in Table 1. The membrane was subsequently washed in TBST and incubated for 1 h with a horseradish peroxidase-linked secondary anti-rabbit antibody (1:2000, 7074S, Cell Signaling Technology, Denver, CO, USA) and anti-mouse antibody (1:2000, 7076C, Cell Signaling Technology, Denver, CO, USA). The enhanced luminol-based chemiluminescent was used for visualization of the proteins and the quantification of the relevant bands was assessed densitometrically using Carestream Molecular Imaging Software (version 5.0, Carestream Health, New Haven, CT, USA) and normalized to GAPDH.

2.7. Statistical Evaluation

Differences between groups were evaluated using one-way analysis of variance (ANOVA) and Bonferroni's multiple comparison test. The Kolmogorov–Smirnov normality test was used to examine whether variables are normally distributed. Data were expressed as means ± standard deviations (SD); $p < 0.05$ was considered to be statistically significant.

Table 1. List of primary antibodies used for the Western blot detection of proteins.

Protein	Dilution	Product Number	Manufacturer of Antibody
Cx43	1:5000	C6219	Sigma-Aldrich, Missouri, USA
$pCx43^{368}$	1:1000	sc-101660	Santa Cruz Biotechnology, Texas, USA
$pCx43^{279}$	1:1000	sc-12900	Santa Cruz Biotechnology, Texas, USA
β-catenin	1:1000	sc-7963	Santa Cruz Biotechnology, Texas, USA
PKCε	1:1000	sc-214	Santa Cruz Biotechnology, Texas, USA
TGF-β1	1:1000	SAB4502954	Sigma-Aldrich, Missouri, USA
SMAD2/3	1:1000	#3102	Cell Signaling Technology, Colorado, USA
Collagen-1	1:1000	ab90395	Abcam, United Kingdom
PKA	1:1000	sc-365615	Santa Cruz Biotechnology, Texas, USA
PKG	1:1000	sc-25429	Santa Cruz Biotechnology, Texas, USA
AktK	1:1000	sc-81436	Santa Cruz Biotechnology, Texas, USA
MAPK42/44	1:1000	137F5	Cell Signaling Technology, Colorado, USA
β1AR	1:1000	ab3442	Abcam, Cambridge, United Kingdom
β2AR	1:1000	bs-0947R	Bioss, Woburn, Massachusetts, USA
β3AR	1:1000	bs-1063R	Bioss, Massachusetts, USA
GAPDH	1:1000	sc-25778	Santa Cruz Biotechnology, Texas, USA

3. Results

3.1. Biometric, Blood Samples and Cardiac Left Ventricular Tissue Parameters of Experimental Rats

Comparing to the systolic blood pressure in normotensive WKY rats (114.2 ± 17 mmHg in males and 99.5 ± 11 mmHg in females), this parameter was significantly elevated in either sex of the wild type SHR (190.0 ± 7 mmHg in males and 179.6 ± 12 mmHg in females) and in hairless SHR^M (182.9 ± 6 mmHg in males and 169.7 ± 9 mmHg in females). As shown in Table 2, wild type SHR males exhibited higher body weights when compared to SHR^M, whereas there was no difference in body weights between these two strains in the females. The heart and left ventricular weights did not noticeably differ between SHR and SHR^M, regardless of the sex, but both cardiac parameters were significantly higher when compared to normotensive WKY. Hyperthyroidism resulted in a decrease in body weight in SHR males and, to a lesser extent, in SHR^M males but not in females. There were no significant differences in the myocardial TBARS levels between wild type SHR and hairless SHR^M or in comparing to WKY rats, but there was a tendency to increase due to hyperthyroidism in both SHR strains. The myocardial levels of GSH did not differ significantly among the experimental groups, regardless of the strain and sex.

As shown in Table 2, the serum total T_3 and T_4 concentrations were similar in wild type SHR and hairless SHR^M males, but in SHR females, the T_3 concentrations were a bit higher in comparison with SHR^M females. While comparing to WKY rats, the serum T_3 concentrations were significantly higher in both SHR strains, regardless of the sex. The serum triglycerides, total cholesterol, HDL and LDL cholesterol did not significantly differ in SHR vs. SHR^M, but these parameters were lower in both SHR strains compared to WKY rats. There were no significant differences in the atherogenic index, expressed as the TC/HDL ratio, among the groups.

Table 2. General characteristics of the experimental rats.

MALES—Variables	WKY	SHR	SHR TH	SHR HY	SHR^M	SHR^M TH	SHR^M HY
BW (g)	410 ± 1.41	398.33 ± 21.08	286 ± 1.41 [b]	383.5 ± 7.78	359 ± 42.76	307.33 ± 31.37	365.25 ± 38.46
HW (g)	1.05 ± 0.01	1.53 ± 0.08 [a]	1.43 ± 0.05	0.87 ± 0.21 [b]	1.59 ± 0.13 [a]	2.01 ± 0.14 [e]	1.14 ± 0.25 [e]
LVW (g)	0.75 ± 0.01	1.19 ± 0.08 [a]	1.06 ± 0.05	0.71 ± 0.08 [b]	1.21 ± 0.13 [a]	1.46 ± 0.11	0.89 ± 0.18 [e]
TBARS (nmol/mg)	1.75 ± 0.29	1.66 ± 0.45	2.06 ± 0.11	1.99 ± 0.21	2.00 ± 0.10	2.37 ± 0.45	1.75 ± 0.22
GSH (μmol/g)	1.99 ± 1.13	1.97 ± 0.27	2.06 ± 0.33	2.43 ± 0.08	2.14 ± 0.15	2.27 ± 0.24	2.39 ± 0.11
T3 (nmol/L)	0.84 ± 0.04	1.12 ± 0.02 [a]	2.84 ± 0.11 [b]	0.48 ± 0.06 [b]	1.08 ± 0.03 [a]	4.05 ± 0.16 [e]	0.59 ± 0.01 [e]
T4 (nmol/L)	55.0 ± 0.8	44.0 ± 6.9 [a]	62.3 ± 1.7 [b]	12.4 ± 1.9 [b]	46.2 ± 2.2 [a]	69.0 ± 1.0 [e]	16.1 ± 0.7 [e]
TG (mmol/L)	1.60 ± 0.20	0.66 ± 0.04 [a]	0.96 ± 0.02 [b]	0.43 ± 0.03	0.70 ± 0.13 [a]	0.80 ± 0.06	0.51 ± 0.08
TC (mmol/L)	1.88 ± 0.10	1.12 ± 0.02 [a]	1.70 ± 0.24 [b]	1.98 ± 0.07 [b]	1.18 ± 0.25 [a]	1.40 ± 0.06	2.42 ± 0.34 [e]
HDL (mmol/L)	1.33 ± 0.03	0.81 ± 0.04 [a]	1.38 ± 0.23 [b]	1.40 ± 0.05 [b]	0.83 ± 0.14 [a]	1.12 ± 0.06	1.77 ± 0.33 [e]
LDL (mmol/L)	0.24 ± 0.09	0.17 ± 0.03	0.15 ± 0.04	0.53 ± 0.10	0.20 ± 0.10	0.13 ± 0.04	0.58 ± 0.21 [e]
TC/HDL (ratio)	1.43 ± 0.10	1.39 ± 0.04	1.24 ± 0.03	1.40 ± 0.05	1.41 ± 0.09	1.26 ± 0.04	1.39 ± 0.20
FEMALES—Variables	**WKY**	**SHR**	**SHR TH**	**SHR HY**	**SHR^M**	**SHR^M TH**	**SHR^M HY**
BW (g)	241.41 ± 2.83	201.50 ± 1.26	197.21 ± 29.70	200.67 ± 1.15	210.60 ± 13.67 [a]	212.50 ± 9.19	185.50 ± 33.74
HW (g)	0.752 ± 0.07	0.827 ± 0.06 [a]	1.116 ± 0.25 [b]	0.679 ± 0.05	0.917 ± 0.10 [a]	1.101 ± 0.25	0.802 ± 0.14
LVW (g)	0.544 ± 0.04	0.639 ± 0.02 [a]	0.884 ± 0.22 [b]	0.512 ± 0.01 [b]	0.624 ± 0.23 [a]	0.737 ± 0.12	0.540 ± 0.16
TBARS (nmol/mg)	2.25 ± 0.27	2.33 ± 0.25	2.90 ± 0.07	2.85 ± 0.41	2.56 ± 0.43	3.51 ± 0.35 [e]	3.85 ± 0.27
GSH (μmol/g)	1.87 ± 0.07	1.98 ± 0.01	1.88 ± 0.18	2.21 ± 0.46	1.98 ± 0.15	1.97 ± 0.19	2.40 ± 0.31
T3 (nmol/L)	0.92 ± 0.02	1.29 ± 0.02 [a]	2.83 ± 0.11	0.38 ± 0.06 [b]	1.08 ± 0.04 [b]	2.75 ± 0.05 [e]	0.51 ± 0.02 [e]
T4 (nmol/L)	47.6 ± 2.1	42.9 ± 1.8 [a]	58.3 ± 1.0[b]	14.8 ± 1.5 [b]	43.8 ± 0.7 [a]	62.7 ± 0.9 [e]	18.2 ± 0.3 [e]
TG (mmol/L)	1.07 ± 0.26	1.28 ± 0.02	1.05 ± 0.34	0.56 ± 0.05 [b]	1.03 ± 0.19	1.16 ± 0.14	0.39 ± 0.06 [e]
TC (mmol/L)	2.28 ± 0.32	1.61 ± 0.04 [a]	1.86 ± 0.21	1.40 ± 0.10	1.37 ± 0.06 [a]	1.59 ± 0.07	2.70 ± 0.07 [e]
HDL (mmol/L)	1.85 ± 0.31	1.19 ± 0.02 [a]	1.47 ± 0.36	1.01 ± 0.16	1.07 ± 0.08 [a]	1.31 ± 0.05	2.07 ± 0.17 [e]
LDL (mmol/L)	0.21 ± 0.07	0.17 ± 0.02	0.14 ± 0.04	0.28 ± 0.7	0.11 ± 0.10	0.11 ± 0.06	0.63 ± 0.04 [e]
TC/HDL (ratio)	1.26 ± 0.05	1.36 ± 0.02	1.28 ± 0.14	1.41 ± 0.13	1.29 ± 0.14	1.24 ± 0.09	1.36 ± 0.05

BW—body weight, HW—heart weight, LVW—left ventricular weight, TBARS—thiobarbituric acid reactive substances in the left ventricle, GSH—reduced glutathione in the left ventricle, T_3—total triiodothyronine, T_4—total thyroxine, TG—plasma triglycerides, TC—total cholesterol, HDL—high-density lipoprotein-cholesterol, LDL—low-density lipoprotein-cholesterol, WKY—Wistar Kyoto normotensive control rats, SHR—wild type spontaneously hypertensive rats, TH—hyperthyroid rats, HY—hypothyroid rats, SHR^M—hairless spontaneously hypertensive rats, n = 5 per group. Data are presented as means ± SD, ^{a} $p < 0.05$ vs. WKY, ^{b} $p < 0.05$ vs. SHR and ^{e} $p < 0.05$ vs. SHR^M. One-way ANOVA and Bonferroni's multiple comparison tests were used as the statistical method.

3.2. Myocardial Histology and Capillary Enzyme Histochemistry

Hematoxylin–eosin staining (Figure 1) revealed that, compared to WKY rats, the left ventricular tissue of both wild type SHR and SHR^M of either sex exhibited focal areas infiltrated with polymorphonuclears. This histopathological feature persisted regardless of the hyper- or hypothyroid status. Besides, Van Gieson staining for collagen deposition (Figure 2) revealed interstitial and perivascular fibrosis in both SHR strains males and females but to lesser extent in SHR^M. Of note, the representative microscopic images indicate the most pronounced area of fibrosis detected in each experimental group of rats. The hypothyroid status enhanced the myocardial fibrosis in both SHR strains.

Histochemical determination of the activity of alkaline phosphatase (AP) that points out the function and myocardial density of the arterial part of capillaries is demonstrated in Figure 3. Alterations of the capillary network AP activity may reflect the myocardial adaptation to hypertension to maintain the function of a structurally remodeled and fibrotic heart. Compared to normotensive WKY rats, the AP activity was significantly increased in both wild type SHR and SHR^M, regardless of the sex, and was enhanced due to hypothyroidism in males and females of both SHR strains, as assessed by a quantitative image analysis.

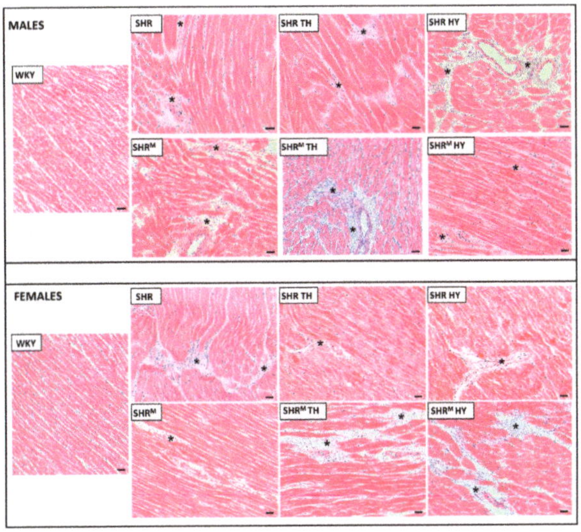

Figure 1. Hematoxylin–eosin staining revealed the infiltration of polymorphonuclears (asterisks) in all SHR groups. WKY—Wistar Kyoto normotensive control rats, SHR—wild type spontaneously hypertensive rats, TH—hyperthyroid rats, HY—hypothyroid rats and SHR^M—hairless spontaneously hypertensive rats, n = 5 per group. Scale bar represents 200 µm.

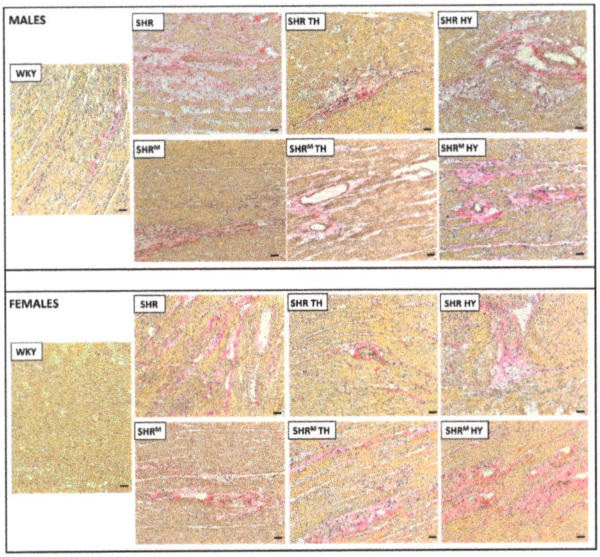

Figure 2. Van Gieson staining shows the collagen deposition (pink) of various degrees in all SHR groups with massive patchy fibrosis in hypothyroid rats. WKY—Wistar Kyoto euthyroid control rats, SHR—wild type spontaneously hypertensive rats, TH—hyperthyroid rats, HY—hypothyroid rats and SHR^M—hairless spontaneously hypertensive rats, n = 5 per group. Scale bar represents 200 µm.

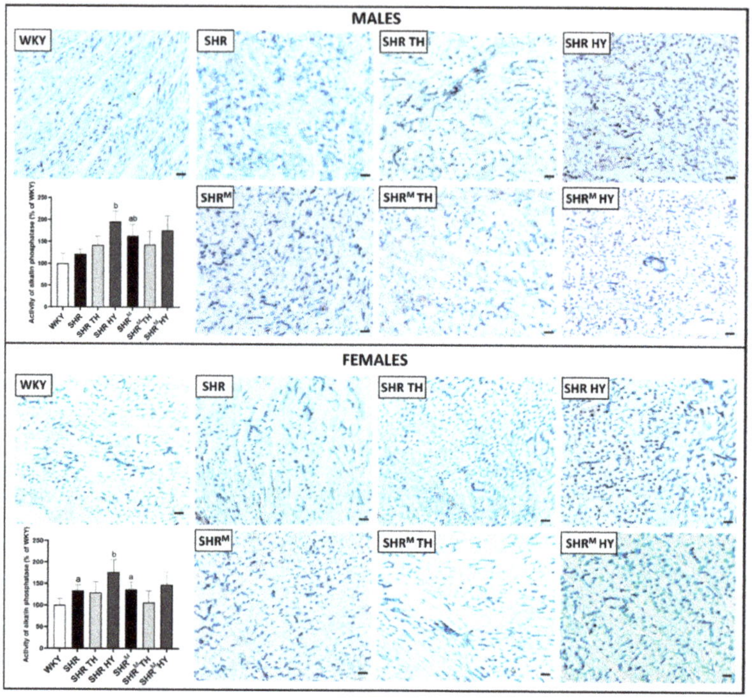

Figure 3. Histochemical demonstration of the alkaline phosphatase (AP) activity (blue) in endothelial cells of the arterial portion of capillaries and quantification of the intensity of the reaction. Note the enhanced AP activity in SHR^{M} males compared to the wild type strain, as well as in response to the hypothyroid status. WKY—Wistar Kyoto normotensive control rats, SHR—wild type spontaneously hypertensive rats, TH—hyperthyroid rats, HY—hypothyroid rats and SHR^{M}—hairless spontaneously hypertensive rats, n = 5 per group. Scale bar represents 200 µm. Data are presented as means ± SD, ^{a} $p < 0.05$ vs. WKY and ^{b} $p < 0.05$ vs. SHR. One-way ANOVA and Bonferroni's multiple comparison test were used as the statistical method.

The activity of dipeptidyl peptidase-4 (DPP4), that points out the function and density of the venous part of capillaries, is demonstrated in Figure 4. Enhanced DPP4 activity is considered detrimental in pathophysiological conditions due to the implication in collagen metabolism and proinflammatory signaling. A quantitative image analysis revealed that the DPP4 activity was significantly lower in SHR^{M} regardless of the sex when compared to normotensive WKY rats or wild type SHR. Hyperthyroidism resulted in a decrease of DPP4 activity, while an increase was observed in hypothyroidism in both wild type SHR and SHR^{M}, regardless of the sex.

3.3. Myocardial Protein Levels of Cx43 and Its Variants

The expression level of the Cx43 protein was significantly reduced in the left ventricle of SHR males and females, as well as in SHR^{M} males (Figure 5), compared to normotensive WKY rats. However, compared to wild type SHR males and females, the protein levels of Cx43 were significantly higher in sex-matched SHR^{M}. The hypothyroid status enhanced the Cx43 levels in wild type SHR, as well as in SHR^{M}, and the increase was more pronounced in females than in males of both SHR strains. The hyperthyroid status resulted in a decrease of Cx43 in SHR^{M} males while not in wild type SHR males, but an increase was demonstrated in SHR females and. to a lesser extent, in SHR^{M} females.

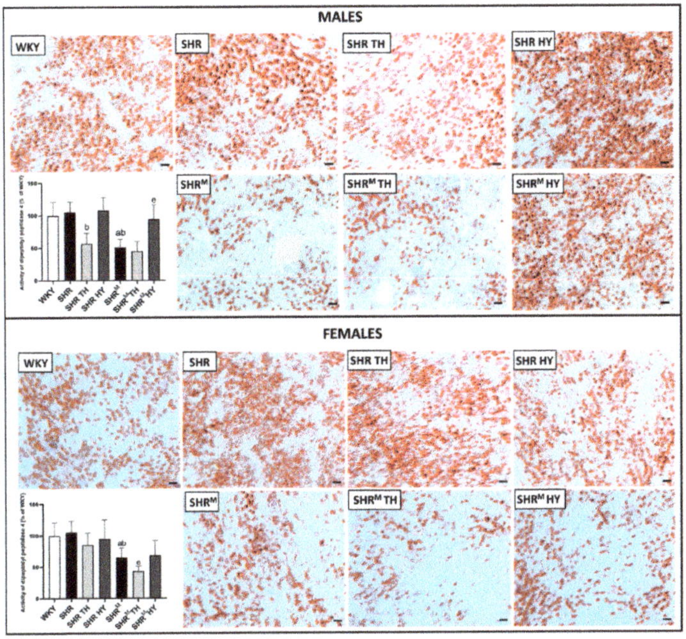

Figure 4. Histochemical demonstration of dipeptidyl peptidase-4 (DPP4) activity (red) in endothelial cells of the venous portion of the capillary network and quantification of the intensity of the reaction. Note the significant decrease of DPP4 activity in SHR^{M} compared to SHR, regardless of the sex, as well as in response to the hyperthyroid status. WKY—Wistar Kyoto normotensive control rats, SHR—wild type spontaneously hypertensive rats, TH—hyperthyroid rats, HY—hypothyroid rats and SHR^{M}—hairless spontaneously hypertensive, n = 5 per group. Scale bar represents 200 µm. Data are presented as means ± SD, ^{a} $p < 0.05$ vs. WKY, ^{b} $p < 0.05$ vs. SHR, ^{e} $p < 0.05$ vs. SHR^{M}. One-way ANOVA and Bonferroni's multiple comparison test were used as the statistical method.

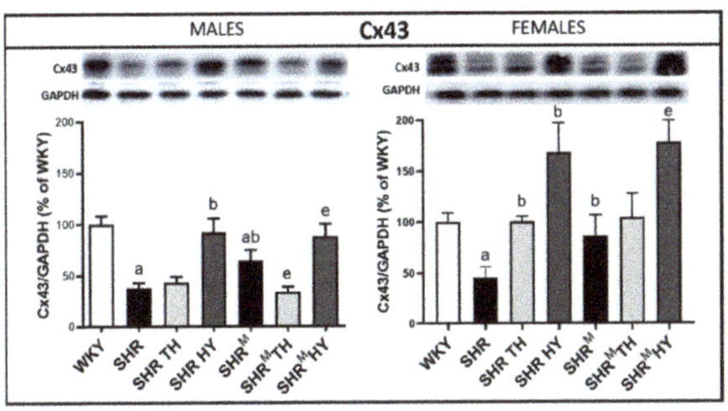

Figure 5. Myocardial protein levels of Cx43 assessed by Western blots. Note the significant decrease of the Cx43 levels in the hypertensive strains compared to normotensive WKY but, to a lesser extent, in hairless SHR^{M} males and females. The hypothyroid status enhanced the Cx43 protein levels in both SHR strains. WKY—Wistar Kyoto normotensive control rats, SHR—wild type spontaneously hypertensive rats, TH—hyperthyroid rats, HY—hypothyroid rats and SHR^{M}—hairless spontaneously hypertensive rats, n = 5 per group. Data are presented as means ± SD, ^{a} $p < 0.05$ vs. WKY, ^{b} $p < 0.05$ vs. SHR, ^{e} $p < 0.05$ vs. SHR^{M}. One-way ANOVA and Bonferroni's multiple comparison test were used as the statistical method.

The levels of expression of the Cx43 variant phosphorylated at serine368 ($pCx43^{368}$), which, is associated with a reduced channel conductivity and the Cx43 variant phosphorylated at serine279 ($pCx43^{279}$) that hampers channel conductivity are shown in Figure 6. Compared to WKY rats, the expression of $pCx43^{368}$ was significantly decreased in wild type SHR males and females but increased in SHR^{M}, apparently in males. The hypothyroid status enhanced the $pCx43^{368}$ expression only in wild type SHR, regardless of the sex, while not in SHR^{M}. On the contrary, the hyperthyroid status reduced the $pCx43^{368}$ protein levels only in SHR^{M} males. There were no significant changes in the $pCx43^{279}$ protein abundance either in wild type SHR or SHR^{M}, regardless of the sex, when compared to normotensive WKY rats. Both the hyperthyroid and hypothyroid status reduced the $pCx43^{279}$ protein levels significantly in SHR^{M} males but not in females.

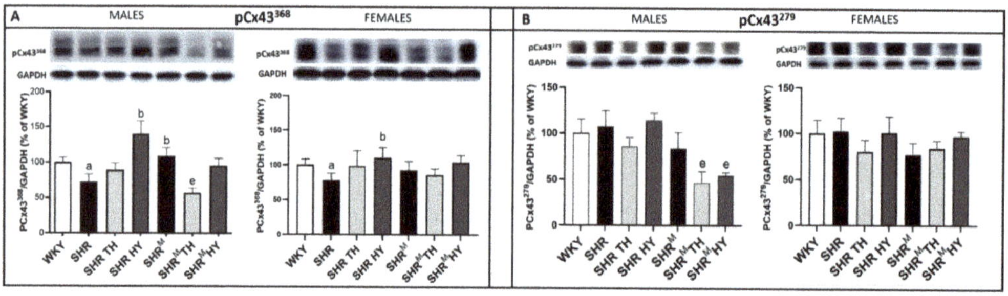

Figure 6. Protein levels of Cx43 variants $pCx43^{368}$ (**A**) and $pCx43^{279}$ (**B**) assessed by the Western blot analysis. Note the reduced levels of $pCx43^{368}$ in wild type SHR males compared to WKY but not in hairless SHR^{M} males. WKY—Wistar Kyoto normotensive control rats, SHR—wild type spontaneously hypertensive rats, TH—hyperthyroid rats, HY—hypothyroid rats and SHR^{M}—hairless spontaneously hypertensive rats, n = 5 in each group. Data are presented as means ± SD, ^{a} $p < 0.05$ vs. WKY, ^{b} $p < 0.05$ vs. SHR and ^{e} $p < 0.05$ vs. SHR^{M}. One-way ANOVA and Bonferroni's multiple comparison test were used as the statistical method.

3.4. Myocardial Topology of Cx43 and Quantification of Its Abnormal Distribution

The cardiomyocyte distribution of Cx43 in experimental rats is demonstrated in Figure 7. The obvious prevalent distribution of Cx43 is detected at the gap junction plaques of the intercalated discs in normotensive WKY rats. Besides this obvious end-to-end pattern, there was enhanced immunofluorescence labeling of Cx43 at the lateral sides of the cardiomyocytes (side-to-side pattern) in both SHR strains, regardless of the sex and thyroid status. A quantitative image analysis of Cx43-positive labeling revealed a significant increase of Cx43 at the lateral sides of the cardiomyocytes in wild type SHR males and females but to a lesser extent in SHR^{M}, regardless of the sex. Moreover, there was a tendency to increase the lateral distribution of Cx43 due to the hyperthyroid status, apparently in females of both SHR strains. In the context of lateral Cx43 distribution, the use of Triton X-100 soluble and Triton X-100 insoluble portions analysis [24] could provide information about the integrity of the gap junction plaques.

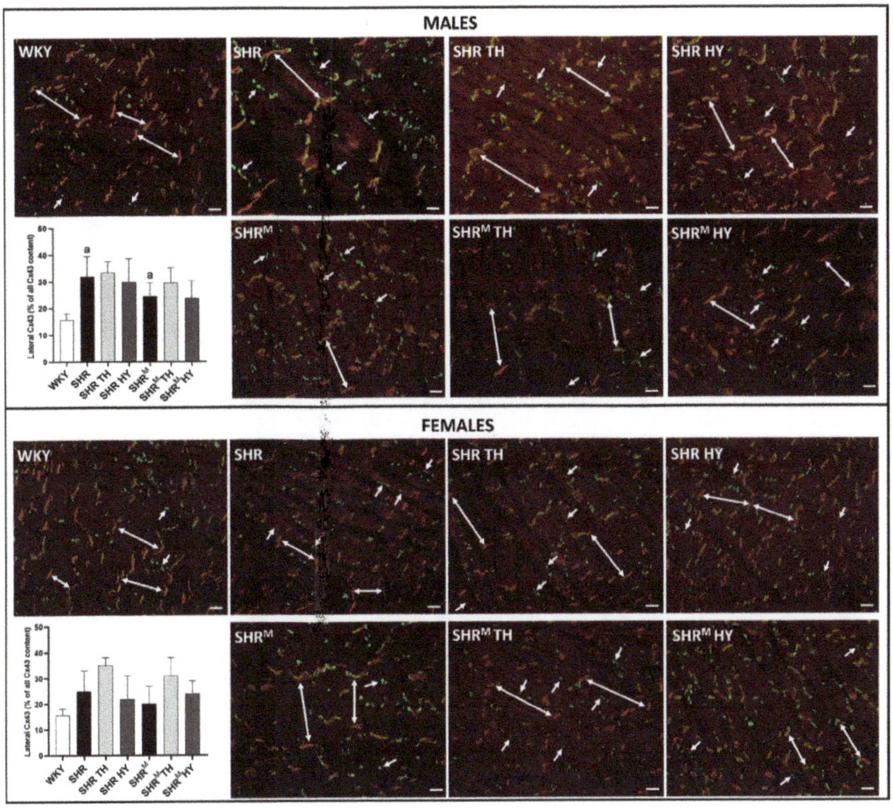

Figure 7. Detection of the myocardial topology of Cx43 using immunofluorescence labeling. Double arrows represent the colocalization of Cx43 (green) with cadherin (red) at the intercalated discs of the cardiomyocytes, and simple arrows show Cx43 localization on the lateral sides of the cardiomyocytes. The graphs represent the quantification of lateral Cx43. Note the abnormal lateral localization is suppressed in male SHR^M compared to SHR. WKY—Wistar Kyoto normotensive control rats, SHR—wild type spontaneously hypertensive rats, TH—hyperthyroid rats, HY—hypothyroid rats and SHR^M—hairless spontaneously hypertensive rats, n = 5 per group. Scale bar represents 400 μm. Data are presented as means ± SD, ^{a} $p < 0.05$ vs. WKY. One-way ANOVA and Bonferroni's multiple comparison test were used as the statistical method.

3.5. Myocardial Protein Levels and Topology of Cx43 Interacting Protein, β-Catenin

β-catenin is an adhesive junction protein interacting with Cx43 and impacting channel assembly and function [25]. Immunolabeling of β-catenin is confined to the intercalated disc junctions (end-to-end pattern), as shown in Figure 8. There was no difference in the topology of β-catenin among the experimental groups of rats. While the Western blotting analysis revealed (Figure 8) that, compared to WKY rats, the protein levels of β-catenin exhibited a tendency to be lower in wild type SHR but higher in SHR^M (significantly in females). The hyperthyroid status increased the β-catenin protein levels in wild type SHR males and females but not in SHR^M. The hypothyroid status exhibited a tendency to decrease in the β-catenin protein levels in males while not in females of both SHR strains. Nevertheless, regarding the Cx43 function, it would be useful to explore active vs. inactive forms of β-catenin.

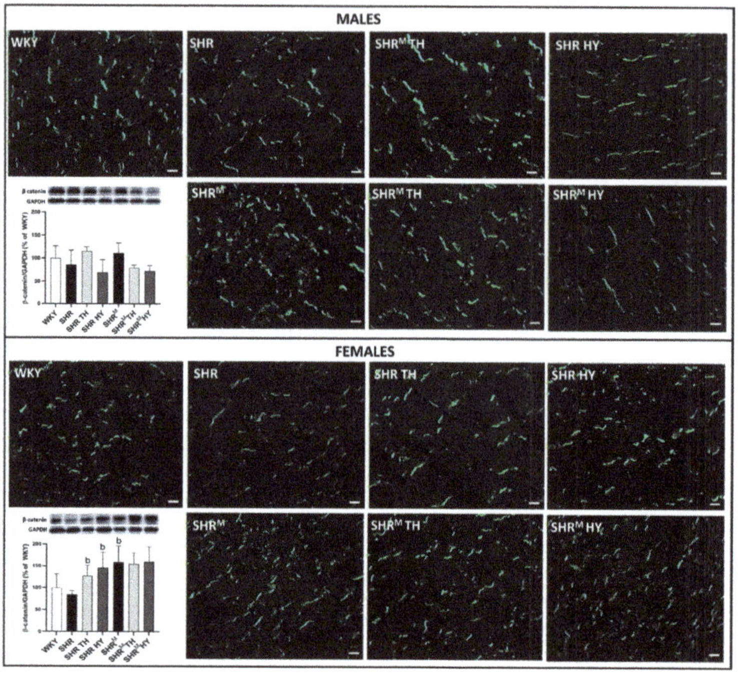

Figure 8. Detection of the myocardial topology in β-catenin (green) using immunofluorescence labeling. Graphs represent quantification of protein levels of β-catenin determined by Western blot analysis. Note the enhanced β-catenin in hairless SHR^M males and, to a lesser extent, in females compared to the wild type strain. WKY—Wistar Kyoto normotensive control rats, SHR—wild type spontaneously hypertensive rats, TH—hyperthyroid rats, HY—hypothyroid rats, SHR^M—hairless spontaneously hypertensive rats, n = 5 per group. Scale bar represents 400 µm. Data are presented as means ± SD, ^{b} $p < 0.05$ vs. SHR. One-way ANOVA and Bonferroni's multiple comparison test were used as the statistical method.

3.6. Myocardial Expression of Cx43 Interacting Protein Kinases

Cx43 interacting protein kinases by the phosphorylation of the Cx43 impact channel's function and inter-myocyte communication [26,27]. Accordingly, PKCε, PKG and MAPK42/44 attenuate, while PKA and Akt kinase facilitate the Cx43 channel conductivity. The expression levels of the assessed protein kinases are demonstrated in Figure 9. Compared to wild type SHR, the protein levels of PKCε were increased in SHR^M males, as well as due to the hypothyroid status in both SHR^M and wild type SHR, regardless of the sex, while the expression of PKG and MAPK42/44 was not significantly altered in SHR^M males vs. wild type SHR or in response to the altered thyroid status. However, the MAPK42/44 expression was increased in SHR^M females compared to wild type SHR. The expression of PKA was lower in SHR^M males vs. wild type SHR but enhanced due to the hypothyroid or hyperthyroid status. There were no changes in PKA expression among the experimental rat groups in the females. Moreover, there was no significant difference in the expression of Akt kinase between the SHR^M and wild type SHR groups, while an increase of Akt kinase protein due to the hypothyroid status was detected in males but not in females of both SHR strains. Demonstrated changes of the protein kinase expression may have an impact on Cx43 phosphorylation and function. Therefore, their interaction with Cx43 (coimmunoprecipitation), along with its phosphorylated status, is challenging to explore.

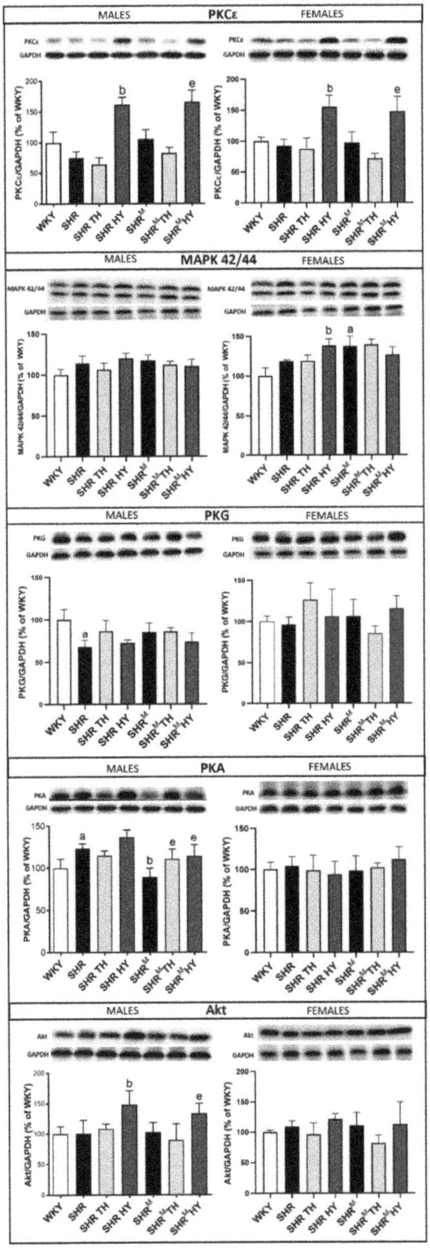

Figure 9. Myocardial levels of the Cx43 interacting protein kinases assessed by Western blotting. Note an increase of PKCε in hairless SHR^M males compared to the wild type strain, as well as in response to a hypothyroid status, regardless of the strain and sex. WKY—Wistar Kyoto normotensive control rats, SHR—wild type spontaneously hypertensive rats, TH—hyperthyroid rats, HY—hypothyroid rats and SHR^M—hairless spontaneously hypertensive rats, n = 5 per group. Data are presented as means ± SD, ^{a} $p < 0.05$ vs. WKY, ^{b} $p < 0.05$ vs. SHR and ^{e} $p < 0.05$ vs. SHR^M. One-way ANOVA and Bonferroni's multiple comparison test were used as the statistical method.

3.7. Myocardial Level of Profibrotic Markers TGF-β1, SMAD2/3, Collagen-1 and Hydroxyproline

As shown in Figure 10, the protein expression of TGF-β1 was significantly higher in wild type SHR males but not in SHR^{M}, compared to normotensive WKY rats. While TGF-β1 expression was not altered in females of both SHR strains when compared to WKY rats. A significant elevation of TGF-β1 was observed only in SHR^{M} females in response to the hypothyroid status.

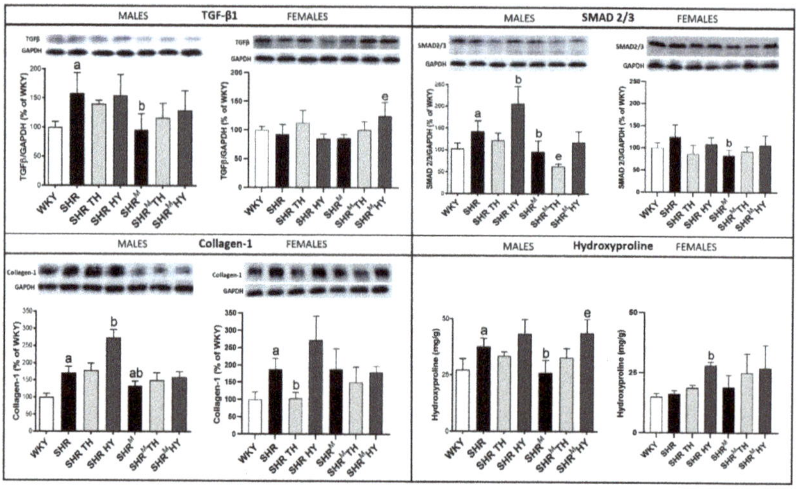

Figure 10. Myocardial protein levels of profibrotic markers TGF-β1, SMAD2/3 and Collagen-1 determined by Western blotting, as well as the levels of hydroxyproline. Note the reduced levels of these proteins in hairless SHR^{M} compared to wild type SHR, apparently in males. WKY—Wistar Kyoto normotensive control rats, SHR—wild type spontaneously hypertensive rats, TH—hyperthyroid rats, HY—hypothyroid rats and SHR^{M}—hairless spontaneously hypertensive rats, n = 5 per group. Data are presented as means ± SD, ^{a} $p < 0.05$ vs. WKY, ^{b} $p < 0.05$ vs. SHR and ^{e} $p < 0.05$ vs. SHR^{M}. One-way ANOVA and Bonferroni's multiple comparison tests were used as the statistical method.

In parallel, there was an increase of SMAD2/3 expression in wild type SHR (significantly in males) but not in SHR^{M}, regardless of the sex, when compared to WKY rats (Figure 10). The hypothyroid status increased the SMAD2/3 expression in wild type males (SHR), while the hyperthyroid status decreased it in SHR^{M}.

Moreover, there was a significant increase of collagen-1 (Figure 10) in wild type SHR males, as well as females, but not so much in SHR^{M} when compared to WKY rats. The hypothyroid status highly increased the content of collagen-1 in wild type SHR males and females but much less in SHR^{M}.

The hydroxyproline levels (Figure 10) were significantly increased in wild type SHR males but not in SHR^{M} when compared to WKY rats. However, the hypothyroid status highly increased the hydroxyproline content in both SHR strains, regardless of the sex.

3.8. Myocardial Protein Levels of β1, β2 and β3-Adrenergic Receptors (AR)

The adrenergic stimulation is important for cold acclimation-elicited non-shivering thermogenesis and the cardiac-specific overexpression of β3-AR hampers hypertrophy and myocardial fibrosis [28,29]. According to our analyses (Figure 11), the expression of cardiac-dominant β1-AR was significantly decreased in wild type SHR, as well as in SHR^{M} males and females, when compared to normotensive WKY rats. There was a tendency to increase the expression of β1-AR in response to the hypothyroid status in wild type SHR, as well as in SHR^{M} males. Like β1-AR, the expression of β2-AR was decreased in males regardless of the SHR strain when compared to WKY rats, while the hyperthyroid status resulted in an increase of β2-AR expression in wild type SHR, as well as in SHR^{M} males. Compared to WKY rats, there were no differences in β3-AR expression among the males, unlike hairless SHR^{M} females that exhibited an increased expression of β3-AR.

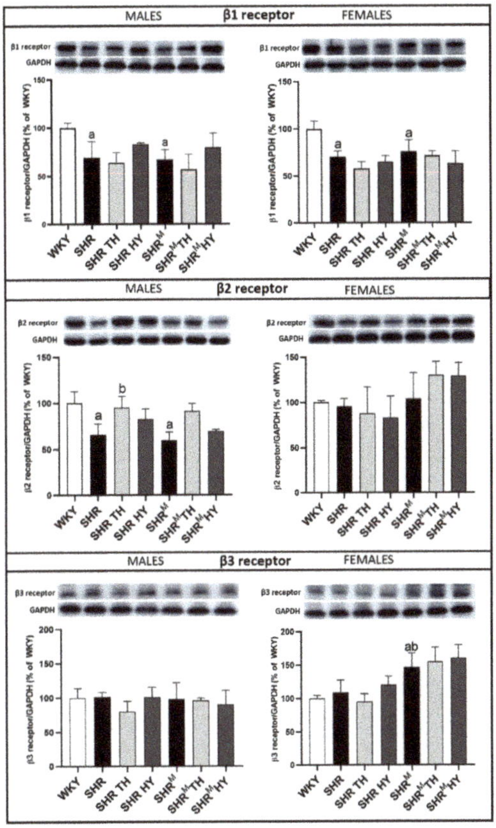

Figure 11. Myocardial protein levels of the β-adrenergic receptors (AR) determined by Western blotting. Note the reduced protein levels of β1-AR and β2-AR in both the SHR and SHR^{M} strains, compared to normotensive WKY rats, as well as an increase of the β3-AR protein in hairless SHR^{M} females compared to wild type SHR. WKY—Wistar Kyoto normotensive control rats, SHR—wild type spontaneously hypertensive rats, TH—hyperthyroid rats, HY—hypothyroid rats and SHR^{M}—hairless spontaneously hypertensive rats, n = 5 per group. Data are presented as means ± SD, ^{a} $p < 0.05$ vs. WKY, ^{b} $p < 0.05$ vs. SHR. One-way ANOVA and Bonferroni's multiple comparison test were used as the statistical method.

4. Discussion

In the current study, we characterized hairless SHR (SHR^{M}) males and females adapted to cold and compared them with sex- and age-matched wild type SHR. While WKY rats were used as a reference normotensive strain to hypertensive rats, there were no significant differences in the biometric, registered blood samples and cardiac tissue parameters between the hairless SHR^{M} and wild type SHR. However, compared to WKY rats, SHR exhibited significantly lower triglycerides and cholesterol but higher circulating T_3, as well as higher heart and left ventricular weight, likewise previously reported [17].

Of interest, the enzyme histochemistry revealed an increase of alkaline phosphatase (AP) activity in the arterial portion of the heart capillary network in males and females of either SHR strain but more pronounced in hairless SHR^{M} males when compared to WKY rats. Vascular AP is involved in purinergic metabolism and the production of adenosine, which is a potent vasodilator and anti-aggregator. However, enhanced purinergic signaling may be implicated in the adverse modulation of myocardial Cx43 channel functions in a pathological setting [9]. In contrast to AP, the activity of dipeptidyl peptidase-4 (DPP4), confined to the venous part of heart capillaries, was significantly reduced in hairless SHR^{M}, regardless of the sex, but not in wild type SHR when compared to WKY rat hearts. Vascular DPP4 is a transmembrane serine protease that degrades vasoactive peptides and hormones and is also involved in collagen metabolism [30] and proinflammatory processes [31], while the inflammation deteriorates the Cx43 channel function [32]. Lower DPP4 activity attenuated capillary rarefaction and myocardial inflammation and contributed to a favorable metabolism in an ischemic heart [33]. Taken together, it appears that hairless SHR^{M} may benefit comparing to wild type SHR.

Besides the apparent hypertrophy, the left ventricular tissue of males and females of both SHR strains was focally infiltrated with polymorphonuclears and exhibited perivascular and interstitial fibrosis. Such structural remodeling underlies the arrhythmogenic substrate that increased the propensity of the heart toward VF [13,22,34]. In this context, it should be emphasized that, compared to wild type SHR, the assessed extracellular matrix proteins collagen-1 and hydroxyproline were significantly lower in hairless SHR^{M} males. In parallel, the markers of profibrotic pathways TGF-β1 and SMAD2/3 were lower as well. Unlike SHR males, the females of both SHR strains did not exhibit significant alterations of hydroxyproline, TGF-β1 and SMAD2/3 when compared to WKY rats. Our findings imply that ECM remodeling is less pronounced in hairless SHR^{M} males, as well as in females of both SHR strains, when compared to sex-matched WKY rat hearts. Altogether, it appears that the hearts of hairless SHR^{M} males and females might be less prone to developing malignant arrhythmias.

In addition to the arrhythmogenic substrate, one of the key factors impacting cardiac arrhythmias occurrence is disorders of Cx43, as previously reviewed [8,9,11]. Structural remodeling is linked with the abnormal and proarrhythmic topology of Cx43, resulting from the redistribution of Cx43 from the intercalated disc to the lateral sides of the hypertrophied cardiomyocytes, as shown previously [17], and in the current study as well. Indeed, compared to the WKY rat heart, "the lateralization" was significantly increased in SHR males but to lesser extent in hairless SHR^{M}, as well as in females of either SHR strain. Apart from this, the levels of the total Cx43 protein and its variant phosphorylated at serine368 (that reduces Cx43 channel conduction and permeability [35]) were significantly decreased in SHR vs. WKY rat hearts (as was also previously demonstrated [17]) but not in hairless SHR^{M} males and females. Moreover, the expression of the Cx43 variant phosphorylated at serine279, that also hampers Cx43 channel communication [26], exhibited a tendency to decrease in hairless SHR^{M} males and females when compared to WKY rat hearts, while no changes were observed in wild type SHR. Finally, a myocardial abundance of β-catenin that impacted the Cx43 channel function [25] was increased in hairless SHR^{M} males and females in contrast to reduced levels in wild type SHR when compared to WKY rat hearts. Both β-catenin and plakoglobin have been indispensable for maintaining mechanoelectrical coupling in the heart to prevent cardiac arrhythmias [36]. Altogether, the myocardial Cx43-related findings indicate that hairless SHR^{M} might be less prone to developing malignant arrhythmias than wild type SHR, namely in males. It is important, because we have previously shown [37] that the threshold to induce ventricular fibrillation was significantly lower in wild type SHR males comparing to normotensive Wistar rats, while females, regardless the strain, are less prone to malignant arrhythmias, in part due to a higher level of Cx43 expression most likely induced by estrogen [38].

Considering the fact that the phosphorylation of Cx43 is required for Cx43 channel function, we were interested to find whether the myocardial expression of most relevant protein kinases interacting with Cx43 [26,27] differ between SHR rat strains. Indeed, the expression of PKCε was increased in hairless SHR^{M} males and MAPK42/44 in SHR^{M} females vs. sex-matched wild type SHR, while the expression of PKA was lower in hairless

SHR^M males vs. wild type SHR. These differences are challenging for further research, since the overexpression of protein kinases may increase Cx43 phosphorylation and vice versa, thereby impacting the channel function [39].

Adrenergic stimulation is important for the cold acclimation [28,29], and adrenoceptors are directly involved in Cx43 mediated inter-myocyte communication [40]. An increase of β2-AR and β3-AR in normotensive rats exposed to cold acclimation at 8 °C for 5 weeks was reported [7]. Of interest, we revealed a distinct expression of β-adrenoceptors in the SHR vs. WKY strains. There was a decrease of cardiac dominant β1-AR protein in males, as well as females, and a decrease of β2-AR protein in males of both SHR strains when compared to WKY rats. It suggests the possible downregulation of these receptors and, hence, the attenuation of their undesirable impact on Cx43 in both SHR strains. There was no difference in β3-AR expression among WKY, wild type SHR and hairless SHR^M males, unlike hairless SHR^M females, that exhibited an increased expression of β3-AR comparing to WKY or wild type SHR. Increased β3-AR may be beneficial due its anti-fibrotic and anti-hypertrophic signaling [29].

Taken into consideration another hormonal factor impacting thermogenesis [3], such as thyroid hormones, we were interested in registering potential differences in the responses of hairless SHR^M vs. wild type SHR to alter the thyroid status. The latter is known to affect the propensity of the heart to arrhythmias, in part due to modulation of the myocardial expression of Cx43 [41].

We demonstrated that the expression of Cx43 and its variant phosphorylated at serine368, as well as PKCε, were increased in hairless SHR^M; similar to wild type SHR males and females in response to their hypothyroid status. It might contribute to a lower susceptibility of the hypothyroid SHR heart to malignant arrhythmias, as it was reported in hypothyroid normotensive rats [42]. On the other hand, an undesirable response to the hypothyroid status was an increase of the myocardial matrix proteins, collagen and hydroxyproline in males of both SHR strains and, to a lesser extent, in SHR^M females. Of interest, the hyperthyroid status did not affect the markers of the extracellular matrix in either SHR strain but resulted in the suppression of Cx43 and its serine368 variant in hairless SHR^M males, as well as in wild type SHR males. It may predict a higher susceptibility to arrhythmias due to hyperthyroidism, as reported in normotensive rats and humans [41,42].

5. Conclusions

The results of this study indicate differences in the levels of the cardiac pro-arrhythmia factors between the hearts of hairless (in particular, in males) vs. wild type SHR while not in the response to an induced hyperthyroid or hypothyroid status. The main findings, i.e., upregulation of myocardial Cx43, along with the suppression of fibrotic factors, suggests that hairless SHR^M are more likely to benefit from adaption to the cold most due to suppression of a malignant arrhythmia risk. Increased thermogenesis induced by the cold might be a possible cause for observed cardiac differences in the SHR^M and wild type SHR. The findings are a challenge to explore the mechanisms by which adaptation to cold triggers a cardiac response.

Limitations

We did not assess either the function of the hairless SHR^M heart or its susceptibility to malignant arrhythmias. This should be explored in the next study to provide the evidence that the adaptation of hypertensive rats to the cold may attenuate their higher susceptibility to lethal cardiac arrhythmias. It might be challenging for clinical trials and the innovative management of hypertensive patients.

Author Contributions: Conceptualization, N.T. and S.P.; methodology, K.A., B.S.B., M.S., H.R. and S.P.; validation, N.T., K.A., B.S.B. and S.P.; investigation, K.A., B.S.B., M.S. and H.R.; resources, N.T.; data curation, K.A., B.S.B., H.R., S.P. and N.T.; writing—original draft preparation, K.A. and N.T.; writing—review and editing, N.T., K.A., B.S.B., M.S., H.R. and S.P.; visualization, N.T., K.A. and S.P.;

supervision, N.T. and S.P.; project administration, S.P. and N.T. and funding acquisition, N.T. All authors have read and agreed to the published version of the manuscript.

Funding: This study was supported by VEGA grants 2/0002/20 and 2/0158/19; Slovak Research and Development Agency under the Contract no. APVV-21-0410, APVV-18-0548 and APVV-19-0317 and EU Structural Fund ITMS 26230120009 from the Ministry of Education.

Institutional Review Board Statement: The maintenance and handling of animals were performed in accordance with the "Guide for the Care and Use of Laboratory Animals" published by US National Institutes of Health (NIH publication No 85-23, revised 1996) and approved by the Ethical Committee of the Institute of Physiology v.v.i., (approval code—49/17/45.0, approved date—16 February 2017) Academy of Sciences of the Czech Republic, Prague.

Informed Consent Statement: Not applicable.

Data Availability Statement: Not applicable.

Acknowledgments: The authors thank Ing. Michal Pravenec and his team for providing the hairless SHR, as well as for the excellent technical assistance to Karla Bohunova from the Institute of Physiology, v.v.i., AS CR, Prague.

Conflicts of Interest: There are no conflict of interest.

References

1. Morrison, S.F. Central neural pathways for thermoregulation. *Front. Biosci.* **2011**, *16*, 74. [CrossRef] [PubMed]
2. Horie, T.; Nakao, T.; Miyasaka, Y.; Nishino, T.; Matsumura, S.; Nakazeki, F.; Ide, Y.; Kimura, M.; Tsuji, S.; Rodriguez, R.R.; et al. microRNA-33 maintains adaptive thermogenesis via enhanced sympathetic nerve activity. *Nat. Commun.* **2021**, *12*, 843. [CrossRef] [PubMed]
3. Yau, W.W.; Yen, P.M. Thermogenesis in adipose tissue activated by thyroid hormone. *Int. J. Mol. Sci.* **2020**, *21*, 3020. [CrossRef]
4. Tsibulnikov, S.; Maslov, L.; Voronkov, N.; Oeltgen, P. Thyroid hormones and the mechanisms of adaptation to cold. *Hormones* **2020**, *19*, 329–339. [CrossRef] [PubMed]
5. Sentis, S.C.; Oelkrug, R.; Mittag, J. Thyroid hormones in the regulation of brown adipose tissue thermogenesis. *Endocr. Connect.* **2021**, *10*, R106–R115. [CrossRef]
6. Kralova Lesna, I.; Rychlikova, J.; Vavrova, L.; Vybiral, S. Could human cold adaptation decrease the risk of cardiovascular disease? *J. Therm. Biol.* **2015**, *52*, 192–198. [CrossRef]
7. Tibenska, V.; Benesova, A.; Vebr, P.; Liptakova, A.; Hejnová, L.; Elsnicová, B.; Drahota, Z.; Hornikova, D.; Galatík, F.; Kolar, D.; et al. Gradual cold acclimation induces cardioprotection without affecting β-adrenergic receptor-mediated adenylyl cyclase signaling. *J. Appl. Physiol.* **2020**, *128*, 1023–1032. [CrossRef]
8. Rodríguez-Sinovas, A.; Sánchez, J.A.; Valls-Lacalle, L.; Consegal, M.; Ferreira-González, I. Connexins in the heart: Regulation, function and involvement in cardiac disease. *Int. J. Mol. Sci.* **2021**, *22*, 4413. [CrossRef]
9. Andelova, K.; Benova, T.E.; Bacova, B.S.; Sykora, M.; Prado, N.J.; Diez, E.R.; Hlivak, P.; Tribulova, N. Cardiac connexin-43 hemichannels and pannexin1 channels: Provocative antiarrhythmic targets. *Int. J. Mol. Sci.* **2021**, *22*, 260. [CrossRef]
10. Tribulova, N.; Szeiffova Bacova, B.; Benova, T.; Viczenczova, C. Can we protect from malignant arrhythmias by modulation of cardiac cell-to-cell coupling? *J. Electrocardiol.* **2015**, *48*, 434–440. [CrossRef]
11. Dhein, S.; Salameh, A. Remodeling of cardiac gap junctional cell–cell coupling. *Cells* **2021**, *10*, 2422. [CrossRef] [PubMed]
12. Jansen, J.A.; Van Veen, T.A.B.; De Jong, S.; Van Der Nagel, R.; Van Stuijvenberg, L.; Driessen, H.; Labzowski, R.; Oefner, C.M.; Bosch, A.A.; Nguyen, T.Q.; et al. Reduced Cx43 expression triggers increased fibrosis due to enhanced fibroblast activity. *Circ. Arrhythmia Electrophysiol.* **2012**, *5*, 380–390. [CrossRef] [PubMed]
13. Egan Benova, T.; Szeiffova Bacova, B.; Viczenczova, C.; Diez, E.; Barancik, M.; Tribulova, N. Protection of cardiac cell-to-cell coupling attenuate myocardial remodeling and proarrhythmia induced by hypertension. *Physiol. Res.* **2016**, *65*, S29–S42. [CrossRef] [PubMed]
14. Saitongdee, P.; Milner, P.; Becker, D.L.; Knight, G.E.; Burnstock, G. Increased connexin43 gap junction protein in hamster cardiomyocytes during cold acclimatization and hibernation. *Cardiovasc. Res.* **2000**, *47*, 108–115. [CrossRef]
15. Fedorov, V.V.; Li, L.; Glukhov, A.; Shishkina, I.; Aliev, R.R.; Mikheeva, T.; Nikolski, V.P.; Rosenshtraukh, L.V.; Efimov, I.R. Hibernator Citellus undulatus maintains safe cardiac conduction and is protected against tachyarrhythmias during extreme hypothermia: Possible role of Cx43 and Cx45 up-regulation. *Heart Rhythm* **2005**, *2*, 966–975. [CrossRef]
16. Trnovská, J.; Šilhavý, J.; Zídek, V.; Šimáková, M.; Mlejnek, P.; Landa, V.; Eigner, S.; Eigner Henke, K.; Škop, V.; Oliyarnyk, O.; et al. Gender-related effects on substrate utilization and metabolic adaptation in hairless spontaneously hypertensive rat. *Physiol. Res.* **2015**, *64*, 51–60. [CrossRef]
17. Sykora, M.; Bacova, B.S.; Benova, T.E.; Barancik, M.; Zurmanova, J.; Rauchova, H.; Weismann, P.; Pavelka, S.; Kurahara, L.H.; Slezak, J.; et al. Cardiac cx43 and ECM responses to altered thyroid status are blunted in spontaneously hypertensive versus normotensive rats. *Int. J. Mol. Sci.* **2019**, *20*, 3758. [CrossRef]

18. Behuliak, M.; Vavrəínová, A.; Bencze, M.; Polgárová, K.; Ergang, P.; Kunesə, J.; Vaneəcəková, I.; Zicha, J. Ontogenetic changes in contribution of calcium sensitization and calciumentry to blood pressure maintenance ofWistar-Kyoto and spontaneously hypertensive rats. *J. Hypertens.* **2015**, *33*, 2443–2454. [CrossRef]
19. Pavelka, S. Development of radiometric assays for quantification of enzyme activities of the key enzymes of thyroid hormones metabolism. *Physiol. Res.* **2014**, *63*, S133–S140. [CrossRef]
20. Rauchová, H.; Vokurková, M.; Pavelka, S.; Vaněčková, I.; Tribulová, N.; Soukup, T. Red palm oil supplementation does not increase blood glucose or serum lipids levels in wistar rats with different thyroid status. *Physiol. Res.* **2018**, *67*, 307–315. [CrossRef]
21. Lojda, Z. Studies on dipeptidyl(amino)peptidase IV (glycyl-proline naphthylamidase). *Histochemistry* **1979**, *59*, 153–166. [CrossRef] [PubMed]
22. Bacova, B.S.; Viczenczova, C.; Andelova, K.; Sykora, M.; Chaudagar, K.; Barancik, M.; Adamcova, M.; Knezl, V.; Benova, T.E.; Weismann, P.; et al. Antiarrhythmic effects of melatonin and omega-3 are linked with protection of myocardial cx43 topology and suppression of fibrosis in catecholamine stressed normotensive and hypertensive rats. *Antioxidants* **2020**, *9*, 546. [CrossRef] [PubMed]
23. Kesava Reddy, G.; Enwemeka, C.S. A simplified method for the analysis of hydroxyproline in biological tissues. *Clin. Biochem.* **1996**, *29*, 225–229. [CrossRef]
24. Trease, A.J.; Capuccino, J.M.V.; Contreras, J.; Harris, A.L.; Sorgen, P.L. Intramolecular signaling in a cardiac connexin: Role of cytoplasmic domain dimerization. *J. Mol. Cell. Cardiol.* **2017**, *111*, 69–80. [CrossRef] [PubMed]
25. Zheng, Q.; Chen, P.; Xu, Z.; Li, F.; Yi, X.P. Expression and redistribution of β-catenin in the cardiac myocytes of left ventricle of spontaneously hypertensive rat. *J. Mol. Histol.* **2013**, *44*, 565–573. [CrossRef] [PubMed]
26. Axelsen, L.N.; Calloe, K.; Holstein-Rathlou, N.-H.; Nielsen, M.S. Managing the complexity of communication: Regulation of gap junctions by post-translational modification. *Front. Pharmacol.* **2013**, *4*, 130. [CrossRef]
27. Solan, J.L.; Lampe, P.D. Spatio-temporal regulation of connexin43 phosphorylation and gap junction dynamics. *Biochim. Biophys. Acta-Biomembr.* **2018**, *1860*, 83–90. [CrossRef]
28. Dubois-Deruy, E.; Gelinas, R.; Beauloye, C.; Esfahani, H.; Michel, L.Y.M.; Dessy, C.; Bertrand, L.; Balligand, J.L. Beta 3 adrenoreceptors protect from hypertrophic remodelling through AMP-activated protein kinase and autophagy. *ESC Heart Fail.* **2020**, *7*, 920–932. [CrossRef]
29. Hermida, N.; Michel, L.; Esfahani, H.; Dubois-Deruy, E.; Hammond, J.; Bouzin, C.; Markl, A.; Colin, H.; Van Steenbergen, A.; De Meester, C.; et al. Cardiac myocyte β3-adrenergic receptors prevent myocardial fibrosis by modulating oxidant stress-dependent paracrine signaling. *Eur. Heart J.* **2018**, *39*, 888–897. [CrossRef]
30. Lambeir, A.M.; Durinx, C.; Scharpé, S.; De Meester, I. Dipeptidyl-peptidase IV from bench to bedside: An update on structural properties, functions, and clinical aspects of the enzyme DPP IV. *Crit. Rev. Clin. Lab. Sci.* **2003**, *40*, 209–294. [CrossRef]
31. Wang, S.C.; Wang, X.Y.; Liu, C.T.; Chou, R.H.; Chen, Z.B.; Huang, P.H.; Lin, S.J. The Dipeptidyl Peptidase-4 Inhibitor Linagliptin Ameliorates Endothelial Inflammation and Microvascular Thrombosis in a Sepsis Mouse Model. *Int. J. Mol. Sci.* **2022**, *23*, 3065. [CrossRef] [PubMed]
32. Andelova, K.; Bacova, B.S.; Sykora, M.; Hlivak, P.; Barancik, M.; Tribulova, N. Mechanisms Underlying Antiarrhythmic Properties of Cardioprotective Agents Impacting Inflammation and Oxidative Stress. *Int. J. Mol. Sci.* **2022**, *23*, 1416. [CrossRef] [PubMed]
33. Suda, M.; Shimizu, I.; Yoshida, Y.; Hayashi, Y.; Ikegami, R.; Katsuumi, G.; Wakasugi, T.; Yoshida, Y.; Okuda, S.; Soga, T.; et al. Inhibition of dipeptidyl peptidase-4 ameliorates cardiac ischemia and systolic dysfunction by up-regulating the FGF-2/EGR-1 pathway. *PLoS ONE* **2017**, *12*, e182422. [CrossRef] [PubMed]
34. Handa, B.S.; Li, X.; Baxan, N.; Roney, C.H.; Shchendrygina, A.; Mansfield, C.A.; Jabbour, R.J.; Pitcher, D.S.; Chowdhury, R.A.; Peters, N.S.; et al. Ventricular fibrillation mechanism and global fibrillatory organization are determined by gap junction coupling and fibrosis pattern. *Cardiovasc. Res.* **2021**, *117*, 1078–1090. [CrossRef] [PubMed]
35. Bao, X.; Reuss, L.; Altenberg, G.A. Regulation of Purified and Reconstituted Connexin 43 Hemichannels by Protein Kinase C-mediated Phosphorylation of Serine 368. *J. Biol. Chem.* **2004**, *279*, 20058–20066. [CrossRef]
36. Swope, D.; Cheng, L.; Gao, E.; Li, J.; Radice, G.L. Loss of Cadherin-Binding Proteins β-Catenin and Plakoglobin in the Heart Leads to Gap Junction Remodeling and Arrhythmogenesis. *Mol. Cell. Biol.* **2012**, *32*, 1056–1067. [CrossRef]
37. Egan Benova, T.; Viczenczova, C.; Szeiffova Bacova, B.; Zurmanova, J.; Knezl, V.; Andelova, K.; Tribulova, N. Omacor Protects Normotensive and Hypertensive Rats Exposed to Continuous Light from Increased Risk to Malignant Cardiac Arrhythmias. *Mar. Drugs* **2021**, *19*, 659. [CrossRef]
38. Grohé, C.; Kahlert, S.; Löbbert, K.; Stimpel, M.; Karas, R.H.; Vetter, H.; Neyses, L. Cardiac myocytes and fibroblasts contain functional estrogen receptors. *FEBS Lett.* **1997**, *416*, 107–112. [CrossRef]
39. Zheng, L.; Trease, A.J.; Katsurada, K.; Spagnol, G.; Li, H.; Shi, W.; Duan, B.; Patel, K.P.; Sorgen, P.L. Inhibition of Pyk2 and Src activity improves Cx43 gap junction intercellular communication. *J. Mol. Cell. Cardiol.* **2020**, *149*, 27–40. [CrossRef]
40. Salameh, A.; Dhein, S. Adrenergic control of cardiac gap junction function and expression. *Naunyn. Schmiedebergs. Arch. Pharmacol.* **2011**, *383*, 331–346. [CrossRef]
41. Tribulova, N.; Kurahara, L.H.; Hlivak, P.; Hirano, K.; Bacova, B.S. Pro-arrhythmic signaling of thyroid hormones and its relevance in subclinical hyperthyroidism. *Int. J. Mol. Sci.* **2020**, *21*, 2844. [CrossRef] [PubMed]
42. Bačová, B.S.; Vinczenzová, C.; Žurmanová, J.; Kašparová, D.; Knezl, V.; Beňová, T.E.; Pavelka, S.; Soukup, T.; Tribulová, N. Altered thyroid status affects myocardial expression of connexin-43 and susceptibility of rat heart to malignant arrhythmias that can be partially normalized by red palm oil intake. *Histochem. Cell Biol.* **2017**, *147*, 63–73. [CrossRef] [PubMed]

biomedicines

MDPI

Article

Beneficial Effects of Empagliflozin Are Mediated by Reduced Renal Inflammation and Oxidative Stress in Spontaneously Hypertensive Rats Expressing Human C-Reactive Protein

Hana Malínská [1], Martina Hüttl [1], Irena Marková [1], Denisa Miklánková [1], Silvie Hojná [2], František Papoušek [2], Jan Šilhavý [2], Petr Mlejnek [2], Josef Zicha [2], Jaroslav Hrdlička [2], Michal Pravenec [2] and Ivana Vaněčková [2,*]

[1] Institute for Clinical and Experimental Medicine, 14220 Prague, Czech Republic
[2] Institute of Physiology, Czech Academy of Sciences, 14220 Prague, Czech Republic
* Correspondence: ivana.vaneckova@fgu.cas.cz; Tel.: +420-241062592

Abstract: Gliflozins (inhibitors of sodium-glucose cotransporter 2) show many beneficial actions beyond their antidiabetic effects. The underlying mechanisms of these additional protective effects are still not well understood, especially under non-diabetic conditions. Therefore, we analyzed the effects of empagliflozin in young (3-month-old) and adult (12-month-old) male spontaneously hypertensive rats (SHR) expressing human C-reactive protein (CRP) in the liver. SHR-CRP rats are a non-diabetic model of metabolic syndrome, inflammation, and organ damage. Empagliflozin was given in a daily dose of 10 mg/kg body weight for 8 weeks. Both age groups of SHR-CRP rats treated with empagliflozin had lower body weight, decreased weight of fat depots, reduced ectopic fat accumulation in the liver and kidneys, and decreased levels of plasma insulin and β-hydroxybutyrate. Empagliflozin effectively reduced ectopic renal fat accumulation, and was associated with decreased inflammation. Exclusively in young rats, decreased microalbuminuria after empagliflozin treatment was accompanied by attenuated oxidative stress. In adult animals, empagliflozin also improved left ventricle function. In conclusion, in young animals, the beneficial renoprotective effects of empagliflozin could be ascribed to reduced lipid deposition in the kidney and the attenuation of oxidative stress and inflammation. In contrast, hepatic lipid metabolism was ameliorated in adult rats.

Keywords: SHR-CRP; SGLT-2 inhibitor; gene expression; age; lipid metabolism

Citation: Malínská, H.; Hüttl, M.; Marková, I.; Miklánková, D.; Hojná, S.; Papoušek, F.; Šilhavý, J.; Mlejnek, P.; Zicha, J.; Hrdlička, J.; et al. Beneficial Effects of Empagliflozin Are Mediated by Reduced Renal Inflammation and Oxidative Stress in Spontaneously Hypertensive Rats Expressing Human C-Reactive Protein. *Biomedicines* **2022**, *10*, 2066. https://doi.org/10.3390/biomedicines10092066

Academic Editor: Albrecht Piiper

Received: 1 July 2022
Accepted: 19 August 2022
Published: 24 August 2022

1. Introduction

Diabetes is among the leading causes of mortality worldwide. As a chronic disease, it is associated with both micro- and macrovascular complications including coronary artery disease, stroke, diabetic nephropathy, neuropathy, and retinopathy. Traditional antidiabetic drugs, such as sulphonylureas or insulin, often have undesirable side effects, including hypoglycemia and weight gain; some of them potentially increase cardiovascular risk [1]. In contrast, new antidiabetics, in particular gliflozins, inhibitors of sodium-glucose cotransporter 2 (SGLT-2), show many beneficial actions beyond their antidiabetic effects. Their effects are mediated through the selective SGLT-2 inhibition at the renal proximal tubule, promoting glucose and sodium excretion, thus leading to significant improvement of glucose control together with the lowering of blood pressure and body weight [2]. The mechanisms underlying the reduction of cardiovascular events and renal protective effects are still poorly understood. Chronic low-grade inflammation is being increasingly recognized as a key feature associated with type 2 diabetes mellitus and its complications [3]. Experimental findings suggest that part of the renoprotective effects of SGLT-2 inhibition may be related to anti-inflammatory actions at the kidney level. The underlying mechanisms to explain this anti-inflammatory effect are multiple, and may involve body weight loss, reduction in adipose tissue inflammation, as well as decreased ectopic fat accumulation in the kidney or attenuation of oxidative stress.

Major large-scale clinical trials such as EMPA-REG, CANVAS, DECLAIRE-TIMI 58, and CREDENCE [4–7] demonstrated positive cardiovascular effects (decrease of the composite endpoints consisting of cardiovascular death, non-fatal myocardial infarction, and non-fatal stroke) of empagliflozin, canagliflozin and dapagliflozin in several thousand diabetic patients. Furthermore, in the DAPA-CKD trial, SGLT-2 inhibitors were also shown to offer renal protection in non-diabetic patients with proteinuria [8]. A recent systematic review and meta-analysis of the DAPA-HF and EMPEROR trials showed improvements in the composite renal endpoint regardless of the presence of diabetes or baseline estimated glomerular filtration rate [9]. Renal benefit has been attributed to metabolic and hemodynamic effects, including lowering of blood pressure, attenuation of glomerular hyperfiltration, body weight loss, and decreased plasma volume, as well as reduced renal hypoxia.

Systemic and renal inflammation is involved in the initiation and progression of diabetic and non-diabetic kidney disease. It has been previously demonstrated in diabetic animal models that the anti-inflammatory potential of SGLT-2 inhibitors is associated with decreased glomerular and tubulo-interstitial damage [10–12]. However, the underlying molecular mechanisms are not completely understood, and only limited information is available on SGLT-2 iinhibitor-mediated anti-inflammatory effects under experimental non-diabetic conditions. In fact, several studies demonstrated cardioprotective effects in experimental non-diabetic models [13,14]. The beneficial effects of gliflozins on renal function were observed in spontaneously hypertensive rats with heart failure [15] and also partly in rats after 5/6 nephrectomy [16]. Our studies in non-diabetic models, namely in hypertensive Ren-2 transgenic rats (TGR) [17], a model of angiotensin II-dependent hypertension, as well as in a model of metabolic syndrome and prediabetes—hereditary hypertriglyceridemic rats (HHTG) [18], showed the beneficial effects of empagliflozin treatment without a glucose-lowering effect. In these models, SGLT-2 inhibition led to a reduction in body weight and visceral adiposity, a decrease of insulin and leptin levels, and an improvement of hepatic metabolism, as well as the attenuation of oxidative stress and inflammation. There were also distinct effects depending on the model used, i.e., blood pressure decrease of TGR, and attenuation of oxidative stress and cell senescence in HHTG rats.

In the current study, we tested the metabolic and renoprotective effects of empagliflozin in spontaneously hypertensive rats expressing human C-reactive protein (SHR-CRP rats), a non-diabetic model of metabolic syndrome with severe hypertension, systemic inflammation, metabolic and hemodynamic disturbances, and target organ injury [19]. The expression of transgenic human CRP in these rats is associated with increased hepatic and renal oxidative tissue damage, increased plasma levels of interleukin 6, and a marked increase of microalbuminuria, which is accompanied by renal histopathologic changes such as fibrosis and inflammatory cellular infiltrates in the interstitium of the kidney [19]. To determine the mechanisms of empagliflozin beyond its antidiabetic effect, we focused our attention on its metabolic and renal effects in this non-diabetic model, both in young rats and in adult animals with the established systemic inflammation. In addition, we analyzed several genes involved in the inflammatory processes in the kidneys.

2. Materials and Methods

2.1. Animals

The SHR-CRP/OlaIpcv transgenic rats (referred to as SHR-CRP) were derived by microinjections of zygotes with a construct containing the cDNA for human CRP under control of the apolipoprotein E promoter with the objective of driving expression of the CRP transgene in the liver, where CRP is normally produced [19]. To investigate the effects of empagliflozin on the metabolic parameters and on kidney injury associated with human CRP, we randomized male SHR-CRP transgenic rats at the age of 3 months (young) and

12 months (adult) into two groups, with or without empagliflozin treatment. In each group, we studied 8 males. The rats were fed ad libitum a standard laboratory diet or the same diet containing empagliflozin in a daily dose of 10 mg/kg body weight for 8 weeks. The rats were housed in an air-conditioned animal facility. All experiments were performed in compliance with the Animal Protection Law of the Czech Republic and were approved by the Ethics Committee of the Institute of Physiology, Czech Academy of Sciences, Prague (Protocol Nr. 47/2019), and conformed to the European Convention on Animal Protection and Guidelines on Research Animal Use (Directive 2010/63/EU).

2.2. Metabolic Parameters in Epididymal Adipose Tissue and Myocardium

As a marker of visceral adipose tissue insulin sensitivity, basal and insulin-stimulated lipid syntheses were determined ex vivo in isolated epididymal fat pad by measuring the incorporation of ^{14}C-U glucose into lipids, as described previously [20]. Basal and adrenaline-stimulated lipolysis in the epididymal adipose tissue were measured ex vivo and evaluated as the release of non-esterified fatty acids (NEFA). Glucose and palmitate oxidation in the myocardium were measured ex vivo in heart tissue sections determined as the incorporation of ^{14}C-U glucose and ^{14}C-palmitate into CO_2, respectively.

2.3. Tissue Triglycerides and Cholesterol Measurements

To determine triglycerides and cholesterol in the liver, kidneys and heart, tissues were powdered under liquid N_2 and extracted in a chloroform:methanol mixture, after which, 2% KH_2PO_4 was added and the solution was centrifuged. The organic phase was collected and evaporated under N_2. The resulting pellet was dissolved in isopropyl alcohol, and lipid concentrations were determined by an enzymatic assay (Erba-Lachema, Brno, Czech Republic).

2.4. Biochemical Analyses

Serum levels of glucose, triglycerides, total and HDL-cholesterol were measured by commercially available kits (Erba-Lachema, Brno, Czech Republic). NEFA levels were determined using an acyl-CoA oxidase-based colorimetric kit (Roche Diagnostics GmbH, Mannheim, Germany). Serum insulin, glucagon, leptin, adiponectin, MCP-1, TNFα, and IL-6 concentrations were determined using a rat insulin ELISA kit (Mercodia, Uppsala, Sweden; MyBioSource, San Diego, CA, USA; eBioscience-Beder, Wien, Austria; BioVendor, Brno, Czech Republic). Rat serum CRP and human serum CRP were also analyzed by ELISA kits (Alpha Diagnostics International, San Antonio, CA, USA). Serum and renal β-hydroxybutyrate (BHB) concentrations were determined by a colorimetric assay kit (Sigma-Aldrich, Saint Louis, MO, USA).

2.5. Parameters of Oxidative Stress

Oxidative stress was evaluated according to the activities of antioxidant enzymes, concentrations of reduced and oxidized glutathione, and levels of lipoperoxidation products TBARS and conjugated dienes [20]. The activities of superoxide dismutase (SOD), glutathione peroxidase (GSH-Px), and glutathione reductase (GR) were analyzed using Cayman Chemicals assay kits (Ann Arbor, MI, USA). Catalase (CAT) activity was measured spectrophotometrically. Lipoperoxidation products were determined by assaying the reaction with thiobarbituric acid. The levels of conjugated dienes were analyzed by extraction in the media (heptane:isopropanol = 2:1) and measured spectrophotometrically in heptane's layer. Concentrations of reduced (GSH) and oxidized (GSSG) forms of glutathione were determined using an HPLC diagnostic kit with fluorescence detection (ChromSystems, Gräfelfing, Germany).

2.6. Urine Collection and Microalbuminuria

Rats were placed in metabolic cages for 16 h to obtain urine samples for the analysis of urinary excretion of albumin, creatinine, sodium, and glucose. The level of albumin in urine was analyzed by HPLC method with UV-VIS detection and adjusted for creatinine concentration. Urinary glucose levels were analyzed by glucose oxidase assay (Erba-Lachema, Brno, Czech Republic). Urinary protein was measured using the Folin method with bovine serum albumin as a standard [21].

2.7. Cardiac Function and Blood Pressure Measurements

In a separate group of adult rats, cardiac function was evaluated at the beginning (week 0) and at the end of experiment (week 7). Gross morphology and function of the left ventricle was assessed by non-invasive ultrasound technique (transthoracic echocardiography, GE Vingmed System Seven with a 14 MHz linear matrix probe) in anesthetized rats (2% isoflurane). Within the baseline echocardiographic evaluation, the following diastolic and systolic dimensions of the left ventricle (LV) were measured: anterior wall thickness (AWTd, AWTs), posterior wall thickness (PWTd, PWTs), and left ventricular cavity diameter (LVDd, LVDs), after which the fractional shortening (FS, %) was calculated [22]. At the end of the study, blood pressure and heart rate (HR) were recorded using a pressure transducer and a multichannel recorder (AD Instruments, Bella Vista, Australia) in conscious rats that were cannulated one day prior to the experiment. Briefly, a polyethylene cannula was implanted under 2.5% isoflurane anesthesia (PE 50 for BP measurement in the left carotid artery), and exteriorized in the interscapular region.

2.8. Quantitative PCR (qPCR)

1 µg total RNA was used to synthesize cDNA using SuperScript IV reverse transcriptase (ThermoFisher, Waltham, MA, USA) according to the manufacturer's protocol. The resulting cDNAs were then used as templates in quantitative real-time PCR (qPCR) reactions. Primers for qPCR reactions were designed using PrimerBLAST [23] to span at least one exon-exon junction: amplicon size was set to 70–150 bp. Each qPCR reaction contained equivalent of 8 ng of input RNA, 300 nM of each forward and reverse primer and 1× Power-up SYBRGreen master mix (ThermoFisher) and was amplified in 7900HT (Applied Biosystems, San Francisco, CA, USA). For the qPCR experiment, cycle threshold (Ct) values of selected genes were normalized relative to the expression of the peptidylprolyl isomerase A (*Ppia*, cyclophilin) gene (for kidney cortex), which served as the internal control, with results being determined in triplicates. Relative quantification was performed using the $\Delta\Delta$Ct method.

2.9. Histology

Kidneys (n = 5) from each group were cut along the longitudinal axis and processed for paraffin embedding. Multiple 4 µm thick sections were cut and stained with Hematoxylin-Eosin, PAS and Azan-Mallory trichrome stain for observation under light microscopy. Slides were observed and pictures acquired with a digitalized camera by an experienced pathologist in a blinded manner. Renal damage, as determined using the glomerulosclerosis index and tubulointerstitial injury, was examined, as described previously [24,25], using the Nikon NIS-Elements AR 3.1 morphometric program (Nikon, Tokyo, Japan).

2.10. Statistical Analysis

All data are expressed as means ± S.E.M. Differences between experimental groups were analyzed by two-way ANOVA with adjustments for multiple comparisons by Holm–Sidak testing. Statistical significance was defined as $p < 0.05$.

3. Results

3.1. Effects of Empagliflozin on Body Weight, Weights of Fat Depots, Cardiac Function and Blood Pressure

As can be seen in Table 1, in both age groups of transgenic SHR-CRP rats, empagliflozin administration reduced body weight as well as adiposity, as evidenced by the significantly decreased relative weights of epididymal and perirenal fat depots. Left ventricular function expressed as fractional shortening (FS) tended to be reduced in adult SHR-CRP during the study, suggesting the deterioration of heart function in these ageing rats (Table 2). Moreover, anterior and posterior diastolic left ventricle diameter (AWTd and PWTd) substantially increased in untreated SHR-CRP rats while it remained stable in empagliflozin-treated animals throughout the study. Developing concentric hypertrophy is seen also in relative wall thickness, which tended to be higher in untreated rats as compared with treated animals, which corresponds to a decrease of relative heart weight at the end of the study. Nevertheless, no effect on blood pressure or heart rate was demonstrated in either age group of empagliflozin-treated rats (Table 1).

Table 1. Basal parameters in experimental groups during the study.

	Young		Adult				
	Control	**Empa**	**Control**	**Empa**	P_{AGE}	$P_{TREATMENT}$	$P_{INTER\text{-}ACTION}$
Body weight (g)	347 ± 6	316 ± 7 **	379 ± 6 #	368 ± 5 #	<0.001	<0.01	n.s.
Epididymal fat (g/100 g BW)	0.58 ± 0.02	0.56 ± 0.01	0.52 ± 0.03	0.39 ± 0.02 ***	<0.001	<0.01	n.s.
Perirenal fat (g/100 g BW)	0.43 ± 0.04	0.33 ± 0.01	0.57 ± 0.06	0.41 ± 0.03 *	<0.01	<0.01	n.s.
Liver weight (g/100 g BW)	3.05 ± 0.11	3.30 ± 0.04 *	4.03 ± 0.09 #	3.68 ± 0.04 **#	<0.001	n.s.	<0.001
Heart weight (g/100 g BW)	0.36 ± 0.01	0.34 ± 0.01	0.41 ± 0.01 #	0.39 ± 0.01 *#	<0.001	<0.05	n.s.
Kidney weight (g/100 g BW)	0.66 ± 0.03	0.70 ± 0.01	0.74 ± 0.01	0.75 ± 0.01	<0.01	n.s.	n.s.
Mean arterial pressure (mm Hg)	192 ± 11	197 ± 7	190 ± 4	197 ± 6	n.s.	n.s.	n.s.
Heart rate (bpm)	373 ± 8	380 ± 5	377 ± 8	355 ± 8 #	n.s.	n.s.	n.s.

P_{AGE} denotes the significance of young vs. adult SHR-CRP rats (age effect); $P_{TREATMENT}$ denotes the significance of empagliflozin treatment (treatment effects); $P_{INTERACTION}$ denotes the significance of different empagliflozin effects in various strains (treatment vs. strain comparison). For multiple comparisons (empagliflozin treatment vs. non-treated controls) Fisher's LSD post-hoc test was used; * denotes $p < 0.05$; ** denotes $p < 0.01$; *** denotes $p < 0.001$; n.s. denotes not significant, Data are means ± SEM; n = 7–8 for each group.

Table 2. Echocardiography parameters in adult animals at the end of the study.

	SHR-CRP			
	Baseline		End of Study	
	Control	Empagliflozin	Control	Empagliflozin
AWTd (mm)	1.96 ± 0.02	1.93 ± 0.03	2.14 ± 0.03 @	1.96 ± 0.04 *
LVDd (mm)	7.75 ± 0.14	7.20 ± 0.09	7.72 ± 0.11	7.41 ± 0.27
PWTd (mm)	1.95 ± 0.02	1.95 ± 0.03	2.08 ± 0.02 @	1.92 ± 0.05 *
AWTs (mm)	2.87 ± 0.06	2.86 ± 0.06	2.99 ± 0.06 @	2.80 ± 0.04 *
LVDs (mm)	4.75 ± 0.09	4.79 ± 0.07	4.35 ± 0.05	4.58 ± 0.20
PWTs (mm)	2.87 ± 0.04	2.88 ± 0.06	3.00 ± 0.04 @	2.80 ± 0.04 *
FS (%)	39.1 ± 0.47	39.5 ± 0.26	37.8 ± 0.31	38.3 ± 0.53
HR (bpm)	329 ± 9	334 ± 5	316 ± 9	306 ± 11

SHR-CRP—spontaneously hypertensive rat (SHR) expressing human C-reactive protein (CRP), AWTd—anterior wall thickness diastolic, LVDd—left ventricular diameter diastolic, PWTd—posterior wall thickness diastolic, AWTs—anterior wall thickness systolic, LVDs—left ventricular diameter systolic, PWTs—posterior wall thickness systolic, FS—fractional shortening, HR—heart rate, * $p < 0.05$ empagliflozin vs. control SHR-CRP, @ $p < 0.05$ vs. baseline.

3.2. Effects of Empagliflozin on Metabolic Parameters and Insulin Sensitivity

Compared to untreated animals, empagliflozin-treated SHR-CRP rats of both age groups exerted no significant differences in serum lipids—triglycerides, non-esterified fatty acids (NEFA), total and HDL cholesterol (Table 3). However, these parameters were substantially affected by the ageing of the animals; adult rats showed increased levels of serum triglycerides and cholesterol compared to young rats. Both age groups of empagliflozin-treated SHR-CRP rats exhibited markedly reduced serum insulin (Figure 1), while the serum levels of fasting and non-fasting glucose were not significantly changed (Table 3). Moreover, empagliflozin treatment reduced ectopic hepatic triglyceride accumulation and hepatic cholesterol concentration (Figure 1), as well as ectopic triglyceride accumulation in kidneys (Figure 2). β-hydroxybutyrate levels were decreased following empagliflozin treatment, this effect being more prominent in young rats than in old ones (Figure 1). By contrast, there were no significant differences in the sensitivity of adipose tissue to insulin action, measured ex vivo as the incorporation of glucose into adipose tissue lipids (lipogenesis). Basal and adrenaline-stimulated lipolysis were also unchanged after empagliflozin treatment in both age groups (Table 3).

Table 3. Biochemical parameters in the experimental groups at the end of the study.

	Young		Adult				
	Control	Empagliflozin	Control	Empagliflozin	P_{AGE}	$P_{TREATMENT}$	$P_{INTER-ACTION}$
Triglycerides (mmol/L)	0.35 ± 0.04	0.39 ± 0.04	0.72 ± 0.13 #	0.57 ± 0.04 #	<0.01	n.s.	n.s.
Total cholesterol (mmol/L)	1.24 ± 0.19	1.00 ± 0.03	1.33 ± 0.08	1.50 ± 0.05 #	<0.01	n.s.	n.s.
HDL-cholesterol (mmol/L)	1.06 ± 0.13	0.85 ± 0.01 *	1.06 ± 0.05	1.22 ± 0.02 #	<0.05	n.s.	<0.05
NEFA (mmol/L)	0.51 ± 0.02	0.44 ± 0.01 *	0.44 ± 0.02	0.47 ± 0.02	n.s.	n.s.	<0.05
Fasting glucose (mmol/L)	4.3 ± 0.1	4.3 ± 0 1	4.6 ± 0.1	4.5 ± 0.2	n.s.	n.s.	n.s.
Non-fasting glucose (mmol/L)	7.1 ± 0.2	6.8 ± 0.1	7.4 ± 0.3	6.9 ± 0.3	n.s.	n.s.	n.s.
AUC0–120 (OGT test)	842 ± 19	802 ± 11	842 ± 25	905 ± 63	n.s.	n.s.	n.s.
Lipogenesis 0 (nmol/g/2 h)	1397 ± 161	1582 ± 204	1105 ± 82	1206 ± 92	<0.05	n.s.	n.s.
Lipogenesis 250 (nmol/g/2 h)	2007 ± 106	2372 ± 350	1948 ± 313	1705 ± 232	<0.05	n.s.	n.s.
Lipolysis 0 (nmol/g/2 h)	4.70 ± 0.24	3.84 ± 0.32	4.61 ± 0.54	4.78 ± 0.49	n.s.	n.s.	n.s.
Lipolysis 250 (nmol/g/2 h)	6.52 ± 0.34	5.28 ± 0.76	8.06 ± 0.76	8.36 ± 0.98 #	n.s.	n.s.	n.s.
Glycogenesis 0 (nmol/g/2 h)	1366 ± 240	1505 ± 307	1803 ± 128	1843 ± 243	<0.01	n.s.	n.s.
Glycogenesis 250 (nmol/g/2 h)	1642 ± 248	2522 ± 462	1995 ± 214	2059 ± 211	n.s.	n.s.	n.s.

P_{AGE} denotes the significance of young vs. adult SHR-CRP rats (age effect); $P_{TREATMENT}$ denotes the significance of empagliflozin treatment (treatment effects); Interaction denotes the significance of empagliflozin in various strains (treatment vs. strain comparison). For multiple comparisons (empagliflozin treatment vs. non-treated controls) Fisher´s LSD post-hoc test was used; * denotes $p < 0.05$; n.s. denotes not significant, Data are means ± SEM; n = 7–8 for each group. AUC—area under curve from the oral glucose tolerance (OGT) test.

3.3. Effects of Empagliflozin on Inflammatory and Oxidative Stress Parameters

In adult SHR-CRP rats, empagliflozin treatment reduced markers of systemic inflammation. The levels of rat (but not human) CRP, leptin, and MCP-1 were significantly decreased in the serum of SHR-CRP rats, while other inflammatory parameters (TNFα and IL-6) were not affected (Table 4). However, when compared with adult animals, young empagliflozin-treated SHR-CRP rats had decreased serum levels of leptin and human CRP. As shown in Table 5, there was a substantial augmentation of oxidative stress in the kidney cortex of aged SHR-CRP rats. The empagliflozin treatment attenuated oxidative stress parameters, most of these effects being more pronounced in young than in adult rats. The lipoperoxidation products TBARS and conjugated dienes (CD) were significantly decreased and the activities of antioxidant enzymes (glutathione peroxidase, catalase) and the concentration of reduced glutathione were increased.

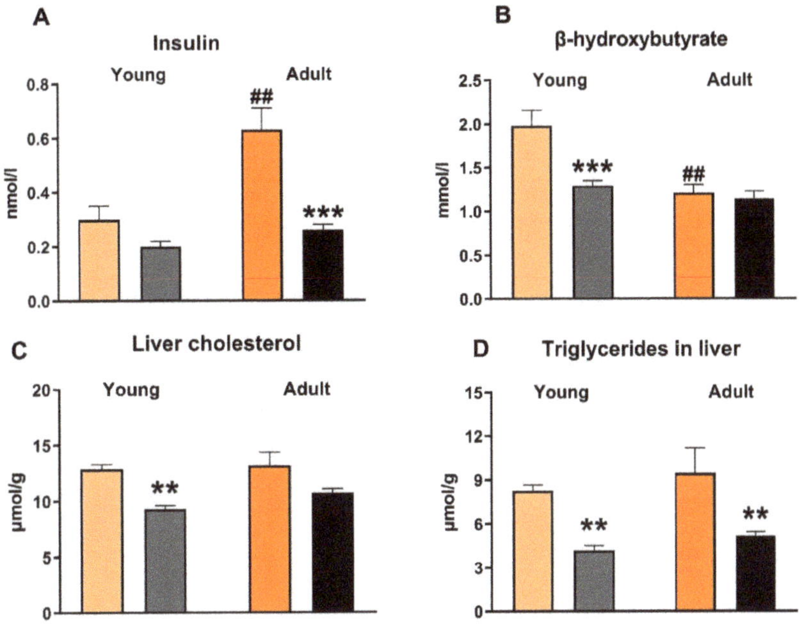

Figure 1. The effect of empagliflozin treatment on plasma insulin (**A**), plasma β-hydroxybutyrate (**B**), liver cholesterol (**C**) and liver triglycerides (**D**). $p < 0.05$ empagliflozin vs. untreated group, $p < 0.05$ vs. respective young group. ## denotes $p < 0.01$, ** denotes $p < 0.01$; *** denotes $p < 0.001$; Data are means ± SEM; n = 7–8 for each group.

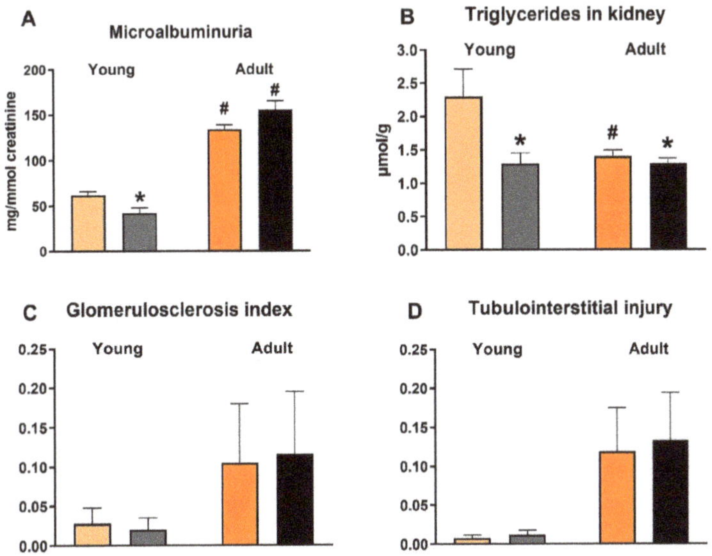

Figure 2. The effect of empagliflozin treatment on microalbuminuria (**A**), triglycerides in kidney (**B**), glomerulosclerosis index (**C**) and tubulointerstitial injury (**D**). * $p < 0.05$ empagliflozin vs. untreated group, # $p < 0.05$ vs. respective young group. Data are means ± SEM; n = 7–8 for each group.

Table 4. Inflammatory parameters in serum in experimental groups at the end of the study.

	Young		Adult				
	Control	Empagliflozin	Control	Empagliflozin	P_{AGE}	$P_{TREATMENT}$	$P_{INTERACTION}$
hs CRP human (mg/mL)	299 ± 59	221 ± 5 *	201 ± 9	227 ± 10	n.s.	n.s.	<0.05
hs CRP rat (mg/mL)	0.82 ± 0.15	0.51 ± 0.15	1.95 ± 0.10 #	1.42 ± 0.13 *#	<0.001	<0.01	n.s.
MCP-1 (ng/mL)	4.36 ± 0.67	3.53 ± 0.17	9.47 ± 0.52 #	7.43 ± 0.17 **#	<0.001	<0.01	n.s.
IL-6 (pg/mL)	51.86 ± 23.17	50.62 ± 17.96	49.75 ± 15.86	50.26 ± 17.23	n.s.	n.s.	n.s.
Leptin (ng/mL)	2.39 ± 0.39	0.73 ± 0.23 ***	1.35 ± 0.03	1.23 ± 0.02	n.s.	<0.001	<0.01

P_{AGE} denotes the significance of young vs. adult SHR-CRP rats (age effect); $P_{TREATMENT}$ denotes the significance of empagliflozin treatment (treatment effects); Interaction denotes the significance of empagliflozin in various strains (treatment vs. strain comparison). For multiple comparisons (empagliflozin treatment vs. non-treated controls) Fisher´s LSD post-hoc test was used; * denotes $p < 0.05$; ** denotes $p < 0.01$; *** denotes $p < 0.001$; n.s. denotes not significant, Data are means ± SEM; n = 7–8 for each group. CRP—C-reactive protein, MCP-1—Monocyte chemoattractant protein-1, IL-6—Interleukin-6.

Table 5. Oxidative stress parameters in kidney cortex in experimental groups at the end of the study.

	Young		Adult				
	Control	Empagliflozin	Control	Empagliflozin	P_{AGE}	$P_{TREATMENT}$	$P_{INTERACTION}$
SOD	0.045 ± 0.003	0.048 ± 0.003	0.046 ± 0.003	0.050 ± 0.002	n.s.	n.s.	n.s.
GSH-Px	148 ± 14	199 ± 12 **	163 ± 14	214 ± 8 **	n.s.	<0.001	n.s.
GR	73 ± 5	67 ± 7	78 ± 4	86 ± 3 #	<0.05	n.s.	n.s.
CAT	626 ± 33	790 ± 32 ***	553 ± 21	636 ± 30 *#	<0.001	<0.001	n.s.
CD	21.7 ± 1.2	17.0 ± 1.8 *	22.6 ± 1.0	16.6 ± 1.2 **	n.s.	<0.001	n.s.
TBARS	0.78 ± 0.03	0.50 ± 0.03 ***	0.65 ± 0.02	0.52 ± 0.03 **	n.s.	<0.001	<0.05
GSH	25.8 ± 5.3	37.8 ± 4.3 ***	13.7 ± 0.7 #	14.3 ± 0.1 #	<0.001	<0.001	<0.001
GSSG	2.4 ± 0.1	2.7 ± 0.1	1.8 ± 0.1 #	1.8 ± 0.1 #	<0.01	n.s.	n.s.
GSH/GSSG	12.1 ± 0.8	14.3 ± 0.6	7.9 ± 0.4 #	7.9 ± 0.2 #	<0.001	n.s.	n.s.

P_{AGE} denotes the significance of young vs. adult SHR-CRP rats (age effect); $P_{TREATMENT}$ denotes the significance of empagliflozin treatment (treatment effects); $P_{INTERACTION}$ denotes the significance of empagliflozin in various strains (treatment vs. strain comparison). For multiple comparisons (empagliflozin treatment vs. non-treated controls) Fisher´s LSD post-hoc test was used; * denotes $p < 0.05$; ** denotes $p < 0.01$; *** denotes $p < 0.001$; n.s. denotes not significant, Data are means ± SEM; n = 7–8 for each group. SOD—superoxide dismutase (U/mg), GSH-Px–glutathione peroxidase (μmol/NADPH/min/mg), GR–glutathione reductase (μmol/NADPH/min/mg), CAT—catalase (μmol/H_2O_2/min/mg), CD—conjugated dienes (nmol/mg), TBARS—thiobarbituric acid reactive substances (μmol/g), GSH—reduced glutathione (μmol/g), GSSG—oxidized glutathione (μmol/g).

3.4. *Effects of Empagliflozin on Renal Expression of Selected Pro-Inflammatory Genes*

The changes in renal expression of selected genes (*Ccl2*, *Il6*, *Tgfb*, *Tnf*) indicated that ageing was related to the substantial aggravation of inflammation in SHR-CRP rats (Figure 3). In addition, empagliflozin treatment reduced the expression of *Ccl2* (to 63% in young and to 66% in adult rats) and *Il6* (to 47% in young and to 61% in adult rats) genes in both age groups, while the expression of *Tgfb* and *Tnf* genes was not affected.

3.5. Effects of Empagliflozin on Renal Function Markers and Histological Analysis

In both age groups, urinary glucose excretion was more than twenty times higher, and sodium excretion two times higher, in empagliflozin-treated SHR-CRP rats (data not shown), suggesting an effective blockade of the SGLT-2 transporter. Microalbuminuria was substantially aggravated in adult SHR-CRP rats as compared with young animals, which is in accordance with the established phase of the incipient nephropathy associated with transgenic CRP expression at the age of one year. This was also confirmed by slightly increased histological markers found in old rats (Figures 2 and 4) and also supported by increased gene expression levels of pro-inflammatory markers MCP-1 and IL-6 (Figure 3). Unexpectedly, empagliflozin treatment decreased microalbuminuria only in young, but not in adult animals. However, this was not paralleled by a decrease of glomerulosclerosis index or tubulointerstitial injury index. Both control and empagliflozin-treated adult SHR-CRP rats exhibited moderate changes with focal segmental glomerular sclerosis accompanied by smaller areas of tubular atrophy when compared to empagliflozin-treated rats. Contrary to adult rats, young SHR-CRP rats exhibited substantially reduced histopathological changes in the kidney and showed no effects of empagliflozin treatment (Figure 4).

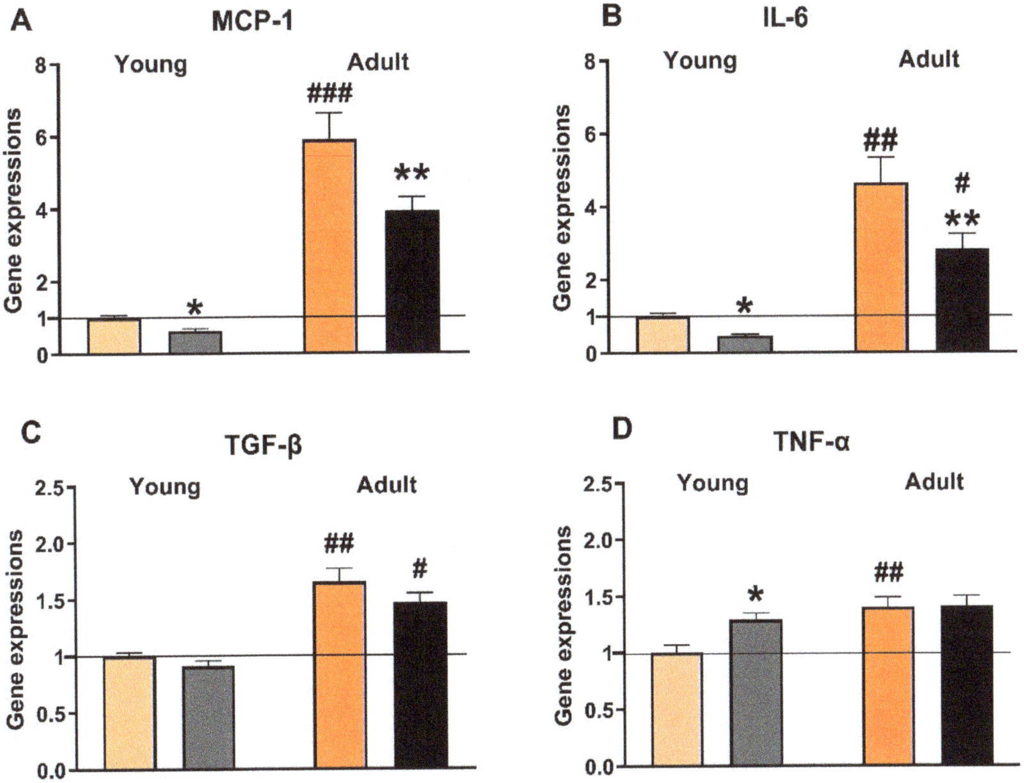

Figure 3. Relative mRNA expression of selected pro-inflammatory genes in the kidneys—monocyte chemoattractant protein-1 (MCP-1; (**A**)), Interleukin-6 (IL-6; (**B**)), Transforming growth factor-β (TGF-β; (**C**)), and Tumor necrosis factor-α (TNF-α; (**D**)). # $p < 0.05$ vs. respective young group. * denotes $p < 0.05$; ** denotes $p < 0.01$; ## denotes $p < 0.01$; ### denotes $p < 0.001$. Data are means ± SEM; n = 7–8 for each group.

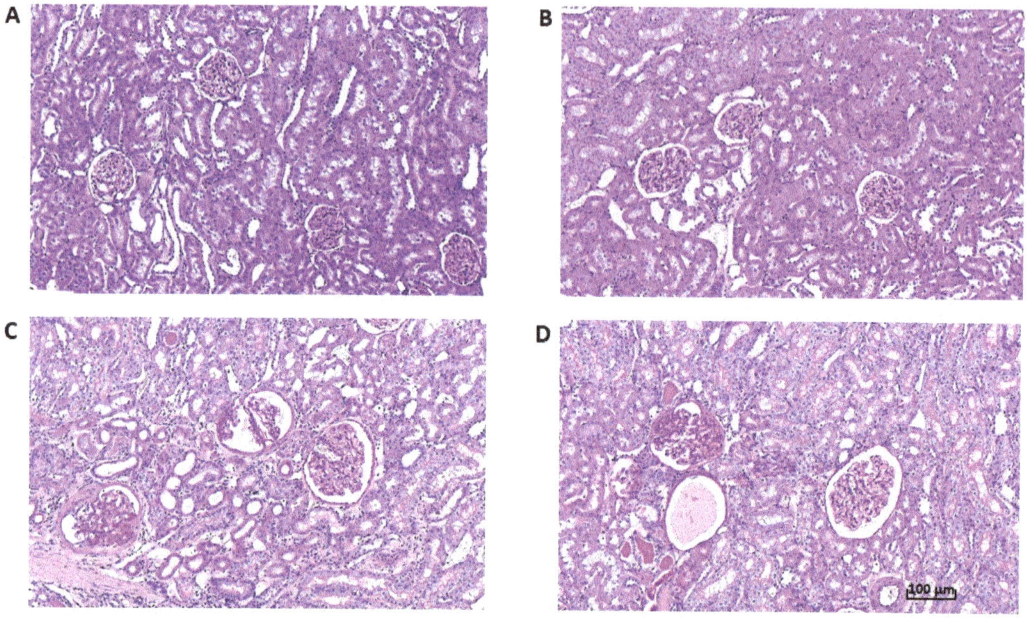

Figure 4. Representative histological images of renal cortex (PAS, 200×) (PAS, original objective 20×) of young (**A**,**B**) and adult (**C**,**D**) untreated (**A**,**C**) and empagliflozin-treated SHR-CRP rats (**B**,**D**). A scale bar is shown in Figure 4D.

4. Discussion

4.1. Renoprotective Effects of Empagliflozin Are Associated with Reduced Ectopic Fat Accumulation and Lower Inflammation and Oxidative Stress

The results of the current study demonstrated the significant renoprotective effects of empagliflozin in SHR-CRP rats. Specifically, reduced ectopic fat accumulation in the kidney was associated with decreased inflammation and oxidative stress and reduced microalbuminuria in young rats. Moreover, we confirmed that in the spontaneously hypertensive rats expressing human C-reactive protein (a model of metabolic syndrome, inflammation and organ damage), there is an age-dependent increase of insulin resistance, inflammatory markers in kidneys (MCP-1, TGF-β, TNF-α and IL-6), and deterioration of kidney function [19].

The reduced weight of renal fat following empagliflozin treatment was also associated with a reduced tissue expression of genes coding for enzymes that regulate inflammation (MCP-1 and IL-6) in both young and adult rats, which is consistent with the reduced serum levels of these pro-inflammatory markers. Conversely, the effects of empagliflozin treatment on gene expression of other pro-inflammatory cytokines TGF-β, TNF-α were not demonstrated. However, we and others found ambiguous effects of gliflozins therapy on distinct pro-inflammatory markers in non-diabetic rat strains—sometimes affecting more TNF-α [17], MCP-1 [18], or IL-1β and NF-κB [18]. Moreover, it could not be excluded that only those pro-inflammatory parameters which were upregulated by the insertion of the CRP transgene (IL-6, but not TNF-α) [19], could be influenced by empagliflozin therapy. The higher albuminuria found in adult control and treated animals is compatible with the long-term effect of high blood pressure, as well as the pro-hypertensive effect of CRP transgene in these animals. Thus, the absence of an empagliflozin effect on the reduction of albuminuria in adult rats could be explained either by the existing long-term changes in the kidneys of ageing animals or by the higher susceptibility of young animals to pharmacological interventions [26]. Similarly, our previous study performed in another non-diabetic hypertensive model—adult Ren-2 transgenic rats—did not show any effect of empagliflozin on proteinuria or albuminuria [17]; while a reduction of albuminuria was disclosed in a pre-diabetic rat model—hereditary hypertriglyceridemic rats [18]. Conversely, a reduction of albuminuria in young rats was associated with substantial attenuation of several parameters of oxidative stress (reduced levels of lipoperoxidation products TBARS and conjugated dienes, increased activity of glutathione peroxidase, increased glutathione levels, etc.) suggesting that intervention applied at an early age (critical developmental window) could be more effective [26]. Altogether, these results support the theory that empagliflozin treatment protects against the incipient nephropathy associated with the overexpression of the CRP transgene by reducing oxidative stress and inflammation either directly or as a consequence of reduced ectopic fat accumulation in the kidney.

In fact, in several animal models, including streptozotocin-induced diabetic mice and rats [27,28], Akita mice [29], OVE26 mice [30], and db/db mice [12,30], it was reported that kidney disease was associated with lipid accumulation and increased activity of pro-inflammatory cytokines, resulting in albuminuria, glomerular mesangial expansion, and tubulointerstitial fibrosis. Ectopic fat accumulation was observed in kidney biopsies of humans with type 2 diabetes mellitus [31,32]. Recently, Wang et al. [12] reported that SGLT-2 inhibition by JNJ-39933673 in db/db mice was associated with decreased renal lipid accumulation and prevention of the development of nephropathy. In addition, Hosokawa et al. [33] showed that ipragliflozin decreased ectopic lipid accumulation in tubular cells in diabetic mice. Although further studies are needed, it is plausible that, in addition to the well-known anti-inflammatory, anti-proliferative, and anti-fibrotic effects of SGLT-2 inhibitors, the reduction of tubular lipid deposition could contribute to the renoprotective mechanism of these molecules [34].

The attenuation of renal inflammation following empagliflozin treatment can lead to a decreased permeability of endothelial cells and subsequent alterations in the hemodynamics of the kidney [35], contributing to the improvement of renal function. The evidence for the anti-inflammatory potential of SGLT-2 inhibitors has been previously demonstrated in diabetic animal models [36]. Moreover, in human proximal tubular cells, SGLT-2 inhibitors (tofogliflozin) attenuated the expression of pro-inflammatory markers [37]. In diabetic patients, SGLT-2 inhibitors were found to reduce systemic levels of pro-inflammatory markers IL-6 and TNFα [8]. The SGLT-2 inhibitor-mediated anti-inflammatory effects, however, have not been shown under normoglycemic conditions.

Consistent with its anti-inflammatory properties, empagliflozin also attenuated oxidative stress—another key pathway known to cause kidney impairment. In the present study, empagliflozin markedly improved renal oxidative stress with the effect also being more pronounced in young rats. Empagliflozin regulates oxidative stress in the kidney cortex by stimulating antioxidant enzyme activity GSH-Px and catalase via the upregulation of Nrf2 rather than the direct inactivation of free radicals. In addition, increased GSH-Px activity can play a role in decreasing lipid peroxidation by participating in the removal of lipoperoxidation products. In the kidney cortex, increased GSH-Px activity following empagliflozin treatment was also linked to increased glutathione levels, a sensitive marker of oxidative damage. A markedly alleviated renal oxidative stress can be one of the pleiotropic metabolic mechanisms of empagliflozin that can also contribute to the improvement of renal function.

4.2. Effects of Empagliflozin on Insulin, β-Hydroxybutyrate, NEFA and Leptin Concentrations

In our study, empagliflozin markedly reduced hyperinsulinemia, although no significant changes in insulin sensitivity were observed in muscle and adipose tissue, as well as in fasting or non-fasting glucose levels. Thus, other factors contributing to the improved insulin sensitivity following empagliflozin treatment such as decreased NEFA and leptin levels can play a role. Although Nishimura et al. [38] have shown that SGLT-2 inhibitors promote systemic NEFA mobilization associated with the induction of ketone bodies as alternative substrate for energy metabolism, we observed no increase of NEFA or β-hydroxybutyrate (BHB) after empagliflozin administration. On the contrary, empagliflozin even reduced circulating BHB concentration. A number of possible mechanisms of SGLT-2 inhibitors are implicated in their cardioprotective effects, which go beyond their diuretic and antihyperglycemic effects. The increase of ketone bodies is only one of the proposed mechanisms, which may not always apply. In addition, increased ketone bodies after SGLT-2 inhibitors are rather related to systemic alterations in substrate utilization and reduced glucose oxidation in preference for fatty acid oxidation. Thus, empagliflozin in a non-diabetic model with genetic hypertension and chronic inflammation affects heart function by other mechanisms than ketone bodies utilization. In our previous study with prediabetic animal model with vascular complications, we observed increased BHB in serum and the heart, but not increased BHB utilization in the heart [20]. In a study with diabetic obese rats, Abdurrachim et al. [39] reported that empagliflozin decreased ketone bodies utilization in the heart despite increasing circulating levels of BHB.

The markedly decreased circulating leptin levels observed in this study may not only contribute to improved insulin resistance, but may also alleviate inflammation and cardiovascular damage. In addition to the effect on insulin sensitivity, leptin may have a pathophysiological role in sodium regulation, as well as in cardiac and renal inflammation and fibrosis [40]. Circulating leptin levels presage the development of heart failure in elderly people and a decline in the glomerular filtration rate in longitudinal studies [41,42]. Although the mechanism/s of leptin action in heart failure are not fully understood, one potential mechanism might be its effect on epicardial adipose expansion and on calcium handling in cardiomyocytes leading to impaired myocardial relaxation. SGLT-2 inhibitors lead to decreases in serum aldosterone, reduction in the activity of NHE1 (sodium-hydrogen exchanger isoform 1), and can probably directly suppress leptin secretion and its paracrine actions on the heart and kidneys to promote fibrosis [40]. All these effects may underlie the action of SGLT-2 inhibitors to ameliorate cardiac and renal injury. Although a positive effect on plasma leptin levels has been observed in other studies with empagliflozin [43,44], it is unclear whether empagliflozin action directly impacts on adipose tissue function or whether it is associated with visceral fat loss. Leptin is secreted by epicardial and perirenal adipose tissue, thus the reduction of these fat depots can also contribute to decreased leptin levels. Although adipose tissue insulin sensitivity was not affected, empagliflozin treatment reduced the weight of visceral adipose tissue, and it is possible that reduced adiposity positively influenced secretion of other adipocytokines, in addition to leptin.

4.3. *Effects of Empagliflozin on Liver Triglycerides and Cholesterol*

Although circulating lipids were not affected in the present study, empagliflozin treatment significantly decreased hepatic triglycerides, as well as cholesterol content. According to our previous studies, empagliflozin modulates genes related to lipid synthesis and fatty acid metabolism, while it has no effect on genes involved in lipid oxidation and transport [20]. Thus, we speculate that reduced lipid accumulation in the liver is probably associated with the inhibited lipogenesis. The suppression of SCD1—the main lipogenic enzyme—was observed in our [20], as well as in another animal study with obese mice [45]. The reduced ectopic hepatic lipid deposition can also ameliorate insulin resistance. The intrahepatic accumulation of fatty acids and lipotoxic intermediates interferes with intracellular signaling pathways like insulin signaling, and can induce endoplasmic reticulum stress; both can contribute to the development of insulin resistance [46]. The hepatic lipid accumulation and impaired lipid metabolism in the liver are independent risk factors for cardiovascular events, thus, the reduced hepatic lipid accumulation can contribute to the cardioprotective effect of empagliflozin.

5. Conclusions

It can be concluded that treatment of SHR-CRP rats with empagliflozin is associated with reduced renal lipid accumulation, inflammation, and oxidative stress, resulting in the attenuation of renal damage; these beneficial effects being more pronounced in young rats. By contrast, the metabolic effects of empagliflozin prevailed in adult rats.

Author Contributions: I.V. conceived and designed the project, S.H., H.M., M.H., I.M., P.M. and J.Š. performed the experiments, F.P. and J.H. performed the echocardiography, I.V., I.M., H.M. and J.Z. analyzed data and interpreted the results, D.M. did the statistical analysis, I.V., H.M and M.P. wrote the manuscript, J.Z. edited the manuscript. All authors have read and agreed to the published version of the manuscript.

Funding: This work was supported by grant of Czech Science Foundation, project number 19-06199S and the Ministry of Health of the Czech Republic—conceptual development of research organisations (Institute for Clinical and Experimental Medicine—IKEM, IN 00023001) and by the project National Institute for Research of Metabolic and Cardiovascular Diseases (Programme EXCELES, ID Project No. LX22NPO5104)—Funded by the European Union—Next Generation EU. This study was also supported by institutional support of the Institute of Physiology, Czech Academy of Sciences, grant Nr. RVO 67985823.

Institutional Review Board Statement: The study was conducted according to the guidelines of the Declaration of Helsinki, and approved by the Institute of Physiology, Czech Academy of Sciences (Protocol Nr. 47/2019).

Informed Consent Statement: Not applicable.

Data Availability Statement: All data arising from this study are contained within the article.

Acknowledgments: The technical assistance of Zdeňka Kopecká and Alena Charvátová is highly appreciated. The authors are grateful to Petr Kujal fo the histological pictures and to Bob Kotanchik for his help in editing the manuscript.

Conflicts of Interest: The authors declare no conflict of interest.

References

1. Aldossari, K.K. Cardiovascular outcomes and safety with antidiabetic drugs. *Int. J. Health Sci.* **2018**, *12*, 70–83.
2. Hediger, M.A.; Rhoads, D.B. Molecular physiology of sodium glucose cotransporters. *Physiol. Rev.* **1994**, *74*, 993–1026. [CrossRef] [PubMed]
3. Prattichizzo, F.; De Nigris, V.; Spiga, R.; Mancuso, E.; La Sala, L.; Antonicelli, R.; Testa, R.; Procopio, A.D.; Olivieri, F. Inflammageing and metaflammation: The yin and yang of type 2 diabetes. *Ageing Res. Rev.* **2018**, *41*, 1–17. [CrossRef] [PubMed]
4. Zinman, B.; Wanner, C.; Lachin, J.M.; Fitchett, D.; Bluhmki, E.; Hantel, S.; Mattheus, M.; Devins, T.; Johansen, O.E.; Woerle, H.J.; et al. Empagliflozin, Cardiovascular Outcomes, and Mortality in Type 2 Diabetes. *N. Engl. J. Med.* **2015**, *373*, 2117–2128. [CrossRef]
5. Wanner, C.; Inzucchi, S.E.; Lachin, J.M.; Fitchett, D.; von Eynatten, M.; Mattheus, M.; Johansen, O.E.; Woerle, H.J.; Broedl, U.C.; Zinman, B.; et al. Empagliflozin and Progression of Kidney Disease in Type 2 Diabetes. *N. Engl. J. Med.* **2016**, *375*, 323–334. [CrossRef]
6. Perkovic, V.; Jardine, M.J.; Neal, B.; Bompoint, S.; Heerspink, H.J.L.; Charytan, D.M.; Edwards, R.; Agarwal, R.; Bakris, G.; Bull, S.; et al. Canagliflozin and Renal Outcomes in Type 2 Diabetes and Nephropathy. *N. Engl. J. Med.* **2019**, *380*, 2295–2306. [CrossRef]
7. Cannon, C.P.; Perkovic, V.; Agarwal, R.; Baldassarre, J.; Bakris, G.; Charytan, D.M.; de Zeeuw, D.; Edwards, R.; Greene, T.; Heerspink, H.J.L.; et al. Evaluating the Effects of Canagliflozin on Cardiovascular and Renal Events in Patients with Type 2 Diabetes Mellitus and Chronic Kidney Disease According to Baseline HbA1c, Including Those with HbA1c <7%: Results from the CREDENCE Trial. *Circulation* **2020**, *141*, 407–410. [CrossRef]
8. Heerspink, H.J.L.; Jongs, N.; Chertow, G.M.; Langkilde, A.M.; McMurray, J.J.V.; Correa-Rotter, R.; Rossing, P.; Sjöstrom, C.D.; Stefansson, B.V.; Toto, R.D.; et al. Effect of dapagliflozin on the rate of decline in kidney function in patients with chronic kidney disease with and without type 2 diabetes: A prespecified analysis from the DAPA-CKD trial. *Lancet Diabetes Endocrinol.* **2021**, *9*, 743–754. [CrossRef]
9. Mima, A. Sodium-Glucose Cotransporter 2 Inhibitors in Patients with Non-Diabetic Chronic Kidney Disease. *Adv. Ther.* **2021**, *38*, 2201–2212. [CrossRef]
10. Benetti, E.; Mastrocola, R.; Vitarelli, G.; Cutrin, J.C.; Nigro, D.; Chiazza, F.; Mayoux, E.; Collino, M.; Fantozzi, R. Empagliflozin protects against diet-induced NLRP-3 inflammasome activation and lipid accumulation. *J. Pharmacol. Exp. Ther.* **2016**, *359*, 45–53. [CrossRef]

11. Gembrandt, F.; Bartaun, C.; Jarzebska, N.; Mayoux, E.; Todorov, V.T.; Hohenstein, B.; Hugo, C. The SGLT2 inhibitor empagliflozin ameliorates early features of diabetic nephropathy in BTBR ob/ob type 2 diabetic mice with and without hypertension. *Am. J. Physiol. Renal. Physiol.* **2014**, *307*, F317–F325. [CrossRef] [PubMed]
12. Wang, X.X.; Levi, J.; Luo, Y.; Kopp, J.B.; Rosenberg, A.Z.; Levi, M. SGLT2 Protein expression is increased in human diabetic nephropathy: SGTL2 protein inhibition decreases renal lipid accumulation, inflammation, and the development of nephropathy in diabetic mice. *J. Biol. Chem.* **2017**, *292*, 5335–5348. [CrossRef] [PubMed]
13. Yurista, S.R.; Silljé, H.H.W.; Oberdorf-Maass, S.U.; Schouten, E.M.; Pavez Giani, M.G.; Hillebrands, J.L.; van Goor, H.; van Veldhuisen, D.J.; de Boer, R.A.; Westenbrink, B.D. Sodium-glucose co-transporter 2 inhibition with empagliflozin improves cardiac function in non-diabetic rats with left ventricular dysfunction after myocardial infarction. *Eur. J. Heart Fail.* **2019**, *21*, 862–873. [CrossRef] [PubMed]
14. Connelly, K.A.; Zhang, Y.; Desjardins, J.F.; Nghiem, L.; Visram, A.; Batchu, S.N.; Yerra, V.G.; Kabir, G.; Thai, K.; Advani, A.; et al. Load-independent effects of empagliflozin contribute to improved cardiac function in experimental heart failure with reduced ejection fraction. *Cardiovasc. Diabetol.* **2020**, *19*, 13. [CrossRef] [PubMed]
15. Lee, H.C.; Shiou, Y.L.; Jhuo, S.J.; Chang, C.Y.; Liu, P.L.; Jhuang, W.J.; Dai, Z.K.; Chen, W.Y.; Chen, Y.F.; Lee, A.S. The sodium-glucose co-transporter 2 inhibitor empagliflozin attenuates cardiac fibrosis and improves ventricular hemodynamics in hypertensive heart failure rats. *Cardiovasc. Diabetol.* **2019**, *18*, 45. [CrossRef]
16. Yang, C.C.; Chen, Y.T.; Wallace, C.G.; Chen, K.H.; Cheng, B.C.; Sung, P.H.; Li, Y.-C.; Ko, S.-F.; Chang, H.-W.; Yip, H.-K. Early administration of empagliflozin preserved heart function in cardiorenal syndrome in rat. *Biomed. Pharmacother.* **2019**, *109*, 658–670. [CrossRef]
17. Hojná, S.; Rauchová, H.; Malínská, H.; Marková, I.; Hüttl, M.; Papoušek, F.; Behuliak, M.; Miklánková, D.; Vaňourková, Z.; Neckář, J.; et al. Antihypertensive and metabolic effects of empagliflozin in Ren-2 transgenic rats, an experimental non-diabetic model of hypertension. *Biomed. Pharmacother.* **2021**, *144*, 112246. [CrossRef]
18. Hüttl, M.; Markova, I.; Miklankova, D.; Zapletalova, I.; Poruba, M.; Haluzik, M.; Vaněčkova, I.; Malinska, H. In a Prediabetic Model, Empagliflozin Improves Hepatic Lipid Metabolism Independently of Obesity and before Onset of Hyperglycemia. *Int. J. Mol. Sci.* **2021**, *22*, 11513. [CrossRef]
19. Pravenec, M.; Kajiya, T.; Zídek, V.; Landa, V.; Mlejnek, P.; Simáková, M.; Silhavý, J.; Malínská, H.; Oliyarnyk, O.; Kazdová, L.; et al. Effects of human C-reactive protein on pathogenesis of features of the metabolic syndrome. *Hypertension* **2011**, *57*, 731–737. [CrossRef]
20. Malinska, H.; Hüttl, M.; Oliyarnyk, O.; Bratova, M.; Kazdova, L. Conjugated linoleic acid reduces visceral and ectopic lipid accumulation and insulin resistance in chronic severe hypertriacylglycerolemia. *Nutrition* **2015**, *31*, 1045–1051. [CrossRef]
21. Lowry, O.H.; Rosebrough, N.J.; Farr, A.L.; Randall, R.J. Protein measurement with the Folin phenol reagent. *J. Biol. Chem.* **1951**, *193*, 265–275. [CrossRef]
22. Hrdlička, J.; Neckář, J.; Papoušek, F.; Husková, Z.; Kikerlová, S.; Vaňourková, Z.; Vernerová, Z.; Akat, F.; Vašinová, J.; Hammock, B.D.; et al. Epoxyeicosatrienoic Acid-Based Therapy Attenuates the Progression of Postischemic Heart Failure in Normotensive Sprague-Dawley but Not in Hypertensive Ren-2 Transgenic Rats. *Front. Pharmacol.* **2019**, *10*, 159. [CrossRef]
23. Ye, J.; Coulouris, G.; Zaretskaya, I.; Cutcutache, I.; Rozen, S.; Madden, T.L. Primer-BLAST: A tool to design target-specific primers for polymerase chain reaction. *BMC Bioinform.* **2012**, *13*, 134. [CrossRef]
24. Nakano, Y.; Hirano, T.; Uehara, K.; Nishibayashi, S.; Hattori, K.; Aihara, M.; Yamada, Y. New rat model induced by anti-glomerular basement membrane antibody shows severe glomerular adhesion in early stage and quickly progresses to end-stage renal failure. *Pathol. Int.* **2008**, *58*, 361–370. [CrossRef] [PubMed]
25. Saito, T.; Sumithran, E.; Glasgow, E.F.; Atkins, R.C. The enhancement of aminonucleoside nephrosis by the co-administration of protamine. *Kidney Int.* **1987**, *32*, 691–699. [CrossRef]
26. Zicha, J.; Kunes, J. Ontogenetic Aspects of Hypertension Development: Analysis in the Rat. *Physiol. Rev.* **1999**, *79*, 1227–1282. [CrossRef]
27. Jiang, T.; Wang, X.X.; Scherzer, P.; Wilson, P.; Tallman, J.; Takahashi, H.; Li, J.; Iwahashi, M.; Sutherland, E.; Arend, L.; et al. Farnesoid X receptor modulates renal lipid metabolism, fibrosis, and diabetic nephropathy. *Diabetes* **2007**, *56*, 2485–2493. [CrossRef]
28. Sun, L.; Halaihel, N.; Zhang, W.; Rogers, T.; Levi, M. Role of sterol regulatory element-binding protein 1 in regulation of renal lipid metabolism and glomerulosclerosis in diabetes mellitus. *J. Biol. Chem.* **2002**, *277*, 18919–18927. [CrossRef]
29. Proctor, G.; Jiang, T.; Iwahashi, M.; Wang, Z.; Li, J.; Levi, M. Regulation of renal fatty acid and cholesterol metabolism, inflammation, and fibrosis in Akita and OVE26 mice with type 1 diabetes. *Diabetes* **2006**, *55*, 2502–2509. [CrossRef]
30. Wang, Z.; Jiang, T.; Li, J.; Proctor, G.; McManaman, J.L.; Lucia, S.; Chua, S.; Levi, M. Regulation of renal lipid metabolism, lipid accumulation, and glomerulosclerosis in FVBdb/db mice with type 2 diabetes. *Diabetes* **2005**, *54*, 2328–2335. [CrossRef] [PubMed]
31. Wang, X.X.; Edelstein, M.H.; Gafter, U.; Qiu, L.; Luo, Y.; Dobrinskikh, E.; Lucia, S.; Adorini, L.; D'Agati, V.D.; Levi, J.; et al. G protein-coupled bile acid receptor TGR5 activation inhibits kidney disease in obesity and diabetes. *J. Am. Soc. Nephrol.* **2016**, *27*, 1362–1378. [CrossRef] [PubMed]
32. Herman-Edelstein, M.; Scherzer, P.; Tobar, A.; Levi, M.; Gafter, U. Altered renal lipid metabolism and renal lipid accumulation in human diabetic nephropathy. *J. Lipid. Res.* **2014**, *55*, 561–572. [CrossRef] [PubMed]

33. Hosokawa, K.; Takata, T.; Sugihara, T.; Matono, T.; Koda, M.; Kanda, T.; Taniguchi, S.; Ida, A.; Mae, Y.; Yamamoto, M.; et al. Ipragliflozin Ameliorates Endoplasmic Reticulum Stress and Apoptosis through Preventing Ectopic Lipid Deposition in Renal Tubules. *Int. J. Mol. Sci.* **2019**, *21*, 190. [CrossRef] [PubMed]
34. Tanaka, S.; Sugiura, Y.; Saito, H.; Sugahara, M.; Higashijima, Y.; Yamaguchi, J.; Inagi, R.; Suematsu, M.; Nangaku, M.; Tanaka, T. Sodium-glucose cotransporter 2 inhibition normalizes glucose metabolism and suppresses oxidative stress in the kidneys of diabetic mice. *Kidney Int.* **2018**, *94*, 912–925. [CrossRef]
35. Charlton, A.; Garzarella, J.; Jandeleit-Dahm, K.A.M.; Jha, J.C. Oxidative stress and inflammation in renal and cardiovascular complications of diabetes. *Biology* **2020**, *10*, 18. [CrossRef]
36. Pirklbauer, M. Anti-inflammatory potencial of empagliflozin. *Inflammopharmacology* **2021**, *29*, 573–576. [CrossRef]
37. Ishibashi, Y.; Matsui, T.; Yamagishi, S. Tofogliflozin, a highly selective inhibitor of SGLT2 blocks proinflammatory and proapoptotic effects of glucose overload on proximal tubular cells partly by suppressing oxidative stress generation. *Horm. Metab. Res.* **2016**, *48*, 191–195. [CrossRef]
38. Nishimura, R.; Tanaka, Y.; Koiwai, K.; Ishida, K.; Salsali, A. Effect of empagliflozin on free fatty acids and ketone bodies in Japanese patients with type 2 diabetes mellitus: A randomized controlled trial. *Adv. Ther.* **2019**, *36*, 2769–2782. [CrossRef]
39. Abdurrachim, D.; Teo, X.Q.; Woo, C.C. Empagliflozin reduces myocardial ketone utilization while preserving glucose utilization in diabetic hypertensive heart disease: A hyperpolarized 13-C magnetic resonance spectroscopy study. *Diabetes Obes. Metab.* **2019**, *21*, 357–365. [CrossRef]
40. Packer, M. Do sodium-glucose co-transporter-2 inhibitors prevent heart failure with a preserved ejection fraction by counterbalancing the effects of leptin? A novel hypothesis. *Diabetes Obes. Metab.* **2018**, *20*, 1361–1366. [CrossRef]
41. Lieb, W.; Sullivan, L.M.; Harris, T.B. Plasma leptin levels and incidence of heart failure, cardiovascular disease, and total mortality in elderly individuals. *Diabetes Care* **2009**, *32*, 612–616. [CrossRef] [PubMed]
42. Pedone, C.; Roshanravan, B.; Scarlata, S.; Patel, K.V.; Ferrucci, L.; Incalzi, R.A. Longitudial association between serum leptin concetration and glomerular filtration rate in humans. *PLoS ONE* **2015**, *10*, e0117828. [CrossRef] [PubMed]
43. Tentolouris, A.; Vlachakis, P.; Tzeravini, E.; Eleftheriadou, I.; Tentolouris, N. SGLT2 inhibitor: A review of their antidiabetic and cardioprotective effects. *Int. J. Environ. Res. Public Health* **2019**, *16*, 2965. [CrossRef]
44. Vickers, S.P.; Cheetham, S.C.; Headland, K.R. Combination of the sodium-glucose cotransporter-2 inhibitor empagliflozin with orlistat or sibutramine further improves the body-weight reduction and glucose homeostasis of obese rats fed a cafeteria diet. *Diabetes Metab. Syndr. Obes.* **2014**, *7*, 265–275. [CrossRef] [PubMed]
45. Nasiri-Ansari, N.; Nikolopoulou, C.; Papoutsi, K.; Kyrou, I.; Mantzoros, C.S.; Kyriakopoulos, G.; Chatzigeorgiou, A.; Kalotychou, V.; Randeva, M.S.; Chatha, K.; et al. Empagliflozin Attenuates Non-Alcoholic Fatty Liver Disease (NAFLD) in High Fat Diet Fed ApoE$^{(-/-)}$ Mice by Activating Autophagy and Reducing ER Stress and Apoptosis. *Int. J. Mol. Sci.* **2021**, *22*, 818. [CrossRef] [PubMed]
46. Lebeaupin, C.; Vallée, D.; Hazari, Y.; Hetz, C.; Chevet, E.; Bailly-Maitre, B. Endoplasmic reticulum stress signalling and the pathogenesis of non-alcoholic fatty liver disease. *J. Hepatol.* **2018**, *69*, 927–947. [CrossRef]

biomedicines

Article

Empagliflozin Is Not Renoprotective in Non-Diabetic Rat Models of Chronic Kidney Disease

Silvie Hojná [1], Zoe Kotsaridou [1,2], Zdeňka Vaňourková [3], Hana Rauchová [1], Michal Behuliak [1], Petr Kujal [4], Michaela Kadlecová [1], Josef Zicha [1] and Ivana Vaněčková [1,*]

1 Institute of Physiology, Czech Academy of Sciences, 14220 Prague, Czech Republic
2 Department of Biotechnology, Agricultural University, 11855 Athens, Greece
3 Institute for Clinical and Experimental Medicine, 14220 Prague, Czech Republic
4 Department of Pathology, Third Faculty of Medicine, Charles University, 14220 Prague, Czech Republic
* Correspondence: ivana.vaneckova@fgu.cas.cz; Tel.: +420-241062592

Abstract: Gliflozins (sodium-glucose transporter-2 inhibitors) exhibited renoprotective effects not only in diabetic but also in non-diabetic patients with chronic kidney disease (CKD). Controversial results were reported in experimental non-diabetic models of CKD. Therefore, we examined empagliflozin effects in three CKD models, namely, in fawn-hooded hypertensive (FHH) rats, uninephrectomized salt-loaded (UNX + HS) rats, and in rats with Goldblatt hypertension (two-kidney, one-clip 2K1C) that were either untreated or treated with empagliflozin (10 mg/kg/day) for eight weeks. Plethysmography blood pressure (BP) was recorded weekly, and renal parameters (proteinuria, plasma urea, creatinine clearance, and sodium excretion) were analyzed three times during the experiment. At the end of the study, blood pressure was also measured directly. Markers of oxidative stress (TBARS) and inflammation (MCP-1) were analyzed in kidney and plasma, respectively. Body weight and visceral adiposity were reduced by empagliflozin in FHH rats, without a significant effect on BP. Experimentally induced CKD (UNX + HS and 2K1C) was associated with a substantial increase in BP and relative heart and kidney weights. Empagliflozin influenced neither visceral adiposity nor BP in these two models. Although empagliflozin increased sodium excretion, suggesting effective SGLT-2 inhibition, it did not affect diuresis in any experimental model. Unexpectedly, empagliflozin did not provide renoprotection because proteinuria, plasma urea, and plasma creatinine were not lowered by empagliflozin treatment in all three CKD models. In line with these results, empagliflozin treatment did not decrease TBARS or MCP-1 levels in either model. In conclusion, empagliflozin did not provide the expected beneficial effects on kidney function in experimental models of CKD.

Keywords: SGLT-2 inhibition; proteinuria; uninephrectomized salt-loaded; two-kidney; one-clip hypertension; fawn-hooded hypertensive rat

Citation: Hojná, S.; Kotsaridou, Z.; Vaňourková, Z.; Rauchová, H.; Behuliak, M.; Kujal, P.; Kadlecová, M.; Zicha, J.; Vaněčková, I. Empagliflozin Is Not Renoprotective in Non-Diabetic Rat Models of Chronic Kidney Disease. *Biomedicines* **2022**, *10*, 2509. https://doi.org/10.3390/biomedicines10102509

Academic Editor: Ramón C. Hermida

Received: 9 September 2022
Accepted: 5 October 2022
Published: 7 October 2022

1. Introduction

The inhibition of the sodium-glucose transporter-2 (SGLT-2) by gliflozins promotes glucose and sodium excretion. Positive renoprotective effects were demonstrated not only in diabetic patients who were treated with gliflozins in addition to a standard antidiabetic therapy [1,2] but also in non-diabetic patients [3]. The CREDENCE trial [4] demonstrated the renoprotective effects of canagliflozin in diabetic patients, while the DAPA-CKD trial [5] also showed the efficacy of dapagliflozin in non-diabetic patients with chronic kidney disease. Moreover, the DAPA-HF and EMPEROR-reduced trials also showed improved cardiovascular outcomes in heart failure patients [6]. Several mechanisms were suggested for their beneficial effects, from hemodynamic and natriuretic effects (the reduction in BP; to the modulation of tubulo-glomerular feedback, and the reduction in sympathetic hyperactivity); to the metabolic factors due to their glycosuric effects (the improvement of insulin sensitivity and the lowering of plasma triglycerides and uric acid); and to the attenuation of oxidative stress, inflammation, and fibrosis. The beneficial renoprotective effects of different gliflozins were demonstrated in various diabetic and prediabetic rat and mice models [7–11]. However, only limited and sometimes contradictory information is available in non-diabetic animals with chronic kidney disease. While Zhang et al. [12], Rajasekaran et al. [13], and Li et al. [14] found no renoprotective effects of gliflozins in rats after 5/6 nephrectomy, Kim et al. [15] and Wan et al. [16] reported on partial benefits in uninephrectomized rats, and Zeng et al. [17] demonstrated renoprotective effects due to reduced fibrosis in the same model. The anti-fibrotic effects of gliflozins in kidneys were also found in the ureteral obstruction model [18] and the angiotensin II-induced model [19]. In the latter model, the additive renoprotective effects of empagliflozin treatment combined with losartan were shown by Reyes-Pardo [20]. Moreover, Ali et al. [18] reported on the amelioration of adenine-induced CKD with canagliflozin due to a reduction in renal inflammation and oxidative stress.

Our previous studies in three non-diabetic hypertensive models (hypertensive Ren-2 transgenic rats, hereditary hypertriglyceridemic rats, and spontaneously hypertensive rats expressing the human CRP gene) [21–23] demonstrated the beneficial effects of SGLT-2 inhibition without a glucose-lowering effect. In all these models, empagliflozin treatment led to a reduction in body weight and visceral adiposity, while the additional effects depended on the model used—an improvement of metabolic parameters and hepatic metabolism (hHTG), a reduction in blood pressure (TGR), and the attenuation of oxidative stress and inflammation in kidneys (SHR-CRP).

Three different models of chronic kidney disease with different pathophysiological backgrounds were selected for the present study—a genetic model of hypertension-associated renal failure, namely, a fawn-hooded hypertensive (FHH) rat, in which hypertension, proteinuria, and focal glomerulosclerosis already develops at young age [24]. Moreover, two models of kidney deterioration induced by experimental procedures were used in this study: either a uninephrectomy combined with increased sodium intake (UNX + HS) or a stenosis of renal artery–two-kidney, one-clip (2K1C) Goldblatt hypertension [25].

We hypothesized that empagliflozin would have renoprotective effects in these three models of chronic kidney disease with different pathological backgrounds, i.e., in rats genetically predisposed to renal impairment (FHH) and in two experimentally induced CKD models (UNX + HS or 2K1C).

2. Materials and Methods

2.1. Animals

Animals were housed at 23o C under a 12 h light/dark cycle, fed maintenance diet for rats and mice (Altromin-1320, Lage, Germany, 11% fat, 24% protein, 65% carbohydrates, 0.45 % NaCl), and given tap water ad libitum. Empagliflozin (Jardiance, Boehringer, Ingelheim, Germany), at a dose of 10 mg/kg/day, was mixed into the Altromin diet (so that the composition of both diets was the same) and given for 8 weeks. The amount of empagliflozin added to the diet was calculated according to our previous studies [21–23]. All procedures and experimental protocols were approved by the Ethical Committee of the Institute of Physiology, Czech Academy of Sciences (Protocol Nr. 47/2019), and conformed to the European Convention on Animal Protection and Guidelines on Research Animal Use (Directive 2010/63/EU). Three different models of chronic kidney disease were used within the study.

2.1.1. Fawn-Hooded Hypertensive Rats

Male fawn-hooded hypertensive rats were treated with empagliflozin from the age of 12 weeks, for eight weeks.

2.1.2. Two-Kidney One-Clip (2K1C) Goldblatt Hypertension

Male Wistar rats underwent a sham operation or clipping of the right renal artery (2K1C) at the age of 10 weeks [25]. Eight weeks later, empagliflozin treatment was started and administered for eight weeks.

2.1.3. Uninephrectomized Rats on High-Salt Intake

Male normotensive Hannover Sprague Dawley rats underwent uninephrectomy (UNX) at the age of 7 weeks [25], and control rats were sham operated. One week later, high-salt diet feeding (2% NaCl) and empagliflozin treatment were started. This regimen was applied for eight weeks (Figure 1).

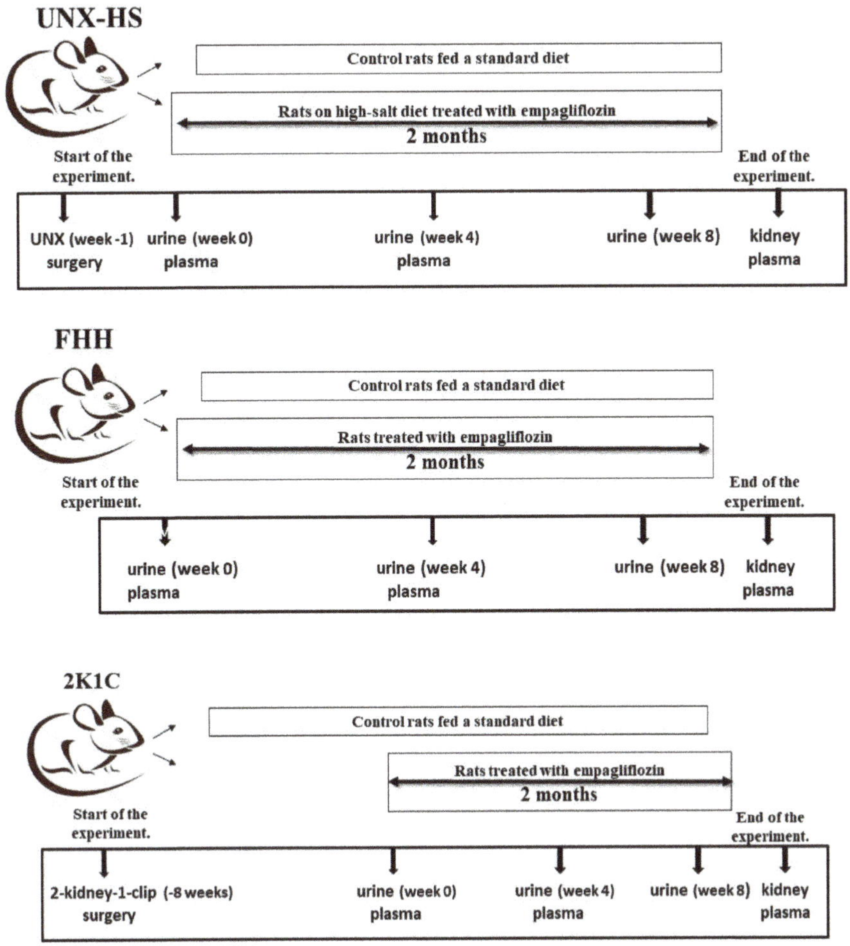

Figure 1. Scheme of the experiment in Fawn-hooded hypertensive rats (FHH), 2-kidney-one-clip (2K1C), and uninephrectomized high salt-fed (UNX-HS) rats.

Body weight was monitored throughout the whole experiment and measured once a week. Systolic blood pressure was measured weekly during the experiment using tail plethysmography (HatterasInstruments, Cary, NC, USA). Food consumption was monitored throughout the experiment.

2.2. Biochemical Analysis

Three times during the experiment (at weeks 0, 4, and 7), the animals were placed in individual metabolic cages for the evaluation of kidney function using 24-h urine collection. Urinary proteinuria, sodium, and glucose were determined. Urinary protein was measured using the Folin method, with bovine serum albumin as a standard [26]. Plasma creatinine was measured by the FUJI DRI-CHEM analyzer using appropriate slides for creatinine CRE-P III (Fujifilm Corp, Tokyo, Japan). Plasma urea were determined using kit (Erba Lachema, Brno, Czech Republic). Lipoperoxidation products were assessed based on levels of thiobarbituric acid-reactive substances (TBARS) by assaying the reaction with thiobarbituric acid. MCP-1 levels were determined in urine using ELISA kit (Invitrogen, Carlsbad, CA, USA). At the end of the study, direct BP measurement was performed under light isoflurane anesthesia [27]. Then, the animals were sacrificed and plasma and tissue samples were collected and stored at −80 °C for further biochemical analysis.

2.3. Statistical Analysis

All data are expressed as means ± SEM. Statistical analysis was done by one-way analysis of variance (ANOVA) with Bonferoni test, using the Graph-Pad Prism software (Graph Pad Software, San Diego, CA, USA). The differences were considered significant at the $p < 0.05$.

3. Results

3.1. Effects of Empagliflozin on Body Weight, Weights of Fat Depots, and Blood Pressure

All three rat models gained weight during the study. However, there was decrease in body weight in uninephrectomized Hannover Sprague Dawley (HanSD) rats on high-salt diet (UNX + HS + empa), while renal artery stenosis (2K1C + empa) in Wistar rats had no effect on body weight. Empagliflozin (empa) decreased body weight in hypertensive fawn-hooded rats (FHH), while it had no effect on body weight in uninephrectomized rats on a high-salt diet or in rats with renal artery stenosis (Figure 2A,C,E and Tables 1–3). The relative food consumption was not different between untreated and empagliflozin-treated animals (14.3 ± 1.4 vs. 13.8 ± 1.2 in FHH rats; 15.2 ± 1.6 vs. 16.8 ± 2.1 in UNX+HS rats, and 12.8 ± 1.6 vs. 14.3 ± 1.4 in 2K1C rats, NS; g/100 g BW).

Table 1. Body and organ weights, as well as blood pressure and TBARS levels, in fawn-hooded hypertensive (FHH) rats treated with empagliflozin (empa).

	FHH Untreated	FHH + empa
Body weight (g)	351 ± 5	322 ± 7 #
Relative heart weight (g/100 g BW)	0.287 ± 0.007	0.286 ± 0.007
Relative kidneys weight (g/100 g BW)	0.767 ± 0.028	0.900 ± 0.036 #
Relative weight of epididymal fat (g/100 g BW)	0.724 ± 0.016	0.603 ± 0.025 #
Relative weight of perirenal fat (g/100 g BW)	0.289 ± 0.018	0.176 ± 0.018 #
Systolic BP (mm Hg)	180 ± 7	173 ± 5
TBARS in kidneys	36.8 ± 1.4	39.3 ± 1.3

* denotes $p < 0.05$ vs. control group; # denotes $p < 0.05$ vs. untreated group; data are means ± SEM; and n = 7–10 for each group.

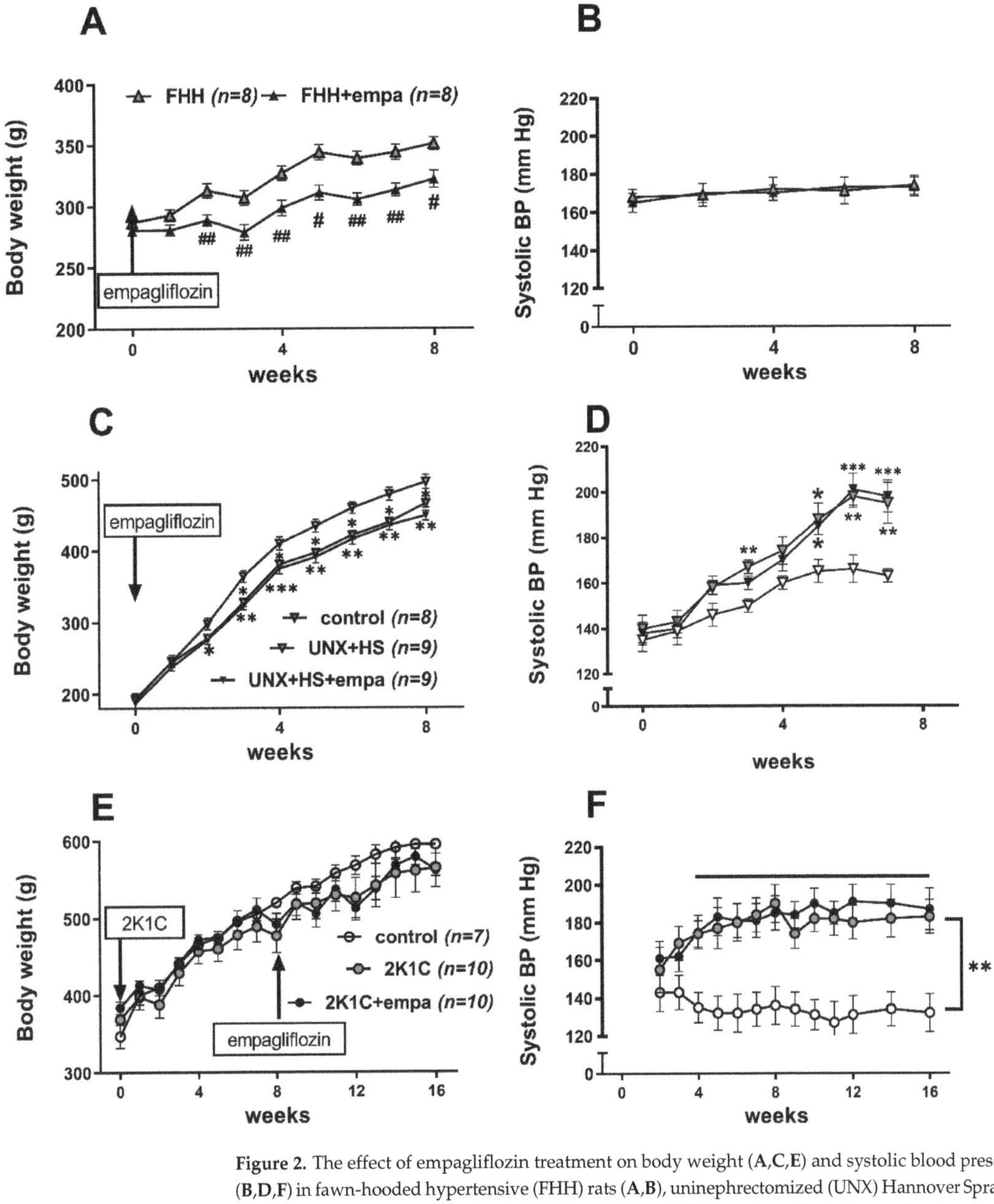

Figure 2. The effect of empagliflozin treatment on body weight (**A**,**C**,**E**) and systolic blood pressure (**B**,**D**,**F**) in fawn-hooded hypertensive (FHH) rats (**A**,**B**), uninephrectomized (UNX) Hannover Sprague Dawley (HanSD) rats on a high-salt (HS) diet (**C**,**D**), and Wistar rats subjected to renal artery stenosis (Goldblatt 2K1C hypertension) (**E**,**F**). * $p < 0.05$ vs. control group, # $p < 0.05$ vs. untreated group. * denotes $p < 0.05$; ** and ## denotes $p < 0.01$; *** denotes $p < 0.001$; data are means ± SEM. Horizontal line in (**F**) depicts time points with significantly different BP levels.

Table 2. Body and organ weights as well as blood pressure and TBARS levels in uninephrectomized salt-loaded (UNX + HS) Hannover Sprague Dawley (HanSD) rats treated with empagliflozin (empa).

	HanSD Control	UNX + HS	UNX + HS + empa
Body weight (g)	472 ± 10	461 ± 6	447 ± 7
Relative heart weight (g/100 g BW)	0.27 ± 0.02	0.33 ± 0.02 *	0.32 ± 0.02 *
Relative left ventricle (g/100g)	0.212 ± 0.005	0.264 ± 0.007 *	0.261 ± 0.009 *
Relative left kidney weight (g/100 g BW)	0.37 ± 0.01	0.74 ± 0.94 *	0.94 ± 0.04 *,#
Relative epididymal fat (g/100 g BW)	1.356 ± 0.081	0.891 ± 0.056 *	0.812 ± 0.028 *
Relative retroperitoneal fat (g/100 g BW)	1.349 ± 0.076	0.795 ± 0.077 *	0.678 ± 0.051 *
Systolic BP (mm Hg)	127 ± 57	166 ± 8 *	179 ± 8 *
TBARS in kidney	33 ± 3	44 ± 4 *	45 ± 2 *

* denotes $p < 0.05$ vs. control group; # denotes $p < 0.05$ vs. untreated group; data are means ± SEM; and $n = 7$–10 for each group.

Table 3. Body and organ weights, as well as blood pressure and TBARS levels, in Wistar rats subjected to renal artery stenosis (Goldblatt 2K1C hypertension) treated with empagliflozin (empa).

	Wistar Control	2K1C	2K1C + empa
Body weight (g)	595 ± 12	565 ± 26	562 ± 13
Relative heart weight (g/100 g BW)	0.212 ± 0.004	0.269 ± 0.017 *	0.269 ± 0.011 *
Relative left ventricle weight (g/100 g BW)	0.173 ± 0.003	0.226 ± 0.017 *	0.223 ± 0.010 *
Relative left kidney weight (g/100 g BW)	0.303 ± 0.006	0.180 ± 0.032 *	0.220 ± 0.030 *
Relative right kidney weight (g/100 g BW)	0.307 ± 0.007	0.419 ± 0.022 *	0.491 ± 0.028 *
Relative epididymal fat (g/100 g BW)	1.47 ± 0.06	1.33 ± 0.17	1.33 ± 0.13
Relative perirenal fat (g/100 g BW)	1.29 ± 0.10	1.15 ± 0.24	1.20 ± 0.17
Systolic BP (mm Hg)	120 ± 4	163 ± 9 *	157 ± 10 *
TBARS in kidney	31 ± 3	20 ± 2	24 ± 3

* denotes $p < 0.05$ vs. control group; # denotes $p < 0.05$ vs. untreated group; data are means ± SEM; and $n = 7$–10 for each group.

There was no change in blood pressure (measured by tail-cuff plethysmography) in FHH rats during the experiment (Tables 1–3). Moreover, empagliflozin treatment did not modify blood pressure and relative heart weight in this strain. In contrast, blood pressure substantially increased (by about 60 mm Hg in HanSD-UNX+HS and 40 mm Hg in Wistar-2K1C) following surgical procedures in both experimentally induced CKD models (Figure 2B,D,F). This was followed by an increase in their relative heart weights. Empagliflozin had no effect on tail-cuff BP in either experimental group. This was also confirmed by a direct BP measurement at the end of the study. Visceral adiposity was decreased by empagliflozin treatment only in FHH rats. Relative kidney weight was increased in empagliflozin-treated FHH and in uninephrectomized salt-loaded rats, while it was unchanged in 2-kidney-1-clip rats.

3.2. *Effects of Empagliflozin on Renal Parameters*

Urine production was not changed in FHH during the experiment, but as expected it was substantially decreased following nephrectomy, while renal artery stenosis had no effect on this parameter (Figure 3A,C,E). Empagliflozin treatment did not affect diuresis in either of three examined models of CKD, although sodium excretion was increased in FHH and in uninephrectomized HanSD rats, suggesting an effective SGLT-2 blockade (Figure 3B,D). In Wistar-2K1C, increased sodium excretion was observed with no further effect on empagliflozin administration.

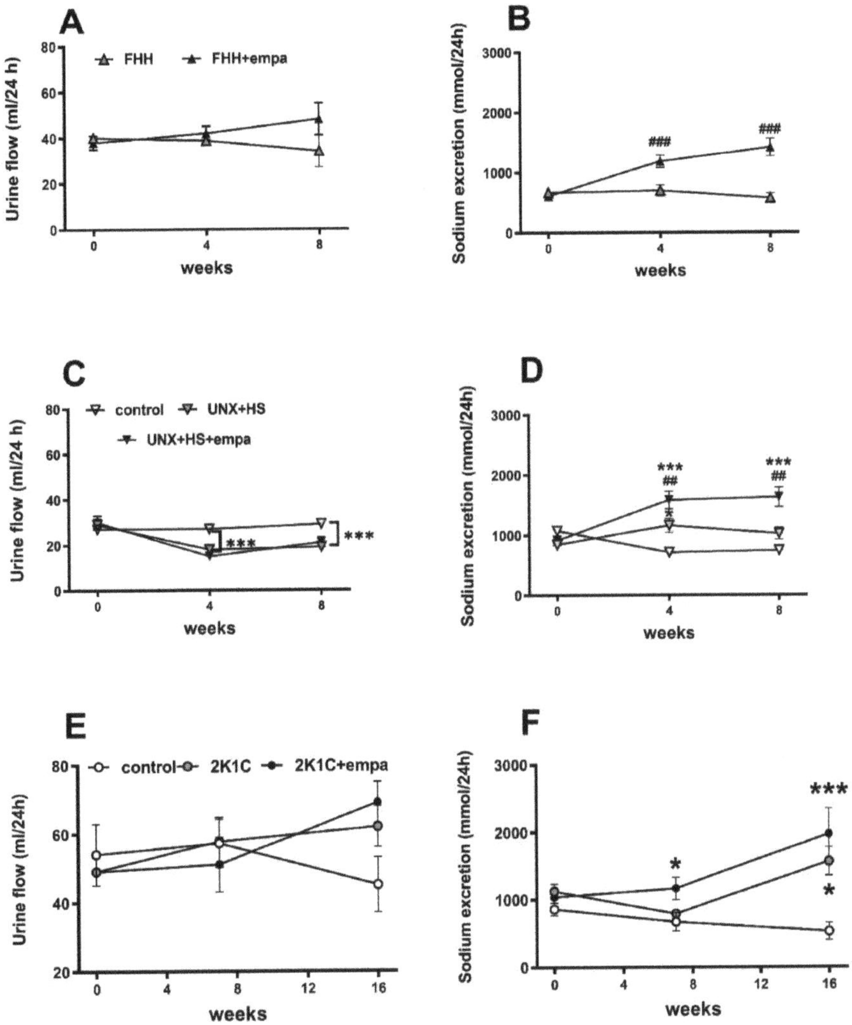

Figure 3. The effect of empagliflozin treatment on diuresis (**A**,**C**,**E**) and sodium excretion (**B**,**D**,**F**) in fawn-hooded hypertensive (FHH) rats (**A**,**B**), uninephrectomized (UNX) Hannover Sprague Dawley (HanSD) rats on high-salt (HS) diet (**C**,**D**), and Wistar rats subjected to renal artery stenosis (Goldblatt 2K1C hypertension) (**E**,**F**). * $p < 0.05$ vs. control group, # $p < 0.05$ vs. untreated group. * denotes $p < 0.05$; ## denotes $p < 0.01$; and ### and *** denotes $p < 0.001$; data are means ± SEM.

Proteinuria increased in all three CKD models during the study (Figure 4). Unexpectedly, empagliflozin treatment not only did not prevent the increase in proteinuria in CKD models but even enhanced it, with this effect being statistically significant in FHH rats. In line with this finding, the plasma urea level was increased in this particular rat strain at the end of the study. In contrast, similar values of plasma urea were found in HanSD-UNX + HS, with a profound increase in this particular rat strain at week 4 and lower values in empagliflozin-treated 2K1C Wistar rats.

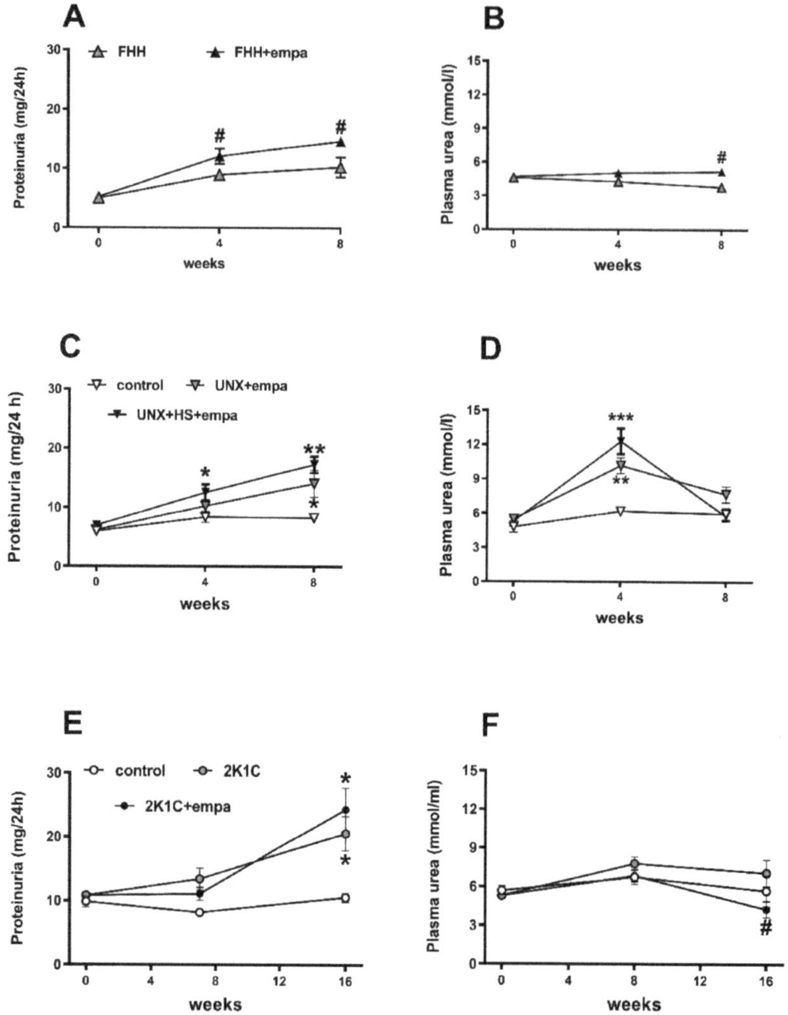

Figure 4. The effect of empagliflozin treatment on proteinuria (**A**,**C**,**E**) and plasma urea (**B**,**D**,**F**) in fawn-hooded hypertensive (FHH) rats (**A**,**B**), uninephrectomized (UNX) Hannover Sprague Dawley (HanSD) rats on high-salt (HS) diet (**C**,**D**), and Wistar rats subjected to renal artery stenosis (Goldblatt 2K1C hypertension) (**E**,**F**). * $p < 0.05$ vs. control group, # $p < 0.05$ vs. untreated group. * denotes $p < 0.05$; ** denotes $p < 0.01$; and *** denotes $p < 0.001$; data are means ± SEM.

At the end of the study, plasma creatinine was decreased in HanSD-UNX rats, while it was not changed in Wistar-2K1C and in FHH following empagliflozin treatment (Figure 5A,C,E).

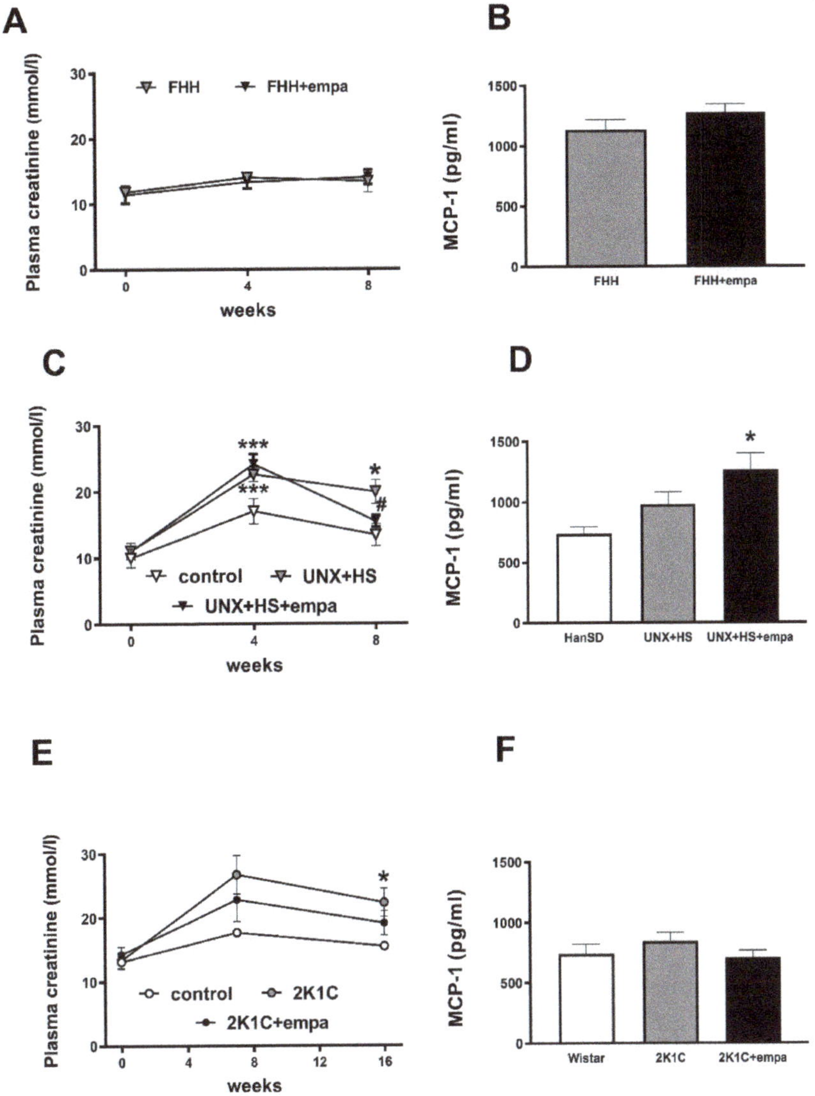

Figure 5. The effect of empagliflozin treatment on plasma creatinine (**A**,**C**,**E**) and on plasma monocyte attractant protein-1 (MCP-1) (**B**,**D**,**F**) in fawn-hooded hypertensive rats (**A**,**B**), uninephrectomized (UNX) Hannover Sprague Dawley (HanSD rats on high-salt (HS) diet (**C**,**D**), and Wistar rats subjected to renal artery stenosis (Goldblatt 2K1C hypertension) (**E**,**F**). * $p < 0.05$ vs. control group, # $p < 0.05$ vs. untreated group. * denotes $p < 0.05$ and *** denotes $p < 0.001$; data are means ± SEM.

3.3. *Effects of Empagliflozin on Oxidative Stress and ROS Production*

At the end of the study, elevated plasma MCP-1 levels (a marker of advanced CKD) were found in uninephrectomized empagliflozin-treated HanSD, while empagliflozin had no effect in other groups (Figure 5B,D,F). In the kidneys, lipoperoxidation products TBARS were increased only due to uninephrectomy, with no effect of empagliflozin treatment on this parameter in either of the three experimental models (Tables 1–3).

4. Discussion

In the present study, empagliflozin did not provide any evidence for renoprotection in three non-diabetic models of chronic kidney disease with a different pathophysiological background. In fact, empagliflozin did not improve proteinuria, plasma urea or plasma creatinine, and it did not decrease oxidative stress or inflammation in either model. This lack of kidney protection in our experimental models of CKD is in sharp contrast to the renoprotective effects of SGLT-2 inhibition demonstrated in several big clinical trials (EMPA-REG, DAPA-CKD) [1,5] in non-diabetic patients. The question arises whether these negative results could be ascribed to the relatively short duration of our studies, the gliflozin type and dose administered, the rat strain used, or a relatively early stage of kidney disease in our models. However, experimental studies of other investigators provide rather conflicting results ranging from beneficial effects [17–19,28], through disputable findings [15,16], to completely negative results [12–14].

Nevertheless, the pleiotropic effects of gliflozins were shown under non-diabetic conditions in many human and experimental studies, including ours. Our previous studies in several hypertensive non-diabetic models demonstrated that their benefits were mainly due to their effects on metabolic and hemodynamic parameters, with less effect on renal or cardiac parameters. Thus, a substantial blood-pressure lowering effect of empagliflozin treatment, together with decreasing body weight and adiposity, was achieved in Ren-2 transgenic rats, a model of angiotensin II-dependent hypertension, but there was little effect on renal parameters [22] and no effect on cardiac function (unpublished results). Similarly, we did not demonstrate the improvement of cardiac function in hereditary hypertriglyceridemic rats, in which beneficial effects were mediated mainly through the improvement of hepatic lipid metabolism [21]. In contrast, in SHR-CRP rats, empagliflozin benefit was based on reduced renal inflammation and oxidative stress [23], which is in line with the findings of Ali et al. [18], who also demonstrated a reduction in renal inflammation and ROS production in adenine-induced CKD. In addition, several studies ascribed renoprotection to the anti-fibrotic effects of gliflozins [17,19,28].

Based on the existing experimental studies, no general conclusion could be reached, suggesting that better results could be expected if the treatment would be prolonged. The problem is that there are no experimental studies evaluating gliflozin effects in the long-term setting, i.e., for months instead of weeks. Thus, the longest study, administering dapagliflozin for 12 weeks [12] or TA-1887 for 10 weeks [14], did not produce positive results, while 10-week treatment with a very low dose of empagliflozin [17] was renoprotective. In humans or in experimental animals with focal segmental glomerulosclerosis [13], eight-week dapagliflozin treatment did not modify renal hemodynamics or did not attenuate proteinuria. In contrast, one- or two-week treatments with empagliflozin or luseogliflozin demonstrated both beneficial and unfavorable effects [16,19,28].

It is also difficult to evaluate the sensitivity of the rat strain to gliflozin treatment. However, it seems that studies with Sprague Dawley (SD) rats demonstrated more negative results than those performed on Wistar rats. In fact, there is only one study demonstrating the beneficial effect of gliflozin in SD rats [19] compared to four negative studies [12–15], while the opposite is true for Wistar rats [16–18,20,28]. This might be due to the higher sensitivity of SD rats to the development of proteinuric kidney disease following nephrectomy (Kala et al., unpublished results), which could be further aggravated by a high-salt diet.

Gliflozin type and dose used should also be taken into consideration. Empagliflozin seems to be used more frequently than dapagliflozin or canagliflozin, or a newly synthetized luseogliflozin. The dose of empagliflozin used in experimental studies is usually 10 mg/kg/day (similar to humans), which, however, yielded inconsistent results. Even a very low dose of empagliflozin (0.6 mg/kg/day) reduced renal and cardiac fibrosis in 5/6NX Wistar rats on a high-salt diet, with these effects being comparable to those of angiotensin receptor blocker telmisartan [17]. In addition, Ali et al. [18] demonstrated the beneficial dose-dependent effects of canagliflozin (10 and 25 mg/kg/day) on adenine-induced CKD, demonstrating the improvement of renal parameters (albuminuria, creatinine clearance, and plasma urea) together with the attenuation of inflammation and oxidative stress. We cannot offer a plausible explanation for a lack of any positive effects of the generally accepted dose of empagliflozin in our experimental CKD models, although positive effects were obtained in our previous studies with hypertensive rat strains [22,23]. However, we cannot exclude the possibility that eight weeks following nephrectomy or 2K1C were insufficient for appropriate CKD development, although both of the other researchers and our previous experiments demonstrated a significant increase in proteinuria and a maximal decrease in creatinine clearance after the procedure [29–32]. Moreover, according to our studies in non-diabetic animals, it seems that the rat models with signs of metabolic disturbances respond to gliflozin therapy more effectively than those without them. This question remains to be analyzed in future studies.

Apart from the antiproteinuric effects of gliflozins in clinical trials in non-diabetic patients, blood pressure lowering was reported to be only 4-6 mm Hg [5]. Unexpectedly, we did not demonstrate any effect on blood pressure (monitored either by tail plethysmography or by direct BP measurement at the end of the study) in either of the examined models. In contrast, BP was reduced by about 25 mm Hg in Ren-2 transgenic rats [22]. Similar BP reduction was shown by Kim et al. [15] or Rajasekeran et al. [13] in uninephrectomized SD rats, while Zhang et al. [12] or Li et al. [14] did not find any blood pressure effect in the same model. Wan et al. [16] reported the blood-pressure-lowering effects of luseogliflozin measured by telemetry in uninephrectomized Wistar rats kept on high-salt intake. In this context, a comparative study evaluating the same treatment and dose in different rat strains would be rational.

Similar to our [21–23] and other investigations [13,19], we have found that relative kidney weight was increased in FHH and UNX-HS rats. Whether this increase is associated with the dilatation of tubular lumen due to SGLT-2-induced diuresis or due to tubular cell hypertrophy remains to be determined. In any case, it is not restricted only to empagliflozin treatment [19,21–23], as was also demonstrated with dapagliflozin administration [13].

5. Limitations of the Study

There are several limitations of our study concerning the translation of negative results into clinical practice. First of all, our study, similar to other investigators [13,17,20], evaluated the effects of empagliflozin treatment for a relatively short time (8 weeks), which is a substantially shorter duration than that applied in big clinical trials. Thus, our study could be negatively affected by the known effect of SGLT-2 inhibitors on the decreasing glomerular filtration rate and other renal functional parameters in the first weeks of gliflozin treatment. However, other studies performed by our group for the same time of empagliflozin administration [21–23] demonstrated the beneficial effects of empagliflozin, relying on different mechanisms of gliflozins in different non-diabetic hypertensive rat strains—metabolic mechanisms in hereditary hypertriglyceridemic rats [21], antihypertensive mechanisms in Ren-2 transgenic [22], or renoprotective mechanisms in spontaneously hypertensive rats expressing C-reactive protein [23]. Second, empagliflozin was the only SGLT-2 inhibitor used, so that the results cannot be simply generalized to all gliflozins. Moreover, it seems that newer gliflozins, such as luseogliflozin, ipragliflozin or ertugliflozin, could have better outcomes than the older ones (dapagliflozin, empagliflozin, and canagliflozin) or the combined SGLT1/2 inhibitors.

On the other hand, the analysis of Certikova Chabova and Cervenka [33] dealt with the opposite situation—the positive experimental results of dual RAS therapy (combined ACEi and ARB blockade) could not be translated into clinical studies using this drug combination in the treatment of cardiovascular diseases. They discussed not only the role of the variability of disease features in humans versus its relative stability in animals (including the genetic uniformity of animals) but also the selection of animal species for experimental studies. There is no doubt that rats and mice are very different from humans, and therefore pigs or primates would be more appropriate for translational research.

However, there are strong limitations concerning animal welfare that almost exclude the performance of such studies.

6. Conclusions

In our experimental models of chronic kidney disease, empagliflozin did not provide the expected beneficial effects on kidney function. Whether this lack of kidney protection is due to the short duration of the study, the models used, or the gliflozin administered remains to be determined. However, this finding provides further evidence about the cautious usage of this class of drugs in humans suffering from renal impairment.

Author Contributions: I.V. conceived and designed the project; S.H., Z.K., Z.V., H.R., M.B., P.K., and M.K. performed the experiments; S.H., I.V., and J.Z. analyzed data and interpreted the results; I.V. wrote the manuscript; and S.H., M.B., and J.Z. edited the manuscript. All authors have read and agreed to the published version of the manuscript.

Funding: This work was supported by institutional support of the Institute of Physiology, Czech Academy of Sciences, grant Nr. RVO 67985823 and by the Czech Science Foundation, grant Nr. 19-06199S.

Institutional Review Board and Statement: The study was conducted according to the guidelines of the Declaration of Helsinki and approved by the Institute of Physiology, the Czech Academy of Sciences (Protocol Nr. 47/2019).

Informed Consent Statement: Not applicable.

Data Availability Statement: All data arising from this study are contained within the article.

Acknowledgments: The technical assistance of Zdeňka Kopecká and Alena Charvátová is highly appreciated.

Conflicts of Interest: The authors declare no conflict of interest.

References

1. Zinman, B.; Wanner, C.; Lachin, J.M.; Fitchett, D.; Bluhmki, E.; Hantel, S.; Mattheus, M.; Devins, T.; Johansen, O.E.; Woerle, H.J.; et al. EMPA-REG OUTCOME Investigators, Empagliflozin, cardiovascular outcomes, and mortality in type 2 diabetes. *N. Engl. J. Med.* **2015**, *373*, 2117–2128. [CrossRef] [PubMed]
2. Neal, B.; Perkovic, V.; Mahaffey, K.W.; de Zeeuw, D.; Fulcher, G.; Erondu, N.; Shaw, W.; Law, G.; Desai, M.; Matthews, D.R.; et al. CANVAS Program Collaborative Group, Canagliflozin and cardiovascular and renal events in type 2 diabetes. *N. Engl. J. Med.* **2017**, *377*, 644–657. [CrossRef] [PubMed]
3. El Din, U.A.A.S.; Salem, M.M.; Abdulazim, D.O. Sodium-glucose cotransporter 2 inhibitors as the first universal treatment of chronic kidney disease. *Nefrologia* **2021**, *42*, 390–403. [CrossRef]
4. Cannon, C.P.; Perkovic, V.; Agarwal, R.; Baldassarre, J.; Bakris, G.; Charytan, D.M.; De Zeeuw, D.; Edwards, R.; Greene, T.; Heerspink, H.J.; et al. Evaluating the Effects of Canagliflozin on Cardiovascular and Renal Events in Patients With Type 2 Diabetes Mellitus and Chronic Kidney Disease According to Baseline HbA1c, Including Those With HbA1c <7%: Results From the CREDENCE Trial. *Circulation* **2020**, *141*, 407–410. [CrossRef]
5. Wheeler, D.C.; Stefansson, B.V.; Batiushin, M.; Bilchenko, O.; Cherney, D.Z.I.; Chertow, G.M.; Douthat, W.; Dwyer, J.P.; Escudero, E.; Pecoits-Filho, R.; et al. The dapagliflozin and prevention of adverse outcomes in chronic kidney disease (DAPA-CKD) trial: Baseline characteristics. *Nephrol. Dial. Transplant.* **2020**, *35*, 1700–1711. [CrossRef] [PubMed]
6. Zannad, F.; Ferreira, J.P.; Pocock, S.J.; Anker, S.D.; Butler, J.; Filippatos, G.; Brueckmann, M.; Ofstad, A.P.; Pfarr, E.; Jamal, W.; et al. SGLT2 inhibitors in patients with heart failure with reduced ejection fraction: A meta-analysis of the EMPEROR-Reduced and DAPA-HF trials. *Lancet* **2020**, *396*, 819–829. [CrossRef]
7. Ojima, A.; Matsui, T.; Nishino, Y.; Nakamura, N.; Yamagishi, S. Empagliflozin, an Inhibitor of Sodium-Glucose Cotransporter 2 Exerts Anti-Inflammatory and Antifibrotic Effects on Experimental Diabetic Nephropathy Partly by Suppressing AGEs-Receptor Axis. *Horm. Metab. Res.* **2015**, *47*, 686–692. [CrossRef]
8. Shin, S.J.; Chung, S.; Kim, S.J.; Lee, E.-M.; Yoo, Y.-H.; Kim, J.-W.; Ahn, Y.-B.; Kim, E.-S.; Moon, S.-D.; Kim, M.-J.; et al. Effect of Sodium-Glucose Co-Transporter 2 Inhibitor, Dapagliflozin, on Renal Renin-Angiotensin System in an Animal Model of Type 2 Diabetes. *PLoS ONE* **2016**, *11*, e0165703. [CrossRef] [PubMed]

9. Steven, S.; Oelze, M.; Hanf, A.; Kröller-Schön, S.; Kashani, F.; Roohani, S.; Welschof, P.; Kopp, M.; Gödtel-Armbrust, U.; Xia, N.; et al. The SGLT2 inhibitor empagliflozin improves the primary diabetic complications in ZDF rats. *Redox Biol.* **2017**, *13*, 370–385. [CrossRef] [PubMed]
10. Gallo, L.A.; Ward, M.S.; Fotheringham, A.K.; Zhuang, A.; Borg, D.J.; Flemming, N.B.; Harvie, B.M.; Kinneally, T.L.; Yeh, S.-M.; McCarthy, D.A.; et al. Once daily administration of the SGLT2 inhibitor, empagliflozin, attenuates markers of renal fibrosis without improving albuminuria in diabetic db/db mice. *Sci. Rep.* **2016**, *26*, 26428. [CrossRef]
11. Tang, L.; Wu, Y.; Tian, M.; Sjöström, C.D.; Johansson, U.; Peng, X.-R.; Smith, D.M.; Huang, Y. Dapagliflozin slows the progression of the renal and liver fibrosis associated with type 2 diabetes. *Am. J. Physiol. Metab.* **2017**, *313*, E563–E576. [CrossRef] [PubMed]
12. Zhang, Y.; Thai, K.; Kepecs, D.M.; Gilbert, R.E. Sodium-Glucose Linked Cotransporter-2 Inhibition Does Not Attenuate Disease Progression in the Rat Remnant Kidney Model of Chronic Kidney Disease. *PLoS ONE* **2016**, *11*, e0144640. [CrossRef] [PubMed]
13. Rajasekeran, H.; Reich, H.N.; Hladunewich, M.A.; Cattran, D.; Lovshin, J.A.; Lytvyn, Y.; Bjornstad, P.; Lai, V.; Tse, J.; Cham, L.; et al. Dapagliflozin in focal segmental glomerulosclerosis: A combined human-rodent pilot study. *Am. J. Physiol. Renal Physiol.* **2018**, *314*, F412–F422. [CrossRef] [PubMed]
14. Li, L.; Konishi, Y.; Morikawa, T.; Zhang, Y.; Kitabayashi, C.; Kobara, H.; Masaki, T.; Nakano, D.; Hitomi, H.; Kobori, H.; et al. Effect of a SGLT2 inhibitor on the systemic and intrarenal renin–angiotensin system in subtotally nephrectomized rats. *J. Pharmacol. Sci.* **2018**, *137*, 220–223. [CrossRef] [PubMed]
15. Kim, S.; Jo, C.H.; Kim, G.-H. Effects of empagliflozin on nondiabetic salt-sensitive hypertension in uninephrectomized rats. *Hypertens. Res.* **2019**, *42*, 1905–1915. [CrossRef] [PubMed]
16. Wan, N.; Fujisawa, Y.; Kobara, H.; Masaki, T.; Nakano, D.; Rahman, A.; Nishiyama, A. Effects of an SGLT2 inhibitor on the salt sensitivity of blood pressure and sympathetic nerve activity in a nondiabetic rat model of chronic kidney disease. *Hypertens. Res.* **2020**, *43*, 492–499. [CrossRef] [PubMed]
17. Zeng, S.; Delic, D.; Chu, C.; Xiong, Y.; Luo, T.; Chen, X.; Gaballa, M.M.; Xue, Y.; Chen, X.; Cao, Y.; et al. Antifibrotic effects of low dose SGLT2 Inhibition with empagliflozin in comparison to Ang II receptor blockade with telmisartan in 5/6 nephrectomised rats on high salt diet. *Biomed. Pharmacother.* **2022**, *146*, 112606. [CrossRef] [PubMed]
18. Ali, B.H.; Al-Salam, S.; Al Suleimani, Y.; Al Za'Abi, M.; Abdelrahman, A.M.; Ashique, M.; Manoj, P.; Adham, S.A.; Hartmann, C.; Schupp, N.; et al. Effects of the SGLT-2 Inhibitor Canagliflozin on Adenine-Induced Chronic Kidney Disease in Rats. *Cell. Physiol. Biochem.* **2019**, *52*, 27–39. [CrossRef] [PubMed]
19. Castoldi, G.; Carletti, R.; Ippolito, S.; Colzani, M.; Barzaghi, F.; Stella, A.; Zerbini, G.; Perseghin, G.; di Gioia, C.R. Renal Anti-Fibrotic Effect of Sodium Glucose Cotransporter 2 Inhibition in Angiotensin II-Dependent Hypertension. *Am. J. Nephrol.* **2020**, *51*, 119–129. [CrossRef] [PubMed]
20. Reyes-Pardo, H.; Bautista, R.; Vargas-Robles, H.; Rios, A.; Sanchez, D.; Escalante, B. Role of sodium/glucose cotransporter inhibition on a rat model of angiotensin II–dependent kidney damage. *BMC Nephrol.* **2019**, *20*, 292. [CrossRef] [PubMed]
21. Huttl, M.; Markova, I.; Miklankova, D.; Oliyarnyk, O.; Trnovska, J.; Kucera, J.; Sedlacek, R.; Haluzik, M.; Malinska, H. Metabolic cardio- and reno-protective effects of empagliflozin in a prediabetic rat model. *J. Physiol. Pharmacol.* **2020**, *71*, 635–645. [CrossRef]
22. Hojná, S.; Rauchová, H.; Malínská, H.; Marková, I.; Hüttl, M.; Papoušek, F.; Behuliak, M.; Miklánková, D.; Vaňourková, Z.; Neckář, J.; et al. Antihypertensive and metabolic effects of empagliflozin in Ren-2 transgenic rats, an experimental non-diabetic model of hypertension. *Biomed. Pharmacother.* **2021**, *144*, 112246. [CrossRef] [PubMed]
23. Malínská, H.; Hüttl, M.; Marková, I.; Miklánková, D.; Hojná, S.; Papoušek, F.; Šilhavý, J.; Mlejnek, P.; Zicha, J.; Hrdlička, J.; et al. Beneficial Effects of Empagliflozin Are Mediated by Reduced Renal Inflammation and Oxidative Stress in Spontaneously Hypertensive Rats Expressing Human C-Reactive Protein. *Biomedicines* **2022**, *10*, 2066. [CrossRef] [PubMed]
24. Provoost, A.P. Spontaneous glomerulosclerosis: Insights from the fawn-hooded rat. *Kidney Int. Suppl.* **1994**, *45*, S2–S5.
25. Drábková, N.; Hojná, S.; Zicha, J.; Vaněčková, I. Contribution of Selected Vasoactive Systems to Blood Pressure Regulation in Two Models of Chronic Kidney Disease. *Physiol. Res.* **2020**, *69*, 405–414. [CrossRef]
26. Lowry, O.H.; Rosebrough, N.J.; Farr, A.L.; Randall, R.J. Protein measurement with the Folin phenol reagent. *J. Biol. Chem.* **1951**, *193*, 265–275. [CrossRef]
27. Vaněčková, I.; Dobešová, Z.; Kuneš, J.; Zicha, J. The effects of repeated delivery of angiotensin II AT1 receptor antisense on distinct vasoactive systems in Ren-2 transgenic rats: Young vs. adult animals. *Hypertens. Res.* **2012**, *35*, 761–768. [CrossRef]
28. Abbas, N.A.T.; El Salem, A.; Awad, M.M. Empagliflozin, SGLT-2 inhibitor, attenuates renal fibrosis in rats exposed to unilateral ureteric obstruction: Potential role of klotho expression. *Naunyn Schmiedebergs Arch Pharmacol.* **2018**, *391*, 1347–1360. [CrossRef]
29. Cruz, C.; Correa-Rotter, R.; Sánchez-González, D.J.; Hernández-Pando, R.; Maldonado, P.D.; Martínez-Martínez, C.M.; Medina-Campos, O.N.; Tapia, E.; Aguilar, D.; Chirino, Y.I.; et al. Renoprotective and antihypertensive effects of S-allylcysteine in 5/6 nephrectomized rats. *Am. J. Physio.l Renal Physiol.* **2007**, *293*, F1691–F1698. [CrossRef]
30. Kujal, P.; Chábová, V.Č.; Vernerová, Z.; Walkowska, A.; Kompanowska-Jezierska, E.; Sadowski, J.; Vaňourková, Z.; Husková, Z.; Opočenský, M.; Skaroupková, P.; et al. Similar renoprotection after renin-angiotensin-dependent and –independent antihypertensive therapy in 5/6-nephrectomized Ren-2 transgenic rats: Are there blood pressure-independent effects? *Clin. Exp. Pharmacol. Physiol.* **2010**, *37*, 1159–1169. [CrossRef]
31. Hao, L.; Kanno, Y.; Fukushima, R.; Watanabe, Y.; Ishida, Y.; Suzuki, H. Effects of eplerenone on heart and kidney in two-kidney, one-clip rats. *Am. J. Nephrol.* **2004**, *24*, 54–60. [CrossRef] [PubMed]

32. Imamura, A.; Mackenzie, H.S.; Lacy, E.R.; Hutchison, F.N.; Fitzgibbon, W.R.; Ploth, D.W. Effects of chronic treatment with angiotensin converting enzyme inhibitor or an angiotensin receptor antagonist in two-kidney, one-clip hypertensive rats. *Kidney Int.* **1995**, *47*, 1394–1402. [CrossRef] [PubMed]
33. Chábová, V.C.; Červenka, L. The dilemma of dual renin-angiotensin system blockade in chronic kidney disease: Why beneficial in animal experiments but not in the clinic? *Physiol. Res.* **2017**, *66*, 181–192. [CrossRef] [PubMed]

biomedicines

MDPI

Article

Disparate Roles of Oxidative Stress in Rostral Ventrolateral Medulla in Age-Dependent Susceptibility to Hypertension Induced by Systemic L-NAME Treatment in Rats

Yung-Mei Chao [1], Hana Rauchová [2] and Julie Y. H. Chan [1,*]

[1] Institute for Translational Research in Biomedicine, Kaohsiung Chang Gung Memorial Hospital, Kaohsiung 833, Taiwan
[2] Institute of Physiology, Czech Academy of Sciences, 14200 Prague, Czech Republic
* Correspondence: jchan@adm.cgmh.org.tw; Tel.: +886-77338415

Abstract: This study aims to investigate whether tissue oxidative stress in the rostral ventrolateral medulla (RVLM), where sympathetic premotor neurons reside, plays an active role in age-dependent susceptibility to hypertension in response to nitric oxide (NO) deficiency induced by systemic L-NAME treatment, and to decipher the underlying molecular mechanisms. Systolic blood pressure (SBP) and heart rate (HR) in conscious rats were recorded, along with measurements of plasma and RVLM level of NO and reactive oxygen species (ROS), and expression of mRNA and protein involved in ROS production and clearance, in both young and adult rats subjected to intraperitoneal (i.p.) infusion of L-NAME. Pharmacological treatments were administered by oral gavage or intracisternal infusion. Gene silencing of target mRNA was made by bilateral microinjection into RVLM of lentivirus that encodes a short hairpin RNA (shRNA) to knock down gene expression of NADPH oxidase activator 1 (*Noxa1*). We found that i.p. infusion of L-NAME resulted in increases in SBP, sympathetic neurogenic vasomotor activity, and plasma norepinephrine levels in an age-dependent manner. Systemic L-NAME also evoked oxidative stress in RVLM of adult, but not young rats, accompanied by augmented enzyme activity of NADPH oxidase and reduced mitochondrial electron transport enzyme activities. Treatment with L-arginine via oral gavage or infusion into the cistern magna (i.c.), but not i.c. tempol or mitoQ_{10}, significantly offset the L-NAME-induced hypertension in young rats. On the other hand, all treatments appreciably reduced L-NAME-induced hypertension in adult rats. The mRNA microarray analysis revealed that four genes involved in ROS production and clearance were differentially expressed in RVLM in an age-related manner. Of them, *Noxa1*, and *GPx2* were upregulated and *Duox2* and *Ucp3* were downregulated. Systemic L-NAME treatment caused greater upregulation of *Noxa1*, but not *Ucp3*, mRNA expression in RVLM of adult rats. Gene silencing of *Noxa1* in RVLM effectively alleviated oxidative stress and protected adult rats against L-NAME-induced hypertension. These data together suggest that hypertension induced by systemic L-NAME treatment in young rats is mediated primarily by NO deficiency that occurs both in vascular smooth muscle cells and RVLM. On the other hand, enhanced augmentation of oxidative stress in RVLM may contribute to the heightened susceptibility of adult rats to hypertension induced by systemic L-NAME treatment.

Keywords: nitric oxide; L-NAME; hypertension; rostral ventrolateral medulla; aging; reactive oxygen species; NADPH oxidase activator 1; mitochondria

Citation: Chao, Y.-M.; Rauchová, H.; Chan, J.Y.H. Disparate Roles of Oxidative Stress in Rostral Ventrolateral Medulla in Age-Dependent Susceptibility to Hypertension Induced by Systemic L-NAME Treatment in Rats. *Biomedicines* **2022**, *10*, 2232. https://doi.org/10.3390/biomedicines10092232

Academic Editor: Marie-Louise Bang

Received: 20 July 2022
Accepted: 6 September 2022
Published: 8 September 2022

1. Introduction

Hypertension, a chronic condition of elevation in blood pressure (BP), is a well-known risk factor for the development of cardiovascular diseases (CVDs), including heart failure, coronary heart diseases, myocardial fibrosis, and infarction [1], atherosclerosis [1,2], stroke [1,3], and kidney failure [4]. The pathogenesis of hypertension is multifaceted, and one of the well-characterized causes is endothelial dysfunction resulting from nitric oxide (NO) deficiency [5–7]. Circulatory NO, derived primarily from endothelial NO synthase (eNOS) in vascular endothelial cells [8], participates in BP regulation via maintenance of vascular tone through the classical NO/soluble guanyl cyclase/cyclic guanosine monophosphate signaling to decrease intracellular calcium in vascular smooth muscle cells and evoke vasodilatation [9,10]. In addition, NO regulates cardiovascular functions by exhibiting anti-inflammatory [11], anti-platelet aggregatory [12], and anti-proliferative [13] actions. Accordingly, deficiency in NO production and/or availability leads to impaired vascular relaxation, increased vascular resistance, augmented platelet aggregation, and vascular inflammation or proliferation; all of which have been demonstrated to underpin the pathophysiology of hypertension [6,14,15].

A rat model of hypertension induced by the systemic administration of N^{ω}-Nitro-L-arginine methyl ester hydrochloride (L-NAME), a NOS inhibitor, is widely used to mimic hypertension in human [16]. This treatment downregulates NOS expression in blood vessels [17], and reduces plasma NO levels [18], resulting in systemic vasoconstriction, increased vascular resistance, and hypertension [17–19]. L-NAME further aggravates the development of hypertension via antagonization of the anti-inflammatory, anti-platelet aggregatory, and/or anti-proliferative actions of NO [20,21]. In addition to its peripheral actions, L-NAME reportedly crosses the blood–brain barrier (BBB) to decrease NOS expression and inhibit NOS activity in brain [22–24], leading to brain-initiated hypertension. The neural mechanism(s) that underlies the pathogenesis of this form of hypertension induced by systemic L-NAME treatment is not fully understood.

The rostral ventrolateral medulla (RVLM), where sympathetic premotor neurons reside [25], is one of the major sites of action in the brain stem where NO participates in the neural control of cardiovascular functions [6,26–28]. Emerging evidence from preclinical studies in animal models suggests that NO in RVLM exhibits a predominantly sympathoinhibitory action in the regulation of sympathetic nerve activity (SNA) under physiological conditions [28,29]. Deficiency in tissue NO availability under various cardiovascular conditions, including hypertension [29], chronic heart failure [30], and metabolic syndrome [31,32], may lead to an increase in SNA because of the blunted sympathoinhibitory action of NO in RVLM. Moreover, NO participates in the neural control of cardiovascular systems via interactions with reactive oxygen species (ROS), in particular superoxide anion ($O_2^{\bullet-}$) [6,26,32]. Under physiological conditions, NO is generated from its substrate L-arginine through coupling of NOS [33]. This coupling capacity is reduced under hypertensive conditions, wherein NOS removes an electron from its cofactor, NADPH, and donates it to an oxygen molecule to generate $O_2^{\bullet-}$ rather than NO [32–34]. The causal role of ROS at RVLM in systemic L-NAME treatment-induced hypertension, nonetheless, has not been fully elucidated.

Epidemiological studies indicate aging is a predominant risk factor for CVDs. The prevalence of hypertension increases markedly with aging, attributed primarily to alterations in the structure, responsiveness, function, and rigidity of vessel walls [35], as well as dysregulation of the autonomic nervous system [36]. Several theories have so far been proposed to explain the etiology of biological aging [37]; among them, tissue oxidative stress was postulated to be a common denominator [38]. Indeed, a wide array of studies suggests the engagement of ROS in age-related CVDs, including hypertension, atherosclerosis, atrial fibrillation, and stroke [38]. Despite evidence suggesting causal links between oxidative stress, aging, and CVDs, there is a paucity in the literature on the molecular mechanisms that underlie age-related dysregulation of redox homeostasis and its contribution to the development of CVDs.

Redox signaling is critically involved in cardiovascular pathophysiology, and oxidative stress in RVLM has been demonstrated unambiguously to play a pivotal role in neural mechanisms of hypertension via the increase in neurogenic sympathetic outflow [6,14,26–28]. However, there is no current information on whether redox homeostasis in RVLM is affected by systemic L-NAME treatment. Furthermore, exactly how aging changes the redox state in RVLM, leading to its engagement in hypertension induced by systemic L-NAME treatment is unknown. This study therefore aims to investigate whether tissue oxidative stress in RVLM plays an active role in age-dependent susceptibility to hypertension in response to systemic L-NAME treatment, and to decipher the underlying molecular mechanisms. Our findings suggest that redox homeostasis in RVLM is tolerated in young (8 weeks of age) normotensive rats subjected to systemic L-NAME treatment; and NO deficiency, both in the vascular smooth muscle cells and RVLM, may likely be the key underlying molecular mechanisms. When animals turn adult (20 weeks of age), tissue ROS level in RVLM is augmented, leading to an enhanced increase in central sympathetic outflow that exacerbates hypertension induced by L-NAME.

2. Materials and Methods

2.1. Animals

Experiments were carried out in young (4 weeks old, $n = 51$) and adult (12 weeks old, $n = 63$) male normotensive Wistar-Kyoto (WKY) rats purchased from BioLASCO, Taipei, Taiwan. They were housed in animal rooms under temperature control (24 ± 0.5 °C) and 12 h light/dark (5 am to 5 pm) cycle. Standard laboratory rat chow (PMI Nutrition International, Brentwood, MO, USA) and tap water were available ad libitum. All animals were allowed to acclimatize for 14 days (young group) or housed to the age of 20 weeks (adult group) in an AAALAC-International accredited animal holding facility in Kaohsiung Chang Gung Memorial Hospital, Taiwan, before experimental manipulations. All experiments were performed in accordance with the guidelines for animal experimentation approved by our institutional animal care and use committee (no. 2018051701) as adopted and promulgated by the U.S. National Institutes of Health.

2.2. Implantation of Osmotic Minipump

Implantation of osmotic minipump into the peritoneal cavity or the cisterna magna was carried out according to previously reported procedures [32]. Briefly, for intraperitoneal (i.p.) implantation, animals were anesthetized with pentobarbital sodium (50 mg/kg, i.p.) and a midline incision of the abdominal cavity was made, followed by implantation of an Alzet® osmotic minipump (model 2002, DURECT Co., Cupertino, CA, USA) into the peritoneal cavity. The incision was closed with layered sutures. Some animals also received implantation of an additional micro-osmotic minipump into the cistern magna. For intracisternal (i.c.) implantation, we performed a midline incision of the dorsal neck, and dissected the underneath muscle layers to expose the dura mater between the foramen magnum and C1 lamina. The dura mater was perforated with a 22-gauge steel needle, followed by insertion of a polyvinylchloride tubing into the cistern magna using an Alzet® brain infusion kit. Observation of presence of cerebrospinal fluid at the outer end of the catheter assured the patency of the implantation. Tissue glue was used to seal the catheter to the dura mater and the incision was closed with layered sutures. A micro-osmotic minipump (Alzet® 1007D) was positioned subcutaneously in the neck, and was connected to the outer end of the catheter. No obvious neurological signs were observed after minipump implantation. All animals received postoperative intramuscular injection of procaine penicillin (1000 IU). Only animals that showed recovery and progressive weight gain after the operation were used in subsequent experiments.

2.3. Blood Pressure and Heart Rate Measurement

Systolic blood pressure (SBP) and heart rate (HR) were determined in conscious animals with a tail-cuff sphygmomanometer (MK-2000; Momuroki Kikai Co., Tokyo, Japan). Baseline values were measured on days 3 and 1 prior to, on the day (day 0) before osmotic pump implantation, and then on days 1, 3, 5, 7, 9, 11, and 14 following i.p. L-NAME (10 mg/kg/day; Sigma-Aldrich, MA, USA) or 0.9% saline (vehicle control) treatment. During the recording sessions (at 14:00–16:00), rats were placed in restraint holders and the tail warmed on a warming pad for 10–15 min to increase blood flow and improve data acquisition. A full recording session consisted of 5 acclimatization cycles to optimize data acquisition, followed by 5 data acquisition cycles during which SBP and HR were recorded. The mean SBP and HR calculated for each time point were used for statistical analyses. It should be noted that SBP obtained by tail-cuff plethysmography in conscious rats has been validated to be comparable to those measured by radiotelemetry [32].

2.4. Evaluation of Sympathetic Vasomotor Activity

Sympathetic vasomotor activity was measured at the end of i.p. L-NAME infusion in young and adult animals from blood pressure recorded from a cannulated femoral artery in animals anesthetized via an anesthesia mask with isoflurane (5% for induction and 2% for maintenance). Each recording session lasted 30 min, and took place between 14:00 and 16:00. Temporal fluctuations in the low-frequency (LF, 0.25–0.8 Hz) component of the SBP signals were detected by continuous online, real-time spectral analysis based on fast Fourier transform using an arterial blood pressure analyzer (APR31a, Notocord, Le Pecq, France). The power density of the LF band was used as our experimental index to reflect sympathetic vasomotor tone [32].

2.5. Measurement of Plasma Norepinephrine

We measured plasma norepinephrine (NE) level by the o-phthalaldehyde (OPA) method using high-performance liquid chromatography (HPLC) with fluorescence detection. Plasma sample was mixed with ice-cold trichloroacetic acid, and centrifuged at room temperature. The supernatant was mixed with 4-fold methanol after being filtered through a syringe filter (0.22 μm; Chroma Technology Corp., Bellows Falls, VT, USA), centrifuged again at room temperature, and kept at −80 °C until analyses. NE was measured by HPLC as previously described [39], and comparing the area under the curve of each sample against standard NE solutions of known concentrations was used to compute the concentration (μg/μL). Each sample was analyzed in triplicates and the mean was used for statistical analyses.

2.6. Measurement of Serum Nitric Oxide

Serum NO concentration was assessed indirectly by measuring the levels of nitrate and nitrite, the stable end-product of NO, with a NO colorimetric assay kit (Arbor assay kit, Ann Arbor, MI, USA) according to manufacturer's instructions. The absorbance of the solution was read on a microplate reader at 540 nm (ThermoFisher Scientific Inc., Waltham, MA, USA). Each sample was analyzed in triplicates and the mean was used for statistical analyses.

2.7. Measurement of Plasma Malondialdehyde

Assays for lipid peroxidation are commonly used for estimation of oxidative status. Levels of lipid peroxidation in plasma were measured by a malondialdehyde (MDA; a primary indicator of lipid peroxidation) assay kit (Biovision, Milpitas, CA, USA), following the protocol provided by the manufacturer. Briefly, plasma samples were reacted with thiobarbituric acid (TBA) at 95 °C for 60 min. A microplate spectrophotometer (ThermoFisher Scientific) was used to determine the level of MDA-TBA adduct with colorimetric absorbance read at 532 nm. Each sample was analyzed in triplicates and the mean was used for statistical analyses.

2.8. Measurement of Plasma Proinflammatory Cytokines

The levels of proinflammatory cytokines, including interleukin-1 β (IL-1β), IL-6, and tumor necrosis factor-alpha (TNF-α), in plasma, were measured using anti-rat ELISA Kits (ThermoFisher Scientific) according to the manufacturer's specification. Plasma was collected and centrifuged for 10 min at 4 °C. The supernatants were used immediately to measure the concentrations of proinflammatory cytokines. Positive and negative controls were included on each plate. The final concentration of the cytokines was calculated by converting the optical density readings against a standard curve. Each sample was analyzed in triplicate and the mean was used for statistical analyses.

2.9. Tissue Collection and Protein Extraction from Rostral Ventrolateral Medulla

Animals were deeply anesthetized at the end of the experiments with an overdose of pentobarbital sodium (100 mg/kg, i.p.), followed by intracardial infusion with 500 mL of warm (37 °C) normal saline. The brain stem was rapidly removed and immediately frozen on ice. Using a rodent brain matrix (World Precision Instruments, Sarasota, FL, USA) and based on the atlas of Watson and Paxinos [40], the medulla oblongata covering RVLM was blocked between 0.5 and 1.5 mm rostral to the obex. Tissue from bilateral RVLM was collected by micropunches, and was stored at −80 °C until use.

For total protein extraction from RVLM, tissue micropunches were homogenized using a Dounce grinder with a tight pestle in ice-cold lysis buffer mixed with a cocktail of protease inhibitors (Sigma-Aldrich, St. Louis, MO, USA) to prevent protein degradation. Solubilized proteins were centrifuged at 20,000× *g* at 4 °C for 15 min, and the total protein in supernatant was quantified by the Bradford assay with a protein assay kit (Bio-Rad, Hercules, CA, USA).

2.10. Measurement of Reactive Oxygen Species in RVLM

RVLM tissues were homogenized in sodium phosphate buffer (20 mM), centrifuged and the supernatant was collected for ROS measurement by electron paramagnetic resonance (EPR) spin trapping technique, as described previously [41]. EPR spectra were captured using a Brucker EMXplus spectrometer (Bruker, Ettlingen, Germany). Typical parameters were set at microwave power: 20 mW, modulation frequency: 100 kHz; modulation amplitude: 2 G; time constant: 655.36 ms; conversion time: 656 ms; sweep time: 335.87 s. We added a membrane-permeable superoxide dismutase (SOD; 350 U/mL) into the incubation medium to determine ROS specificity. Spectra represented the average of 6 scans. Each sample was analyzed in triplicate and the mean was used for statistical analyses.

2.11. Measurement of Nitric Oxide in RVLM

For measurement of NO in RVLM tissue, tissue micropunches were homogenized in lysis buffer, centrifuged, and the supernatant was stored at −80 °C until use after being deproteinized using a Centricon-30 filtrator (Microcon YM-30, Bedford, MA, USA). Tissue NO levels were determined based on chemiluminescence reaction with the purge system of a NO analyzer (Sievers NOA 280^{TM}, Boulder, CO, USA) [32]. Each sample was analyzed in triplicate and the mean was used for statistical analyses.

2.12. Measurement of Nitric Oxide Synthase Activity in RVLM

Tissue NOS activity in RVLM was measured according to previously reported procedures [32]. RVLM tissues were lysed and centrifuged to obtain the supernatants, which were then used for detection of the enzyme activity following the manufacturer's instructions of a NOS activity assay kit (Merck KGaA, Darmstadt, Germany). After colorimetric reaction, the optical density was read using a microplate spectrophotometer (ThermoFisher Scientific) at an absorbance wavelength of 540 nm. Each sample was analyzed in triplicate and the mean was used for statistical analyses.

2.13. Measurement of NADPH Oxidase Activity in RVLM

NADPH oxidase activity of protein samples from RVLM was measured using the lucigenin-derived chemiluminescence method [32]. Briefly, the luminescence assay was performed in phosphate buffer with NADPH as the substrate. After dark adaptation, a tissue homogenate (100 μg protein) was added, and the chemiluminescence value was recorded. $O_2^{\bullet -}$ production was measured with the addition of NADPH, in the presence or absence of an NADPH oxidase, diphenyleneiodonium. All measurements were conducted in the dark room with temperature maintained at 22–24 °C. Light emission was recorded by a Sirius Luminometer (Berthold, Germany). Protein concentrations were determined using a Bio-Rad protein assay kit (Bio-Rad Laboratories). Each sample was analyzed in triplicate and the mean was used for statistical analyses.

2.14. Measurement of Total Antioxidant Activity in RVLM

Tissue antioxidant activity in RVLM was measured by a total antioxidant capacity assay kit (Sigma-Aldrich), following the protocol provided by the manufacturer. RVLM tissues were homogenized in lysis buffer, centrifuged, and the supernatant was used for analysis. The reaction was based on Cu^{2+} reduction by the small molecule antioxidants and the reduced Cu^{+} ion chelates with a colorimetric probe that was read with a standard 96-well spectrophotometric microplate reader at 570 nm. Antioxidant capacity was determined by comparison with Trolox, a water-soluble vitamin E analog that serves as an antioxidant standard. Each sample was analyzed in triplicate and the mean was used for statistical analyses.

2.15. Measurement of Mitochondrial Respiratory Enzyme Activity in RVLM

The mitochondrial fraction from RVLM tissue was isolated following procedures reported previously [42]. Purity of the mitochondrial-rich fraction was verified by the expression of the mitochondrial cytochrome *c* oxidase. Activities of mitochondrial respiratory chain enzymes were measured immediately after mitochondrial isolation, according to procedures reported previously [42] using a thermostatically regulated spectrophotometer (ThermoFisher Scientific). Enzyme activity was expressed in nmol/mg protein/min.

For the measurement of nicotinamide adenine dinucleotide (NADH) cytochrome *c* reductase (NCCR; enzyme for electron transport between ETC Complex I and Complex III) activity, mitochondrial fraction was incubated in a mixture containing K_2HPO_4 buffer, KCN, β-NADH, and rotenone at 37 °C for 2 min. After the addition of cytochrome *c* (50 μM), the reduction of oxidized cytochrome *c* was measured as the difference in the presence or absence of rotenone at 550 nm for 3 min at 37 °C.

For the determination of succinate cytochrome *c* reductase (SCCR; enzyme for electron transport between ETC Complex II and Complex III) activity, mitochondrial fraction was performed in the same buffer solution supplemented with succinate. After a 5 min equilibration at 37 °C, cytochrome *c* (50 μM) was added and the reaction was monitored at 550 nm for 3 min at 37 °C.

For the determination of cytochrome *c* oxidase (CCO, marker enzyme for ETC Complex IV) activity, mitochondria fraction was pre-incubated at 30 °C for 5 min in K_2HPO_4 buffer, then 45 μM ferrocytochrome *c* was added to start the reaction, which was monitored at 550 nm for 3 min at 30 °C. In all measurements, experiments were performed in triplicate, and the mean was used for statistical analyses.

2.16. Measurement of ATP Levels in RVLM

RVLM tissues were centrifuged at 10,000× *g* for 10 min after homogenization in a protein extraction solution (Pierce, Rockford, IL, USA), ATP concentration in the supernatant was determined by an ATP colorimetric assay kit (AbCam, Waltham, MA, USA) using a microplate reader (ThermoFisher Scientific). The ATP level was normalized to the protein concentration of the sample. Each measurement was performed in triplicate and the mean was used for statistical analyses.

2.17. Western Blot Analysis

We determined the expression levels of $gp91^{phox}$, $p22^{phox}$, $p67^{phox}$, and $p47^{phox}$ subunits of NADPH oxidase, manganese dismutase (SOD2), nuclear factor erythroid 2-related factor 2 (Nrf2), nNOS, iNOS, eNOS, NADPH oxidase activator 1 (Noxa1), uncoupling protein 3 (UCP3), and GAPDH in total protein extracted from RVLM by Western blot analysis [43]. In brief, 8–12% SDS-polyacrylamide gel electrophoresis was used for protein separation, and a Bio-Rad miniprotein-III wet transfer unit (Bio-Rad) was employed to transfer samples onto polyvinylidene difluoride transfer membranes (Immobilon-P membrane; Millipore, Bedford, MA) for 1.5 h at 4 °C. The transfer membranes were then incubated with a blocking solution (5% nonfat dried milk dissolved in Tris-buffered saline-Tween buffer) for 1 h at room temperature. Primary antisera used in this study included goat polyclonal, rabbit polyclonal or monoclonal, or mouse monoclonal antiserum against gp91^{phox} (1:5000; BD Biosciences, Sparks, MD, USA), p22^{phox}, p67^{phox}, and p47^{phox} (1:5000; Santa Cruz Biotechnology, Santa Cruz, CA, USA), SOD2 (1:3000; Stressgen, San Deigo, CA, USA), Nrf2 (1:1000; Santa Cruz), nNOS, iNOS, and eNOS (1:1000; BD Biosciences), Noxa1 (1:1000, Santa Cruze), UCP3 (1:1000, Santa Cruz), and GAPDH (1:10,000; Merck). Membranes were subsequently washed three times with TBS-t buffer, followed sequentially by incubation with the secondary antibodies (1:10,000; Jackson ImmunoResearch, West Grove, PA, USA) for 1 h and horseradish peroxidase-conjugated goat anti-rabbit IgG or goat anti-mouse IgG (Jackson ImmunoResearch). An enhanced chemiluminescence Western blot detection system (GE Healthcare Bio-Sciences Corp., Piscataway, NJ, USA) was used to detect specific antibody–antigen complex. For the detection of eNOS or nNOS dimerization, nondenaturing, low-temperature sodium dodecyl sulfate polyacrylamide gel electrophoresis was used [32,34]. During the electrophoresis process and transfer of proteins to nitrocellulose membrane, buffers were placed in an ice-water bath and the whole apparatus was kept at 4 °C. ImageJ software (NIH, Bethesda, MD, USA) was used to quantify the number of detected proteins, which was expressed as the ratio to loading control (GAPDH).

2.18. Generation of Lentiviral Vector

NOXa1 shRNA lentiviral particles (sc-150038-V, Santa Cruz) were used in gene silencing experiments. These transduction-ready viral particles contain a target-specific construct that encodes a 19–25 nt (plus hairpin) shRNA designed to knock down gene expression of *Noxa1*. Each vial contains 200 µL frozen stock of 1.0×10^6 infectious units of virus (IFU) in Dulbecco's Modified Eagle's Medium with HEPES pH 7.3 (25 mM). Control shRNA lentiviral particles (sc-108080, Santa Cruz) contain an shRNA construct that encodes a scrambled sequence that will not lead to the specific degradation of any known cellular mRNA.

2.19. Microinjection of Lentiviral Vectors into RVLM

Microinjection of the Lv-Noxa1-shRNA, or scramble (Lv-scr-RNA), was carried out stereotaxically and sequentially into the bilateral RVLM of rats that were anesthetized with sodium pentobarbital (50 mg/kg, i.p.). Adequate anesthesia of animals was confirmed by observations of unresponsive to paw pinch and no corneal withdrawal reflex. The animals were placed into a stereotaxic head holder (Kopf, Tujunga, CA, USA) on a thermostatically controlled heating pad. Bilateral microinjection of the viral vectors was carried out, as described previously [32,44]. In brief, a glass micropipette (external tip diameter: 50–80 µm), connected to a 0.5-µL Hamilton microsyringe, was positioned into RVLM. A total of eight injections (4 on each side) of undiluted viral particles (200 nl total volume on each side) were made at two rostro-caudal levels at stereotaxic coordinates of 4.5–5.0 mm posterior to lambda, 1.8–2.1 mm lateral to the midline, and 8.0–8.5 mm below the dorsal surface of cerebellum. These coordinates cover the confines of RVLM within which sympathetic premotor neurons reside [25,40]. After the lentivirus injection, the wound was closed in layers, and animals were allowed to recover in individual cages with free access to food and water.

2.20. Reverse Transcription and Quantitative Polymerase Chain Reaction

Total RNA from RVLM tissues was isolated with TRIzol reagent (Invitrogen, Carlsbad, CA, USA) according to the manufacturer's protocol. All RNA isolated was quantified by spectrophotometry and the optical density 260/280 nm ratio was determined. Reverse transcriptase (RT) reaction was performed using a SuperScript Preamplification System (Invitrogen) for the first-strand cDNA synthesis.

Noxa1 and *Ucp3* mRNA levels were analyzed by quantitative polymerase chain reaction (qPCR) using SYBR Green and normalized to the GAPDH mRNA signal as described [44]. The following primers were used: *Noxa1*: 5′-TTA CTC TGC CCC TGA AGG TC-3′ (forward) and 5′-CTC GGG CTT TGT TGA AC-3′ (reverse); *Ucp3*: 5′-TTC CTG GGG GCC GGC ACT G-3′ (forward) and 5′-CAT GGT GGA TCC GAG CTC GGT AC-3′ (reverse) [45]; and *GAPDH*: 5′-AGA CAG CCG CAT CTT CTT GT-3′ (forward), 5′-CTT GCC GTG GGT AGA GTC AT-3′ (reverse). Noxa1 and *Ucp3* mRNA were amplified under the following conditions: 95 °C for 3 min, followed by 50 cycles consisting of 95 °C for 10 s, 50 °C for 20 s, and 72 ° C for 2 s, and finally a 10 min extension at 40 °C. GAPDH was amplified under identical conditions, with the exception of a 55 °C primer annealing temperature. All samples were analyzed in triplicate. All qPCR reactions were followed by dissociation curve analysis. Relative quantification of gene expression was performed using the $2^{\Delta\Delta CT}$ method.

For amplification of oxidative stress-related mRNA, RT^2 Profiler PCR Arrays (Qiagen GmbH, Hilden, Germany) were employed following the manufacturer's protocol. The microarrays include primer assays for 84 oxidative stress-focused genes, 5 housekeeping genes, a genomic DNA control, 3 wells containing reverse-transcription controls, and 3 wells containing a positive PCR control (Supplementary Figure S1). Total RNA from RVLM tissues was converted into first-strand cDNA using the RT^2 First Strand Kit. The cDNA was next mixed with an appropriate RT^2 SYBR® Green Mastermix. This mixture was then aliquoted into the wells of the RT^2 Profiler PCR Array. PCR was performed, and the relative expression was determined using data from the real-time cycler and the $2^{\Delta\Delta CT}$ method.

2.21. Experimental Design

Figure 1 illustrates the experimental design of the present study. The first group of young and adult rats (n = 6 per group) was used to evaluate the effect of i.p. L-NAME on SBP, HR, LF component of SAP signals, plasma NE and serum NO levels, plasma MAD and proinflammatory cytokine, and tissue ROS levels and expression of proteins for the production and degradation of ROS in RVLM. The hemodynamic parameters were recorded on days 3 and 1 prior to, and on the day (day 0) before, osmotic pump implantation and days 1, 3, 5, 7, 9, 11, and 14 following i.p. infusion of L-NAME or 0.9% saline. NE and NO were measured on day 3 before, and days 0, 7, and 14 following L-NAME treatment. At the end of the 14-day infusion, the power density of the LF component of SAP signals was determined before animals were killed to collect blood for proinflammatory cytokine measurements, and collect RVLM tissue for measurements of ROS and protein expressions.

The protocol was repeated in a second group of young and adult animals to evaluate various treatments (n = 5 per group; see below) on hemodynamic and/or biochemical changes induced by i.p. L-NAME infusion. The pharmacological manipulations, including oral gavage or i.c. infusion, were performed from days 7 to 14 during the 14-day L-NAME treatment period. BP was recorded on day 3 before, and on days 0, 3, 7, 9, 11, and 14 following L-NAME infusion, and power density of LF component, tissue levels of ROS and NO in RVLM were determined at the end of the treatment period.

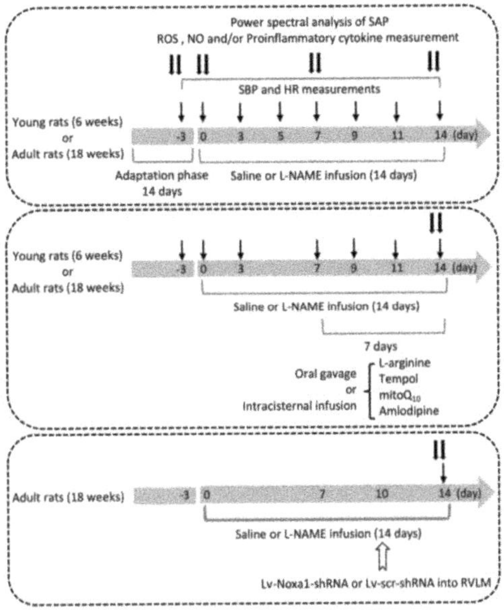

Figure 1. Experimental design of the present study. The first group of young and adult animals ($n = 6$ per group) was used to assess systolic blood pressure (SBP), heart rate (HR), and power density of low-frequency component in the SBP spectrum at various time intervals (arrows) before and after i.p. infusion of L-NAME for 14 days. Nitric oxide (NO), reactive oxygen species (ROS), and/or proinflammatory cytokine in plasma and/or tissue of rostral ventrolateral medulla (RVLM) were measured at the end of the 14-day L-NAME treatment (double arrows). The protocol was repeated in a second group of young and adult rats ($n = 5$ per group) to evaluate the effect of various treatments, delivered via oral gavage or intracisternal infusion during days 7–14, on L-NAME-induced changes in SBP, HR, LF power, and NO and ROS levels in RVLM. The third group of adult animals ($n = 6$ per group) was used to examine the effect of gene silencing NADPH oxidase activator 1 (*Noxa1*) in RVLM on changes in SBP, HR, LF power, and NO and ROS levels in RVLM induced by systemic L-NAME. Lentiviral vector contains a target-specific construct that encodes a short hairpin RNA (shRNA) to knock down gene expression of *Noxa1* (Lv-Noxa1-shRNA) or control scrambled shRNA (Lv-scr-shRNA) was microinjected into the bilateral RVLM (open arrow) on day 10 after L-NAME infusion.

The third group of young and adult animals was used to identify candidate genes discriminately expressed in RVLM of adult animals ($n = 3$ per group), and the functional significance of the identified mRNA in susceptibility to hypertension induced by L-NAME ($n = 6$ per group) in adult animals. Gene manipulation was carried out via bilateral microinjection of lentiviral vectors into RVLM on day 10 following L-NAME infusion, and mRNA and protein expressions, as well as SBP were measured at the end of the 14-day infusion.

Treatments employed in the present study included i.p. infusion of L-NAME (10 mg/kg/day; Sigma-Aldrich, MA, USA), oral intake via gavage of L-arginine (2%, Sigma-Aldrich), 4-hydroxy-2,2,6,6-tetramethylpiperidine-1-oxyl (tempol; 100 μmol/kg; Sigma-Aldrich), or amlodipine (10 mg/kg; Sigma-Aldrich); or i.c. infusion of L-arginine (2 μg/kg/day), tempol (1 μmol//h/μL), or mitoQ_{10} (2.5 μmol//h/μL; Sigma-Aldrich); or microinjection bilaterally into RVLM of Lv-Noxa1-shRNA (1×10^5 IFU per animal) or Lv-scr-shRNA (1×10^5 IFU per animal). Control infusion of 0.9% saline (for i.p. or oral gavage treatment) or artificial CSF (aCSF; for i.c. infusion) served as the volume and vehicle control. The composition of aCSF was (mM): NaCl 117, $NaHCO_3$ 25, Glucose 11, KCl 4.7, $CaCl_2$ 2.5, $MgCl_2$ 1.2, and NaH_2PO_4.

2.22. Statistical Analysis

All data were presented as mean ± standard deviation (SD). The normality of the data distribution was checked before all the statistical analyses using Shapiro–Wilk test to confirm that the data complied with normal distribution. Differences in SBP and HR to various treatments were analyzed with a two-way analysis of variance (ANOVA) with repeated measures, followed by the Tukey multiple comparisons test using time and treatment group as the main factors. All the other differences in mean values were analyzed by one-way ANOVA with Tukey's multiple comparisons tests. Statistical differences between experimental groups in young and adult animals were evaluated using unpaired Student's *t*-tests. All the data were analyzed by GraphPad Prism software (version 6.0; GraphPad Software Inc., La Jolla, CA, USA). $p < 0.05$ was considered statistically significant.

3. Results

3.1. Age-Dependent Changes in Blood Pressure, Heart Rate, Sympathetic Vasomotor Activity, and Plasma NE Levels in Response to Systemic NO Deficiency

Our first set of experiments evaluated the age-dependent hemodynamic responses to systemic NO deficiency, a well-established animal model for the study of human hypertension [16]. In young (at age of 8 weeks) normotensive WKY rats, i.p. infusion of L-NAME (10 mg/kg/day) for 14 days resulted in gradual increases in SBP, power density of the LF component of the SAP signals, our experimental index for neurogenic sympathetic vasomotor activity [32,43], and plasma NE levels, but not HR, which became statistically significant on postinfusion days 7−14 (Figure 2A−D). Similar cardiovascular responses were observed in adult animals (at age of 20 weeks). Of note, baseline SBP (105 ± 4.4 versus 90.7 ± 5.4 mmHg, $n = 6$, $p < 0.05$), as well as temporal increases in SBP (139.8 ± 3.5 versus 125.5 ± 6.5 mmHg, $n = 6$; $p < 0.05$), LF power density (2.76 ± 0.26 versus 2.39 ± 0.24 mmHg^2, $n = 6$, $p < 0.05$), and plasma NE levels (7.16 ± 0.33 versus 6.40 ± 0.46 ng/mL, $n = 6$, $p < 0.05$) measured on day 14 postinfusion, were significantly greater in adult animals when compared to young rats. L-NAME infusion evoked similar decreases in serum NO (nitrite and nitrate) levels, detected on days 7 and 14 postinfusion in both age groups (Figure 2E). The same treatment, on the other hand, exerted no effect on plasma Il-1β, IL-6, and TNF-α levels measured at the end of L-NAME infusion (Table 1). These data suggest an age-dependent vulnerability in hemodynamic changes associated with systemic NO deficiency.

Table 1. Changes in plasma levels of proinflammatory cytokines in young (8 weeks) and adult (20 weeks) rats in response to i.p. infusion of saline or L-NAME.

	Saline		**L-NAME**	
	Young	**Adult**	**Young**	**Adult**
IL-1β (ng/mL)	0.58 ± 0.11	0.47 ± 0.21	0.69 ± 0.31	0.61 ± 0.34
IL-6 (pg/mL)	208 ± 42	186 ± 39	223 ± 68	217 ± 56
TNF-α (pg/mL)	89 ± 38	105 ± 43	93 ± 54	119 ± 72

Saline or L-NAME (10 mg/kg/day) was infused into the peritoneal cavity for 14 days. Data are presented as mean ± SD, $n = 6$ per group. No significant difference exists between groups in One-Way ANOVA. IL-1β, interleukin 1-β; IL-6, interleukin 6; TNF-α, tumor necrosis factor α.

3.2. Effect of Systemic L-NAME Treatment on Expression of NOS Isoforms and Activity in RVLM

L-NAME reportedly crosses the BBB to alter NOS expression and inhibit NOS activity in brain [22–24]. At the end of 14-day i.p. infusion of L-NAME, protein expression of eNOS, but not nNOS or iNOS, isoform in RVLM was significantly decreased (Figure 3A), alongside a notable suppression of NOS activity (Figure 3B) in both young and adult animals. Of note, systemic L-NAME treatment resulted in comparable suppression in eNOS expression and total NOS activity in RVLM of both age groups.

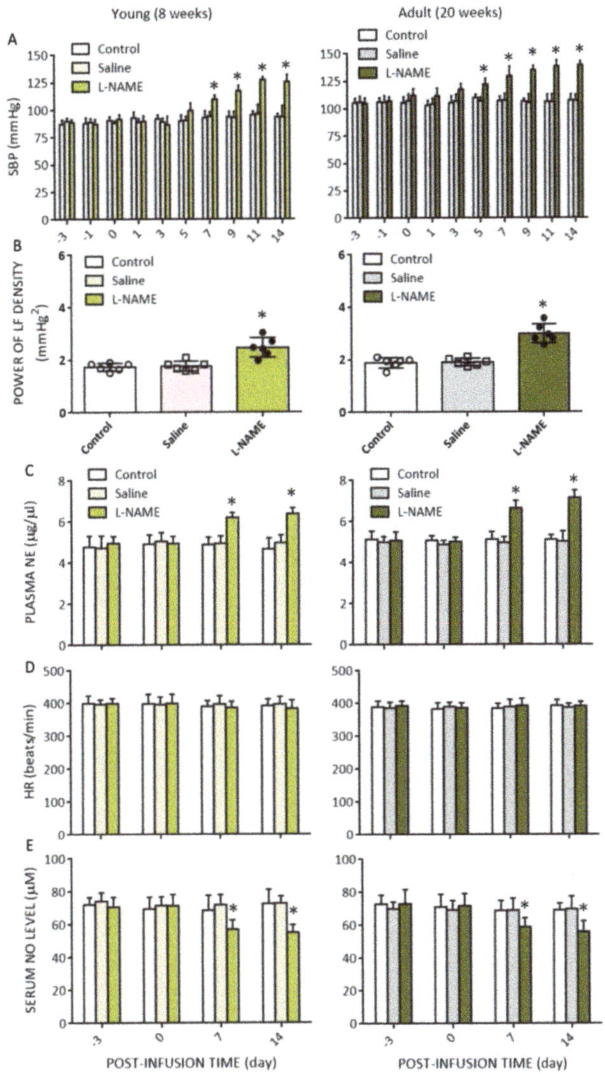

Figure 2. Temporal changes in hemodynamic parameters and serum NO (nitrite and nitrate) levels in response to intraperitoneal infusion of L-NAME (10 mg/kg/day) for 14 days. Changes in (**A**) systolic blood pressure (SBP); (**B**) power density of low-frequency (LF) component of SBP signal; (**C**) plasma norepinephrine (NE) levels; (**D**) heart rate (HR); as well as (**E**) serum NO levels detected at different time points in the untreated group, or animals treated with i.p. infusion of saline or L-NAME. Data are presented as mean ± SD, $n = 6$ per group at each time interval. * $p < 0.05$ versus saline-treated group (pink or gray bars) in post hoc Tukey's multiple-range test.

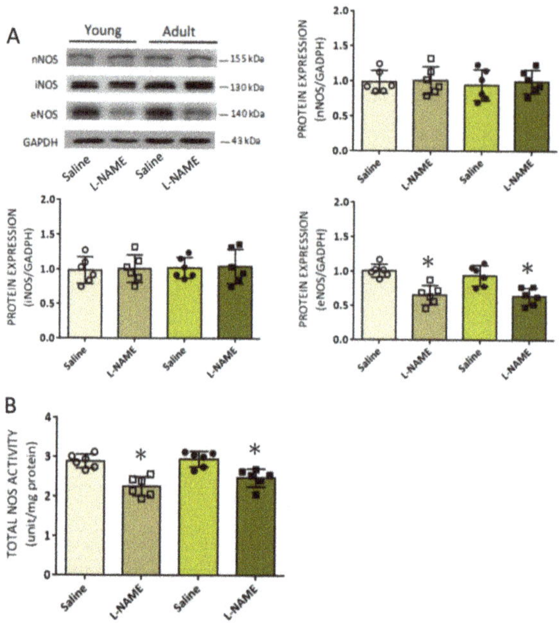

Figure 3. Effect of systemic L-NAME treatment on the expression of NOS isoforms and NOS activity in RVLM: (**A**) Representative gels (insets) and densitometric analysis of results from Western blot analysis showing changes in protein expression of nNOS, iNOS, and eNOS and (**B**) enzyme activity of NOS in RVLM 14 days after i.p. infusion of saline or L-NAME (10 mg/kg/day) in young (8 weeks, open circles or squares) or adult (20 weeks, filled circles or sqaures) rats. Data on protein expression were normalized to the respective saline control value, which is set to 1.0, and are presented as mean ± SD, n = 6 per group. * $p < 0.05$ versus corresponding saline-treated group in unpaired Student's t-test.

3.3. Differential Effect of Systemic L-NAME Treatment on Tissue ROS Levels in RVLM of Animals at Different Ages

A series of studies from our laboratory [32,39,42,43] suggest that the LF component of SAP signals originates from the RVLM, and tissue oxidative stress in RVLM augments sympathetic vasomotor activity and BP [28,32,39,41–44]. We therefore investigated whether the differential effect of L-NAME infusion on hemodynamic parameters at different ages is the consequence of disparate tissue oxidative stress in RVLM. As shown in Figure 4A, baseline tissue ROS levels were higher, albeit statistically insignificant, in RVLM of adult animals. Moreover, i.p. L-NAME (10 mg/kg/day) infusion for 14 days resulted in further increases in tissue ROS levels in RVLM of adult, but not young rats. On the other hand, there were no detectable increases in plasma MDA levels (a biomarker of oxidative stress) in both groups (Figure 4B).

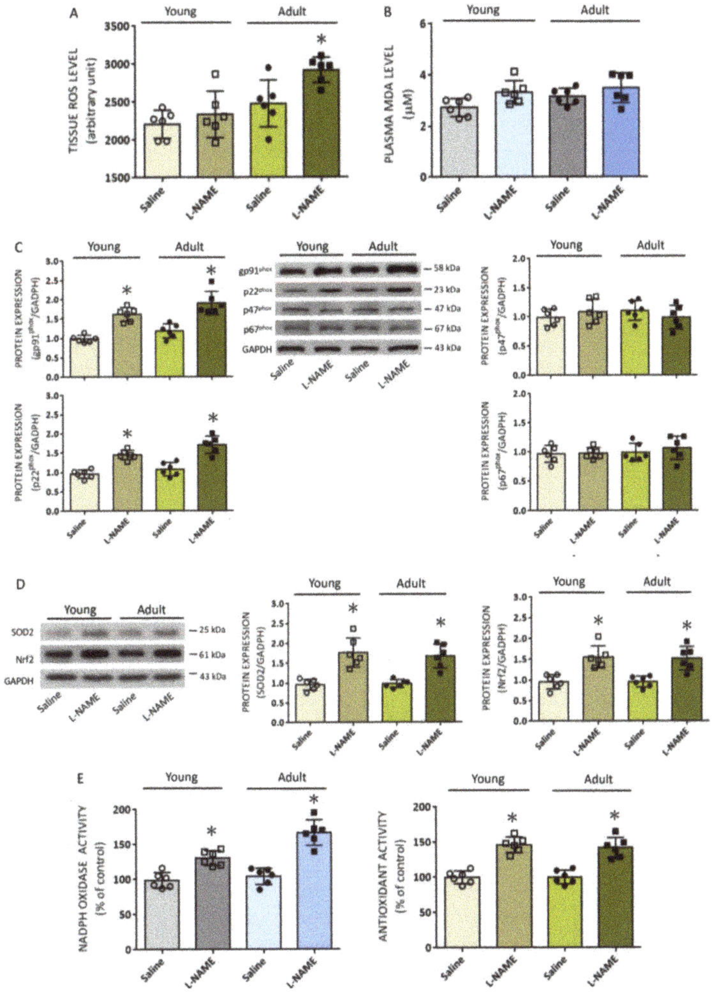

Figure 4. Effect of systemic L-NAME treatment on reactive oxygen species (ROS) levels and expression of proteins involved in ROS production and degradation. Data showing ROS levels in RVLM (**A**) and MDA levels in plasma (**B**) of young (8 weeks, open circles or squares) or adult (20 weeks, filled circles or squares) rats after i.p. infusion of saline or L-NAME (10 mg/kg/day) for 14 days. Also shown are representative gels (insets) and densitometric analysis of results from Western blot changes in protein expression of pg91^{phox}, p22^{phox}, p47^{phox}, and p67^{phox} (**C**) or SOD2 and Nrf2 (**D**), as well as enzyme activity of NADPH oxidase and total antioxidants (**E**) in RVLM 14 days after saline or L-NAME treatment. Data on protein expression are normalized to the respective saline control value, which is set to 1.0. Data are presented as mean $\pm$ SD, $n = 6$ per group. * $p < 0.05$ versus corresponding saline-treated group in unpaired Student's t-test.

Dysregulated redox homeostasis because of an imbalance in ROS production over degradation leads to tissue oxidative stress [6,26,28,33]. In RVLM, we reported previously that increases in the protein expression of NADPH oxidase subunits [43], and decreases in the expression of antioxidants [46], contribute to oxidative stress that results in sympathoexcitation and hypertension in spontaneously hypertensive rats and normotensive animals treated with angiotensin II (Ang II). We therefore examined the expression of NADPH oxidase subunits and antioxidants in RVLM of animals that were subjected to i.p. L-NAME infusion. In RVLM of WKY rats at age of 8 weeks, the protein expression of pg91^{phox} and p22^{phox}, but not p47^{phox} or p67^{phox} subunit of NADPH oxidase (Figure 4C), determined on day 14 following i.p. infusion of L-NAME, was appreciably increased. Interestingly, the protein expression of two key antioxidants, SOD2 and Nrf2 (Figure 4D), was also increased at the same postinfusion time point. Similar results were found in RVLM of L-NAME-treated animals at the age of 20 weeks. As shown in Figure 4E, systemic L-NAME treatment also significantly augmented the enzyme activity of NADPH oxidase and total antioxidant capacity in RVLM, measured on day 14 postinfusion in both age groups. Notably, the increase in NADPH oxidase activity in RVLM was significantly greater in the adult animals (+66.5 ± 13.8% versus +30.5 ± 10.2 %, $n = 6$, $p < 0.05$).

Mitochondria are considered another important cellular source of ROS. In RVLM, impairment of enzyme activity of the mitochondrial electron transport chain (ETC) for oxidative phosphorylation contributes to cellular oxidative stress, leading to sympathetic hyperactivity and neurogenic hypertension [42]. In young normotensive rats, systemic L-NAME treatment increased the enzyme activity of CCO (electron transport in Complex IV), but not NCCR (enzyme for electron transport between Complexes I and III) or SCCR (enzyme for electron transport between Complexes II and III) in RVLM (Figure 5A), accompanied by a mild increase in tissue ATP content (15.8 ± 2.1 versus 17.8 ± 1.3 pmol/μg, $n = 6$, $p = 0.071$). On the other hand, the enzyme activity of both NCCR and CCO was notably depressed in RVLM of control adult rats, and remained reduced following L-NAME treatment; together with a moderate decrease in tissue ATP content (15.4 ± 1.3 versus 13.2 ± 1.3 pmol/μg, $n = 6$, $p < 0.05$).

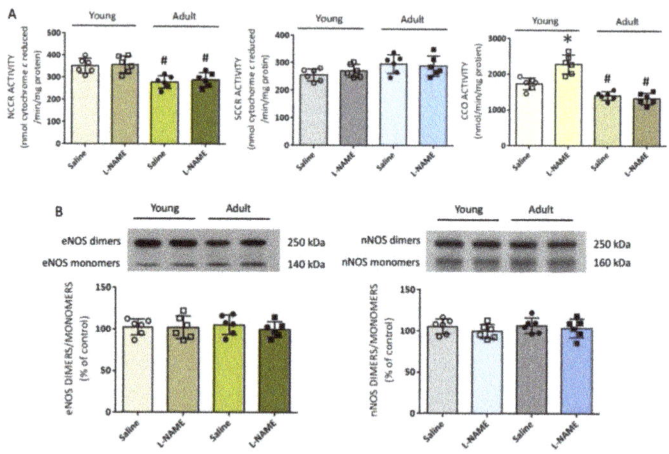

Figure 5. Effect of systemic L-NAME treatment on mitochondrial electron transport chain enzyme activity and NO synthase (NOS) uncoupling in RVLM: (**A**) Enzyme activities of NADH cytochrome *c* reductase (NCCR, marker for coupling capacity between complexes I and III), succinate cytochrome *c* reductase (SCCR, marker for coupling capacity between complexes II and III), and cytochrome *c* oxidase (CCO, marker for complexes IV) in RVLM of young (8 weeks, open circles or squares) or adult (20 weeks, filled circles or squares) rats after i.p. infusion of saline or L-NAME (10 mg/kg/day). (**B**) Representative gels (insets) and densitometric analysis of results from Western blot showing changes in the ratio of dimers versus monomers of eNOS and nNOS protein in RVLM of young and adult rats after saline or L-NAME treatment. Data on protein expression were normalized to the respective saline control value, which is set to 100%. Data are presented as mean ± SD, $n = 6$ per group. * $p < 0.05$ versus corresponding saline-treated group, and # $p < 0.05$ versus saline-treated young rats in unpaired Student's *t*-test.

The increase in ROS production in RVLM could also result from NOS uncoupling [32]. In RVLM of the L-NAME-treated animals, the ratio of eNOS monomer over dimer, an experimental index of NOS uncoupling, remained unchanged in both age groups (Figure 5B); although, eNOS expression, NOS activity, and tissue NOx levels were suppressed. Similarly, the same treatment did not affect nNOS coupling, which has been reported to promote tissue oxidative stress in RVLM [32], in both age groups.

We interpret our observations that tissue ROS levels, particularly after i.p. L-NAME infusion, were significantly augmented in RVLM of adult rats to suggest an active role of ROS in RVLM in age-dependent exacerbation of hemodynamic responses in this NO deficiency model of hypertension. The elevated protein expression and enzyme capacity of the antioxidants may explain why redox homeostasis in RVLM of young rats is maintained despite the increase in both protein expression and enzyme activity of NADPH oxidase after L-NAME treatment. On the other hand, the enhanced augmentation of NADPH oxidase activity may account for the further increase in tissue ROS levels observed in RVLM of adult animals. The reduced mitochondrial bioenergetics because of impaired enzyme activity of NCCR and CCO may also give rise to the heightened ROS levels in RVLM of adult animals after systemic L-NAME treatment.

3.4. Causal Involvement of Tissue Oxidative Stress in RVLM in Age-Dependent Exacerbation of Hemodynamic Responses to Systemic NO Deficiency

To ascertain a causal disparate role of NO deficiency and ROS production in RVLM in age-dependent augmentation of hemodynamic responses to i.p. L-NAME infusion, young and adult animals were randomly divided into six groups to receive, respectively, an oral intake or i.c. infusion of L-arginine, a NOS precursor; tempol, a ROS scavenger; mitoQ_{10}, a mitochondrial-targeted SOD mimetic, or amlodipine, a vasodilator of dihydropyridine class of calcium channel blockers; from days 7 to 14 following i.p. infusion of L-NAME (10 mg/kg/day). In young rats, oral intake of L-arginine (2%) or amlodipine (10 mg/kg), or i.c. infusion of L-arginine (2 μg/kg/day), but not oral intake of tempol (100 μmol/kg) or i.c. infusion of tempol (1 μmol//h/μL) or mitoQ_{10} (2.5 μmol//h/μL), significantly attenuated the increase in SBP in the L-NAME-treated animals (Figure 6A). On the other hand, when L-arginine, tempol, or mitoQ was microinfused into the cisterna magna of adult animals, one week after i.p. L-NAME infusion, the treatments discernibly diminished the increases in SBP induced by systemic L-NAME treatment (Figure 6B). Microinfusion of tempol or mitoQ_{10}, but not L-arginine, into the cisterna magna, at the same time, restored tissue ROS in RVLM of the L-NAME-treated rats to saline-control levels (Figure 6C). Infusion of tempol or mitoQ_{10} into the cisternal magna had no apparent effect on tissue NO levels in RVLM (Figure 6D), and i.c. infusion of tempol or mitoQ_{10} infusion did not affect the reduced serum NO levels following L-NAME infusion.

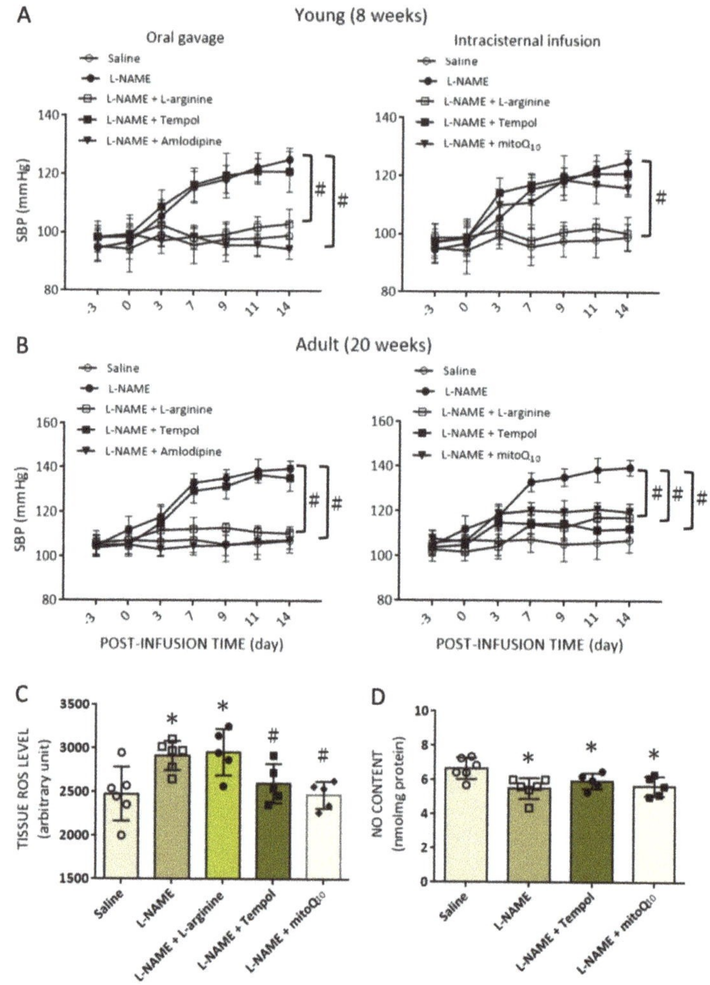

Figure 6. Effect of NO donor and antioxidants on changes in SBP, and tissue ROS and NO levels in RVLM of young and adult rats induced by systemic L-NAME treatment. Temporal changes in SBP at different postinfusion time points after i.p. infusion of L-NAME, alone or with additional oral intake or i.c. infusion of various pharmacological treatments in young (**A**) and (**B**) adult rats. Also shown are tissue levels of ROS (**C**) and NO (**D**) in RVLM, measured at day 14 after i.p. infusion of L-NAME, alone or with additional i.c. infusion of pharmacological treatments. The pharmacological treatments included i.p. infusion of L-NAME (10 mg/kg/day), oral intake via gavage of L-arginine (2%), tempol (100 µmol/kg) or amlodipine (10 mg/kg), or i.c. infusion of L-arginine (2 µg/kg/day), tempol (1 µmol//h/µL), or mitoQ$_{10}$ (2.5 µmol//h/µL). Control infusion of 0.9% saline (for i.p. or oral gavage treatment) or artificial CSF (aCSF; for i.c. infusion) served as the volume and vehicle control. Data are presented as mean ± SD, n = 5–6 per group, and * $p < 0.05$ versus the corresponding saline-treated group, and # $p < 0.05$ versus the L-NAME group in post hoc Tukey's multiple comparisons tests or unpaired Student's t-test. Data on saline and L-NAME treatments from Figure 2 are adopted for comparison.

These results are interpreted to suggest that in response to systemic NO deficiency, a predominant increase in vasomotor tone because of vascular constriction and sympathetic outflow from RVLM may underline the increase in SBP of the L-NAME-treated young animals. When animals become older, tissue oxidative stress in RVLM is actuated to further increase sympathetic vasomotor activity and promote greater hemodynamic responses in the systemic L-NAME treatment model of hypertension.

3.5. Identification of Additional Age-Dependent Redox Homeostasis-Related Genes in RVLM in the Systemic NO-Deficiency Model of Hypertension

Our observations that the increase in tissue ROS levels was greater in RVLM of the L-NAME-treated adult animals (cf. Figure 4E) prompted the speculation that additional oxidative stress-related genes on top of those reported previously are upregulated in this brain stem site. To address this issue, we performed a whole genome microarray analysis of RVLM tissue using a Qiagen RT^2 Profiler^{TM} PCR array (Qiagen). As shown in the supplementary data (Supplementary Figure S1), four differentially expressed genes, whose expression levels are at least two times different from young rats, were identified in RVLM of adult animals. These included two upregulated genes, NADPH oxidase activator 1 (*Noxa1*) and glutathione peroxidase 2 (*GPx2*), and two downregulated genes, dual oxidase 2 (*Duox2*) and uncoupling protein 3 (*Ucp3*). Given the limitation in the RVLM sample volume, we decided to verify the accuracy of our microarray analysis by real-time qPCR only on the candidate transcriptomes that exhibited the highest (Noxa1) and lowest (*Ucp3*) changes. Our RT-qPCR results confirmed the microarray data for *Noxa1* and *Ucp3* mRNA in RVLM (Figure 7A). Moreover, their expressions were further upregulated in the L-NAME-treated animals in both age groups. Similar patterns were found in protein expression of Noxa1 and UCP3 in RVLM of both groups of animals (Figure 7B). Of note, the upregulation of *Noxa1* (+1.9 ± 0.2 versus +1.1 ± 0.5 fold, $n = 6$, $p < 0.05$), but not a change in *Ucp3* mRNA (+1.7 ± 0.4 versus +1.8 ± 0.3 fold, $n = 6$, $p > 0.05$), was significantly greater in RVLM of adult animals when compared to young rats. Together these results suggest that age-dependent alterations in oxidative stress-related gene transcription, including the upregulation of a ROS-producing enzyme, *Nox1a*, and the downregulation of an antioxidant, *Ucp3*, may contribute to the elevated ROS levels in RVLM following i.p. L-NAME infusion.

3.6. Silencing Nox1a mRNA in RVLM Ameliorates Oxidative Stress and Attenuates Hemodynamic Responses to Systemic NO Deficiency in Adult Rats

Our final series of experiments was performed to validate the functional significance of the newly identified *Noxa1* gene in RVLM on exacerbated hemodynamic responses in the L-NAME-treated adult animals. In situ gene silencing via bilateral microinjection into RVLM of lentiviral vectors encoding shRNA targeting *Noxa1* (Lv-Noxa1-shRNA; 1 × 10^5 IFU) was performed on day 10 after the onset of i.p. L-NAME infusion. Effective transduction of the viral vectors into RVLM was confirmed by qPCR, which showed a significant decrease in *Noxa1* mRNA and protein, measured on day 4 after vector transduction (Figure 8A). Compared to their scramble shRNA control, silencing *Nox1a* in RVLM with its shRNA resulted in notable attenuation of hypertension as well as the increase in sympathetic vasomotor activity and plasma NE levels (Figure 8B) in the L-NAME-treated adult animals. Gene silencing of *Noxa1* also significantly alleviated the elevated ROS levels (Figure 8C) in RVLM, but not the reduced plasma NO levels (Figure 8D). These data provide evidence to suggest an active role of *Noxa1* in RVLM on age-dependent susceptibility to hypertension induced by systemic L-NAME treatment in adult rats.

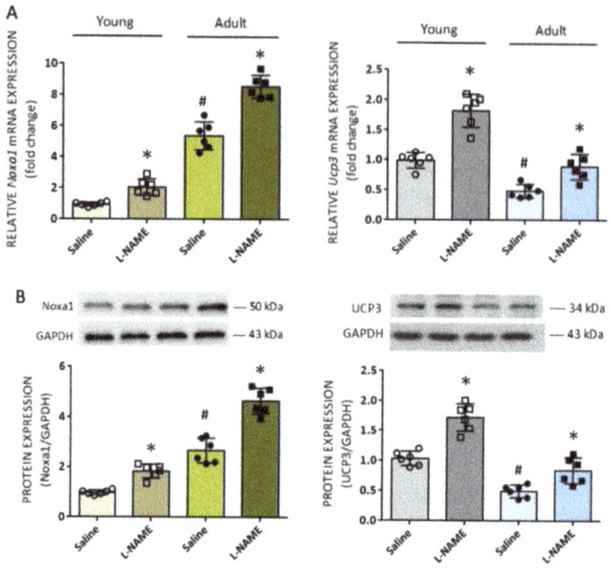

Figure 7. Age-dependent expression of ROS-related proteins in RVLM and the effect of systemic L-NAME treatment on their expressions: (**A**) Relative expression of *Noxa1* and *Ucp3* mRNA quantified by RT-qPCR in RVLM tissues from young (open circles or squares) and adult (filled circles or squares) rats on 14 days after i.p. infusion of saline or L-NAME. (**B**) Representative gels (insets) and densitometric analysis of results from Western blot changes in protein expression of Noxa1 and UCP3 in RVLM of young and adult 14 days after systemic L-NAME treatment. Data on protein expression were normalized to the respective saline control value, which is set to 1.0. Data are presented as mean ± SD, $n = 6$ per group. * $p < 0.05$ versus corresponding saline-treated groups, and # $p < 0.05$ versus saline-treated young rats in unpaired Student's t-test.

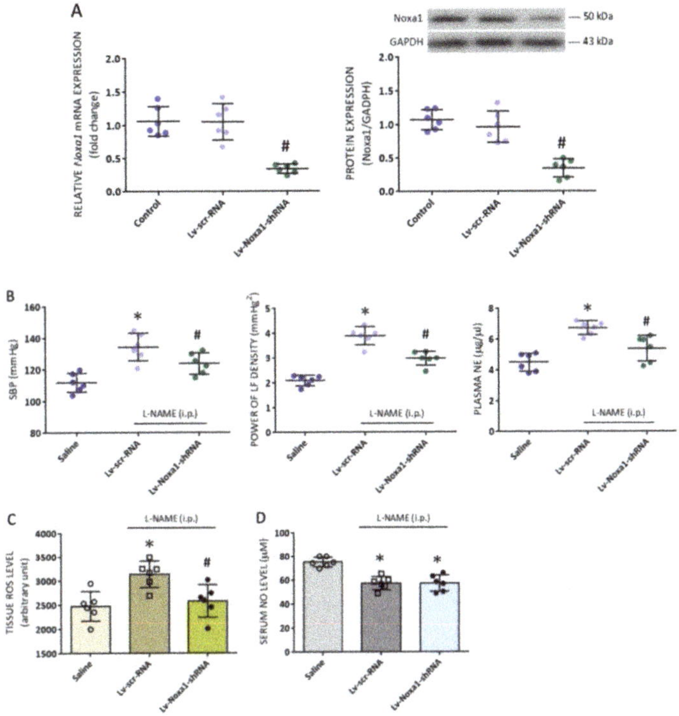

Figure 8. Effect of manipulations of *Noxa1* gene in RVLM on mRNA and protein expression, and changes in cardiovascular responses and levels of ROS in RVLM and serum NO of adult rats induced by systemic L-NAME treatment: (**A**) Changes in mRNA transcription of Noxa1 and representative gels (insets) and densitometric analysis of results from Western blot showing changes in protein expression of Noxa1 in RVLM tissues 4 days after bilateral microinjection into RVLM of lentiviral vectors (Lv) containing short hairpin interfering RNA (shRNA) targeting the rat *Noxa1* sequence (Lv-Noxa1-shRNA) or a scrambled (scr) control shRNA. (**B**) SBP, power density of the LF component of SBP signals, and plasma NE levels; and (**C**) tissue ROS and (**D**) serum NO levels, determined on day 14 after i.p. infusion of saline or L-NAME (10 mg/kg/day) in adult rats that received additional treatment with bilateral microinjection into RVLM of Lv-scr-RNA or Lv-Noxa1-shRNA on day 10 after L-NAME treatment. Data on mRNA transcription and protein expression are normalized to the respective saline control value, which is set to 1.0. Data are presented as mean ± SD, $n = 6$ per group. * $p < 0.05$ versus corresponding saline-treated groups, and # $p < 0.05$ versus Lv-scr-RNA-treated groups in unpaired Student's *t*-test.

4. Discussion

The present study was designed to explore the role of oxidative stress in RVLM on age-dependent susceptibility to hypertension in response to systemic NO deficiency, and to decipher the underlying molecular mechanisms. There are four major findings. First, i.p. infusion of L-NAME evoked oxidative stress in RVLM in adult, but not young, normotensive rats, accompanied by augmented enzyme activity of NADPH oxidase and reduced mitochondrial NCCR and CCO enzyme activities. Second, treatment with L-arginine via oral gavage or infusion into the cistern magna, but not i.c. tempol or mitoQ_{10}, significantly offset the L-NAME-induced hypertension in young rats. On the other hand, all treatments appreciably reduced L-NAME-induced hypertension in adult rats. Third, four genes involved in ROS production and clearance were differentially expressed in RVLM in an age-related manner. Of them, *Noxa1* and *GPx2* were upregulated and *Duox2* and

Ucp3 were down-regulated. Systemic L-NAME treatment caused greater upregulation of *Noxa1*, but not *Ucp3*, mRNA expression in RVLM of adult rats. Fourth, gene silencing of *Noxa1* in RVLM effectively alleviated oxidative stress and protected adult rats against L-NAME-induced hypertension. These data together suggest that hypertension induced by systemic L-NAME treatment in young rats is mediated primarily by NO deficiency that occurs both in vascular smooth muscle cells and RVLM. On the other hand, enhanced augmentation of oxidative stress in RVLM contributes to a heightened susceptibility of adult rats to hypertension induced by systemic L-NAME treatment.

NO deficiency is a well-characterized trait in human hypertension [5–7], and NOS inhibition by L-NAME is commonly used to establish NO deficiency in animal models of human hypertension [16]. L-NAME exerts various pharmacological effects on cardiovascular functions and molecular activities that are dependent on dose (1–50 mg/kg/day), route (oral, subcutaneous, i.p., or cerebral ventricle), and mode (acute, daily bolus, or continuous infusion) of administration, as well as the duration of treatment (minutes, hours, days, or weeks). In this study, we employed a relatively low dose for i.p. infusion of L-NAME (10 mg/kg/day) for 2 weeks to establish the expected hemodynamic responses, with minimal concomitant oxidative and inflammatory actions (cf. Table 1) to avoid their confounding influences on hypertension development. At a higher dosage and/or longer duration of L-NAME treatment, ROS production in vascular smooth muscle cells [47], kidney [48], and heart [21], as well as pro-inflammatory cytokines in kidney [49] and heart [21], have been demonstrated to mediate hypertension and associated cardiovascular complications such as cardiac hypertrophy and renal injury. The findings that the plasma level of NE was increased and serum NO level was reduced, and that oral intake of L-arginine and amlodipine (cf. Figure 6A,B) conferred protection against L-NAME-induced hypertension suggest that an increase in SNA and vasoconstriction may contribute to the observed hemodynamic changes. We did not find significant changes in HR following L-NAME administration; although, other studies demonstrated reduction [50] or augmentation [18] in HR. The exact reason behind these diverse findings is not clear, but might simply be the consequence of differences in dose and duration of L-NAME treatment. In this regard, L-NAME given at 1.5 times our dose for 2 weeks decreases [50], and at 4 times our dose for 5 weeks increases, HR [18].

To date, cardiovascular responses to L-NAME treatment have primarily focused on its effects on NOS expression and NO bioavailability in the vasculature, with little attention on the role of L-NAME in the central nervous system. As such, one of the major findings of the present study is the identification of the suppressive effects of systemic L-NAME treatment on eNOS protein expression, NOS activity, and tissue NO level in RVLM of both young and adult rats. L-NAME has been reported to cross the BBB to reach brain tissues [22–24]. All three NOS isoforms are constitutively present in RVLM [51], and their roles in RVLM on neural control of cardiovascular functions, have been extensively reviewed in the literature [26]. In the present study, we found that the expression of eNOS, but not nNOS or iNOS, protein in RVLM was suppressed by i.p. L-NAME infusion. Since constitutive eNOS tonically inhibits RVLM neuronal activity and sympathetic outflow [52], a diminished eNOS expression and eNOS-derived NO availability in RVLM may therefore contribute to the observed increases in sympathetic vasomotor activity and plasma NE levels in rats subjected to systemic L-NAME treatment. These findings are in concordance with the observations that inhibition of eNOS by L-NAME evokes central sympathoexcitation, leading to increased SNA in experimental animals [19] and healthy men [53]. In addition, diminished eNOS expression and eNOS-derived NO bioavailability in the hypothalamic paraventricular nucleus (PVN) by L-NAME contributes to sympathoexcitation and hypertension associated with heart failure [54]. We reported previously a concentration-dependent action of NO in RVLM on neural control of cardiovascular functions. Whereas nNOS-derived NO is responsible for sympathoexcitation, iNOS-induced NO elicits sympathoinhibition [51]. In contrast, a sympathoinhibitory action of NO derived from nNOS in RVLM has also been reported [29]. In the present study, neither nNOS nor iNOS expression in RVLM was affected by systemic L-NAME treatment. These results are at variance with a previous study [24] that shows nNOS mRNA in RVLM is downregulated in young (4 weeks), but upregulated in adult (10 weeks) rats, following oral intake of L-NAME (50 mg/kg/day) for 6 weeks. Such discrepancies may again reflect differences in dose, route, and duration of L-NAME treatment.

In addition to diminished eNOS/NO signaling, we found in the present study that systemic L-NAME treatment also affected proteins involved in ROS production and clearance in RVLM. Of the major sources for the production of ROS, we found the protein expression of gp91^{phox} and p22^{phox} subunits was increased in both age groups, alongside elevated enzyme activity of the NADPH oxidase. In addition, the enzyme activity of mitochondrial CCO was increased in young, but decreased in adult, rats, and NCCR activity was also decreased in adult rats. On the contrary, NOS uncoupling, which is considered a secondary source of ROS production [6], was not affected by systemic L-NAME treatment. The NADPH oxidase family is the most important enzymatic source of ROS in the cardiovascular system [6,26]. In RVLM, augmented ROS production resulting from increases in pg91^{phox} and P22^{phox} subunits initiates a series of molecular events, leading to tissue oxidative stress and sympathoexcitation that contribute to the neural mechanism of hypertension [43]. Expression of p47^{phox} and p67^{phox} proteins, the other two NADPH oxidase subunits that play active roles in the redox-associated neural mechanism of hypertension [43], on the other hand, were not affected by L-NAME treatment; this might be related to the susceptibility of individual subunits to NO deficiency. The mechanisms underpinning the increase in gp91^{phox} and p22^{phox} protein expressions induced by L-NAME are not immediately clear, but might be the consequence of transcriptional upregulation of these subunits [14,24].

Another major cellular source of ROS production is the mitochondrial ETC in association with oxidative phosphorylation for ATP synthesis [55]. The effects of NO on mitochondrial functions and metabolism are mediated mainly through their interactions at specific sites in the ETC enzyme complexes. In this regard, NO, at subnanomolar amounts, inhibits Complex IV via interactions with the ferrous heme iron or oxidized copper at the heme iron:copper binuclear center of the enzyme [56]. At high concentrations, NO inhibits Complex I via oxidation or *S*-nitrosation of specific thiols [57]. Accordingly, an increase in CCO activity observed in RVLM of L-NAME-treated young rats could result from the withdrawal of an inhibitory effect of NO on Complex IV. On the other hand, the mechanism that underlies the lower NCCR and CCO activity in RVLM of adult rats and their reduced responsiveness to NO deficiency is unclear. Nonetheless, it is noteworthy that aging selectively downregulates genes encoding Complex I and III of the mitochondria ETC both in rat and human hearts [58]. Functionally, impairment of both NCCR and CCO activity in RVLM has been reported to increase mitochondrial ROS production that contributes to hypertension in SHR or Ang II treatment in normotensive rats [46].

Under the condition of oxidative stress, NOS may remove an electron from NADPH and donate it to an oxygen molecule for generation of $O_2^{\bullet-}$ rather than NO [6,14,59]. In RVLM, tissue oxidative stress causes an uncoupling of eNOS during hypertension [28], further depleting the levels of NO and aggravating hypertension progression. In addition, a redox-sensitive feedforward mechanism of nNOS uncoupling in RVLM contributes to sympathoexcitation and hypertension associated with metabolic disorders [32]. In the present study, the ratio between dimmers over monomers of either eNOS or nNOS was not affected (cf. Figure 5B) by i.p. L-NAME infusion, suggesting a negligible role of NOS uncoupling in ROS production in RVLM in response to systemic NO deficiency.

Redox homeostasis depends on the balance between the production and degradation of the oxidants. At the same time, antioxidant treatments offset the development of L-NAME-induced hypertension by a reduction in ROS production during NOS inhibition [50]. In young and adult rats subjected to systemic L-NAME treatment, we found that the protein expression of two key antioxidants, SOD2 and Nrf2, was significantly increased, alongside an increase in total antioxidant activity in RVLM. SOD2, or manganese SOD, is one of the most well-characterized antioxidant defensive mechanisms for the elimination of cellular oxidants, particularly $O_2^{\bullet -}$. Transcribed from *sod2* and synthesized in the cytoplasm, SOD2 is subsequently relocated to the mitochondrial matrix, endowed with the responsibility to scavenge $O_2^{\bullet -}$ produced by the mitochondrial ETC [60]. In RVLM, transcriptional upregulation of *sod2* protects against mitochondrial oxidative stress and hypertension in Ang II treatment in normotensive rats [61]. SOD2 also participates in the protection again hypertension and cardiovascular complications conferred by the mitochondria-target antioxidants [62]. Nrf2 is the master regulator of antioxidant genes and, hence, of antioxidant status. Nrf2 has been demonstrated to be a key to redox homeostasis in RVLM; targeted ablation of Nrf2 in RVLM leads to hypertension [41]. The increase in the expression of these antioxidants may explain why redox homeostasis in RVLM of young rats was maintained despite significant elevations in ROS production by mitochondria and NADPH oxidase induced by systemic NO deficiency. These data, at the same time, suggest that the oxidant responsiveness to systemic L-NAME treatment is well tolerated in RVLM of young rats but may turn remitted when animals become older.

Aging is associated with an increase in ROS production, which together with a decline in antioxidant defense efficiency significantly contributes to the manifestation of an oxidative stress state [38]. Compared to young rats, we found in this study greater increases in NADPH oxidase activity and augmented ROS accumulation in RVLM of adult rats in response to systemic NO deficiency. These intriguing findings prompted us to search for additional candidate molecules that are associated with age-dependent oxidative stress in RVLM. Based on microarray analysis of redox signal-related genes, we identified four genes whose expression levels are at least two times up- or downregulated in RVLM of adult animals. We found that *Noxa1* and *Gpx2* mRNA were upregulated, whereas *Duox2* and *Ucp3* mRNA were downregulated. Among them, upregulation of the antioxidant Gpx2 could be an antioxidant defense mechanism to compensate for tissue oxidative stress, and Duox2 is a $p22^{phox}$-independent isoform that is not important in cardiovascular pathophysiology [6]. We therefore focused on Noxa1 and *Ucp3* mRNA to further interrogate their roles in the augmented ROS levels in RVLM of adult rats. First, we confirmed that expression of *Noxa1* mRNA was higher, whereas *Ucp3* mRNA was lower, in RVLM of adult rats. An age-dependent decrease in Ucp3 expression in male mice has recently been reported [63]. Second, expression of *Noxa1*, but not *Ucp3*, mRNA was upregulated by systemic L-NAME treatment, suggesting that transcriptional regulation of ROS signal-related genes as a consequence of tissue NO deficiency is target specific. Third, gene silencing of *Nox1* appreciably alleviated the augmented ROS levels in RVLM, indicating the age-dependent accumulation of ROS in RVLM may be attributed to an upregulation of *Noxa1* transcription. Finally, the functional significance of the newly identified *Noxa1* mRNA in RVLM on age-dependent susceptibility of cardiovascular responses to tissue NO deficiency is validated by our findings that bilateral microinjection into RVLM of Lv-Noxa1-shRNA appreciably ameliorated hypertension, the exaggerated sympathetic vasomotor activity, and plasma NE levels evoked by L-NAME treatment in adult animals. Noxa1 is a critical functional homolog of p67phox for NADPH oxidase activation in vascular smooth muscle cells [64]. NOX1 (a $p22^{phox}$-dependent oxidase) interacts with p67phox homolog Noxa1, causing constitutive production of $O_2^{\bullet -}$ [65]. Conversely, genetic deletion of *Noxa1* reduces basal and Ang II-induced hypertension and renal oxidative stress [66]. The observations of comparable changes in Ucp3 expression in RVLM of young and adult animals to systemic L-NAME treatment (cf. Figure 7) are interpreted to suggest a minor role of mitochondrial Ucp3 in RVLM on age-related susceptibility to hypertension in adult rats to systemic NO deficiency. This suggestion, nonetheless, waits for further validation.

Treatments targeting NO and ROS signals in the periphery and RVLM were employed to further verify the differential roles of NO and ROS in RVLM on age-dependent cardiovascular responses induced by L-NAME. In young rats, both oral intake and i.c. infusion of L-arginine, but not i.c. application of tempol or mitoQ_{10}, significantly reduced hypertension induced by systemic NO deficiency. These results indicate that L-NAME-induced cardiovascular responses in young animals may mainly be the result of the NO deficiency that occurs both in the smooth muscle cells and RVLM. The engagement of tissue oxidative stress in RVLM on cardiovascular responses to L-NAME in adult animals was unveiled by observations that apart from L-arginine, i.c. infusion of tempol and mitoQ_{10} significantly diminished hypertension. It is noteworthy that i.c. infusion of tempol or mitoQ_{10} had no effect on the reduced NO levels in RVLM, indicating the protective actions of tempol and mitoQ_{10} are not secondary to changes in tissue NO contents. Moreover, the results that i.c. infusion of L-arginine had no effect on the augmented ROS levels in RVLM of adult rats (cf. Figure 6C) suggesting that aging-associated oxidative stress may be related to changes in ROS signals but not NOS activity or tissue NO bioavailability in RVLM. This notion of a minor role of NO in ROS production in RVLM of adult animals is further supported by findings that L-NAME had no effect on the reduced enzyme activity of mitochondrial NCCR and CCO in RVLM of adult rats (cf. Figure 5A). Finally, the observations that oral intake of L-arginine and amlodipine protected both young and adult rats from L-NAME-induced hypertension suggest that the observed cardiovascular changes are likely the final outcomes of vasoconstriction in response to systemic NO deficiency.

There are several limitations to our study. First, the present findings were made from animals that were subjected to a low-dose L-NAME treatment. As discussed above, in view of the disparity of cardiovascular responses that are dependent on doses of L-NAME, the notion of an interplay between NOS and ROS in the pathogenesis of hypertension induced by systemic NO deficiency should be taken with caution. Second, since the present study focused only on RVLM, the roles of NOS and ROS in other "pre-autonomic" neurons, such as the nucleus tractus solitarii (NTS) and PVN, in neural mechanisms of the L-NAME-induced cardiovascular complications remain to be delineated. In this regard, both NOS and ROS signaling in the NTS and PVN have been reported to play pivotal roles in hypertension induced by systemic L-NAME treatment [6,20,26,30,31,67]. Third, an increase in Ang II release along with depressed NO production is considered the principal culprit in hemodynamic and structural alterations in L-NAME-treated rats [68]. In RVLM, Ang II induces ROS production via activation of NADPH oxidase [6,26,32]. In addition, NO deficiency differentially affects the expression of Ang II receptors in RVLM of young versus adult rats [24]. Therefore, it would be of interest to further investigate the role of Ang II in RVLM in the interplay between NOS and ROS on age-dependent susceptibility to hypertension induced by systemic L-NAME treatment. Fourth, in the present study, we used a commercially available microarray kit to screen the oxidative stress-related genes that are differentially expressed in RVLM of young versus adult rats. The genes provided in the kit are far from complete; consequently, the identified genes could be underestimated. In a recent study, out of 47 genes that are involved in ROS metabolism, 39 are downregulated and 8 upregulated in the aged (24 months) versus adult (6 months) rat heart [58].

5. Conclusions

In conclusion, our findings reveal that disparate mechanisms underlie the increase in SNA and BP in rats subjected to systemic L-NAME treatment in an age-dependent manner. In young rats, cardiovascular responses to L-NAME are mediated mainly by NO deficiency, both in the vascular smooth muscle cells and RVLM. When animals become older, additional ROS generation from both mitochondrial (reduction in enzyme activity of NCCR and CCO) and extra-mitochondrial (transcriptional upregulation of Noxa1) pathways may contribute to the enhanced susceptibility to sympathoexcitation and hypertension induced by systemic L-NAME treatment. This information provides novel insights into potential targets involved in the responsiveness to systemic NO deficiency during aging that could be manipulated to prevent age-associated deterioration in cardiovascular functions. Moreover, recognizing the functional significance of aging on the transcription of genes encoding ROS signaling molecules may help to identify novel targets that can be selectively intervened to prevent aging-associated hypertension and cardiovascular complications.

Supplementary Materials: The following supporting information can be downloaded at: https://www.mdpi.com/article/10.3390/biomedicines10092232/s1, Figure S1: Differentially expressed oxidative stress-related genes in RVLM of adult animals.

Author Contributions: Conceptualization, J.Y.H.C. and H.R.; methodology, Y.-M.C.; formal analysis, Y.-M.C. and J.Y.H.C.; investigation, Y.-M.C. and J.Y.H.C.; writing—original draft preparation, Y.-M.C. and J.Y.H.C.; writing—review and editing, H.R. and J.Y.H.C.; visualization, Y.-M.C. and J.Y.H.C.; funding acquisition, J.Y.H.C. All authors have read and agreed to the published version of the manuscript.

Funding: This research was funded in part by the Ministry of Science and Technology, Taiwan (grant number: MOST108-2923-B-182A-001-MY3), Chang Gung Medical Foundation, Taiwan (grant number: OMRPG80011), and International Joint Project of GACR, Czech Republic and MOST, Taiwan: 19-08260J.

Institutional Review Board Statement: The animal study protocol was approved by the Institutional Animal Care and Use Committee (IACUC) of Kaohsiung Chang Gung Memorial Hospital, Taiwan (protocol code: 2018051701, date of approval: 8 June 2018).

Data Availability Statement: The data presented in this study are available on reasonable request from the corresponding author.

Acknowledgments: The authors thank Yen-Hua Hung for her technical assistance in this study and administrative support of the project.

Conflicts of Interest: The authors declare no conflict of interest.

References

1. Bays, H.E.; Kulkarni, A.; German, C.; Satish, P.; Iluyomade, A.; Dudum, R.; Thakkar, A.; Rifai, M.I.; Mehta, A.; Thobani, A.; et al. Ten things to know about ten cardiovascular disease risk factors—2022. *Am. J. Prev. Cardiol.* **2022**, *10*, 100342. [CrossRef] [PubMed]
2. Wong, N.D.; Budoff, M.J.; Ferdinand, K.; Graham, I.M.; Michos, E.D.; Reddy, T.; Shapiro, M.D.; Toth, P.P. Atherosclerotic cardiovascular disease risk assessment: An American Society for Preventive Cardiology clinical practice statement. *Am. J. Prev. Cardiol.* **2022**, *10*, 100335. [CrossRef] [PubMed]
3. Cipolla, M.J.; Liebeskind, D.S.; Chan, S.L. The importance of comorbidities in ischemic stroke: Impact of hypertension on the cerebral circulation. *J. Cereb. Blood. Flow. Metab.* **2018**, *38*, 2129–2149. [CrossRef] [PubMed]
4. Fu, X.; Ren, H.; Xie, J.; Wang, Y.; Li, Y.; Gao, P.; Chen, N. Association of nighttime masked uncontrolled hypertension with left ventricular hypertrophy and kidney function among patients with chronic kidney disease not receiving dialysis. *JAMA Netw. Open* **2022**, *5*, e2214460. [CrossRef]
5. Wu, Y.; Ding, Y.; Ramprasath, T.; Zou, M.H. Oxidative stress, GTPCH1, and endothelial nitric oxide synthase uncoupling in hypertension. *Antioxid. Redox. Signal.* **2021**, *34*, 750–764. [CrossRef]
6. Griendling, K.K.; Camargo, L.L.; Rios, F.; Alves-Lopes, R.; Montezano, A.C.; Touyz, R.M. Oxidative stress and hypertension. *Circ. Res.* **2021**, *128*, 993–1020. [CrossRef]
7. Bernatova, I. Endothelial dysfunction in experimental models of arterial hypertension: Cause or consequence? *Biomed. Res. Int.* **2014**, *2014*, 598271. [CrossRef]
8. Palmer, R.M.J.; Ferrige, A.G.; Moncada, S. Nitric oxide release accounts for the biological activity of endothelium-derived relaxing factor. *Nature* **1987**, *327*, 524–526. [CrossRef]
9. Furchgott, R.F.; Zawadzki, J.V. The obligatory role of endothelial cells in the relaxation of arterial smooth muscle by acetylcholine. *Nature* **1980**, *288*, 373–376. [CrossRef]
10. Francis, S.H.; Busch, J.L.; Corbin, J.D. cGMP-dependent protein kinases and cGMP phosphodiesterases in nitric oxide and cGMP action. *Pharmacol. Rev.* **2010**, *62*, 525–563. [CrossRef]
11. Iwata, M.; Inoue, T.; Asai, Y.; Hori, K.; Fujiwara, M.; Matsuo, S.; Tsuchida, W.; Suzuki, S. The protective role of localized nitric oxide production during inflammation may be mediated by the heme oxygenase-1/carbon monoxide pathway. *Biochem. Biophys. Rep.* **2020**, *23*, 100790. [CrossRef] [PubMed]
12. Knowles, R.B.; Warner, T.D. Anti-platelet drugs and their necessary interaction with endothelial mediators and platelet cyclic nucleotides for therapeutic efficacy. *Pharmacol. Ther.* **2019**, *193*, 83–90. [CrossRef] [PubMed]
13. Nunokawa, Y.; Takana, S. Interefon-γ inhibits proliferation of vascular smooth muscle cells by nitric oxide generation. *Biochem. Biophys. Res. Comm.* **1992**, *188*, 409–415. [CrossRef]
14. Daiber, A.; Kröller-Schön, S.; Oelze, M.; Hahad, O.; Li, H.; Schulz, R.; Steven, S.; Münzel, T. Oxidative stress and inflammation contribute to traffic noise-induced vascular and cerebral dysfunction via uncoupling of nitric oxide synthases. *Redox. Biol.* **2020**, *34*, 101506. [CrossRef] [PubMed]

15. Zhang, Y.; Murugesan, P.; Huang, K.; Cai, H. NADPH oxidases and oxidase crosstalk in cardiovascular diseases: Novel therapeutic targets. *Nat. Rev. Cardiol.* **2020**, *17*, 170–194. [CrossRef]
16. Ribeiro, M.O.; Antunes, E.; de Nucci, G.; Lovisolo, S.M.; Zatz, R. Chronic inhibition of nitric oxide synthesis. A new model of arterial hypertension. *Hypertension* **1992**, *20*, 298–303. [CrossRef]
17. Kopincová, J.; Púzserová, A.; Bernátová, I. L-NAME in the cardiovascular system-nitric oxide synthase activator? *Pharmacol. Rep.* **2012**, *64*, 511–520. [CrossRef]
18. Berkban, T.; Boonprom, P.; Bunbupha, S.; Welbat, J.U.; Kukongviriyapan, U.; Kukongviriyapan, V.; Pakdeechote, P.; Prachaney, P. Ellagic acid prevents L-NAME-induced hypertension via restoration of eNOS and $p47^{phox}$ expression in rats. *Nutrients* **2015**, *7*, 5265–5280. [CrossRef]
19. Zambrano, L.I.; Pontes, R.B.; Garcia, M.L.; Nishi, E.E.; Nogueira, F.N.; Higa, E.M.S.; Cespedes, J.G.; Bergamaschi, C.T.; Campos, R.R. Pattern of sympathetic vasomotor activity in a model of hypertension induced by nitric oxide synthase blockade. *Physiol. Rep.* **2019**, *7*, e14183. [CrossRef]
20. Shu, W.; Li, H.; Gong, H.; Zhang, M.; Niu, X.; Ma, Y.; Zhang, X.; Cai, W.; Yang, G.; Wei, M.; et al. Evaluation of blood vessel injury, oxidative stress and circulating inflammatory factors in an L-NAME-induced preeclampsia-like rat model. *Exp. Ther. Med.* **2018**, *16*, 585–594. [CrossRef]
21. Poasakate, A.; Maneesai, P.; Rattanakanokchai, S.; Bunbupha, S.; Tong-Un, T.; Pakdeechote, P. Genistein prevents nitric oxide deficiency-induced cardiac dysfunction and remodeling in rats. *Antioxidants* **2021**, *10*, 237. [CrossRef] [PubMed]
22. Ayers, N.A.; Kapas, L.; Krueger, J.M. The inhibitory effects of N^{ω}-nitro-L-arginine methyl ester on nitric oxide synthase activity vary among brain regions in vivo but not in vitro. *Neurochem. Res.* **1997**, *22*, 81–86. [CrossRef] [PubMed]
23. Jendekova, L.; Kojsova, S.; Andriantsitohaina, R.; Pechanova, O. The time-dependent effect of provinols on brain NO synthase activity in L-NAME-induced hypertension. *Physiol. Res.* **2006**, *55* (Suppl. S1), S31–S37. [CrossRef] [PubMed]
24. Majzúnová, M.; Pakanová, Z.; Kvasnička, P.; Bališ, P.; Čačányiová, S.; Dovinová, I. Age-dependent redox status in the brain stem of NO-deficient hypertensive rats. *J. Biomed. Sci.* **2017**, *24*, 72. [CrossRef] [PubMed]
25. Ross, C.A.; Ruggiero, D.A.; Joh, T.H.; Park, D.H.; Reis, D.J. Rostral ventrolateral medulla: Selective projections to the thoracic autonomic cell column from the region containing C1 adrenaline neurons. *J. Comp. Neurol.* **1984**, *228*, 168–185. [CrossRef] [PubMed]
26. Chan, S.H.H.; Chan, J.Y.H. Brain stem NOS and ROS in neural mechanisms of hypertension. *Antioxid. Redox. Signal.* **2014**, *20*, 146–163. [CrossRef]
27. Guo, Z.L.; Tjen-A-Looi, S.C.; Fu, L.W.; Longhurst, J.C. Nitric oxide in rostral ventrolateral medulla regulates cardiac-sympathetic reflexes: Role of synthase isoforms. *Am. J. Physiol. Heart Circ. Physiol.* **2009**, *297*, H1478–H1486. [CrossRef]
28. Hirooka, Y.; Kishi, T.; Sakai, K.; Takeshita, A.; Sunagawa, K. Imbalance of central nitric oxide and reactive oxygen species in the regulation of sympathetic activity and neural mechanisms of hypertension. *Am. J. Physiol. Regul. Integr. Comp. Physiol.* **2011**, *300*, R818–R826. [CrossRef]
29. Ally, A.; Powell, I.; Ally, M.M.; Chaitoff, K.; Nauli, S. Role of neuronal nitric oxide synthase on cardiovascular functions in physiological and pathophysiological states. *Nitric Oxide* **2020**, *102*, 52–73. [CrossRef]
30. Sharma, N.M.; Zheng, H.; Mehta, P.P.; Li, Y.F.; Patel, K.P. Decreased nNOS in the PVN leads to increased sympathoexcitation in chronic heart failure: Role for CAPON and Ang II. *Cardiovasc. Res.* **2011**, *92*, 342–357. [CrossRef]
31. Zheng, H.; Mayhan, W.G.; Bidasee, K.R.; Patel, K.P. Blunted nitric oxide-mediated inhibition of sympathetic nerve activity within the paraventricular nucleus in diabetic rats. *Am. J. Physiol. Regul. Integr. Comp. Physiol.* **2006**, *290*, R992–R1002. [CrossRef] [PubMed]
32. Wu, K.L.H.; Chao, Y.M.; Tsay, S.J.; Chen, C.H.; Chan, S.H.H.; Dovinova, I.; Chan, J.Y.H. Role of nitric oxide synthase uncoupling at rostral ventrolateral medulla in redox-sensitive hypertension associated with metabolic syndrome. *Hypertension* **2014**, *64*, 815–824. [CrossRef]
33. Tejero, J.; Shiva, S.; Gladwin, M.T. Sources of vascular nitric oxide and reactive oxygen species and their regulation. *Physiol. Rev.* **2019**, *99*, 311–379. [CrossRef]
34. Zweier, J.L.; Chen, C.A.; Druhan, L.J.; Zweier, J.L. S-glutathionylation reshapes our understanding of endothelial nitric oxide synthase uncoupling and nitric oxide/reactive oxygen species-mediated signaling. *Antioxid. Redox. Signal.* **2011**, *14*, 1769–1775. [CrossRef] [PubMed]
35. Xu, X.; Wang, B.; Ren, C.; Hu, J.; Greenberg, D.A.; Chen, T.; Xie, L.; Jin, K. Age-related impairment of vascular structure and functions. *Aging Dis.* **2017**, *8*, 590–610. [CrossRef] [PubMed]
36. Liu, W.L.; Lin, Y.Y.; Mündel, T.; Chou, C.C.; Liao, Y.H. Effects of acute interval exercise on arterial stiffness and cardiovascular autonomic regulatory responses: A narrative review of potential impacts of aging. *Front. Cardiovasc. Med.* **2022**, *9*, 864173. [CrossRef] [PubMed]
37. Poznyak, A.V.; Sadykhov, N.K.; Kartuesov, A.G.; Borisov, E.E.; Sukhorukov, V.N.; Orekhov, A.N. Aging of vascular system is a complex process: The cornerstone mechanisms. *Int. J. Mol. Sci.* **2022**, *23*, 6926. [CrossRef]
38. Ungvari, Z.; Tarantini, S.; Donato, a.J.; Galvan, V.; Csiszar, A. Mechanisms of vascular aging. *Circ. Res.* **2018**, *123*, 849–867. [CrossRef]
39. Tsai, P.C.; Chao, Y.M.; Chan, J.Y.H. Sympathetic activation of splenic T-lymphocytes in hypertension of adult offspring programmed by maternal high fructose exposure. *Chin. J. Physiol.* **2020**, *63*, 263–275. [CrossRef]

40. Paxinos, G.; Watson, C. *The Rat Brain in Stereotaxic Coordinates*, 7th ed.; American Press: New York, NY, USA, 2014; pp. 133–137.
41. Gao, L.; Zimmerman, M.C.; Biswal, S.; Zucker, I.H. Selective Nrf2 Gene Deletion in the Rostral Ventrolateral Medulla Evokes Hypertension and Sympathoexcitation in Mice. *Hypertension* **2017**, *69*, 1198–1206. [CrossRef]
42. Chan, S.H.H.; Wu, K.L.H.; Chang, A.Y.W.; Tai, M.H.; Chan, J.Y.H. Oxidative impairment of mitochondrial electron transport chain complexes in rostral ventrolateral medulla contributes to neurogenic hypertension. *Hypertension* **2009**, *53*, 217–227. [CrossRef] [PubMed]
43. Chan, S.H.H.; Hsu, K.S.; Huang, C.C.; Wang, L.L.; Ou, C.C.; Chan, J.Y.H. NADPH oxidase-derived superoxide anion mediates angiotensin II-induced pressor effect via activation of p38 mitogen-activated protein kinase in the rostral ventrolateral medulla. *Cir. Res.* **2005**, *97*, 772–780. [CrossRef]
44. Chao, Y.M.; Wu, K.L.N.; Tsai, P.C.; Tain, Y.L.; Leu, S.; Lee, W.C.; Chan, J.Y.H. Anomalous AMPK-regulated angiotensin AT_1R expression and SIRT1-mediated mitochondrial biogenesis at RVLM in hypertension programming of offspring to maternal high fructose exposure. *J. Biomed. Sci.* **2020**, *27*, 68. [CrossRef] [PubMed]
45. Hirasaka, K.; Lago, C.U.; Kenoston, M.A.; Fathe, K.; Nowinski, S.M.; Nikawa, T.; Mills, E.M. Identification of a redox-modulatory interaction between uncoupling protein 3 and thioredoxin 2 in the mitochondrial intermembrane space. *Antioxid. Redox Signal.* **2011**, *15*, 2465–2661. [CrossRef] [PubMed]
46. Chan, S.H.H.; Tai, M.H.; Li, C.Y.; Chan, J.Y.H. Reduction in molecular synthesis or enzyme activity of superoxide dismutases and catalase contributes to oxidative stress and neurogenic hypertension in spontaneously hypertensive rats. *Free Radic. Biol. Med.* **2006**, *40*, 2028–2039. [CrossRef]
47. Idris-Khodja, N.; Ouerd, S.; Trindade, M.; Gornitsky, J.; Rehman, A.; Barhoumi, T.; Offermanns, S.; Gonzalez, F.J.; Neves, M.F.; Paradis, P.; et al. Vascular smooth muscle cell peroxisome proliferator-activated receptor γ protects against endothelin-1-induced oxidative stress and inflammation. *J. Hypertens.* **2017**, *35*, 1390–1401. [CrossRef]
48. Rincón, J.; Correia, D.; Arcaya, J.L.; Finol, E.; Fernández, A.; Pérez, M.; Yaguas, K.; Talavera, E.; Chávez, M.; Summer, R.; et al. Role of Angiotensin II type 1 receptor on renal NAD(P)H oxidase, oxidative stress and inflammation in nitric oxide inhibition induced-hypertension. *Life Sci.* **2015**, *124*, 81–90. [CrossRef]
49. Sorriento, D.; De Luca, N.; Trimarco, B.; Iaccarino, G. The antioxidant therapy: New insights in the treatment of hypertension. *Front. Physiol.* **2018**, *9*, 258. [CrossRef]
50. Chia, T.Y.; Murugaiyah, V.; Khan, N.A.K.; Sattar, M.A.; Abdulia, M.H.; Johns, E.J.; Ahmad, A.; Hassan, Z.; Kaur, G.; Mei, H.Y.; et al. Inhibition of L-NAME-induced hypertension by combined treatment with apocynin and catalase: The role of Nox 4 expression. *Physiol. Res.* **2021**, *70*, 13–26. [CrossRef]
51. Chan, S.H.H.; Wang, L.L.; Wang, S.H.; Chan, J.Y.H. Differential cardiovascular responses to blockade of nNOS or iNOS in rostral ventrolateral medulla of the rat. *Br. J. Pharmacol.* **2001**, *133*, 606–614. [CrossRef]
52. Kishi, T.; Hirooka, Y.; Sakai, K.; Shigematsu, H.; Shimokawa, H.; Takeshita, A. Overexpression of eNOS in the RVLM causes hypotension and bradycardia via GABA release. *Hypertension* **2001**, *38*, 896–901. [CrossRef] [PubMed]
53. Young, C.N.; Fisher, J.P.; Gallagher, K.M.; Whaley-Connell, A.; Chaudhary, K.; Victor, R.G.; Thomas, G.D.; Fadel, P.J. Inhibition of nitric oxide synthase evokes central sympatho-excitation in healthy humans. *J. Physiol.* **2009**, *587*, 977–4986. [CrossRef]
54. Biancardi, V.C.; Bergamaschi, C.T.; Lopes, O.U.; Campos, R.R. Sympathetic activation in rats with L-NAME-induced hypertension. *Braz. J. Med. Biol. Res.* **2007**, *40*, 401–408. [CrossRef] [PubMed]
55. Rubattu, S.; Pagliaro, B.; Pierelli, G.; Santolamazza, C.; Castro, S.D.; Mennuni, S.; Volpe, M. Pathogenesis of target organ damage in hypertension: Role of mitochondrial oxidative stress. *Int. J. Mol. Sci.* **2014**, *16*, 823–839. [CrossRef] [PubMed]
56. Mason, M.G.; Nicholls, P.; Wilson, M.T.; Cooper, C.E. Nitric oxide inhibition of respiration involves both competitive (heme) and noncompetitive (copper) binding to cytochrome c oxidase. *Proc. Natl. Acad. Sci. USA* **2006**, *103*, 708–713. [CrossRef]
57. Clementi, E.; Brown, G.C.; Feelisch, M.; Moncada, S. Persistent inhibition of cell respiration by nitric oxide: Crucial role of S-nitrosylation of mitochondrial complex I and protective action of glutathione. *Proc. Natl. Acad. Sci. USA* **1998**, *95*, 7631–7636. [CrossRef]
58. Rizvi, F.; Preston, C.C.; Emelyanova, L.; Yousufuddin, M.; Viqar, M.; Dakwar, O.; Ross, G.R.; Faustino, R.S.; Holmuhamedov, E.L.; Jahangir, A. Effects of aging on cardiac oxidative stress and transcriptional changes in pathways of reactive oxygen species generation and clearance. *J. Am. Heart Assoc.* **2021**, *10*, e019948. [CrossRef]
59. Daiber, A.; Xia, N.; Steven, S.; Oelze, M.; Hanf, A.; Kröller-Schön, S.; Münzel, T.; Li, H. New therapeutic implications of endothelial nitric oxide synthase (eNOS) function/dysfunction in cardiovascular disease. *Int. J. Mol. Sci.* **2019**, *20*, 187. [CrossRef]
60. Fukai, T.; Ushio-Fukai, M. Superoxide dismutases: Role in redox signaling, vascular function, and diseases. *Antioxid. Redox. Signal.* **2011**, *15*, 1583–1606. [CrossRef]
61. Chan, S.H.H.; Wu, C.A.; Wu, K.L.H.; Ho, Y.H.; Chang, A.Y.W.; Chan, J.Y.H. Transcriptional upregulation of mitochondrial uncoupling protein 2 protects against oxidative stress-associated neurogenic hypertension. *Circ. Res.* **2009**, *105*, 886–896. [CrossRef]
62. Dikalova, A.; Mayorov, V.; Xiao, L.; Panov, A.; Amarnath, V.; Zagol-Ikapitte, I.; Vergeade, A.; Ao, M.; Yermalitsky, V.; Nazarewicz, R.R.; et al. Mitochondrial isolevuglandins contribute to vascular oxidative stress and mitochondria-targeted cavenger of isolevuglandins reduces mitochondrial dysfunction and hypertension. *Hypertension* **2020**, *76*, 1980–1991. [CrossRef] [PubMed]
63. Moschinger, M.; Hilse, K.E.; Rupprecht, A.; Zeitz, U.; Erben, R.G.; Rülicke, T.; Pohl, E.E. Age-related sex differences in the expression of important disease-linked mitochondrial proteins in mice. *Biol. Sex Differ.* **2019**, *10*, 56. [CrossRef] [PubMed]

64. Niu, X.L.; Madamanchi, N.R.; Vendrov, A.E.; Tchivilev, I.; Rojas, M.; Madamanchi, C.; Brandes, R.P.; Krause, K.H.; Humphries, J.; Smith, A.; et al. Nox activator 1: A potential target for modulation of vascular reactive oxygen species in atherosclerotic arteries. *Circulation* **2010**, *121*, 549–559. [CrossRef] [PubMed]
65. Takeya, R.; Ueno, N.; Kami, K.; Taura, M.; Kohjima, M.; Izaki, T.; Nunoi, H.; Sumimoto, H. Novel human homologues of $p47^{phox}$ and $p67^{phox}$ participate in activation of superoxide-producing NADPH oxidases. *J. Biol. Chem.* **2003**, *278*, 25234–25246. [CrossRef]
66. Vendrov, A.E.; Sumida, A.; Canugovi, C.; Lozhkin, A.; Hayami, T.; Madamanchi, N.R.; Runge, M.S. NOXA1-dependent NADPH oxidase regulates redox signaling and phenotype of vascular smooth muscle cell during atherogenesis. *Redox. Biol.* **2018**, *21*, 101063. [CrossRef]
67. Guyenet, P.G.; Stornetta, R.L.; Souza, G.M.P.R.; Abbott, S.B.G.; Brooks, V.L. Neuronal networks in hypertension: Recent advances. *Hypertension* **2020**, *76*, 300–311. [CrossRef]
68. Crowley, S.D. The cooperative roles of inflammation and oxidative stress in the pathogenesis of hypertension. *Antioxid. Redox. Signal.* **2014**, *20*, 102–120. [CrossRef]

biomedicines

MDPI

Article

Angiotensin II Modulates Calcium/Phosphate Excretion in Experimental Model of Hypertension: Focus on Bone

Giovanna Castoldi [1,*], Raffaella Carletti [2], Silvia Ippolito [3], Isabella Villa [4], Biagio Palmisano [5], Simona Bolamperti [4], Alessandro Rubinacci [6], Gianpaolo Zerbini [7], Michela Meani [1,8], Giovanni Zatti [1,8] and Cira R. T. di Gioia [9]

1 Dipartimento di Medicina e Chirurgia, Università degli Studi di Milano-Bicocca, 20900 Monza, Italy
2 Dipartimento di Medicina Traslazionale e di Precisione, Sapienza Università di Roma, 00185 Rome, Italy
3 Laboratorio Analisi Chimico Cliniche, Ospedale San Gerardo, ASST Monza, 20900 Monza, Italy
4 Laboratorio di Endocrinologia e Metabolismo Osseo, Istituto di Endocrinologia e Scienze Metaboliche, IRCCS Ospedale San Raffaele, 20132 Milano, Italy
5 Dipartimento di Medicina Molecolare, Sapienza Università di Roma, 00161 Roma, Italy
6 Osteoporosis and Bone and Mineral Metabolism Unit, IRCCS Ospedale San Raffaele, 20132 Milano, Italy
7 Unita' Complicanze del Diabete, IRCCS Ospedale San Raffaele, 20132 Milano, Italy
8 Clinica Ortopedica, Ospedale San Gerardo, ASST Monza, 20900 Monza, Italy
9 Dipartimento di Scienze Radiologiche, Oncologiche e Anatomopatologiche, Istituto di Anatomia Patologica, Sapienza Università di Roma, 00161 Rome, Italy
* Correspondence: giovanna.castoldi@unimib.it; Tel.: +39+2-64488058

Abstract: A link between hypertension and long-term bone health has been suggested. The aim of this study was to investigate the effects of chronic angiotensin II administration on urinary calcium/phosphate excretion, bone mineral density, bone remodeling and osteoblast population in a well-established experimental model of hypertension, in the absence of possible confounding factors that could affect bone metabolism. Male Sprague–Dawley rats, divided in the following groups: (a) Angiotensin II (Ang II, 200 ng/kg/min, osmotic minipumps, sub cutis, $n = 8$); (b) Ang II+losartan (Los, 50 mg/kg/day, *per os*, $n = 6$); (c) control group (physiological saline, sub cutis, $n = 9$); and (d) control+losartan ($n = 6$) were treated for four weeks. During the experimental period, 24-hour diuresis, urinary calcium, phosphate and sodium excretion were measured prior to the treatment, at two weeks of treatment, and at the end of the treatment. Systolic blood pressure was measured by plethysmography technique (tail cuff method). At the end of the experimental protocol, the rats were euthanized and peripheral quantitative computed tomography at the proximal metaphysis and at the diaphysis of the tibiae and quantitative bone histomorphometry on distal femora were performed. Angiotensin II-dependent hypertension is associated with increased calcium and phosphate excretion. AT1 receptor blockade prevented the increase of blood pressure and phosphate excretion but did not affect the increase of calcium excretion. These changes took place without significantly affecting bone density, bone histology or osteoblast population. In conclusion, in our experimental conditions, angiotensin II-dependent hypertension gave rise to an increased urinary excretion of calcium and phosphate without affecting bone density.

Keywords: angiotensin II; experimental hypertension; bone; rats

Citation: Castoldi, G.; Carletti, R.; Ippolito, S.; Villa, I.; Palmisano, B.; Bolamperti, S.; Rubinacci, A.; Zerbini, G.; Meani, M.; Zatti, G.; et al. Angiotensin II Modulates Calcium/Phosphate Excretion in Experimental Model of Hypertension: Focus on Bone. *Biomedicines* **2022**, *10*, 2928. https://doi.org/10.3390/biomedicines10112928

Academic Editors: Ivana Vaněčková and Josef Zicha

Received: 2 October 2022
Accepted: 8 November 2022
Published: 14 November 2022

1. Introduction

Epidemiologic data suggest a possible link between arterial hypertension, characterized by cardiovascular and renal complications, and a degenerative disease of the bone, known as osteoporosis [1–4]. Both arterial hypertension and osteoporosis show increased frequency in aging, especially in post-menopausal women, and share the same risk factors, such as low physical activity or smoking. These two diseases have similar pathophysiological mechanisms, such as increased oxidative stress, inflammation, and dysfunctions of calcium homeostasis [5–9].

Systemic renin-angiotensin system (RAS) has a fundamental role in the regulation of blood pressure and electrolyte balance [10,11], and Angiotensin II (Ang II) has been historically considered the main effector of RAS. Besides the 'classical' systemic RAS, it has been demonstrated that the components of RAS are expressed in multiple tissues [12]. At the same time, Ang II, in addition to its well-known hemodynamic action, has many cellular effects in different tissues promoting inflammation [13–15], fibrosis [13,16,17], and hypertrophy [17–19], mainly mediated by AT1 receptors.

Bone tissue expresses the components of RAS which are involved in bone remodeling [20–22]. In fact, experimental studies have demonstrated that Ang II increases the activity of osteoclasts, enhancing bone loss in rats affected by osteoporosis caused by estrogen deficiency [23]. A role for bone RAS has been described in age-related osteoporosis in mice [24] and in the reduction of bone formation and in the increase of resorption in obstructive nephropathy [25].

AT1 receptor knockout mice showed an increase in bone mass [26], possibly suggesting a protective role of RAS inhibition in bone. However, the inhibitory effects of captopril on bone RAS components did not improve bone damage caused by type II diabetes in db/db mice [27]. In type 1 diabetes mice (streptozotocin induced) the increased activity of bone RAS was shown to be involved in osteoporosis processes, which was not improved by ACE inhibitors [28]. Finally, in ovariectomized rats, the stimulation of the protective arm of RAS (ACE-2/Ang 1-7/Mas) was shown to mediate the osteo-protective effects of RAS inhibition [21].

These data strongly suggest that the role of RAS in bone remodeling might differ depending on the underlying pathological conditions, such as estrogen deficiency, hypertension, diabetes and osteoporosis.

Nevertheless, the direct effects of angiotensin II dependent hypertension on bone tissue are not completely understood.

In the present study we investigated the effects of chronic angiotensin II administration on urinary calcium/phosphate excretion, bone mineral density, bone remodeling and on osteoblast population—the cells responsible for the synthesis of bone matrix—in a well-established experimental model of hypertension, in the absence of possible confounding factors that could affect bone metabolism.

In this study we demonstrate that angiotensin II-dependent hypertension is associated with a modulation of urinary calcium/phosphate excretion over time leading to an increase of calcium and phosphate excretion. AT1 receptor blockade prevented the increase of blood pressure and phosphate excretion but did not affect the increase of calcium excretion. These changes took place without significantly affecting bone density, bone histology, and osteoblast population.

2. Materials and Methods

2.1. Animal Study Design of Ang II-Dependent Hypertension

Experiments were performed in male Sprague–Dawley rats (body weight 150–175 g) in accordance with the Guide for the Care and Use of Laboratory Animals published by the US National Institutes of Health (NIH Publication No. 85–23, revised 1996). Animal husbandry was in conformity with the institutional guidelines in compliance with national laws and policies (D.L.n. 116, Gazzetta Ufficiale della Repubblica Italiana, suppl.40, 18 February 1992). Rats were individually housed in cages (or metabolic cages as necessary) in a temperature-controlled room with a 12:12 light:dark cycle, with free access to a standard rat chow and tap water. One week before the beginning of the protocol, rats were accustomed to metabolic cages and experimental procedures. Body weight (BW, g) was measured once a week. Systolic blood pressure was measured by plethysmography technique (tail cuff method, average of six recordings using a BP Recorder (Ugo Basile Instruments, Varese, Italy)) at the beginning and at the end of the experimental protocols [29,30]. Ang II was administrated through osmotic minipumps (Alzet 2004, Palo Alto, CA, USA), subcutaneously implanted under sodium pentobarbital anesthesia (40 mg/kg/i.p.), in order to administer Ang II

(200 ng/Kg/min for 4 weeks, $n = 8$, Sigma) or physiological saline in the control group ($n = 9$). Losartan (Los, 50 mg/kg/day), dissolved in drinking water, was administered in control + Los ($n = 6$) and Ang II + Los treated rats ($n = 6$). Glomerular filtration rate (mL/min) was evaluated as creatinine clearance using 24 h diuresis (ml/24 h).

Non-fasting plasma glucose (mg/dL), creatinine (mg/dL), sodium (mEq/L), potassium (mEq/L), calcium (mg/dL), phosphate (mg/dL), alkaline phosphatase (U/L), uric acid (mg/dL), total cholesterol (mg/dL), triglycerides (mg/dL) and 24 h urinary calcium (mg/24 h), phosphate (mg/24 h), and sodium (mEq/24 h) were measured by colorimetric technique on Cobas Roche (Mannheim, Germany) [15].

At the end of the experimental periods, rats were euthanized by an overdose of anesthesia. Hearts and kidneys were excised and weighted. The femurs were excised and fixed with 10% formalin, embedded in paraffin and used for light microscopic examination and morphometric analysis. Contralateral femur with tibia was also dissected, fixed with 10% formalin solution, neutral buffered, and used for peripheral quantitative computed tomography.

2.2. Peripheral Quantitative Computed Tomography (pQCT) Analysis

The measurements were obtained by using a Stratec Research SA+ pQCT scanner (Stratec Medizintechnik GmbH, Pforzheim, Germany) with a voxel size of 70 mm and a scan speed of 3 mm/s. The excised tibiae were hold in place with manufacturer-made plastic holders in order to position the long axes of the specimen parallel to the image planes. The correct longitudinal positioning was resolved by means of a "scout scan". The scans were performed at the proximal metaphysis and at the diaphysis of the tibiae. The scans were analyzed with pQCT software 6.00B using contour mode 2 and peel mode 2. The threshold for the calculation of trabecular and total bone parameters was 500 mg/cm^3 and for cortical bone parameters 710 mg/cm^3.

2.3. Bone Histology and Histomorphometry

For each rat the formalin fixed femur was decalcified in 10% EDTA for 30–45 days at 4 °C with gentle shaking and processed for paraffin embedding. Three-micron-thick paraffin embedded sections were used for standard histology after staining with Hematoxylin-Eosin (H&E) or with Sirius red to visualize collagen fibers with morphometric analysis.

Quantitative bone histomorphometry was conducted on distal femora. Experiments were performed in a blinded fashion. The region of interest (ROI) was identified in the secondary spongiosa of distal femora, starting 500 μm below the growth plate and for a length of 5 mm. Sections were stained with H&E and Sirius red and used to measure trabecular bone volume per tissue volume (BV/TV), osteoblast number per bone surface (N.Ob/BS) and osteoblast surface per bone surface (Ob.S/BS). Ten pictures at 20X were acquired with an optical microscope (Leica DMRB, Leica Biosystems, Muttenz, Switzerland) through a digital camera (Leica LAS EZ, Leica Biosystems, Switzerland) and all histomorphometric analyses were performed using Image J [31].

2.4. Statistical Analysis

Data are reported as means ± standard error of the mean (SEM). Differences among the groups of rats (control, control+losartan, Ang II, Ang II+losartan) for systolic blood pressure, body weight, glomerular filtration rate, heart/body weight, kidney/body weight, blood glucose, alkaline phosphatase, sodium, potassium, calcium, phosphate, uric acid, cholesterol and triglycerides, 24 h diuresis, 24 h urinary calcium, phosphate and sodium excretion, pQCT parameters, trabecular bone volume per tissue volume (BV/TV), osteoblast number per bone surface (N.Ob/BS) and osteoblast surface per bone surface (Ob.S/BS) were assessed using ANOVA followed by Fisher's protected least-significant test for post hoc comparisons. Differences between means were considered significant at $p < 0.05$.

3. Results

3.1. Effects of Ang II and Losartan Administration on Systolic Blood Pressure, Body Weight, Glomerular Filtration Rate, Heart/Body Weight Ratio, Kidney/Body Weight Ratio and Serological Parameters

Ang II administration caused a marked increase in blood pressure, which was prevented by losartan treatment in Ang II treated rats (Table 1). Losartan administration did not significantly modify blood pressure in control rats (Table 1).

Table 1. Systolic blood pressure (SBP, mmHg), body weight (BW, g.), glomerular filtration rate (GFR, ml/min), heart/body weight ratio (mg/g), kidney/body weight ratio (mg/g), non-fasting plasma glucose (mg/dL), alkaline phosphatase (U/L), sodium (mEq/L), potassium (mEq/L), calcium (mg/dL), phosphate (mg/dL), uric acid (mg/dL), total cholesterol (mg/dL), triglycerides (mg/dL) in control, control+los, Ang II, Ang II+los-treated rats at the end of the experimental period (four weeks). Data are means ± SEM. * = $p < 0.05$ vs. control; † = $p < 0.01$ vs. control; ‡ = $p < 0.0001$ vs. control; ° = $p < 0.05$ vs. control+los; δ = $p < 0.01$ vs. control+los; § = $p < 0.0001$ vs. control+los; ^ = $p < 0.05$ vs. Ang II+los; & = $p < 0.01$ vs. Ang II+los; ζ = $p < 0.0001$ vs. Ang II+los.

Parameters	Control	Control + Los	Ang II	Ang II + Los
SBP, mmHg	140.5 ± 3.3	135.0 ± 5.0	202.1 ± 3.0 ‡§ζ	142.5 ± 3.2
BW, g	459.1 ± 13.0	433.0 ± 5.2	409.7 ± 13.6 †	423.8 ± 9.4
GFR, ml/min	4.28 ± 0.23	3.58 ± 0.07	3.35 ± 0.36 *	4.15 ± 0.28
Heart/Body Weight, mg/g	2.81 ± 0.06	2.66 ± 0.13	3.93 ± 0.11 ‡§ζ	2.71 ± 0.05
Kidney/Body Weight, mg/g	3.34 ± 0.05	3.51 ± 0.06	4.06 ± 0.09 ‡δζ	3.42 ± 0.13
Plasma				
Glucose (mg/dL)	139± 4.08	142.1 ± 8.15	150.5 ± 5.03	137.2 ± 7.84
Alkaline phosphatase (U/L)	194.3 ± 13.7	194.4 ± 19.5	207.3 ± 6.97	226.7 ± 21.2
Sodium (mEq/L)	140.7 ± 0.38	141.3 ± 0.56	138.0 ± 0.93 †δ^	141.1 ± 1.03
Potassium (mEq/L)	3.95 ± 0.13	4.59 ± 0.33	3.16 ± 0.13 *δ&	4.75 ± 0.54 *
Calcium (mg/dL)	9.83 ± 0.10	9.96 ± 0.09	9.79 ± 0.17	10.0 ± 0.13
Phosphate (mg/dL)	5.72 ± 0.25	6.27 ± 0.46	4.84 ± 0.38 *°^	6.14 ± 0.45
Uric acid (mg/dL)	0.72 ± 0.15	0.72 ± 0.14	0.95 ± 0.18	0.94 ± 0.20
Cholesterol (mg/dL)	71.88 ± 3.96	70.2 ± 4.23	69.31 ± 3.51	71.14 ± 2.09
Triglycerides (mg/dL)	89.81 ± 4.89	98.6 ± 6.86	87.58 ± 21.32	116.34 ± 14.54

As compared to control rats, Ang II administration caused a significant reduction in BW, which was blunted by losartan treatment (Table 1). Ang II administration caused a slight decrease in glomerular filtration rate as compared to control rats. Losartan treatment prevented the decrease of GFR in Ang II treated rats (Table 1).

Ang II administration caused a significant increase in the heart/BW and kidney/BW ratios as compared to control rats, in line with the possible development of myocardial and kidney hypertrophy/fibrosis [15–17]. Losartan administration prevented the increase of heart and kidney/BW ratio in Ang II treated rats (Table 1).

As shown in Table 1, Ang II administration caused a slight decrease in plasma sodium, potassium and phosphate values as compared to control rats, which was prevented by losartan treatment in Ang II treated rats. No significant changes in plasma alkaline phosphatase, calcium, uric acid and in glyco-metabolic parameters (non-fasting glucose, cholesterol, triglycerides) were observed among the different groups (Table 1).

3.2. Effects of Ang II and Losartan Administration on 24 h Diuresis, Urinary Sodium, Calcium, and Phosphate Excretion

Before starting the treatments, diuresis and urinary sodium, calcium, and phosphate excretion were similar in all the groups (Figures 1 and 2).

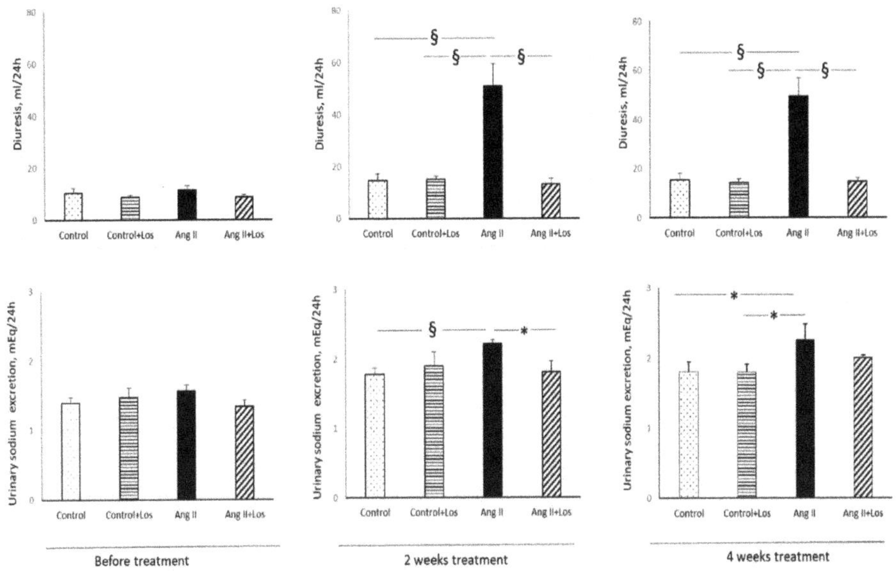

Figure 1. Effects of Ang II and losartan administration on 24 h diuresis and urinary sodium excretion before and after two and four weeks of different treatments. Data are means ± SEM. * = $p < 0.05$; § = $p < 0.01$.

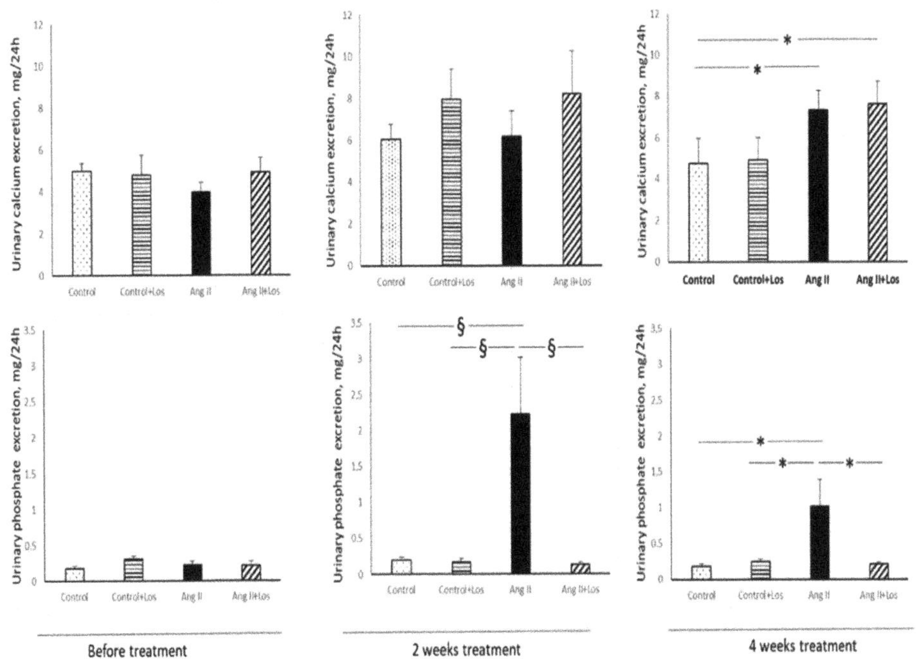

Figure 2. Effects of Ang II and losartan administration on 24 h urinary calcium and phosphate excretion before and after two and four weeks of different treatments. Data are means ± SEM. * = $p < 0.05$; § = $p < 0.01$.

Ang II administration increased diuresis just after two weeks of administration and until the end of the experimental period as compared to other groups (Figure 1). Losartan treatment blocked the increase in diuresis in Ang II treated rats during the entire experimental period (Figure 1).

After two weeks of Ang II administration and until the end of the experimental protocol Ang II administration caused a significant increase in urinary sodium excretion as compared to control rats (Figure 1). Losartan treatment blocked the increase in urinary sodium excretion in Ang II-treated rats (Figure 1).

After two weeks of Ang II administration no significant changes in urinary calcium excretion were observed in Ang II treated rats, as compared to other groups (Figure 2).

At the end of the experimental period, treatment with Ang II, alone or in combination with losartan, induced an increase of urinary calcium excretion as compared to control rats (Figure 2). Ang II administration increased phosphaturia just after two weeks of administration and until the end of the experimental period as compared to control rats (Figure 2). Otherwise, losartan treatment blocked the increase in urinary phosphate excretion in Ang II treated rats just after two weeks of administration until the end of experimental protocol (Figure 2).

3.3. pQCT Analysis of Bone Parameters after Four Weeks of Treatment with Ang II and Losartan

After four weeks of treatment, Ang II did not modify any bone parameters compared to control measured at the proximal metaphysis and at the diaphysis midshaft of the tibiae. Losartan induced an increase in Tb.BMD in rats chronically treated with Ang II and in control rats, although not reaching statistical significance (Table 2, Figure 3).

Table 2. Bone parameters measured by pQCT.

Parameters	Control	Control + Los	Ang II	Ang II + Los
Tb.BMD (mg/cm^3)	257.0 ± 11.3	277.6 ± 19.7	251.9 ± 9.9	291.7 ± 10.2
Tb.Area (mm^2)	20.47 ± 1.08	20.80 ± 1.53	21.66 ± 1.07	20.35 ± 0.71
Ct.BMD (mg/cm^3)	1334.6 ± 4.8	1336.2 ± 8.5	1332.2 ± 6.6	1331.8 ± 5.3
Ct.Area (mm^2)	4.75 ± 0.16	4.67 ± 0.10	4.52 ± 0.18	4.71 ± 0.16
Tot.Area (mm^2)	6.91 ± 0.24	6.75 ± 0.19	6.54 ± 0.25	6.90 ± 0.15

Tb.BMD: trabecular bone mineral density, Tb.Area: trabecular area, Ct.BMD: cortical bone mineral density, Ct.Area: cortical area, Tot.Area: total area. Tb.BMD and Tb.Area were measured at the proximal metaphysis of the tibia, Ct.BMD, Ct.Area and Tot.Area were measured at the diaphyseal midshaft. Data are means ± SEM.

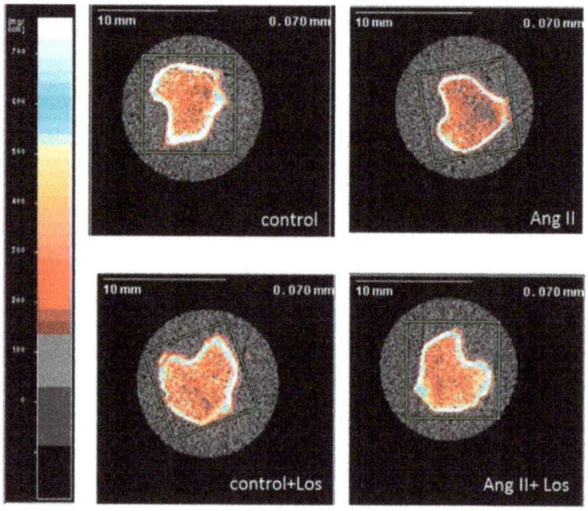

Figure 3. Representative image of the metaphyseal site of the rat tibiae analyzed by pQCT. The bar on the left is the colour scale bar of the density (mg/cm^3): white > 700 mg/cm^3; grey ≤ 100 mg/cm^3.

3.4. Histomorphometric Analysis of Bone Parameters after Four Weeks of Treatment with Ang II and Losartan

Osteoblast parameters, quantified by morphology and histomorphometry and normalized for trabecular bone surfaces (N.Ob/BS and Ob.S/BS) did not differ between Ang II treated rats and control animals. Losartan treatment did not modify osteoblast number/bone surface and osteoblast surface/bone surface in control and Ang II treated rats (Figure 4).

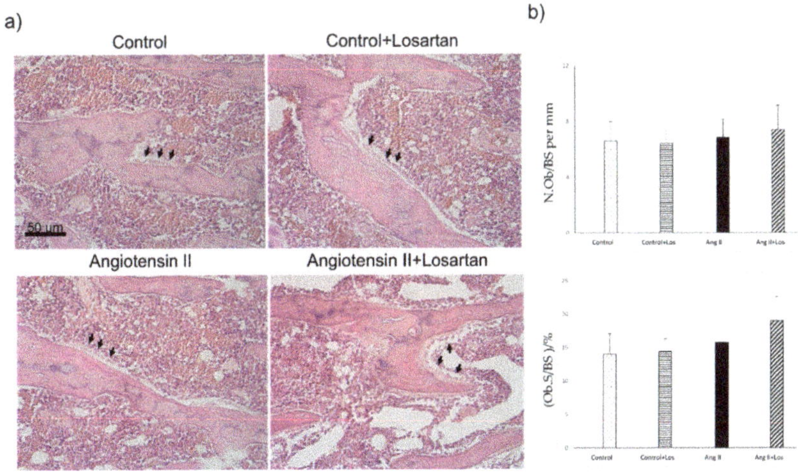

Figure 4. Effects of Ang II and Losartan administration on bone tissue. (**a**) Representative images from H&E-stained sections showing osteoblasts (arrow) and (**b**) quantification of osteoblast parameters in the different groups of rats. Data are means ± SEM. N.Ob: Osteoblast number; BS: bone surface; Ob.S: osteoblast surface; BS: bone surface. Ang II: Angiotensin II; Los: losartan.

Trabecular bone volume of femora in Ang II treated rats did not show significant difference as compared with control groups and Ang II+losartan treated rats (Figure 5).

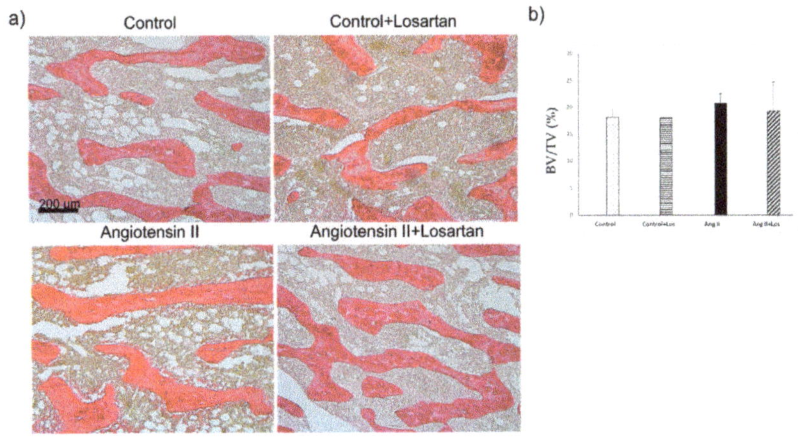

Figure 5. Effects of Ang II and losartan administration on bone tissue. (**a**) Representative histological pictures from Sirius red-stained bone tissue sections (trabecular bone stained in red) and (**b**) quantification of femoral trabecular bone in the different groups of rats. BV: Trabecular bone volume; TV: tissue volume. Data are means ± SEM. Ang II: Angiotensin II; Los: losartan.

4. Discussion

The results of this study demonstrate that chronic Ang II administration caused an early increase in urinary excretion of calcium and phosphate that lasted a long time. The increase was paralleled by a decrease in plasma phosphate levels, while plasma calcium levels did not change.

The blockade of AT1 receptors by losartan treatment prevented the increase of phosphaturia, but not the increase in urinary calcium excretion.

A link between hypertension and bone metabolism has been described in the literature, in particular a relationship between high blood pressure and hypercalciuria, which may promote bone mineral loss [32–34].

In our case the Ang II-induced increase of calcium and phosphate excretion is not associated with changes in bone mass.

The observed enhancement of renal phosphate loss in Ang II treated animals might be primarily related to a direct effect of Ang II on the type II sodium-phosphate co-transporter (NaPi-IIa) that is the critical player in renal Pi regulation. It has indeed been shown that phosphate excretion increased seven fold in rats chronically treated with Ang II with an associated enhancement, likely induced by post-transcriptional mechanisms, of NaPi-IIa protein level with Western blot analysis of the brush border membrane vesicles exposed to Ang II [35]. This effect is potentially mediated by the Ang II type 1 receptor, and, in line with our experimental model, losartan prevented the increase of phosphate excretion in Ang II treated rats.

Beyond the direct effect of Ang II on NaPi-IIa, it should be considered that the endocrine regulators of the renal tubular maximum phosphate reabsorption (TmP) might be affected by RAS activation and subsequently constitute an endocrine milieu favoring kidney phosphate loss. In fact, the increased (sodium-driven) calcium excretion induced by Ang II administration could give rise to a negative calcium balance with secondary activation of parathyroid hormone (PTH). This phenomenon would result, just as in our case, an increased urinary excretion of phosphate and a decrease in plasma phosphate. PTH is in fact a known regulator of TmP and could directly contribute to bone and heart damage. PTH secretion could also be directly enhanced by aldosterone [36] that can in turn be stimulated by Ang II. This assumption is supported by the evidence described in the literature that RAS inhibitors lower PTH [37,38].

Taking into account the extent of the effects on phosphate metabolism in relation to the minor effects on calcium, it is also possible, as an additional hypothesis, that the increased urinary excretion of phosphate and reduced plasma phosphate levels are related to the activation of Fibroblast Growth Factor23 (FGF23) secretion by Ang II and/or aldosterone [39]. It has indeed been shown that Ang II and aldosterone stimulate FGF23 transcription and secretion in cultured osteoblast-like cells [40] and cardiac myocytes [41]. FGF23 is a 251-amino-acid long, primarily bone derived, hormone that is critical to the maintenance of phosphate homeostasis [42]. In the proximal kidney tubule, FGF23 binds to the FGF receptor (FGFR) and its co-receptor klotho and downregulates the membrane availability of the phosphate type II sodium-phosphate co-transporter, NaPi-IIa. This effect, acting in a combined manner with PTH signaling [43], results in increased urinary excretion of phosphate and reduced plasma phosphate levels. The interactions among PTH, FGF23 and RAS signaling on NaPi-IIa might therefore constitute the endocrine milieu which drives the phosphate loss in a synergistic manner, although the specific contribution of each regulatory signal remains unclear. If this was the case, the activation of FGF23 could at the same time participate in cardiac damage. In fact, Ang II inhibition has been shown to reduce FGF23-induced changes in Ca2+ homeostasis and hypertrophy in cardiac cells [44]

Interestingly, in our experimental conditions, an increase in urinary calcium excretion, but not an increase in urinary phosphate excretion, was present in Ang II+losartan treated group, in the absence of both hypertension and increased urinary sodium excretion. Whether this increase is a transitory effect or may last a long time, mediated by other mechanisms not related to the blockade of AT1 receptors or by the increase in blood pressure caused by Ang II, remains to be clarified.

However, in our experimental model of Ang II dependent hypertension we did not find significant changes in bone mass measured by both pQCT and histomorphometric analysis in the different appendicular skeleton sites. In line with our results, no detectable changes in the trabecular and cortical bone parameters, as assessed by micro-CT, were shown to be at the bone sites far from the inflamed joints after Ang II administration for four weeks in male mice with Tumor Necrosis Factor-mediated arthritis [45]. Our findings differ from results previously described in experimental models that used chronic administration of Ang II at the same dosage and for the same time used in our experimental model, with subsequent enhancement of bone loss [20,23]. These differences could depend on the different pathophysiologic conditions of animal model applied. In the present study we administered Ang II to male adult Sprague–Dawley rats in physiological conditions. Conversely, in the osteoporosis animal model, Ang II was administered to female animals that had developed ovariectomy induced bone loss [20,23]. It is therefore likely that the gender and the pre-activation of bone remodeling with a negative bone balance might be both critical determinants of RAS activation in bone. Taking into account that in our experimental model we administered Ang II to healthy male Sprague–Dawley rats for four weeks, we can reasonably hypothesize that the duration of the treatment, which was shown to be sufficient to cause changes to other organs, such as the heart [15,17] and kidney [16], is probably too short to cause an alteration in the bone tissue. In fact, in the absence of a pre-activation of bone remodeling characterized by bone loss, it is likely that, to observe a significant modification of bone density, it should be necessary to prolong the duration of administration of Ang II, and possibly also to increase the dosage. Further experimental studies will be needed to verify these hypotheses.

In conclusion, in our experimental model of Ang II dependent hypertension we found changes in urinary calcium/phosphate excretion, which might be due to direct or endocrine mediated mechanisms. In our experimental model, chronic administration of Ang II and/or increased blood pressure values did not affect bone tissue in the absence of pre-activation of bone loss. However, we cannot exclude that, by further extending the experimental time, a detrimental effect of Ang II on bone could become apparent.

Author Contributions: Conceptualization, G.C., G.Z. (Gianpaolo Zerbini) and C.R.T.d.G.; methodology, G.C., R.C., S.I., M.M. and B.P.; software, R.C. and B.P.; validation, R.C. and S.I.; formal analysis, G.C., M.M., R.C. and S.I.; investigation G.C., R.C., G.C., S.I., I.V., S.B., A.R., M.M. and C.R.T.d.G.; resources, G.Z. (Giovanni Zatti) and C.R.T.d.G.; data curation, G.C., R.C., S.I., G.Z. (Gianpaolo Zerbini), I.V., S.B. and G.Z. (Giovanni Zatti); writing—original draft preparation, G.C., G.Z. (Gianpaolo Zerbini), G.Z. (Giovanni Zatti), S.I., I.V., S.B., A.R. and C.R.T.d.G. writing—reviewing, interpreting the results and editing, G.C., G.Z. (Gianpaolo Zerbini), S.I., A.R., G.Z. (Giovanni Zatti) and C.R.T.d.G.; visualization, R.C., M.M. and B.P.; supervision, G.C., G.Z. (Gianpaolo Zerbini) and C.R.T.d.G.; funding acquisition, G.Z. (Giovanni Zatti). All authors have read and agreed to the published version of the manuscript.

Funding: This research was funded by Fondo Ateneo Ricerca (Giovanni Zatti).

Institutional Review Board Statement: The study was conducted according to the Guidelines of the Declaration of Helsinki and approved by the Ministero della Salute della Repubblica Italiana-Direzione Generale della Sanita' Animale e dei Farmaci Veterinari (n. 1123/2015, 22 October 2015).

Informed Consent Statement: Not applicable.

Data Availability Statement: The data presented in this study are available on request from the corresponding author.

Conflicts of Interest: The authors declare no conflict of interest.

References

1. Farhat, G.N.; Cauley, J.A. The link between osteoporosis and cardiovascular disease. *Clin. Cases Miner. Bone Metab.* **2008**, *5*, 19–34. [PubMed]
2. Canoy, D.; Harvey, N.C.; Prieto-Alhambra, D.; Cooper, C.; Meyer, H.E.; Asvold, B.O.; Nazarzadeh, M.; Rahimi, K. Elevated blood pressure, antihypertensive medications and bone health in the population: Revisiting old hypotheses and exploring future research directions. *Osteoporos. Int.* **2022**, *33*, 315–326. [CrossRef] [PubMed]
3. London, G.M. Soft bone–hard arteries: A link? *Kidney Blood Press. Res.* **2011**, *34*, 203–208. [CrossRef] [PubMed]
4. Doherty, T.M.; Asotra, K.; Fitzpatrick, L.A.; Qiao, J.H.; Wilkin, D.J.; Detrano, R.C.; Dunstan, C.R.; Shah, P.K.; Rajavashisth, T.B. Calcification in atherosclerosis: Bone biology and chronic inflammation at the arterial crossroads. *Proc. Natl. Acad. Sci. USA* **2003**, *100*, 11201–11206. [CrossRef] [PubMed]
5. Singh, M.V.; Chapleau, M.W.; Harwani, S.C.; Abboud, F.M. The immune system and hypertension. *Immunol. Res.* **2014**, *59*, 243–253. [CrossRef]
6. Byon, C.H.; Chen, Y. Molecular mechanisms of vascular calcification in chronic kidney disease: The link between bone and the vasculature. *Curr. Osteoporos. Rep.* **2015**, *13*, 206–215. [CrossRef]
7. Grobbee, D.E.; Hackeng, W.H.L.; Birkenhager, J.C.; Hofman, A. Raised plasma intact parathyroid hormone concentrations in young people with mildly raised blood pressure. *Br. Med. J.* **1998**, *296*, 814–816. [CrossRef]
8. Butt, D.A.; Alharty, R.; Leu, R.; Cheung, A.M. Hypertension, antihypertensive drugs and the risk of fractures. *Clinic. Rev. Bone Mineral Metab.* **2015**, *13*, 160–172. [CrossRef]
9. Ilić, K.; Obradović, N.; Vujasinović-Stupar, N. The relationship among hypertension, antihypertensive medications, and osteoporosis: A narrative review. *Calcif. Tissue Int.* **2013**, *92*, 217–227. [CrossRef]
10. Te Riet, L.; van Esch, J.H.M.; Roks, A.J.M.; van den Meiracker, A.H.; Danser, A.H. Hypertension: Renin-Angiotensin-Aldosterone system alterations. *Circ. Res.* **2015**, *116*, 960–975. [CrossRef]
11. Pugliese, N.R.; Masi, S.; Taddei, S. The renin-angiotensin-aldosterone system: A crossroad from arterial hypertension to heart failure. *Heart Fail. Rev.* **2020**, *25*, 31–42. [CrossRef] [PubMed]
12. Paul, M.; Poyan Mehr, A.; Kreutz, R. Physiology of local renin-angiotensin systems. *Physiol Rev.* **2006**, *86*, 747–803. [CrossRef] [PubMed]
13. Muñoz-Durango, N.; Fuentes, C.A.; Castillo, A.E.; González-Gómez, L.M.; Vecchiola, A.; Fardella, C.E.; Kalergis, A.M. Role of the renin-angiotensin-aldosterone system beyond blood pressure regulation: Molecular and cellular mechanisms involved in end-organ damage during arterial hypertension. *Int. J. Mol. Sci.* **2016**, *17*, 797. [CrossRef] [PubMed]
14. Li, X.; Zhang, Z.; Luo, M.; Cheng, Z.; Wang, R.; Liu, Q.; Lv, D.; Yan, J.; Shang, F.; Luo, S.; et al. NLRP3 inflammasome contributes to endothelial dysfunction in angiotensin II-induced hypertension in mice. *Microvasc. Res.* **2022**, *143*, 104384. [CrossRef] [PubMed]
15. Castoldi, G.; Carletti, R.; Ippolito, S.; Stella, A.; Zerbini, G.; Pelucchi, S.; Zatti, G.; di Gioia, C.R.T. Angiotensin Type 2 and Mas Receptor Activation Prevents Myocardial Fibrosis and Hypertrophy through the Reduction of Inflammatory Cell Infiltration and Local Sympathetic Activity in Angiotensin II-Dependent Hypertension. *Int. J. Mol. Sci.* **2021**, *22*, 13678. [CrossRef]
16. Castoldi, G.; di Gioia, C.; Giollo, F.; Carletti, R.; Bombardi, C.; Antoniotti, M.; Roma, F.; Zerbini, G.; Stella, A. Different regulation of miR-29a-3p in glomeruli and tubules in an experimental model of angiotensin II-dependent hypertension: Potential role in renal fibrosis. *Clin. Exp. Pharmacol. Physiol.* **2016**, *43*, 335–342. [CrossRef]
17. Castoldi, G.; Di Gioia, C.R.T.; Bombardi, C.; Catalucci, D.; Corradi, B.; Gualazzi, M.G.; Leopizzi, M.; Mancini, M.; Zerbini, G.; Condorelli, G.; et al. MiR-133a regulates collagen 1A1: Potential role of miR-133a in myocardial fibrosis in angiotensin II-dependent hypertension. *J Cell Physiol.* **2012**, *227*, 850–856. [CrossRef]
18. Li, N.; Wang, H.-X.; Han, Q.-Y.; Li, W.-J.; Zhang, Y.-L.; Du, J.; Xia, Y.-L.; Li, H.-H. Activation of the cardiac proteasome promotes angiotensin II-induced hypertrophy by down-regulation of ATRAP. *J. Mol. Cell. Cardiol.* **2015**, *79*, 303–314. [CrossRef]
19. Forrester, S.J.; Booz, G.W.; Sigmund, C.D.; Coffman, T.M.; Kawai, T.; Rizzo, V.; Scalia, R.; Eguchi, S. Angiotensin II signal transduction: An update on mechanisms of physiology and pathophysiology. *Physiol. Rev.* **2018**, *98*, 1627–1738. [CrossRef]
20. Zhou, Y.; Guan, X.; Chen, X.; Yu, M.; Wang, C.; Chen, X.; Shi, J.; Liu, T.; Wang, H. Angiotensin II/Angiotensin II receptor blockade affects osteoporosis via AT1/AT2-mediated cAMP-dependent PKA pathway. *Cell Tissues Organs* **2017**, *204*, 25–37. [CrossRef]
21. Abuohashish, H.M.; Ahmed, M.M.; Sabry, D.; Khattab, M.M.; Al-Rejaie, S.S. The ACE-2/Ang 1-7/Mas cascade enhances bone structure and metabolism following angiotensin-II type 1 receptor blockade. *Europ. J. Pharmacol.* **2017**, *807*, 44–55. [CrossRef] [PubMed]
22. Asaba, Y.; Ito, M.; Fumoto, T.; Watanabe, K.; Fukuhara, R.; Takeshita, S.; Nimura, Y.; Ishida, J.; Fukamizu, A.; Ikeda, K. Activation of renin-angiotensin system induces osteoporosis independently of hypertension. *J. Bone Mineral. Res.* **2009**, *24*, 241–250. [CrossRef] [PubMed]
23. Shimizu, H.; Nakagami, H.; Osako, M.K.; Hanayama, R.; Kunugiza, Y.; Kizawa, T.; Tomita, T.; Yoshikawa, H.; Ogihara, T.; Morishita, R. Angiotensin II accelerates osteoporosis by activating osteoclasts. *FASEB J.* **2008**, *22*, 2465–2475. [CrossRef] [PubMed]
24. Gu, S.S.; Zhang, Y.; Li, X.L.; Wu, S.Y.; Diao, T.Y.; Hai, R.; Deng, H.W. Involvement of the skeletal renin-angiotensin system in age-related osteoporosis in ageing mice. *Biosci. Biotechnol. Biochem.* **2012**, *76*, 1367–1371. [CrossRef] [PubMed]
25. Gu, S.S.; Zhang, Y.; Wu, S.Y.; Diao, T.Y.; Gebru, Y.A.; Deng, H.W. Early molecular response of bone to obstructive nephropathy induced by unilateral ureteral obstruction in mice. *Nephrology* **2012**, *17*, 767–773. [CrossRef] [PubMed]

26. Kaneko, K.; Ito, M.; Fumoto, T.; Fukuhara, R.; Ishida, J.; Fukamizu, A.; Ikeda, K. Physiological function of the angiotensin AT1a receptor in bone remodeling. *J. Bone Mineral Res.* **2011**, *26*, 2959–2966. [CrossRef] [PubMed]
27. Zhang, Y.; Li, X.-L.; Sha, N.-N.; Shu, B.; Zhao, Y.-J.; Wang, X.-L.; Xiao, H.-H.; Shi, Q.; Wong, M.-S.; Wang, Y.-J. Differential response of bone and kidney to ACEI in db/db mice: A potential effect of captopril on accelerating bone loss. *Bone* **2017**, *97*, 222–232. [CrossRef]
28. Diao, T.Y.; Pan, H.; Gu, S.S.; Chen, X.; Zhang, F.Y.; Wong, M.S.; Zhang, Y. Effects of angiotensin-converting enzyme inhibitor, captopril, on bone of mice with streptozotocin-induced type 1 diabetes. *J. Bone Miner. Metab.* **2014**, *32*, 261–270. [CrossRef]
29. Castoldi, G.; Carletti, R.; Ippolito, S.; Colzani, M.; Barzaghi, F.; Stella, A.; Zerbini, G.; Perseghin, G.; Zatti, G.; di Gioia, C.R.T. Sodium-glucose cotransporter 2 inhibition prevents renal fibrosis in cyclosporine nephropathy. *Acta Diabetol.* **2021**, *58*, 1059–1070. [CrossRef]
30. Castoldi, G.; di Gioia, C.R.T.; Roma, F.; Carletti, R.; Manzoni, G.; Stella, A.; Zerbini, G.; Perseghin, G. Activation of angiotensin type 2 (AT2) receptors prevents myocardial hypertrophy in Zucker diabetic fatty rats. *Acta Diabetol.* **2019**, *56*, 97–104. [CrossRef]
31. Palmisano, B.; Labella, R.; Donsante, S.; Remoli, C.; Spica, E.; Coletta, I.; Farinacci, G.; Dello Spedale Venti, M.; Saggio, I.; Serafini, M.; et al. GsαR201C and estrogen reveal different subsets of bone marrow adiponectin expressing osteogenic cells. *Bone Res.* **2022**, *10*, 50. [CrossRef] [PubMed]
32. Cappuccio, F.P.; Meilahn, E.; Zmuda, J.M.; Cauley, J.A. High blood pressure and bone-mineral loss in elderly white women: A prospective study. Study of Osteoporotic Fractures Research Group. *Lancet* **1999**, *354*, 971–975. [CrossRef]
33. Cappuccio, F.P.; Kalaitzidis, R.; Duneclift, S.; Eastwood, J.B. Unravelling the links between calcium excretion, salt intake, hypertension, kidney stones and bone metabolism. *J. Nephrol.* **2000**, *13*, 169–177. [PubMed]
34. Ye, Z.; Lu, H.; Liu, P. Association between essential hypertension and bone mineral density: A systematic review and meta-analysis. *Oncotarget* **2017**, *8*, 68916–68927. [CrossRef] [PubMed]
35. Xu, L.; Dixit, M.P.; Chen, R.; Dixit, N.M.; Collins, J.F.; Ghishan, F.K. Effects of angiotensin II on NaPi-IIa-co-trasporter expression and activity in rat renal cortex. *Biochim. Biophys. Acta* **2004**, *1667*, 114–121. [CrossRef] [PubMed]
36. Fischer, E.; Hannemann, A.; Rettig, R.; Lieb, W.; Nauck, M.; Pallauf, A.; Bildingmaier, M.; Beuschlein, F.; Wallaschofski, H.; Reincke, M. A high aldosterone to renin ratio is associated with high serum parathyroid hormone concentrations in the general population. *J. Clin. Endocrinol. Metab.* **2014**, *99*, 965–971. [CrossRef]
37. Koiwa, F.; Komukai, D.; Hirose, M.; Yoshimura, A.; Ando, R.; Sakaguchi, T.; Komatsu, Y.; Shinoda, T.; Inaguma, D.; Joki, N.; et al. Influence of renin-angiotensin system on serum parathyroid hormone levels in uremic patients. *Clin. Exp. Nephrol.* **2012**, *16*, 130–135. [CrossRef]
38. Catena, C.; Colussi, G.L.; Brosolo, G.; Bertin, N.; Novello, M.; Palomba, A.; Sechi, L.A. Salt, Aldosterone, and Parathyroid Hormone: What Is the Relevance for Organ Damage? *Int. J. Endocrinol.* **2017**, *2017*, 4397028. [CrossRef]
39. Radloff, J.; Pagitz, M.; Andrukhova, O.; Oberbauer, R.; Burgener, I.A.; Erben, R.G. Aldosterone is positively associated with circulating FGF23 levels in chronic kidney disease across four species and may drive FGF23 secretion directly. *Front. Physiol.* **2021**, *12*, 649921. [CrossRef]
40. Zhang, B.; Umbach, A.T.; Chen, H.; Yan, J.; Fakhri, H.; Fajol, A.; Salker, M.S.; Spichtig, D.; Daryadel, A.; Wagner, C.A.; et al. Up-regulation of FGF23 release by aldosterone. *Biochem. Biophys. Res. Commun.* **2016**, *470*, 384–390. [CrossRef]
41. Leifheit-Nestler, M.; Kirchhoff, F.; Nespor, J.; Richter, B.; Soetje, B.; Klintschar, M.; Heineke, J.; Haffner, D. Fibroblast growth factor 23 is induced by an activated renin-angiotensin-aldosterone system in cardiac myocytes and promotes the pro-fibrotic crosstalk between cardiac myocytes and fibroblasts. *Nephrol. Dial. Transplant.* **2018**, *33*, 1722–1734. [CrossRef] [PubMed]
42. Portales-Castillo, I.; Simic, P. PTH, FGF-23, Klotho and Vitamin D as regulators of calcium and phosphorus: Genetics, epigenetics and beyond. *Front. Endocrinol.* **2022**, *13*, 992666. [CrossRef] [PubMed]
43. Ide, N.; Ye, R.; Courbebaisse, M.; Olauson, H.; Densmore, M.J.; Larsson, T.E.; Hanai, J.I.; Lanske, B. In vivo evidence for an interplay of FGF23/Klotho/PTH axis on the phosphate handling in renal proximal tubules. *Am. J. Physiol. Renal Physiol.* **2018**, *315*, F1261–F1270. [CrossRef] [PubMed]
44. Mhatre, K.N.; Wakula, P.; Klein, O.; Bisping, E.; Volkl, J.; Pieske, B.; Heinzel, F.R. Crosstalk between FGF23- and angiotensin II-mediated Ca^{2+} signaling in pathological cardiac hypertrophy. *Cell. Mol. Life Sci.* **2018**, *75*, 4403–4416. [CrossRef]
45. Akagi, T.; Mukai, T.; Mito, T.; Kawahara, K.; Tsuji, S.; Fujita, S.; Uchida, H.A.; Morita, Y. Effect of angiotensin II on bone erosion and systemic bone loss in mice with tumor necrosis factor-mediated arthritis. *Int. J. Mol. Sci.* **2020**, *21*, 4145. [CrossRef]

biomedicines

MDPI

Review

Role of Translationally Controlled Tumor Protein (TCTP) in the Development of Hypertension and Related Diseases in Mouse Models

Jeehye Maeng and Kyunglim Lee *

Graduate School of Pharmaceutical Sciences, College of Pharmacy, Ewha Womans University, Seoul 03760, Korea
* Correspondence: klyoon@ewha.ac.kr

Abstract: Translationally controlled tumor protein (TCTP) is a multifunctional protein that plays a wide variety of physiological and pathological roles, including as a cytoplasmic repressor of Na,K-ATPase, an enzyme pivotal in maintaining Na^+ and K^+ ion gradients across the plasma membrane, by binding to and inhibiting Na,K-ATPase. Studies with transgenic mice overexpressing TCTP (TCTP-TG) revealed the pathophysiological significance of TCTP in the development of systemic arterial hypertension. Overexpression of TCTP and inhibition of Na,K-ATPase result in the elevation of cytoplasmic Ca^{2+} levels, which increases the vascular contractility in the mice, leading to hypertension. Furthermore, studies using an animal model constructed by multiple mating of TCTP-TG with apolipoprotein E knockout mice (ApoE KO) indicated that TCTP-induced hypertension facilitates the severity of atherosclerotic lesions in vivo. This review attempts to discuss the mechanisms underlying TCTP-induced hypertension and related diseases gleaned from studies using genetically altered animal models and the potential of TCTP as a target in the therapy of hypertension-related pathological conditions.

Keywords: ApoE KO; apolipoprotein E knockout mice; atherosclerosis; hypertension; Na,K-ATPase; TCTP; TCTP-overexpressing transgenic mice; translationally controlled tumor protein; TCTP-TG

Citation: Maeng, J.; Lee, K. Role of Translationally Controlled Tumor Protein (TCTP) in the Development of Hypertension and Related Diseases in Mouse Models. *Biomedicines* **2022**, *10*, 2722. https://doi.org/10.3390/biomedicines10112722

Academic Editors: Josef Zicha and Ivana Vaněčková

Received: 13 September 2022
Accepted: 24 October 2022
Published: 27 October 2022

1. Introduction

Arterial hypertension, a major cause of premature death, remains the most prevalent risk factor that is implicated in the development of cardiovascular disease (CVD) and its complications [1]. Globally, the prevalence of hypertension in adults is nearly 32% and 34%, in women and men, respectively [2]. Uncontrolled or untreated high blood pressure (BP) over an extended period of time leads to damage to target organs, including the heart, kidneys, eyes, brain, and arterial blood vessels, causing functional and structural alterations, which could be the clinical manifestations of hypertension-related complications, such as atherosclerosis, heart failure, and chronic kidney disease (CKD) [3]. The dysfunction of vascular smooth muscle cells (VSMCs), among others, that shows the dysregulation in calcium signaling, migrative, and contractile properties is responsible for the mechanism underlying the development of arterial hypertension [3]. Therefore, understanding the molecular players and the modalities and modulation of their function has long been an area of extensive investigation in the field of pathophysiology of hypertension.

Translational research using a variety of animal models that reflect the genetic and environmental factors has significantly contributed to the understanding of the pathophysiologic processes of hypertensive disorders. These efforts have led to the discovery of promising anti-hypertensive drug targets. Among those, Na,K-ATPase, also called the sodium pump, and its repressor, translationally controlled tumor protein (TCTP), have long been the theme of our research using TCTP-overexpressing and TCTP-deficient transgenic mouse models. Cumulative evidence has established that dysregulation of the sodium pump is involved in the development of hypertension and related complications, and that inhibitors of Na,K-ATPase can induce the hypertensive status. Prolonged suppression of Na,K-ATPase by ouabain, a well-established extracellular inhibitor of the sodium pump, has been shown to result in the development of hypertension in vivo [4–6]. Analogous to the action of ouabain, transgenic overexpression of TCTP, an intracellular inhibitor of Na,K-ATPase, showed phenotypes of systemic arterial hypertension in vivo, suggesting the causative role of TCTP as an inhibitor of Na,K-ATPase.

This review attempts to describe the current status of our understanding of the role of TCTP in hypertensive disorders and related diseases, such as atherosclerosis and cataracts, as a consequence of its ability to suppress Na,K-ATPase, and to collate what we have learned from the studies using genetically engineered animal models. Therefore, studies that examined the role of TCTP in the pathophysiology of arterial hypertension and potentially related conditions such as atherosclerosis, heart failure, and obesity, if any, were included for the description in the present review. In addition, studies that use genetically engineered mouse models whose TCTP expression was overexpressed or reduced either systemically or tissue-specifically, were included, whereas research that dealt with non-genetic models or genetic models unrelated to TCTP were excluded in this study.

2. Pathophysiological Consequences of Na,K-ATPase Inhibition

Na,K-ATPase is an essential cell surface enzyme, which maintains ion gradients between the intracellular space and extracellular fluids [7]. This pump consists of three distinctive subunits, which appear in several isoforms. These isoforms include four α isoform (α1~4) and three β isoform (β1~3) subunits and an optional third γ subunit, a member of a Phe-X-Tyr-Asp (FXYD) motif-containing FXYD family [8]. The catalytic α and β subunits are required for ion-pumping activity while the β subunit affects the activity of this pump. Members of the FXYD proteins regulate the Na,K-ATPase activity in specific tissues [8]. The catalytic α subunit that consists of three cytoplasmic domains and ten transmembrane (TM) domains offers the binding sites for Na^+, K^+, and ATP as well as for cardiotonic steroids (CTSs), such as ouabain [9]. The high degree of conservation of the ouabain-binding site of this enzyme implies the existence of potential endogenous regulators [8].

Several endogenous "digitalis-like" or "ouabain-like" factors have been identified as a circulatory moiety that inhibits Na,K-ATPase [7]. The identification of endogenous CTS-like activity and the observation that Na,K-ATPase subunits exist in several isoforms imply the hormone-like functions of CTS and receptor-like function of the sodium pump [10]. Numerous investigations have found that CTS in relation with the sodium pump mediates various signaling cascades related to cell growth and division, release of endothelin-1 (ET-1) by endothelial cells, and other effects in cells [10]. Thus far, two distinctive, but coupled, pathways have been identified in the cellular signaling of Na,K-ATPase not only as an ion pump but also as a signaling transducer [10–13]. Therefore, dysregulation of this pump in both pathways has been an area of attention due to its pathophysiologic association with hypertension and the complications thereof [9].

2.1. Inhibition of Pump Activity of Na,K-ATPase by CTS

Na,K-ATPase is a key player in osmotic equilibrium that transports Na^+ in exchange for K^+ by hydrolyzing ATP to establish an electrochemical gradient across the cell membrane. Electrochemical gradients can drive certain types of cotransporters or exchangers, such as the Na^+/Ca^{2+} exchanger (NCX) [8]. In addition, ionic gradients are also indispensable for the regulation of membrane potential, electrical excitability, cell volume, renal reabsorption of ions, and nutrient transport [14]. Therefore, it has been reported that mutations of the sodium pump α1 are related to the development of secondary hypertension, endocrine syndrome, peripheral neuropathy, and neuromuscular disorders [15]. In addition, dysfunction of Na,K-ATPase substantially affects the osmotic equilibrium and several other essential functions that relate to the development of hypertension, cardiac hypertrophy, cataracts, diabetes, and other disorders [16–18].

Moreover, the extracellular domain of the sodium pump serves as a receptor for cardiac glycosides, such as ouabain, and binding of cardiac glycoside to the sodium pump induces inhibition of its enzymatic activity. In smooth muscle cells, astrocytes, and hippocampal neurons of rodents, CTSs such as ouabain inhibit α2- or α3-isoforms of Na,K-ATPase of the cell membrane that is in close proximity to the sarcoplasmic or endoplasmic reticulum (S/ER), and induces a transient elevation of sub-plasmalemma sodium concentration [10]. With the coupling of the Na,K-ATPase to that of NCX, the latter works in its reverse mode, which results in the local increase in cytosolic Ca^{2+} levels. This triggers Ca^{2+} release from the S/ER to the cytoplasm, which in turn initiates Ca^{2+}-mediated intracellular signaling cascades [10]. The subtle elevation of cytosolic calcium may amplify the impact on intracellular Ca^{2+} stores, which induces considerable changes in the contractility of myocardial and smooth muscles [9]. The resultant arteriolar contraction is one of the mechanisms for essential hypertension [10,19].

Additionally, enhanced intracellular Ca^{2+} transients upon stimuli render vasomotor neurons and endothelial cells hyper-responsive. The synergistic action of neurotransmitter release by sympathetic neurons and elevated vascular reactivity increases arterial tone and peripheral vascular resistance, all of which are the traits of essential hypertension [19]. The involvement of the sodium pump in hypertension is corroborated by the fact that the inhibited expression of the smooth muscle-specific α2 subunit elevates basal blood pressure with enhanced sensitivity to angiotensin II (Ang II), whereas overexpression of this subunit reduces the basal BP and its sensitivity to Ang II [4]. In addition, abnormal elevation of sodium in the lens is related to the opacification of the lens cortex in age-related cataracts, and the alteration of Na,K-ATPase activity appears relevant to cataract formation in human and animal models [20].

2.2. *Signal Transduction of Na,K-ATPase by CTS*

It is noteworthy that low-dose ouabain treatment activates the specific signaling without affecting the intracellular ion concentrations [21]. Mounting evidence has shown that Na,K-ATPase is engaged in the cardiac glycoside-mediated signaling cascades [22]. In the context of the signal-transducing functions of Na,K-ATPase, this pump is capable of communicating with signaling molecules by ligand-induced conformational changes within the signalosome [10]. Ouabain binding to the Na,K-ATPase in caveolae activates not only inositol triphosphate receptor (IP_3R)-mediated Ca^{2+} release from S/ER but also Src kinase-mediated signal cascades [8]. CTS-induced conformational changes occurring within the caveolar sodium pump α1 subunit, which is a negative regulator of Src, allow the activation of Src [9].

Currently, Na,K-ATPase complex with Src is regarded as a unique receptor for CTS [9]. Activated Src triggers the execution of downstream signaling cascades, such as epidermal growth factor receptor (EGFR) transactivation-mediated Ras/Raf/mitogen-activated protein kinase (MAPK) kinase (MEK)/extracellular signal-regulated kinase (ERK) 1/2 pathways [23], inducing gene activation [10]. Src is known to be a key player in the Na,K-ATPase-mediated inter-receptor crosstalk with EGFR [24]. Activated as well are phosphoinositide 3-kinase (PI3K) [25] and EGFR-transactivated generation of reactive oxygen species (ROS), and the latter explains the clinical role of Na,K-ATPase as an oxidant amplifier contributing to the progression of cardiac fibrosis [9]. In addition, Src activates the phospholipase C (PLC) pathway [26] that also contributes to the IP_3-induced Ca^{2+} release [10]. The signaling pathways of the sodium transporter are assumed to be implicated in hypertension, renal disease, diabetes, metabolic disease, and cardiovascular diseases. When there are alterations in the signaling pathways, signaling receptor function would change, causing hypertension, cardiac hypertrophy, tissue fibrosis, or cancer [9].

2.3. *Ouabain-Induced Hypertension and Related Diseases*

Ouabain, a well-known positive ionotropic agent in the heart, is also an endogenous adrenal cortical hormone because it can be synthesized in the adrenal glands [27]. It is known that the infusion of ouabain for a prolonged period in rodents [4–6], elevation of endogenous ouabain [28], and ouabain-like factors [29] in the circulation are associated with the development of hypertension. The elevation of endogenous ouabain also facilitates the proliferation of myocardial and smooth muscle cells [27]. Circulating endogenous ouabain inhibits the activity of the ouabain-sensitive Na,K-ATPase α2 subunit [4] and inhibition of the sodium pump by cardiac glycosides, such as ouabain, is associated with the pathogenesis of hypertension [17], cataract [18], and diabetic diseases [16], through the elevation of the intracellular Ca^{2+} concentrations.

3. TCTP, as an Intracellular Na,K-ATPase Suppressor

TCTP, alternatively referred to as fortilin, p23, and histamine-releasing factor (HRF), is a highly conserved multifunctional 172-amino-acid protein that is under a high degree of regulation both at transcriptional and translational levels [30]. TCTP plays fundamental roles in the regulation of cell cycle progression, apoptosis, autophagy, survival, stress responses, growth and development, and cytokine-like activities, among others [30]. Given the multiplicity of roles it plays, dysregulation of TCTP can lead to cardiovascular and metabolic diseases including systemic and pulmonary arterial hypertension [31], atherosclerosis, diabetes, carcinogenesis, and allergic and inflammatory disorders [32].

Our research group has identified TCTP as a cytoplasmic repressor for Na,K-ATPase [33]. In contrast to ouabain that binds to the outer part of the sodium pump, TCTP interacts with the cytosolic domain of Na,K-ATPase [33]. Searching for the cytoplasmic molecules that are potentially associated with the Na,K-ATPase α subunit, we screened a rat skeletal muscle DNA library using a yeast two-hybrid system. We found that the third large cytoplasmic domain (CD3) of Na,K-ATPase α1 and α2 isoforms interacts with the rat TCTP [33].

While studying the interaction of TCTP with the sodium pump in HeLa cells, we discovered the inhibitory effect of TCTP on Na,K-ATPase activity. Interaction of rat TCTP with the sodium pump does not affect the mRNA and protein levels of the Na,K-ATPase α subunit, indicating a relatively short-term regulation by TCTP [33]. TCTP inhibited the sodium pump activity in a dose-dependent manner [33]. In addition, the C-terminal region (102–172 residues) of rat TCTP was the moiety involved in the inhibition of pump activity of Na,K-ATPase [33]. These observations suggest that TCTP is possibly playing a promotive role in hypertension and raise the question of whether TCTP elevates cytosolic Ca^{2+} in vivo [33]. Interaction of the sodium pump with TCTP is now generally accepted to have important roles not only in the biological functions and signaling but also in the pathophysiology of human diseases [34–36].

It has been shown that TCTP plays either protective or causative roles in certain diseases in some conditions, but it is clear that the dysregulation of TCTP contributes to the development of various pathophysiological processes [37], as shown in several genetically altered mouse models (Table 1). Conventional or tissue/cell-specific TCTP-overexpressing transgenic mice (TCTP-TG) were used to study the gain-of-function of TCTP. Conversely, heterozygous TCTP-deficient mice ($TCTP^{+/-}$), in which TCTP gene is deficient causing a loss of function, were studied because of embryonic lethality of homozygous TCTP knockout mice ($TCTP^{-/-}$) [38]. This review focuses on the roles of TCTP in hypertension and related diseases, including atherosclerosis, cataracts, and other metabolic disorders that have been studied in genetically engineered animal models.

Table 1. Genetically modified mouse models that exhibit elevated or reduced expressions of TCTP and its phenotypes and pathophysiological significances in vivo.

Animal Models	Related Disease	Phenotypes	Ref.
		Pathophysiological roles of TCTP	
TCTP-TG	Hypertension	Development of systemic arterial hypertension and increased vascular contractility ↑ RhoA expression and phospho-myosin light chain (p-MLC) in aorta	[39,40]
	Cataract	↑ Incidence of cataract formation in lens ↑ Abnormal eye development	[41]
	Osteoporosis	↓ Bone mass in femurs ↑ Osteoporotic features in femur bones, and osteoclast cell count	[42]
	Rheumatoid arthritis	↑ Inflammatory responses, bone erosion, and cartilage destruction upon collagen-induced arthritis (CIA)	[43]
$TCTP^{+/-}$	Hypertension	↓ RhoA expression and p-MLC in aorta	[40]
	Osteoporosis	↑ Bone mass in femurs ↓ Osteoporotic features in femur bones, and osteoclast cell count	[42]
	Rheumatoid arthritis	↓ Synovial inflammation, bone erosion, cartilage damage, and osteoclastic bone resorption upon CIA	[43]
TCTP-TG/ ApoE-KO	Atherosclerosis	↑ Exacerbation of atherosclerotic lesion formation by high-fat diet without alteration in plasma lipid profiles, compared with ApoE-KO mice	[44]
$TCTP^{+/-}$ $Ldlr^{-/-}Apobec1^{-/-}$	Atherosclerosis	Similar lipid profiles and BP compared with $TCTP^{+/+}$ $Ldlr^{-/-}$ $Apobec1^{-/-}$ mice ↓ Atherosclerotic lesion in aorta and macrophage numbers in lesions, compared to $TCTP^{+/+}$ $Ldlr^{-/-}$ $Apobec1^{-/-}$ mice ↑ Bax expression and apoptosis of peritoneal macrophages in the intima layer, compared with $TCTP^{+/+}$ $Ldlr^{-/-}$ $Apobec1^{-/-}$ mice	[45]
Clara cell-specific TCTP-TG	Allergic asthma	↑ Allergic and asthmatic inflammation with increase in serum and bronchoalveolar lavage (BAL) IgE, interleukin-4 (IL-4), and eosinophil count upon ovalbumin (OVA) challenge ↑ TCTP secretion and macrophage counts in BAL fluids	[46]
		Protective roles of TCTP	
TCTP-TG	Obesity	↑ Metabolic homeostasis under both normal and high-fat diet conditions with enhanced glucose tolerance and insulin sensitivity ↑ Energy expenditure with upregulation of uncoupling protein 1 (UCP1) in the brown adipose tissue (BAT) ↑ Adaptive thermogenesis of BAT after cold exposure	[47]
Tregs-specific TCTP-TG	Diabetes	↑ Forkhead box protein P3(FOXP3) expression and prolonged survival of regulatory T cells (Tregs) ↓ Development of autoimmune diabetes by inhibiting the apoptosis of Tregs	[48]
Cardiomyocyte-specific TCTP-TG	Heart failure	↓ Doxorubicin-induced cardiac dysfunction and Bcl-2 interacting protein 3 (Bnip3) induction ↓ Dihydroartemisinin-induced heart failure and cardiomyocyte death	[49]
Heart-specific TCTP-KO	Heart failure	Mice lacking TCTP in the heart die by 9 weeks of age because of severe heart failure and extensive cardiomyocyte apoptosis	[50]
Liver-specific TCTP-TG	Liver damage	↑ Peroxiredoxin-1 (PRX1) activity in the liver and protection against alcohol, and ROS-mediated liver damage	[51]
Liver-specific TCTP-KO	Liver damage	↓ Endoplasmic reticulum (ER) stress-induced liver failure and death by blocking apoptosis in the liver	[52]

↑ indicates increase; ↓ indicates decrease.

4. Pathophysiology of Hypertension in TCTP-TG

Animal models have been used for conducting etiological studies of genetic hypertension. These include spontaneous hypertensive rats (SHRs), Dahl salt-sensitive rats, and transgenic models generated by overexpression of particular genes [53]. To address the questions raised regarding the pathophysiological consequences of TCTP-mediated Na,K-ATPase inhibition in vivo, we constructed transgenic mice overexpressing TCTP (TCTP-TG). It was established in either C57BL/6×CBA hybrid or C57BL/6N inbred backgrounds using pCAGGS-TCTP cDNA construct consisting of cytomegalovirus enhancer (CMV-IE) and chicken β-actin promotor (Figure 1A).

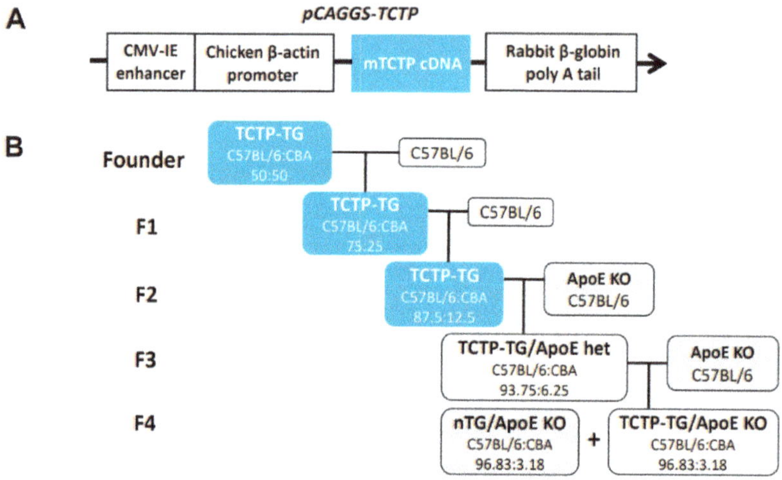

Figure 1. Generation of TCTP-TG and TCTP-TG/ApoE KO mice. (**A**) Schematic representation of the gene construct used for TCTP transgenic overexpression in mice. The TCTP transgene of pCAGGS-TCTP cDNA containing CMV-IE enhancer and chicken β-actin promoter constructed in the transgenic expression vector pCAGGS was used to generate the transgenic mice. (**B**) Strategy of crossbreeding for generation of nTG/ApoE KO and TCTP-TG/ApoE KO mice exhibiting identical genetic backgrounds.

Their phenotypic characteristics are partly affected by the genetic background of TCTP-TG because, for example, cataractogenesis was profound in C57BL/6N background mice [41] while both types of transgenic mice show hypertensive properties regardless of their genetic backgrounds [39,41,44]. Because of the decreased fertility of inbred offspring, TCTP-TG was generated in C57BL/6×CBA hybrid strains in the majority of our investigations. For the study regarding atherogenesis, TCTP-TG was subjected to crossbreeding to generate apolipoprotein (ApoE)-deficient TCTP-TG (ApoE-KO/TCTP-TG) with a hybrid background (Figure 1B).

4.1. Systemic Arterial Hypertension by Na,K-ATPase Inhibition in TCTP-TG

Both C57BL/6×CBA and C57BL/6N mice overexpressing TCTP showed phenotypes of hypertension and hybrid mice were used in the study. The gender-related dimorphism regarding BP was observed, also reported in heterozygous female TGR(mRen2)27 transgenic rats [54], possibly due to the differences in hormonal changes or vascular responsiveness between female and male TCTP-TG mice occurring with ageing [39]. At approximately 6 weeks after birth, systemic arterial hypertension developed in both male and female TCTP-TG but after 19–20 weeks female TCTP-TG did not exhibit a significant difference in BP compared with non-transgenic (nTG) mice [39]. Therefore, heterozygous male TCTP-TG mice were studied for further investigations.

As with arterial BP elevation, an increase in left ventricular systolic pressure (LVSP) was found in TCTP-TG mice [39]. The thickening of the left ventricular posterior walls was observed in 15- to 16-month-old male TG mice, possibly caused by the sustained elevation of BP, but TCTP-TG did not exhibit the significant cardiac dysfunction [39]. TCTP-TG showed the characteristics of hypertension: first, the alteration of vascular functions that exhibit enhanced contractility of aortic rings devoid of endothelial layer, and secondly, augmented sensitivity to vasoconstrictor stimuli such as norepinephrine (NE) and serotonin (5-HT). Moreover, vascular relaxation reactivity following sodium nitroprusside (SNP) treatment after NE-induced contraction was decreased, indicating the reduced sensitivity to vasodilators [39]. Impaired vasodilatory responses accompanied by the hypercontractile profile in the vasculature isolated from TCTP-TG mice aberrantly elevated the BP [39].

As expected, vascular smooth muscle cells (VSMCs) isolated from TCTP-TG showed suppressed activity of Na,K-ATPase $\alpha1$ and $\alpha2$ isoforms that are the predominant Na,K-ATPase in mouse aorta, confirming the TCTP-induced sodium pump inhibition in VSMCs [39]. The expression of $\alpha1$ in VSMCs from TCTP-TG (TG-VSMCs) was elevated, whereas that of $\alpha2$ did not show any significant changes [39]. Because transient TCTP overexpression in HeLa cells did not affect the Na,K-ATPase expression, as observed in our earlier studies [33], elevated $\alpha1$ expression in TCTP-TG mice suggests a potential compensatory mechanism due to the prolonged pump inhibition in vivo [39].

Then, we confirmed that inhibition of Na,K-ATPase activity in TG-VSMCs induces augmented intracellular Ca^{2+} mobilization, possibly through the reverse mode of NCX. Upon 5-HT stimuli, TG-VSMCs exhibited elevated Ca^{2+} transients that were mobilized from the ER [39]. Of interest, ouabain treatment at a concentration that blocks almost all the activity of Na,K-ATPase did not further enhance the 5-HT-evoked Ca^{2+} transients in TG-VSMCs whereas that of non-transgenic mice was elevated by ouabain, compared with untreated cells [39]. This phenomenon can be explained by the fact that ouabain may have negligible impact on TCTP-TG whose homeostasis has already been established by TCTP-induced Na,K-ATPase inhibition [39]. Likewise, VSMCs of TCTP-TG showed increased sensitivity to vasoconstrictor, elevated levels of Ca^{2+} in the resting state, and increased capacity for intracellular Ca^{2+} mobilization, and ultimately enhanced contraction by stimuli, accompanied by reduction in vasorelaxant-induced relaxation [39].

Taking these observations together, we delineated the pathophysiology of TCTP-induced hypertension using TCTP-TG, as depicted in Figure 2. Repression of Na,K-ATPase by TCTP overexpression results in the rise of cytosolic Na^+ level, and subsequent elevation of cytosolic Ca^{2+} mobilization and Ca^{2+} storage, possibly through the reversion action of NCX, ultimately inducing the hyper-contractility of VSMC in TCTP-TG. Clearly, based on the fact that high blood pressure implies the dysfunctions in cardiac and/or arterial parameters [55], non-significant cardiac malfunctions indicate that TCTP-TG-induced hypertension is mediated by vascular malfunctions, as shown by elevated contractility and reduced relaxation in VSMC from TCTP-TG mice. This phenomenon is similar to that of ouabain treatment-induced hypertension exhibiting augmented vascular reactivity to vasopressors in renal artery of rats [56], confirming that TCTP can be viewed as an endogenous ouabain-like entity. Based on the observation that $TCTP^{+/-}$ and wild-type mice did not show a significant difference in systolic BP [45], overexpression of TCTP might be essential in inducing arterial systemic hypertension in vivo.

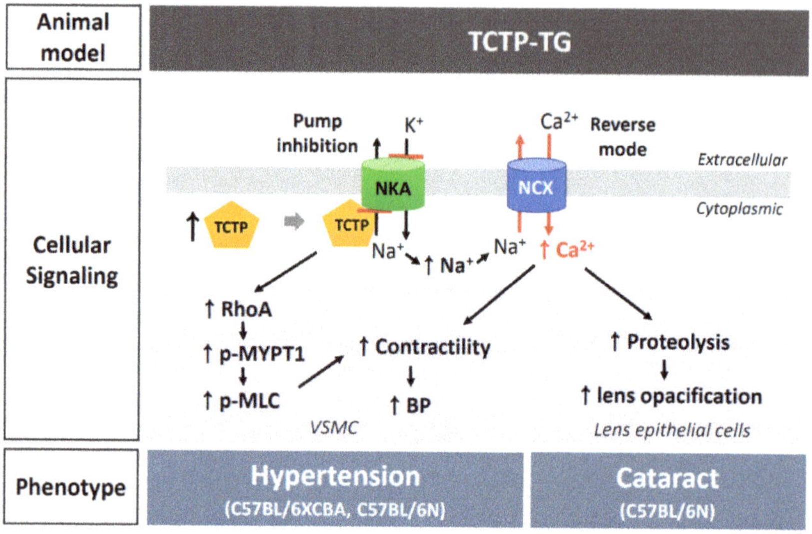

Figure 2. TCTP overexpression-induced consequences in transgenic TCTP-overexpressing (TCTP-TG) mice. Transgenic overexpression of TCTP led to the Na,K-ATPase inhibition, which results in the development of hypertension through elevated cytosolic Ca^{2+} levels and vascular contractility and tone. Such an increase in Ca^{2+} can also occur in the lens epithelial cells of TCTP-TG, at this time, inducing the activation of proteolytic enzyme and lens swelling, all of which affect the cataractogenic processes. RhoA/Rho kinase activation and resultant MYPT-1 phosphorylation in TCTP-TG appears to partly contribute to the increased sensitivity of myofibrils upon constrictor stimuli, thereby increasing the contractility of VSMC in TCTP-TG (↑ indicates increase).

4.2. *Upregulation of RhoA/Rho Kinase Signaling in TCTP-TG*

The contractility of smooth muscle cells is largely determined by the phosphorylation status of the regulatory myosin light chain (MLC) whose activity is under regulation by myosin light chain kinase (MLCK) and myosin light chain phosphatase (MLCP) [57]. Vasoconstrictors like Ang II binding to G-protein coupled receptors (GPCR) activates $G_{q/11}$, which in turn activates phospholipase C (PLC)-β and elevates Ca^{2+} concentration, thereby stimulating MLCK activity and induces $G_{12/13}$ to activate RhoA/Rho kinase and, in turn, inhibits MLCP. A myosin binding subunit, the myosin phosphatase target unit (MYPT-1) of MLCP, is one of the most important substrates of Rho kinase [57]. Both pathways together converge into the MLC phosphorylation and enhance the contraction of VSMCs and myofibrils [57]. As a downstream mediator of Ang II signaling, a small GTP-binding protein, RhoA, which maintains MLC phosphorylation in VSMCs, regulates Ca^{2+} sensitivity of the myofilaments through MLCP inhibition [58,59].

It is generally accepted that Ca^{2+} sensitivity of myofilaments is under regulation by RhoA/Rho kinase-induced MLCP inhibition [58,59] and that dysregulation of RhoA/Rho kinase signaling results in the hypercontractile properties of VSMC, thereby contributing to the pathogenesis of hypertension [57]. Upregulation of RhoA/Rho kinase signaling pathway has been described in cardiovascular diseases, such as systemic and pulmonary hypertension, atherosclerosis, heart failure, and stroke [57,60] and it is deeply involved in cardiovascular and renal pathophysiology [61]. Consistently, dysregulation of RhoA/Rho kinase pathway has been reported in several hypertensive rat models, including SHRs, renal hypertensive rats, and deoxycorticosterone acetate (DOCA) salt-induced hypertensive rats [58].

We found that TCTP-induced RhoA/Rho kinase upregulation is involved in the hypertensive phenotype of TCTP-TG [40]. Increased contraction in the aorta from TCTP-TG under lower concentrations of K^+ [39] implies the potential dysregulation of pathways regulating Ca^{2+} sensitization, such as RhoA/Rho kinase pathway. When we examined these alterations in aortas isolated from TCTP-TG mice, elevated RhoA expression and phosphorylated MLC (p-MLC) were confirmed, whereas aortas obtained from heterozygous mice containing a deleted allele of TCTP ($TCTP^{+/-}$) showed a decrease in RhoA expression and p-MLC in their aortas [40].

Additionally, overexpression of TCTP increased RhoA expression and activated its downstream signaling, including MYPT-1 and MLC phosphorylation in primary cultured VSMC. In contrast, knockdown of TCTP expression in VSMCs showed suppression of RhoA expression as well as Rho kinase signaling pathways [40]. Inhibition of Rho kinase signaling can be an alternative approach in the modulation of TCTP-induced hypertension. It is evident that dysregulation of RhoA-related Ca^{2+} sensitization mechanism partly underlies the hypertensive phenotypes of TCTP-TG (Figure 2).

4.3. Elevation of the Expression of Peroxiredoxin 3 (Prx3) and Heat Shock Protein 25 (Hsp25) in TCTP-TG

A proteomic analysis to delineate the potential pathophysiological changes in TCTP-TG mice of C57BL/6×CBA strain showed differentially expressed proteins involving reactive oxygen species (ROS) metabolism, fatty acids/amino acids metabolism, energy metabolism, and cytoskeletal organization in heart tissue of TCTP-TG compared with those of control mice [62]. Of note, the expression of a mitochondrial antioxidant enzyme, peroxiredoxin 3 (Prx3) showed decrease in 9-week-old TCTP-TG mice, whereas cytosolic Prx2 expression was elevated in 19-week-old TCTP-TG [62]. Because mitochondrial oxidative stress takes part in the pathogenesis of hypertension [63–65], altered expression of these enzymes implies the potential relation of ROS in TCTP-induced hypertension [62]. The protective and anti-inflammatory roles of Prx3 in various cell types were suggested [66] and Prx3 overexpression protects the heart from the ventricular remodeling and heart failure following myocardial infarction by attenuating mitochondrial oxidative stress [67]. Additionally, the higher Hsp25 expression in 9-week-old TCTP-TG mice, also indicates the possible role of ROS in TCTP-induced hypertension. Stress-inducible Hsp25 is upregulated in the kidney isolated from the rats with prolonged Ang II administration, and its expression was reported to be involved in the regulation of BP [68]. Studies on the potential roles of Prx3 and Hsp25 in the heart of TCTP-TG mice are an area of interest for the further studies.

5. Pathophysiology of Cataractogenesis in TCTP-TG

There is a 2.13 times greater risk of having a cataract extraction in hypertensive patients, especially in the 60–69 year age group, compared with non-hypertensives individuals [69]. A meta-analysis also indicates that hypertension enhances the risk of cataracts [70]. In salt-sensitive hypertensive Dahl rats, the incidence of cataracts was 30–35% and it can be predicted by the suppression of Na,K-ATPase, suggesting a defect in this pump in hypertensive rats [71–73].

Cataractogenesis, an impaired lens transparency, results from the dysregulation of ionic and fluid homeostasis [41], which is tightly regulated by Na,K-ATPase in the lens cells [20,74]. Elevated sodium and calcium concentrations or digitalis-like compounds have been found in human cataractous lenses [75]. Sodium pump inhibitors like ouabain induce the elevation of the sodium level with a concomitant decrease in the potassium level in the lens, and such osmotic imbalance leads to the accumulation of fluids in the lens, and then cells swell and degrade, resulting in the formation of fluid droplets, the cause to opacify [76–78]. In addition to fluid-induced cell death, elevated sodium levels due to the Na,K-ATPase repression induce the rise in cytosolic level and transient of Ca^{2+}, which activates Ca^{2+}-dependent proteolytic enzymes, cellular integrity impairment and apoptosis, eventually resulting in the loss of lens transparency [76–78].

Of interest, the elevation of BP is not necessarily the only effect of TCTP-induced Na,K-ATPase inhibition in the TCTP-TG because an elevated incidence rate of cataracts was found in TCTP-TG mice of C57BL/6N inbred strain, compared to that of non-TG (7.38% vs.

1.47%) at 10 weeks after birth [41]. Similar to the majority types of human cataracts, TCTP-TG showed mixed types of nuclear, cortical, and anterior subcapsular opacification [41]. For still unknown reasons, this phenotype was not prominent in C57BL6×CBA hybrid TCTP-TG although both strains of TCTP-TG elicited hypertension as a pathophysiological consequence of Na,K-ATPase inhibition [41]. This mode of cataractogenesis operates in human lens epithelial cells because TCTP overexpression induced an increase in cytosolic Ca^{2+} mobilization upon histamine stimuli via sodium pump inhibition, which might induce the proteolytic activity of enzymes (Figure 2). The transient overexpression of TCTP in lens epithelial cells did not affect the expression of the Na,K-ATPase α-subunit [41].

Interestingly, TCTP appears to affect the normal eye development because some TCTP-TG mice showed abnormal eye development, including unusual sizing of the eyes, and even absence of the eye [41]. This intriguing observation emphasizes the need to identify the fundamental roles of TCTP in the developmental processes of the eyes.

6. Pathophysiology of Atherosclerosis in TCTP-TG and TCTP$^{+/-}$ Mice

6.1. Exacerbation of Atherosclerosis by TCTP-Induced Hypertension in TCTP-TG/ApoE KO

Atherosclerosis is a pathological remodeling of arterial walls with the development of atheromatous plaques involving endothelial dysfunction, migration of SMCs, and accumulation of cholesterol in the lesions [79,80]. Arterial hypertension is a key etiological factor in the development of atherosclerosis and, thus, of cardiovascular disorders [79], through the various pathological links including hemodynamic stress [81]. Ang II- and high salt-induced hypertension facilitated the development of atherogenesis in apolipoprotein E-deficient mice (ApoE KO) [82,83] whereas Ang II-induced atherosclerosis was attenuated by enalapril treatment in vivo because of its anti-atherogenic and anti-inflammatory effects [84]. In clinical settings, angiotensin-converting enzyme inhibitors (ACEIs) and angiotensin receptor blockers (ARBs) retard the pathogenic process of atherosclerosis in hypertensive patients [85].

Genetically modified animal models of atherosclerosis include apolipoprotein E knockout mice (ApoE KO), low-density lipoprotein (LDL) receptor knockout mice (LDLR$^{-/-}$), scavenger receptor class B member 1 knockout mice (SR-BI KO), *db/db* and *ob/ob* mice, Zucker fatty rats, and cholesterol ester transfer protein (CETP) transgenic rats [53]. Among those, the most extensively investigated is the ApoE KO that lacks ApoE, a glycoprotein ligand for receptors that eliminates chylomicron remnants and very low-density lipoproteins (VLDLs), which exhibits spontaneous hypercholesterolemia and the formation of atherosclerotic lesions in vessels [86,87].

Our group investigated whether TCTP overexpression accelerates the process of atherogenesis in animal models [44]. Multiple crossbreeding of TCTP-TG with ApoE-null mice generated nTG/ApoE KO and TCTP-TG/ApoE KO mice having comparable genetic backgrounds (about 97% C57BL/6 and 3% CBA) (Figure 1B). TCTP-TG and nTG mice carrying identical genetic makeup were generated by backcrossing the TCTP-TG C57BL/6×CBA hybrid mice to C57BL/6 mice. The differences in body weight changes and major organ weights were insignificant among the four groups, but the brain, lung, and spleen weights showed an increase in TCTP-TG/ApoE KO compared to TCTP-TG [44]. Following the lipid-enriched western diet of 7-week-old mice for 16 weeks, lipid-laden atherosclerotic plaques were found in the thoracic aorta of ApoE-null mice, but not in that of TCTP-TG [44]. Importantly, the transgenic overexpression of TCTP in ApoE KO showed an exacerbation of atherosclerotic lesions, exhibiting a wider distribution and thicker lesions than that of ApoE KO [44]. The BP elevation of TCTP-TG and TCTP-TG/ApoE KO was comparable [44]. Based on the observation that TCTP-TG shows negligible atherosclerotic lesion formation, hypertension alone was not enough to establish atherosclerotic lesions. Moreover, high-fat diet feeding in TCTP-TG also did not cause atherosclerosis and the animals exhibited normal lipid metabolism [44], suggesting that ApoE genetic malfunction may be a prerequisite for establishing atherosclerosis.

We found that TCTP overexpression in an ApoE KO genetic background did not affect the plasma glucose, albumin, and lipid profiles of ApoE KO mice [44], indicating that atherosclerotic acceleration by TCTP is not mediated by those factors, including increased burden of plasma lipid. Therefore, we speculated that hemodynamic stress or inflammatory responses provoked by high BP [81,88] might be the proatherogenic factors that operate in TCTP-induced atherogenic progression [44]. The potential involvement of ET-1, a potent endothelium-derived vasoactive peptide that is upregulated in atherogenic status to facilitate the proliferation, migration, and contraction of VSMC [89–91], was also proposed. Such a hypothesis can be supported by the findings that Na,K-ATPase inhibition, such as that which results from ouabain treatment, induces ET-1 release from the endothelial cells [92–96] and that TCTP synthesis was upregulated following stimulation of human lung fibroblasts with ET-1 [97], though the interplay between TCTP and ET-1 and its consequences require experimental verification.

Our unpublished data on ligation-induced atherosclerotic mouse model also supports the role of TCTP in the atherogenesis. Two weeks after the ligation operation on the left common carotid artery (LCCA), ligated carotid arteries of TCTP-TG showed a 3.4-fold increase in intimal area compared to that of non-TG, suggesting TCTP-induced exacerbation of atherosclerosis through intimal layer thickening.

6.2. Aggravation of Atherosclerosis by TCTP-Induced Survival of Macrophages in TCTP$^{+/-}$ Mice

Pinkaew et al. described the pathophysiological roles of TCTP in protecting macrophages in atherosclerosis using heterozygous TCTP-deficient mice [45]. This group noted that TCTP, not hypertension per se, can facilitate atherosclerosis by protecting macrophages against apoptosis. They constructed mouse models of TCTP-deficient (TCTP$^{+/-}$) and wild-type (TCTP$^{+/+}$) mice on a hypercholesterolemic genetic background, Ldlr$^{-/-}$ Apobec1$^{-/-}$ whose low-density lipoprotein receptor (*Ldlr*) and apolipoprotein B mRNA editing enzyme, catalytic polypeptide 1 (*Apobec 1*) genes are lacking. Ldlr$^{-/-}$ Apobec1$^{-/-}$ mice have considerably elevated blood cholesterol levels on a normal diet and induce atherogenic processes from fatty streaks through the fibrous cap [45]. Of note, this group tried to find the role of TCTP itself by ruling out the effect of BP as there were no significant differences in BP between TCTP$^{+/-}$ and TCTP$^{+/+}$ [45].

On a normal diet for 10 months, both groups of mice showed comparable BP and lipid profiles, but TCTP$^{+/-}$ had significantly less atherosclerosis in aortas compared to TCTP$^{+/+}$ littermate controls [45], indicating that TCTP deficiency is related to a lesser degree of atherosclerosis. It was found that TCTP$^{+/-}$ mice had fewer macrophages within atherosclerotic plaques and more apoptotic macrophages in the intima than those of TCTP$^{+/+}$ mice [45]. Moreover, peritoneal macrophages isolated from TCTP$^{+/-}$ mice showed elevated expression of Bax as well as apoptosis in the plaques both at baseline and upon oxidized LDL stimuli [45]. TCTP expression in peritoneal macrophages was more enhanced in hypercholesterolemic sera from Ldlr$^{-/-}$ Apobec1$^{-/-}$ mice than that from control mice. Additionally, macrophage colony-stimulating factor (M-CSF), which is elevated in the plasma of patients with coronary artery disease [98], induced TCTP expression [45]. Based on these findings, they concluded that the pro-atherosclerotic microenvironment, including hypercholesterolemia and inflammatory cytokines like M-CSF, allows the induction of TCTP expression in macrophages and this protects macrophages against Bax-induced apoptosis, thereby facilitating the propagation of macrophages and atherosclerosis [45]. Moreover, TCTP levels are abundantly elevated in human atherosclerotic plaques, showing a positive correlation with the severity of lesions from the fatty streak to the fibrous plaques [45].

A recent study using patients' data concluded that elevated TCTP in atherosclerotic lesions might result from the adaptive responses to profound apoptosis [99]. Higher plasma TCTP levels were found in the patients with coronary artery disease (CAD), particularly those with three-vessel disease (3VD) [99]. Furthermore, the degree of TCTP levels is positively correlated with the severity of CAD, suggesting it as a biomarker for CAD [99]. It was speculated that insufficient clearance of extensive apoptotic cells in atherosclerotic lesions might cause the accumulation of apoptotic cells and inflammatory responses that in turn induce the adaptive responses involving TCTP expression to reduce the excessive apoptosis [99].

Several lines of evidence unequivocally show the facilitative role of TCTP overexpression in atherosclerotic plaque formation via TCTP-induced hypertension or by reducing apoptosis in macrophages [44,45]. In this review, TCTP serves as a novel target for the modulation of hypertension-related atherosclerosis.

7. Role of TCTP in Hypertension-Related Diseases

Studies using genetically altered animal models, including conventional or tissue-specific TCTP-overexpressing and TCTP-deficient mice, have shed light on TCTP's potential pathophysiological role in chronic diseases, such as osteoporosis [42], rheumatoid arthritis [43], and allergy [46]. Other studies have indicated the protective roles of TCTP due to its anti-apoptotic and cell-protective functions. Such beneficial effects were found in the studies regarding diabetes [48], heart failure [49,50], and liver damage [51,52]. Additionally, TCTP promotes the energy expenditure and metabolic homeostasis that might be preventive of obesity-related metabolic disorders [47] (Table 1). Here, we briefly discuss the protective roles of TCTP in some conditions, including obesity and heart failure, among others, to help understand the plethora of TCTP functions in cardiovascular and metabolic diseases in vivo.

7.1. Obesity

Obesity, characterized by excessive accumulation of body fats, is a representative of the modifiable risk factors for cardiovascular diseases including hypertension and atherosclerosis [100]. Based on the fortuitous finding that TCTP-TG mice with C57BL/6N background have the phenotypes of relatively lighter body weight than control mice, our group investigated the impact of TCTP on metabolic tissues and systemic energy metabolism [47]. We confirmed that TCTP-TG mice of C57BL/6N strain under normal chaw diet (NCD) showed significantly lower body weight compared to wild-type mice (WT) as early as six weeks after birth despite their comparable amount of food intake [47]. TCTP-TG showed an increase in muscle weight and reduction in fat mass due to inhibition of the hypertrophy of adipocytes in epididymal white adipose tissue (WAT). In addition, TCTP-TG exhibited improvement in hepatic lipid accumulation, plasma lipid profiles, and glucose tolerance as well as overall energy expenditures [47]. When we compared the metabolic homeostasis between TCTP-TG and WT under NCD and high-fat diet (HFD) conditions, TCTP-TG indicated improved metabolic homeostasis under both conditions, with the enhanced glucose tolerance and insulin sensitivity [47].

Upon cold exposure that triggers thermogenesis by the sympathetic nerve system (SNS) activation in the brown adipose tissue (BAT), TCTP-TG showed improved adaptive thermogenesis [47]. TCTP overexpression attenuated systemic metabolic imbalance by upregulating the uncoupling protein 1 (UCP1)-mediated thermogenesis in the BAT, serving as a modulator in the process of energy expenditure [47]. Affluent mitochondria contained in the BAT, where mitochondrial carrier protein, UCP1, mediates thermogenesis by sympathetic stimuli and produces heat by consuming triglycerides [101]. BAT activation is reported to reduce hypercholesterolemia and to exert protective roles from atherogenesis [102]. More importantly, TCTP-TG exhibited resistance to HFD-induced obesity and metabolic disorders [47]. The increase in energy expenditure and thermogenesis of BAT under HFD in TCTP-TG underscores the critical roles of TCTP in metabolic homeostasis through energy expenditure [47]. Therefore, TCTP can be viewed as a rational target for the energy expenditure-related conditions such as obesity and metabolic disorders [47].

7.2. Heart Failure

A recent publication by Cai et al. addressed TCTP's role in cardiomyocyte survival based on the observation that an animal model overexpressing cardiomyocyte-specific

TCTP-TG showed a protective role in heart failure [49]. Cardiomyocyte-specific TCTP overexpression drastically reduced the susceptibility to doxorubicin (DOX)-induced cardiac dysfunction in mice [49]. It also inhibited the induction of Bcl-2/adenovirus E1B 19 kDa-interacting protein 3 (Bnip3), a molecule that mediates the TCTP-loss-induced cardiomyocyte death [49]. In mice with cardiomyocyte-specific TCTP overexpression, treatment with dihydroartemisinin, a pharmacological TCTP inhibitor, did not induce heart failure and cardiomyocyte death, both of which were induced in control mice [49]. Altogether, these observations show that TCTP is essential for cardiomyocyte survival and can be suggested as a therapeutic target that ameliorates DOX-induced heart failure [49].

In addition, studies in an animal model that lacks TCTP in their heart confirm the protective role of TCTP in heart failure [50]. Mice devoid of TCTP expression in the heart showed premature death by 9 weeks of age because of extensive apoptotic cardiomyocytes and severe heart failure [50]. Heart-specific TCTP KO mice showed upregulation of p53 target genes in their hearts and heart-targeted deletion of p53 in those mice prolonged survival of mice from 9 to 18 weeks by preventing apoptosis of cardiomyocytes. Therefore, this group concluded that inappropriate expression of TCTP is related to the pathophysiology of heart failure [50].

8. Conclusions

New perspectives on the pathophysiological roles of TCTP as a multifunctional protein as well as a cytosolic sodium pump inhibitor and as a player in the development of systemic and arterial hypertension, cardiac hypertrophy, cataracts, diabetes, and related disorders via inhibition of the sodium pump in VSMC, lens epithelial, or other relevant cells have emerged from the studies of genetically engineered animal models. Additionally, TCTP accelerates the pathogenesis and severity of atherosclerosis by the mechanisms involving its anti-apoptotic activity on macrophages in atherosclerotic lesions. Conversely, modalities that modulate TCTP could be a promising strategy in the therapeutics for systemic hypertension, hypertension-induced atherosclerosis, and cataracts, all of which are associated with Na,K-ATPase suppression.

Intriguingly, TCTP also showed beneficial effects on heart failure and obesity by protecting the cells from apoptosis and enhancing the metabolic expenditure, respectively, in animal models. Because TCTP has a plethora of physiological functions and is essential for the survival and growth of cells and organisms, a certain extent of TCTP expression in vivo seems indispensable for maintaining normal physiology. However, not only excessive but also deficient expression of TCTP is related to the pathophysiology in specific conditions, as shown in the phenotypes of TCTP-TG and $TCTP^{+/-}$ mice. Further studies regarding hitherto unknown potential mechanisms and roles of TCTP in terms of various cellular functions might delineate its pathophysiological network in the cardiovascular conditions. In this perspective, possible implications of TCTP in the cellular senescence and its contributory mechanism in the arterial hypertension and related disorders can be one of the future studies based on the causative roles of senescence in the majority of cardiovascular diseases [103,104].

It is challenging, at present, to describe the complicated networking between protective and causative roles of TCTP in a unified figure, and disease-specific mechanisms, and regulations affecting the expression and activity of TCTP need to be delineated for future investigations. Clearly, studies in disease-specific animal models and human cases may help us understand how TCTP exerts manifold functions in a certain disease and pave the way for TCTP-targeting therapeutics.

Author Contributions: Conceptualization, J.M. and K.L.; writing—original draft preparation, J.M.; writing—review & editing, K.L.; supervision, K.L.; project administration, K.L.; funding acquisition, K.L. All authors have read and agreed to the published version of the manuscript.

Funding: This research was supported by the Bio and Medical Technology Development Program (2020M3E5E2036808) and (2021R1A2C2003629) by the National Research Foundation of Korea (NRF) grant funded by the Korea government (MSIT).

Institutional Review Board Statement: Not applicable.

Informed Consent Statement: Not applicable.

Data Availability Statement: Not applicable.

Conflicts of Interest: The authors declare no conflict of interest.

References

1. Al Ghorani, H.; Götzinger, F.; Böhm, M.; Mahfoud, F. Arterial Hypertension—Clinical Trials Update 2021. *Nutr. Metab. Cardiovasc. Dis.* **2022**, *32*, 21–31. [CrossRef] [PubMed]
2. Zhou, B.; Carrillo-Larco, R.M.; Danaei, G.; Riley, L.M.; Paciorek, C.J.; Stevens, G.A.; Gregg, E.W.; Bennett, J.E.; Solomon, B.; Singleton, R.K.; et al. NCD Risk Factor Collaboration (NCD-RisC) Worldwide Trends in Hypertension Prevalence and Progress in Treatment and Control from 1990 to 2019: A Pooled Analysis of 1201 Population-Representative Studies with 104 Million Participants. *Lancet* **2021**, *398*, 957–980. [CrossRef]
3. Kućmierz, J.; Frąk, W.; Młynarska, E.; Franczyk, B.; Rysz, J. Molecular Interactions of Arterial Hypertension in Its Target Organs. *Int. J. Mol. Sci.* **2021**, *22*, 9669. [CrossRef] [PubMed]
4. Blaustein, M.P.; Chen, L.; Hamlyn, J.M.; Leenen, F.H.H.; Lingrel, J.B.; Wier, W.G.; Zhang, J. Pivotal Role of A2 Na^+ Pumps and Their High Affinity Ouabain Binding Site in Cardiovascular Health and Diseaysse. *J. Physiol.* **2016**, *594*, 6079–6103. [CrossRef] [PubMed]
5. Yuan, C.M.; Manunta, P.; Hamlyn, J.M.; Chen, S.; Bohen, E.; Yeun, J.; Haddy, F.J.; Pamnani, M.B. Long-Term Ouabain Administration Produces Hypertension in Rats. *Hypertension* **1993**, *22*, 178–187. [CrossRef]
6. Manunta, P.; Rogowski, A.C.; Hamilton, B.P.; Hamlyn, J.M. Ouabain-Induced Hypertension in the Rat: Relationships among Plasma and Tissue Ouabain and Blood Pressure. *J. Hypertens.* **1994**, *12*, 549–560. [CrossRef] [PubMed]
7. Buckalew, V.M. Endogenous Digitalis-like Factors: An Overview of the History. *Front. Endocrinol.* **2015**, *6*, 49. [CrossRef]
8. Lingrel, J.B. The Physiological Significance of the Cardiotonic Steroid/Ouabain-Binding Site of the Na,K-ATPase. *Annu. Rev. Physiol.* **2010**, *72*, 395–412. [CrossRef] [PubMed]
9. Yan, Y.; Shapiro, J.I. The Physiological and Clinical Importance of Sodium Potassium ATPase in Cardiovascular Diseases. *Curr. Opin. Pharmacol.* **2016**, *27*, 43–49. [CrossRef] [PubMed]
10. Scheiner-Bobis, G. The Na^+, K^+-ATPase: More than Just a Sodium Pump. *Cardiovasc. Res.* **2011**, *89*, 6–8. [CrossRef] [PubMed]
11. Cui, X.; Xie, Z. Protein Interaction and Na/K-ATPase-Mediated Signal Transduction. *Molecules* **2017**, *22*, 990. [CrossRef]
12. Xie, Z.; Xie, J. The Na/K-ATPase-Mediated Signal Transduction as a Target for New Drug Development. *Front. Biosci.* **2005**, *10*, 3100–3109. [CrossRef] [PubMed]
13. Kinoshita, P.F.; Orellana, A.M.M.; Nakao, V.W.; de Souza Port's, N.M.; Quintas, L.E.M.; Kawamoto, E.M.; Scavone, C. The Janus Face of Ouabain in Na^+/K^+-ATPase and Calcium Signalling in Neurons. *Br. J. Pharmacol.* **2022**, *179*, 1512–1524. [CrossRef] [PubMed]
14. Gagnon, K.B.; Delpire, E. Sodium Transporters in Human Health and Disease. *Front. Physiol.* **2020**, *11*, 588664. [CrossRef] [PubMed]
15. Biondo, E.D.; Spontarelli, K.; Ababioh, G.; Méndez, L.; Artigas, P. Diseases Caused by Mutations in the Na^+/K^+ Pump A1 Gene ATP1A1. *Am. J. Physiol. Cell Physiol.* **2021**, *321*, C394–C408. [CrossRef] [PubMed]
16. Kjeldsen, K.; Braendgaard, H.; Sidenius, P.; Larsen, J.S.; Nørgaard, A. Diabetes Decreases Na^+-K^+ Pump Concentration in Skeletal Muscles, Heart Ventricular Muscle, and Peripheral Nerves of Rat. *Diabetes* **1987**, *36*, 842–848. [CrossRef] [PubMed]
17. Schoner, W.; Scheiner-Bobis, G. Endogenous and Exogenous Cardiac Glycosides: Their Roles in Hypertension, Salt Metabolism, and Cell Growth. *Am. J. Physiol. Cell Physiol.* **2007**, *293*, C509–C536. [CrossRef] [PubMed]
18. Tao, Q.F.; Hollenberg, N.K.; Graves, S.W. Sodium Pump Inhibition and Regional Expression of Sodium Pump Alpha-Isoforms in Lens. *Hypertension* **1999**, *34*, 1168–1174. [CrossRef]
19. Blaustein, M.P. Endogenous Ouabain: Role in the Pathogenesis of Hypertension. *Kidney Int.* **1996**, *49*, 1748–1753. [CrossRef] [PubMed]
20. Delamere, N.A.; Tamiya, S. Expression, Regulation and Function of Na,K-ATPase in the Lens. *Prog. Retin. Eye Res.* **2004**, *23*, 593–615. [CrossRef] [PubMed]
21. Kaplan, J.H. Biochemistry of Na,K-ATPase. *Annu. Rev. Biochem.* **2002**, *71*, 511–535. [CrossRef] [PubMed]
22. Askari, A. The Other Functions of the Sodium Pump. *Cell Calcium* **2019**, *84*, 102105. [CrossRef] [PubMed]
23. Haas, M.; Askari, A.; Xie, Z. Involvement of Src and Epidermal Growth Factor Receptor in the Signal-Transducing Function of Na^+/K^+-ATPase. *J. Biol. Chem.* **2000**, *275*, 27832–27837. [CrossRef]
24. Haas, M.; Wang, H.; Tian, J.; Xie, Z. Src-Mediated Inter-Receptor Cross-Talk between the Na^+/K^+-ATPase and the Epidermal Growth Factor Receptor Relays the Signal from Ouabain to Mitogen-Activated Protein Kinases. *J. Biol. Chem.* **2002**, *277*, 18694–18702. [CrossRef]

25. Zhou, X.; Jiang, G.; Zhao, A.; Bondeva, T.; Hirszel, P.; Balla, T. Inhibition of Na,K-ATPase Activates PI3 Kinase and Inhibits Apoptosis in LLC-PK1 Cells. *Biochem. Biophys. Res. Commun.* **2001**, *285*, 46–51. [CrossRef]
26. Yuan, Z.; Cai, T.; Tian, J.; Ivanov, A.V.; Giovannucci, D.R.; Xie, Z. Na/K-ATPase Tethers Phospholipase C and IP3 Receptor into a Calcium-Regulatory Complex. *Mol. Biol. Cell* **2005**, *16*, 4034–4045. [CrossRef]
27. Schoner, W.; Scheiner-Bobis, G. Endogenous and Exogenous Cardiac Glycosides and Their Mechanisms of Action. *Am. J. Cardiovasc. Drugs* **2007**, *7*, 173–189. [CrossRef]
28. Manunta, P.; Hamilton, J.; Rogowski, A.C.; Hamilton, B.P.; Hamlyn, J.M. Chronic Hypertension Induced by Ouabain but Not Digoxin in the Rat: Antihypertensive Effect of Digoxin and Digitoxin. *Hypertens. Res.* **2000**, *23*, S77–S85. [CrossRef]
29. Goto, A.; Yamada, K.; Yagi, N.; Yoshioka, M.; Sugimoto, T. Physiology and Pharmacology of Endogenous Digitalis-like Factors. *Pharmacol. Rev.* **1992**, *44*, 377–399.
30. Bommer, U.-A.; Thiele, B.-J. The Translationally Controlled Tumour Protein (TCTP). *Int. J. Biochem. Cell Biol.* **2004**, *36*, 379–385. [CrossRef]
31. Pinkaew, D.; Fujise, K. Fortilin: A Potential Target for the Prevention and Treatment of Human Diseases. *Adv. Clin. Chem.* **2017**, *82*, 265–300. [CrossRef] [PubMed]
32. Bommer, U.-A.; Kawakami, T. Role of TCTP in Cell Biological and Disease Processes. *Cells* **2021**, *10*, 2290. [CrossRef] [PubMed]
33. Jung, J.; Kim, M.; Kim, M.-J.; Kim, J.; Moon, J.; Lim, J.-S.; Kim, M.; Lee, K. Translationally Controlled Tumor Protein Interacts with the Third Cytoplasmic Domain of Na,K-ATPase Alpha Subunit and Inhibits the Pump Activity in HeLa Cells. *J. Biol. Chem.* **2004**, *279*, 49868–49875. [CrossRef]
34. Assrir, N.; Malard, F.; Lescop, E. Structural Insights into TCTP and Its Interactions with Ligands and Proteins. *Results Probl. Cell Differ.* **2017**, *64*, 9–46. [CrossRef]
35. Bommer, U.-A. The Translational Controlled Tumour Protein TCTP: Biological Functions and Regulation. *Results Probl. Cell Differ.* **2017**, *64*, 69–126. [CrossRef]
36. Lee, H.-J.; Song, K.-H.; Oh, S.J.; Kim, S.; Cho, E.; Kim, J.; Park, Y.G.; Lee, K.-M.; Yee, C.; Song, S.-H.; et al. Targeting TCTP Sensitizes Tumor to T Cell-Mediated Therapy by Reversing Immune-Refractory Phenotypes. *Nat. Commun.* **2022**, *13*, 2127. [CrossRef]
37. Bommer, U.-A.; Telerman, A. Dysregulation of TCTP in Biological Processes and Diseases. *Cells* **2020**, *9*, 1632. [CrossRef]
38. Koide, Y.; Kiyota, T.; Tonganunt, M.; Pinkaew, D.; Liu, Z.; Kato, Y.; Hutadilok-Towatana, N.; Phongdara, A.; Fujise, K. Embryonic Lethality of Fortilin-Null Mutant Mice by BMP-Pathway Overactivation. *Biochim. Biophys. Acta* **2009**, *1790*, 326–338. [CrossRef]
39. Kim, M.-J.; Kwon, J.-S.; Suh, S.H.; Suh, J.-K.; Jung, J.; Lee, S.-N.; Kim, Y.-H.; Cho, M.-C.; Oh, G.T.; Lee, K. Transgenic Overexpression of Translationally Controlled Tumor Protein Induces Systemic Hypertension via Repression of Na^+,K^+-ATPase. *J. Mol. Cell. Cardiol.* **2008**, *44*, 151–159. [CrossRef]
40. Maeng, J.; Sheverdin, V.; Shin, H.; Ha, I.; Bae, S.S.; Yang-Yen, H.-F.; Lee, K. Up-Regulation of Rhoa/Rho Kinase Pathway by Translationally Controlled Tumor Protein in Vascular Smooth Muscle Cells. *Int. J. Mol. Sci.* **2014**, *15*, 10365–10376. [CrossRef]
41. Kim, M.-J.; Lyu, J.; Sohn, K.-B.; Kim, M.; Cho, M.-C.; Joo, C.-K.; Lee, K. Over-Expression of Translationally Controlled Tumor Protein in Lens Epithelial Cells Seems to Be Associated with Cataract Development. *Transgenic Res.* **2009**, *18*, 953–960. [CrossRef] [PubMed]
42. Choi, S.-W.; Shin, H.K.; Yang-Yen, H.-F.; Lee, M.S.; Lee, C.H.; Park, S.-J.; Kim, K.-J.; Lee, K.; Kim, S.H. Osteoclastogenic Activity of Translationally-Controlled Tumor Protein (TCTP) with Reciprocal Repression of P21. *FEBS Lett.* **2014**, *588*, 4026–4031. [CrossRef]
43. Kim, M.; Choe, Y.; Lee, H.; Jeon, M.-G.; Park, J.-H.; Noh, H.S.; Cheon, Y.-H.; Park, H.J.; Park, J.; Shin, S.J.; et al. Blockade of Translationally Controlled Tumor Protein Attenuated the Aggressiveness of Fibroblast-like Synoviocytes and Ameliorated Collagen-Induced Arthritis. *Exp. Mol. Med.* **2021**, *53*, 67–80. [CrossRef] [PubMed]
44. Cho, Y.; Maeng, J.; Ryu, J.; Shin, H.; Kim, M.; Oh, G.T.; Lee, M.-Y.; Lee, K. Hypertension Resulting from Overexpression of Translationally Controlled Tumor Protein Increases the Severity of Atherosclerosis in Apolipoprotein E Knock-out Mice. *Transgenic Res.* **2012**, *21*, 1245–1254. [CrossRef] [PubMed]
45. Pinkaew, D.; Le, R.J.; Chen, Y.; Eltorky, M.; Teng, B.-B.; Fujise, K. Fortilin Reduces Apoptosis in Macrophages and Promotes Atherosclerosis. *Am. J. Physiol. Heart Circ. Physiol.* **2013**, *305*, H1519–H1529. [CrossRef] [PubMed]
46. Yeh, Y.-C.; Xie, L.; Langdon, J.M.; Myers, A.C.; Oh, S.-Y.; Zhu, Z.; Macdonald, S.M. The Effects of Overexpression of Histamine Releasing Factor (HRF) in a Transgenic Mouse Model. *PLoS ONE* **2010**, *5*, e11077. [CrossRef]
47. Jeon, Y.; Choi, J.-Y.; Jang, E.-H.; Seong, J.K.; Lee, K. Overexpression of Translationally Controlled Tumor Protein Ameliorates Metabolic Imbalance and Increases Energy Expenditure in Mice. *Int. J. Obes.* **2021**, *45*, 1576–1587. [CrossRef] [PubMed]
48. Yan, Y.; Xiong, Z.; Zhang, S.; Song, J.; Huang, Y.; Thornton, A.M.; Wang, H.; Yang, X.-F. CD25high T Cells with a Prolonged Survival Inhibit Development of Diabetes. *Int. J. Immunopathol. Pharmacol.* **2008**, *21*, 767–780. [CrossRef]
49. Cai, W.; Fujita, T.; Hidaka, Y.; Jin, H.; Suita, K.; Shigeta, M.; Kiyonari, H.; Umemura, M.; Yokoyama, U.; Sadoshima, J.; et al. Translationally Controlled Tumor Protein (TCTP) Plays a Pivotal Role in Cardiomyocyte Survival through a Bnip3-Dependent Mechanism. *Cell Death Dis.* **2019**, *10*, 549. [CrossRef]
50. Chunhacha, P.; Pinkaew, D.; Sinthujaroen, P.; Bowles, D.E.; Fujise, K. Fortilin Inhibits P53, Halts Cardiomyocyte Apoptosis, and Protects the Heart against Heart Failure. *Cell Death Discov.* **2021**, *7*, 310. [CrossRef]
51. Chattopadhyay, A.; Pinkaew, D.; Doan, H.Q.; Jacob, R.B.; Verma, S.K.; Friedman, H.; Peterson, A.C.; Kuyumcu-Martinez, M.N.; McDougal, O.M.; Fujise, K. Fortilin Potentiates the Peroxidase Activity of Peroxiredoxin-1 and Protects against Alcohol-Induced Liver Damage in Mice. *Sci. Rep.* **2016**, *6*, 18701. [CrossRef] [PubMed]

52. Pinkaew, D.; Chattopadhyay, A.; King, M.D.; Chunhacha, P.; Liu, Z.; Stevenson, H.L.; Chen, Y.; Sinthujaroen, P.; McDougal, O.M.; Fujise, K. Fortilin Binds IRE1α and Prevents ER Stress from Signaling Apoptotic Cell Death. *Nat. Commun.* **2017**, *8*, 18. [CrossRef] [PubMed]
53. Leong, X.-F.; Ng, C.-Y.; Jaarin, K. Animal Models in Cardiovascular Research: Hypertension and Atherosclerosis. *BioMed Res. Int.* **2015**, *2015*, 528757. [CrossRef] [PubMed]
54. Cargnelli, G.; Rossi, G.P.; Pessina, A.C.; Luciani, S.; Debetto, P.; Ganten, D.; Peters, J.; Bova, S. Changes of Blood Pressure and Aortic Strip Contractile Responses to ET-1 of Heterozygous Female Transgenic Rats, TGR(MRen2)27. *Pharmacol. Res.* **1998**, *37*, 207–211. [CrossRef]
55. McVeigh, G.E.; Plumb, R.; Hughes, S. Vascular Abnormalities in Hypertension: Cause, Effect, or Therapeutic Target? *Curr. Hypertens. Rep.* **2004**, *6*, 171–176. [CrossRef] [PubMed]
56. Kimura, K.; Manunta, P.; Hamilton, B.P.; Hamlyn, J.M. Different Effects of in Vivo Ouabain and Digoxin on Renal Artery Function and Blood Pressure in the Rat. *Hypertens. Res.* **2000**, *23*, S67–S76. [CrossRef]
57. Wirth, A. Rho Kinase and Hypertension. *Biochim. Biophys. Acta* **2010**, *1802*, 1276–1284. [CrossRef]
58. Seko, T.; Ito, M.; Kureishi, Y.; Okamoto, R.; Moriki, N.; Onishi, K.; Isaka, N.; Hartshorne, D.J.; Nakano, T. Activation of RhoA and Inhibition of Myosin Phosphatase as Important Components in Hypertension in Vascular Smooth Muscle. *Circ. Res.* **2003**, *92*, 411–418. [CrossRef]
59. Uehata, M.; Ishizaki, T.; Satoh, H.; Ono, T.; Kawahara, T.; Morishita, T.; Tamakawa, H.; Yamagami, K.; Inui, J.; Maekawa, M.; et al. Calcium Sensitization of Smooth Muscle Mediated by a Rho-Associated Protein Kinase in Hypertension. *Nature* **1997**, *389*, 990–994. [CrossRef]
60. Shimokawa, H.; Sunamura, S.; Satoh, K. RhoA/Rho-Kinase in the Cardiovascular System. *Circ. Res.* **2016**, *118*, 352–366. [CrossRef]
61. Seccia, T.M.; Rigato, M.; Ravarotto, V.; Calò, L.A. ROCK (RhoA/Rho Kinase) in Cardiovascular-Renal Pathophysiology: A Review of New Advancements. *J. Clin. Med.* **2020**, *9*, 1328. [CrossRef] [PubMed]
62. Cheon, M.S.; Suh, J.-K.; Kim, M.-J.; Kim, S.H.; Lee, K. Identification of Differentially Expressed Proteins in the Heart of Translationally Controlled Tumor Protein Over-Expressing Transgenic Mice. *Biomed. Chromatogr.* **2008**, *22*, 1091–1099. [CrossRef] [PubMed]
63. Dikalov, S.I.; Dikalova, A.E. Contribution of Mitochondrial Oxidative Stress to Hypertension. *Curr. Opin. Nephrol. Hypertens.* **2016**, *25*, 73–80. [CrossRef] [PubMed]
64. Dikalov, S.I.; Ungvari, Z. Role of Mitochondrial Oxidative Stress in Hypertension. *Am. J. Physiol. Heart Circ. Physiol.* **2013**, *305*, H1417–H1427. [CrossRef] [PubMed]
65. Griendling, K.K.; Camargo, L.L.; Rios, F.J.; Alves-Lopes, R.; Montezano, A.C.; Touyz, R.M. Oxidative Stress and Hypertension. *Circ. Res.* **2021**, *128*, 993–1020. [CrossRef]
66. Jeong, S.-J.; Park, J.-G.; Oh, G.T. Peroxiredoxins as Potential Targets for Cardiovascular Disease. *Antioxidants* **2021**, *10*, 1244. [CrossRef]
67. Matsushima, S.; Ide, T.; Yamato, M.; Matsusaka, H.; Hattori, F.; Ikeuchi, M.; Kubota, T.; Sunagawa, K.; Hasegawa, Y.; Kurihara, T.; et al. Overexpression of Mitochondrial Peroxiredoxin-3 Prevents Left Ventricular Remodeling and Failure after Myocardial Infarction in Mice. *Circulation* **2006**, *113*, 1779–1786. [CrossRef]
68. Ishizaka, N.; Aizawa, T.; Ohno, M.; Usui Si, S.; Mori, I.; Tang, S.-S.; Ingelfinger, J.R.; Kimura, S.; Nagai, R. Regulation and Localization of HSP70 and HSP25 in the Kidney of Rats Undergoing Long-Term Administration of Angiotensin II. *Hypertension* **2002**, *39*, 122–128. [CrossRef]
69. Szmyd, L.; Schwartz, B. Association of Systemic Hypertension and Diabetes Mellitus with Cataract Extraction. A Case-Control Study. *Ophthalmology* **1989**, *96*, 1248–1252. [CrossRef]
70. Yu, X.; Lyu, D.; Dong, X.; He, J.; Yao, K. Hypertension and Risk of Cataract: A Meta-Analysis. *PLoS ONE* **2014**, *9*, e114012. [CrossRef]
71. Unakar, N.J.; Johnson, M. Lenticular Alterations in Hypertensive Rats. *Exp. Eye Res.* **1994**, *59*, 645–652. [CrossRef] [PubMed]
72. Rodríguez-Sargent, C.; Cangiano, J.L.; Berríos Cabán, G.; Marrero, E.; Martínez-Maldonado, M. Cataracts and Hypertension in Salt-Sensitive Rats. A Possible Ion Transport Defect. *Hypertension* **1987**, *9*, 304–308. [CrossRef] [PubMed]
73. Rodríguez-Sargent, C.; Estapé, E.S.; Fernández, N.; Irizarry, J.E.; Cangiano, J.L.; Candia, O.A. Altered Lens Short-Circuit Current in Adult Cataract-Prone Dahl Hypertensive Rats. *Hypertension* **1996**, *28*, 440–443. [CrossRef] [PubMed]
74. Delamere, N.A.; Tamiya, S. Lens Ion Transport: From Basic Concepts to Regulation of Na,K-ATPase Activity. *Exp. Eye Res.* **2009**, *88*, 140–143. [CrossRef]
75. Lichtstein, D.; Gati, I.; Samuelov, S.; Berson, D.; Rozenman, Y.; Landau, L.; Deutsch, J. Identification of Digitalis-like Compounds in Human Cataractous Lenses. *Eur. J. Biochem.* **1993**, *216*, 261–268. [CrossRef] [PubMed]
76. Gupta, P.D.; Johar, K.; Vasavada, A. Causative and Preventive Action of Calcium in Cataracto-Genesis. *Acta Pharmacol. Sin.* **2004**, *25*, 1250–1256.
77. Karlsson, J.O.; Andersson, M.; Kling-Petersen, A.; Sjöstrand, J. Proteolysis in Human Lens Epithelium Determined by a Cell-Permeable Substrate. *Invest. Ophthalmol. Vis. Sci.* **1999**, *40*, 261–264.
78. Gupta, J.D.; Harley, J.D. Decreased Adenosine Triphosphatase Activity in Human Senile Cataractous Lenses. *Exp. Eye Res.* **1975**, *20*, 207–209. [CrossRef]

79. Ziegler, T.; Abdel Rahman, F.; Jurisch, V.; Kupatt, C. Atherosclerosis and the Capillary Network; Pathophysiology and Potential Therapeutic Strategies. *Cells* **2019**, *9*, 50. [CrossRef] [PubMed]
80. Libby, P.; Ridker, P.M.; Hansson, G.K. Progress and Challenges in Translating the Biology of Atherosclerosis. *Nature* **2011**, *473*, 317–325. [CrossRef]
81. Hollander, W. Role of Hypertension in Atherosclerosis and Cardiovascular Disease. *Am. J. Cardiol.* **1976**, *38*, 786–800. [CrossRef]
82. Ketonen, J.; Merasto, S.; Paakkari, I.; Mervaala, E.M.A. High Sodium Intake Increases Vascular Superoxide Formation and Promotes Atherosclerosis in Apolipoprotein E-Deficient Mice. *Blood Press.* **2005**, *14*, 373–382. [CrossRef] [PubMed]
83. Weiss, D.; Taylor, W.R. Deoxycorticosterone Acetate Salt Hypertension in Apolipoprotein $E^{-}/^{-}$ Mice Results in Accelerated Atherosclerosis: The Role of Angiotensin II. *Hypertension* **2008**, *51*, 218–224. [CrossRef]
84. da Cunha, V.; Tham, D.M.; Martin-McNulty, B.; Deng, G.; Ho, J.J.; Wilson, D.W.; Rutledge, J.C.; Vergona, R.; Sullivan, M.E.; Wang, Y.-X.J. Enalapril Attenuates Angiotensin II-Induced Atherosclerosis and Vascular Inflammation. *Atherosclerosis* **2005**, *178*, 9–17. [CrossRef] [PubMed]
85. Standridge, J.B. Hypertension and Atherosclerosis: Clinical Implications from the ALLHAT Trial. *Curr. Atheroscler. Rep.* **2005**, *7*, 132–139. [CrossRef] [PubMed]
86. Zhang, S.H.; Reddick, R.L.; Piedrahita, J.A.; Maeda, N. Spontaneous Hypercholesterolemia and Arterial Lesions in Mice Lacking Apolipoprotein, E. *Science* **1992**, *258*, 468–471. [CrossRef]
87. Nakashima, Y.; Plump, A.S.; Raines, E.W.; Breslow, J.L.; Ross, R. ApoE-Deficient Mice Develop Lesions of All Phases of Atherosclerosis throughout the Arterial Tree. *Arterioscler. Thromb. J. Vasc. Biol.* **1994**, *14*, 133–140. [CrossRef] [PubMed]
88. Li, J.-J.; Chen, J.-L. Inflammation May Be a Bridge Connecting Hypertension and Atherosclerosis. *Med. Hypotheses* **2005**, *64*, 925–929. [CrossRef]
89. Barton, M.; Yanagisawa, M. Endothelin: 30 Years From Discovery to Therapy. *Hypertension* **2019**, *74*, 1232–1265. [CrossRef]
90. Ivey, M.E.; Osman, N.; Little, P.J. Endothelin-1 Signalling in Vascular Smooth Muscle: Pathways Controlling Cellular Functions Associated with Atherosclerosis. *Atherosclerosis* **2008**, *199*, 237–247. [CrossRef] [PubMed]
91. Lin, Y.-J.; Juan, C.-C.; Kwok, C.-F.; Hsu, Y.-P.; Shih, K.-C.; Chen, C.-C.; Ho, L.-T. Endothelin-1 Exacerbates Development of Hypertension and Atherosclerosis in Modest Insulin Resistant Syndrome. *Biochem. Biophys. Res. Commun.* **2015**, *460*, 497–503. [CrossRef] [PubMed]
92. Krishnamoorthy, R.R.; Prasanna, G.; Dauphin, R.; Hulet, C.; Agarwal, N.; Yorio, T. Regulation of Na,K-ATPase Expression by Endothelin-1 in Transformed Human Ciliary Non-Pigmented Epithelial (HNPE) Cells. *J. Ocul. Pharmacol. Ther.* **2003**, *19*, 465–481. [CrossRef]
93. Mandal, A.; Shahidullah, M.; Beimgraben, C.; Delamere, N.A. The Effect of Endothelin-1 on Src-Family Tyrosine Kinases and Na,K-ATPase Activity in Porcine Lens Epithelium. *J. Cell. Physiol.* **2011**, *226*, 2555–2561. [CrossRef] [PubMed]
94. Prasanna, G.; Dibas, A.; Hulet, C.; Yorio, T. Inhibition of Na(+)/K(+)-Atpase by Endothelin-1 in Human Nonpigmented Ciliary Epithelial Cells. *J. Pharmacol. Exp. Ther.* **2001**, *296*, 966–971. [PubMed]
95. Scheiner-Bobis, G.; Eva, A.; Kirch, U. Signalling Pathways Involving Sodium Pump Stimulate Endothelin-1 Secretion and Nitric Oxide Production in Endothelial Cells. *Cell. Mol. Biol.* **2006**, *52*, 58–63.
96. Eva, A.; Kirch, U.; Scheiner-Bobis, G. Signaling Pathways Involving the Sodium Pump Stimulate NO Production in Endothelial Cells. *Biochim. Biophys. Acta* **2006**, *1758*, 1809–1814. [CrossRef]
97. Predic, J.; Soskic, V.; Bradley, D.; Godovac-Zimmermann, J. Monitoring of Gene Expression by Functional Proteomics: Response of Human Lung Fibroblast Cells to Stimulation by Endothelin-1. *Biochemistry* **2002**, *41*, 1070–1078. [CrossRef]
98. Saitoh, T.; Kishida, H.; Tsukada, Y.; Fukuma, Y.; Sano, J.; Yasutake, M.; Fukuma, N.; Kusama, Y.; Hayakawa, H. Clinical Significance of Increased Plasma Concentration of Macrophage Colony-Stimulating Factor in Patients with Angina Pectoris. *J. Am. Coll. Cardiol.* **2000**, *35*, 655–665. [CrossRef]
99. Aoyama, M.; Kishimoto, Y.; Saita, E.; Ohmori, R.; Tanimoto, K.; Nakamura, M.; Kondo, K.; Momiyama, Y. High Plasma Levels of Fortilin in Patients with Coronary Artery Disease. *Int. J. Mol. Sci.* **2022**, *23*, 8923. [CrossRef]
100. Piché, M.-E.; Tchernof, A.; Després, J.-P. Obesity Phenotypes, Diabetes, and Cardiovascular Diseases. *Circ. Res.* **2020**, *126*, 1477–1500. [CrossRef]
101. Lowell, B.B.; Spiegelman, B.M. Towards a Molecular Understanding of Adaptive Thermogenesis. *Nature* **2000**, *404*, 652–660. [CrossRef] [PubMed]
102. Berbée, J.F.P.; Boon, M.R.; Khedoe, P.P.S.J.; Bartelt, A.; Schlein, C.; Worthmann, A.; Kooijman, S.; Hoeke, G.; Mol, I.M.; John, C.; et al. Brown Fat Activation Reduces Hypercholesterolaemia and Protects from Atherosclerosis Development. *Nat. Commun.* **2015**, *6*, 6356. [CrossRef] [PubMed]
103. Evangelou, K.; Vasileiou, P.V.; Papaspyropoulos, A.; Hazapis, O.; Petty, R.; Demaria, M.; Gorgoulis, V.G. Cellular Senescence and Cardiovascular Diseases: Moving to the "Heart" of the Problem. *Physiol. Rev.* **2022**. [CrossRef] [PubMed]
104. Hu, C.; Zhang, X.; Teng, T.; Ma, Z.-G.; Tang, Q.-Z. Cellular Senescence in Cardiovascular Diseases: A Systematic Review. *Aging Dis.* **2022**, *13*, 103–128. [CrossRef]

MDPI
St. Alban-Anlage 66
4052 Basel
Switzerland
Tel. +41 61 683 77 34
Fax +41 61 302 89 18
www.mdpi.com

Biomedicines Editorial Office
E-mail: biomedicines@mdpi.com
www.mdpi.com/journal/biomedicines

www.ingramcontent.com/pod-product-compliance
Lightning Source LLC
Chambersburg PA
CBHW061814210926
51850CB00006B/705
* 9 7 8 3 0 3 6 5 6 3 1 8 3 *